5G Wireless

A Comprehensive Introduction

Dr. William Stallings

Addison-Wesley

Boston • Columbus • New York • San Francisco • Amsterdam • Cape Town
Dubai • London • Madrid • Milan • Munich • Paris • Montreal • Toronto • Delhi • Mexico City
São Paulo • Sydney • Hong Kong • Seoul • Singapore • Taipei • Tokyo

For information about buying this title in bulk quantities, or for special sales opportunities (which may include electronic versions; custom cover designs; and content particular to your business, training goals, marketing focus, or branding interests), please contact our corporate sales department at corpsales@pearsoned.com or (800) 382-3419.

For government sales inquiries, please contact governmentsales@pearsoned.com.

For questions about sales outside the U.S., please contact intlcs@pearson.com.

Visit us on the Web: informit.com/aw

Library of Congress Control Number: 2021937463

ISBN-13: 978-0-13-676714-5
ISBN-10: 0-13-676714-1

1 2021

Editor-in-Chief
Mark Taub

Director Product Management
Brett Bartow

Development Editor
Marianne Bartow

Managing Editor
Sandra Schroeder

Technical Reviewers
Toon Norp
Tim Stammers

Senior Project Editor
Lori Lyons

Copy Editor
Catherine D. Wilson

Production Manager
Aswini Kumar/codeMantra

Indexer
Cheryl Ann Lenser

Proofreader
Donna E. Mulder

Editorial Assistant
Cindy Teeters

Cover Designer
Chuti Prasertsith

Compositor
codeMantra

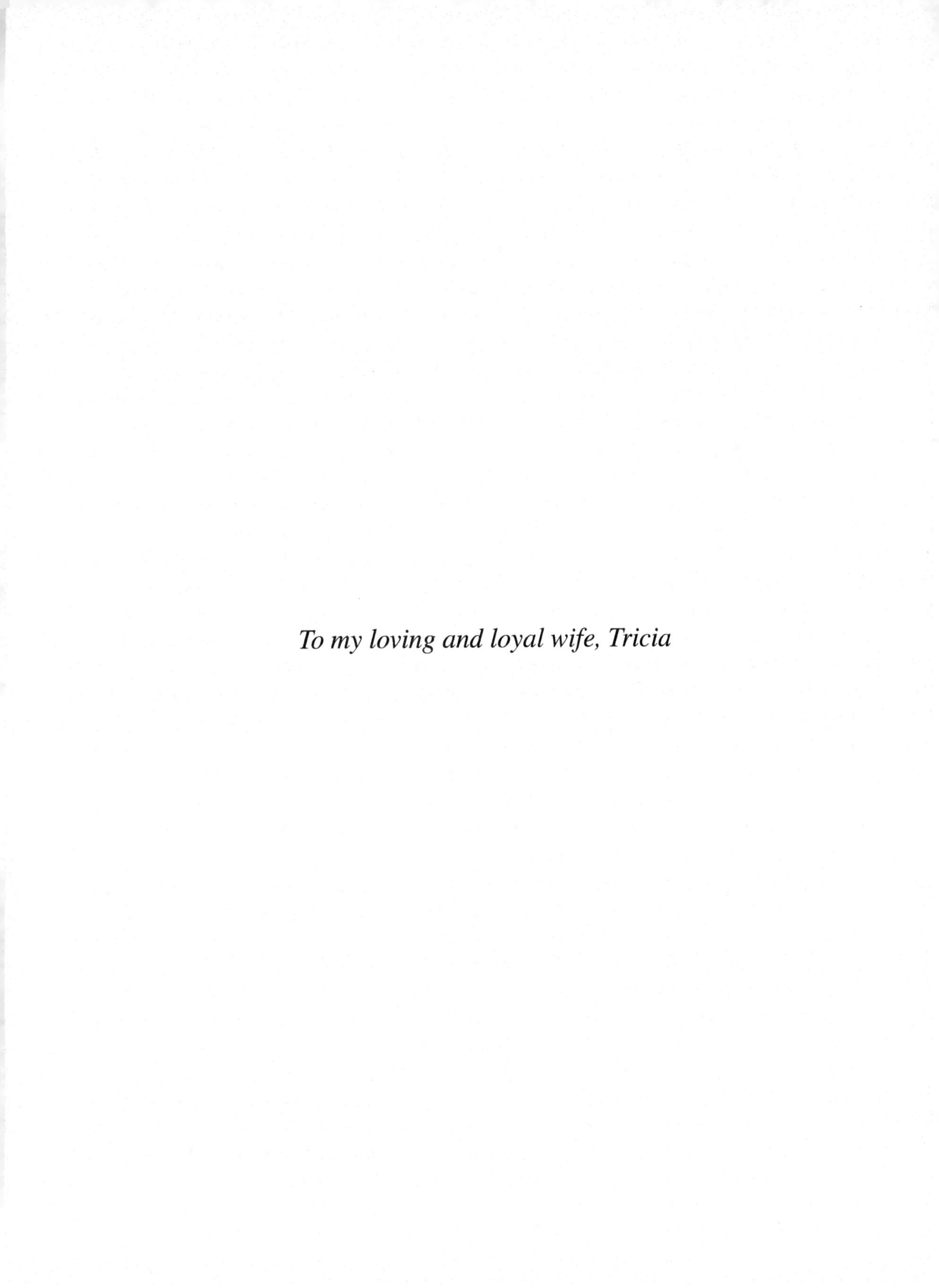

To my loving and loyal wife, Tricia

Contents at a Glance

Contents

About the Author

Dr. William Stallings has made a unique contribution to understanding the broad sweep of technical developments in computer security, computer networking, and computer architecture. He has authored 20 textbooks, and, counting revised editions, more than 75 books on various aspects of these subjects. His writings have appeared in numerous ACM and IEEE publications, including the *Proceedings of the IEEE* and *ACM Computing Reviews*. He has 13 times received the award for the best computer science textbook of the year from the Text and Academic Authors Association.

In over 30 years in the field, he has been a technical contributor, a technical manager, and an executive with several high-technology firms. He has designed and implemented both TCP/IP-based and OSI-based protocol suites on a variety of computers and operating systems, ranging from microcomputers to mainframes. Currently he is an independent consultant whose clients have included computer and networking manufacturers and customers, software development firms, and leading-edge government research institutions.

He created and maintains the Computer Science Student Resource Site at ComputerScienceStudent.com. This site provides documents and links on a variety of subjects of general interest to computer science students (and professionals). He is a member of the editorial board of *Cryptologia*, a scholarly journal devoted to all aspects of cryptology.

Dr. Stallings holds a PhD from M.I.T. in computer science and a B.S. from Notre Dame in electrical engineering.

About the Technical Reviewers

Toon Norp is a Senior Business Consultant at TNO. He joined TNO (former KPN Research) in 1991, where he has since been working on network aspects of mobile communications. Toon advises European operators, government organizations, and others on strategy and architecture related to mobile networks, M2M/IoT, and 5G. He has been involved in standardization of mobile networks for more than 20 years, and as chairman of the 3GPP SA1 service aspects working group, he was responsible for the requirements specification phase of 5G. Toon has been instrumental in getting several new sectors (e.g., public safety, satellite, media, railway, industry) involved in the 3GPP standardization process. Toon is a member of the 5G-PPP association, a joint initiative between the European ICT industry and the European Commission, a reviewer of European R&D projects, and a regular speaker at conferences. Toon holds a master's degree in electrical engineering from the Eindhoven University of Technology, The Netherlands.

Tim Stammers is a Principal Engineer at Cisco Systems. Tim joined Cisco in 2000, where he has since been working on the architecture and development of products for mobile data services. Prior to that, Tim worked at Alcatel in the area of mobile switching as well as at a number of telecommunications startups.

Tim serves as a technical advisor for cellular core technologies across a number of products and services. Tim was directly involved with Tier 1 operators in the commercial launch of 4G data services and is now focused on 5G, IoT, and multi-access technologies; he has an interest in service adjacencies for private cellular opportunities.

Tim is the author of over 50 U.S. patents in the areas of mobile networking and security.

Tim represents Cisco in 5G-ACIA, an industry group promoting 5G in the area of industrial automation. Tim has provided and reviewed material for 3GPP and participated in the core standards development for ANSI-136.

Tim holds a bachelor of science degree in electrical and electronic engineering from the University of Bristol, United Kingdom.

Acknowledgments

This book has benefited from review by a number of people, who gave generously of their time and expertise. I especially thank Nikhil Bhargava, A.H.J (Toon) Norp, and Tim Stammers, who each devoted an enormous amount of time to a detailed review of the entire manuscript. I also thank the people who provided thoughtful reviews of the initial book proposal: Jerome Henry and David Hucaby.

Thanks also to the many people who provided detailed technical reviews of one or more chapters: Mohammad Abbas, Yaser Ahmed, Vladimir Atanasovski, Rashmi Bhattad, Yufei Blankenship, Khishigbayar Dushchuluun, Samuel Edeagu, Mohamed Ghozzi, Prakash Nagarajan, and Apostolos Savvakis.

Finally, I would like to thank the many people at Pearson responsible for the publication of the book. This includes Brett Bartow (director, IT Professional Product Management), Marianne Bartow (development editor), Lori Lyons (senior project editor), Aswini Kumar (production), and Kitty Wilson (copy editor). Thanks also to the marketing and sales staffs at Pearson, without whose efforts this book would not be in front of you.

Preface

5G, the fifth-generation technology for wireless cellular networks, takes a significant technological leap beyond the capabilities of the 4G networks that currently dominate available cellular network services. 5G delivers a substantial increase in peak and average speeds and capacity. A significant increase in download and upload speeds will enhance many existing use cases, including cloud-based storage, augmented reality, and artificial intelligence. 5G will also enable cell sites to communicate with a greater number of devices. Reduced latency and enhanced use of edge computing will transform Internet of Things (IoT) capabilities and application breadth.

This book provides comprehensive coverage of 5G networks, including the technology and the application areas for enterprises, government, and consumers. This is a technical book, intended for readers with some technical background, but it is sufficiently self-contained to be a valuable resource for managers responsible for procuring or offering 5G services, IT managers, and product marketing personnel, in addition to system engineers, network maintenance personnel, and network and protocol designers.

The book is designed to provide:

- Understanding of the overall 5G ecosystem and individual building blocks
- Awareness of the 5G research, development, and standardization roadmaps
- Insights into the specification and standardization process
- Knowledge of 5G applicability in the variety of use cases
- Understanding of the underlying core network technologies
- Understanding of the underlying wireless access and radio transport network technologies

Specifications and Standards

The development of 5G technology and its implementation and deployment depend heavily on international specifications and standards that are universally accepted and that enable the interoperation of equipment from many different sources. Because an understanding of 5G depends on an understanding of the process by which the standards are developed and the content of those standards, this book deals in depth with this subject. Two organizations are responsible for the development of 5G standards. The International Telecommunication Union (ITU) has issued standards, called Recommendations, and other documents, call Reports, that define the overall idea of 5G as well as the technical, performance, and service requirements for 5G. Based on these documents, the misleadingly named 3rd Generation Partnership Project (3GPP), which is a consortium of national standards organizations and government and industry participants, has developed and continues to develop a detailed set of technical specifications for the implementation of 5G. In turn, ITU transforms stable specifications into international standards. In describing the technology and application areas of 5G, this book relies on and explains the work of ITU and 3GPP.

How This Book Is Organized

The book consists of four parts, each of which includes a number of chapters.

Part I, "Overview": This part provides a concise history of the development of cellular networks through 4G. It also introduces 5G, discussing the motivation for its development and its characteristics, and covers 5G technologies.

- **Chapter 1, "Cellular Networks: Concepts and Evolution":** This chapter begins with an overview of basic concepts of cellular networks and cellular data transmission. It also provides summaries of the architecture and functionality of the first four generations of cellular networks. This overview will enable you to better grasp the complexities of 5G networks.
- **Chapter 2, "5G Standards and Specifications":** This chapter goes into some detail, although at a high level, on the requirements, objectives, specifications, and standards for 5G. It provides a useful overview of 5G.
- **Chapter 3, "Overview of 5G Use Cases and Architecture":** This chapter presents a high-level view of the concepts underlying 5G. It begins with a discussion of usage scenarios and use cases that require 5G. It also provides a detailed overview of the 5G architecture, which includes the core network and the radio access network.

Part II, "Use Cases and Applications": ITU defines three broad service areas for 5G, called usage scenarios. A usage scenario dictates various performance and technical requirements for a network. A wide but nevertheless constrained variety of use cases are encompassed by each usage scenario. Part II provides a broad survey of the usage scenarios and use cases for 5G. This part includes a look at the impact of 5G on IoT, cloud computing, and fog computing.

- **Chapter 4, "Enhanced Mobile Broadband":** Of the three usage scenarios defined by ITU, eMBB is the only general-purpose case, and it is the one that is most familiar to current 4G users. In essence, eMBB is an enhanced version of 4G, providing improved performance and an increasingly seamless user experience. Chapter 4 discusses various deployment scenarios for eMBB as well as performance requirements. It then examines three important use cases: smart office, dense urban communications, and communication between train and trackside.
- **Chapter 5, "Massive Machine Type Communications":** The mMTC usage scenario is characterized by a very large number of connected devices, typically transmitting a relatively low volume of non-delay-sensitive data. Devices are required to be low cost and have a very long battery life. This chapter looks at performance requirements and provides an overview of the Internet of Things (IoT). The chapter also discusses smart agriculture and smart cities.
- **Chapter 6, "Ultra-Reliable and Low-Latency Communications":** URLCC has stringent requirements for capabilities such as throughput, latency, and availability. This chapter begins with a discussion of performance requirements and then highlights many use cases categorized by mission criticality and performance requirements. The chapter then surveys Industry 4.0 and unmanned aircraft system traffic management.

Part III, "5G NextGen Core Network": The core network of a cellular network system, also called the backbone network, provides networking services and long-distance interconnection for the wireless access networks through which user devices connect to the cellular network system. Part III provides a detailed survey of the underlying technology and the architecture of 5G core networks.

- **Chapter 7, "Software-Defined Networking":** The two technology underpinnings for the 5G NextGen core network are SDN and NFV. SDN is an approach to designing, building, and operating large-scale networks based on programming the forwarding decisions in routers and switches via software from a central server. This chapter provides detailed treatment of the major aspects of SDN.
- **Chapter 8, "Network Functions Virtualization":** NFV is the virtualization of compute, storage, and network functions by implementing these functions in software and running them on virtual machines. Chapter 8 provides detailed treatment of the major aspects of NFV.
- **Chapter 9, "Core Network Functionality, QoS, and Network Slicing":** This chapter begins with a discussion of the requirements for 5G core networks. It then examines the functional architecture of the core network, covering tunneling, session establishment, and policy control. The chapter also covers quality of service (QoS) and network slicing.
- **Chapter 10, "Multi-Access Edge Computing":** MEC is a distributed information technology architecture in which client data is processed at the periphery of the core network, as close to the originating user as possible. This chapter examines the 5G MEC architecture and examines the relationship between MEC and NFV and network slicing. The chapter also looks at a number of specific use cases.

Part IV: "5G NR Air Interface and Radio Access Network": Beyond the core network, the essential elements of a cellular network are the radio access network (RAN) and the air interface. The RAN is an interconnected set of base stations that provide radio transmission and reception in cells to and from user equipment and also provide connections to the core network. The air interface is the wireless interface between user equipment and the base station. Chapters 11 through 14 cover aspects of the air interface. Chapter 15 surveys the RAN.

- **Chapter 11, "Wireless Transmission":** This chapter provides background on concepts related to wireless transmission, focused on the data rate that can be achieved over a wireless transmission link. The chapter also discusses transmission in the millimeter wavelength region of the wireless spectrum, which is crucial to the success of 5G.
- **Chapter 12, "Antennas":** This chapter provides background on antenna technology, including multiple-input/multiple-output (MIMO) antennas, which are critical to 5G.
- **Chapter 13, "Air Interface Physical Layer":** This chapter examines the two most important aspects of the physical layer of the air interface: modulation and waveform definition. The chapter provides an overview of modulation schemes, as well as treatment of the specific

techniques standardized for 5G. The chapter also surveys various waveform definitions based on orthogonal frequency-division multiplexing (OFDM) and discusses the OFDM techniques specified for 5G.

- **Chapter 14, "Air Interface Channel Coding":** This chapter provides an overview of forward error correction and examines the two techniques specified for 5G: low-density parity-check coding and polar coding.
- **Chapter 15, "5G Radio Access Network":** This chapter provides an overview of key aspects of the 5G RAN. It begins with an overview of RAN architecture, highlighting the roles of various types of RAN nodes. The chapter details the functional split between the RAN and the core network and discusses the protocol architecture and key RAN interfaces. The chapter also provides an overview of the RAN transport network and discusses the concept of integrated architecture and backhaul (IAB).

Supporting Websites

The author maintains a companion website at WilliamStallings.com/5G that includes a list of relevant links and an errata sheet for the book.

WilliamStallings.com/5G
Companion Website

The author also maintains the Computer Science Student Resource Site, at ComputerScienceStudent.com. The purpose of this site is to provide documents, information, and links for computer science students and professionals. Links and documents are organized into seven categories:

computersciencestudent.com
Computer Science Student
Resource Site

- **Math:** Includes a basic math refresher, a queuing analysis primer, a number system primer, and links to numerous math sites.
- **How-to:** Advice and guidance for solving homework problems, writing technical reports, and preparing technical presentations.
- **Research resources:** Links to important collections of papers, technical reports, and bibliographies.
- **Other useful:** A variety of other useful documents and links.
- **Computer science careers:** Useful links and documents for those considering a career in computer science.
- **Writing help:** Help in becoming a clearer, more effective writer.
- **Miscellaneous topics and humor:** You have to take your mind off your work once in a while.

Register Your Book

Register your copy of *5G Wireless* on the InformIT site for convenient access to updates and/or corrections as they become available. To start the registration process, go to informit.com/register and log in or create an account. Enter the product ISBN (9780136767145) and click Submit. Look on the Registered Products tab for an Access Bonus Content link next to this product and follow that link to access any available bonus materials. If you would like to be notified of exclusive offers on new editions and updates, please check the box to receive email from us.

Figure Credits

Figure No.	Page Number	Selection Title	Attribution/Credit Line
FIG01-06a	010	Illustration of landline telephone	Arvind Singh Negi/Red Reef Design Studio. Pearson India Education Services Pvt. Ltd
FIG01-06b; FIG01-11; FIG01-12a; FIG03-08; FIG06-03a; FIG06-06b; FIG15-02; FIG15-05; FIG15-17d; FIG15-19b	010; 027; 029; 094; 172; 175; 003; 006; 026; 029	Modern computer server over white background	Sashkin/Shutterstock
FIG01-06; FIG05-08c; FIG07-04; FIG08-03b; FIG09-03c	010; 139; 210; 243; 278	Server Case on white	Alex Mit/Shutterstock
FIG01-07	013	Illustration of car icon	Victor Metelskiy/ Shutterstock
FIG01-08; FIG02-04; FIG02-05; FIG02-06	017; 047; 048	Communications Tower	Mipan/Shutterstock
FIG01-08; FIG01-11; FIG01-14; FIG09-03	017; 027; 032; 278	Illustration of smart phone	Yukipon/123RF
FIG01-11; FIG01-12b; FIG06-04; FIG06-05; FIG06-06d; FIG09-03b; FIG09-05; FIG09-15; FIG09-17; FIG10-01; FIG12-14; FIG12-15a; FIG15-14; FIG15-17	027; 029; 173; 174; 175; 278; 282; 313; 318; 324; 022; 023; 022; 026;	Vector antenna icon	Yulia Glam/123RF
FIG01-14a; FIG10-10	032; 341	Tablet PC	Oleksiy Mark/ Shutterstock
FIG02-03a	042	Cityscape connected line, technology concept, internet of things conceptual	Vasin Leenanuruksa/123RF
FIG02-03b	042	Smart home. Flat design style vector illustration concept of smart house technology system with centralized control. Editable vector icons for video, mobile apps, Web sites and print projects	MaDedee/Shutterstock
FIG02-03c	042	LCD high definition flat screen TV-set over a white background	AXL/Shutterstock
FIG02-03d	042	Microphone on white background. Computer generated image	Alex Kalmbach/ Shutterstock

Figure No.	Page Number	Selection Title	Attribution/Credit Line
FIG02-06	048	Tower with radio waves	Valdis torms/ Shutterstock
FIG04-05; FIG05-08b; FIG06-07e	117; 139; 177	Server. Server rack, half open door	Jojje/Shutterstock
FIG05-08a	139	Woman-programmer near computer	Ganna Rassadnikova/123RF
FIG05-11	154	Perfect face (with lines) - A so-called perfect face, the expected result of cosmetic surgery. The lines show the perfect proportions of the human face	4634093993/ Shutterstock
FIG06-03b	172	Bathroom scale. Vector.	Alhovik/Shutterstock
FIG06-03c; FIG06-04a	172; 173	Yellow robot arm for industry isolated included clipping path	Baloncici/Shutterstock
FIG06-04b	173	Industry 4.0 Robot concept .Engineers are using virtual AR to maintain and check the work of human robot in the 4.0 Smart Factory.	PaO_STUDIO/ Shutterstock
FIG06-05a	174	Soccer stadium	Alexander Chaikin/ Shutterstock
FIG06-05; FIG06-06c	174; 175	VR virtual reality headset half turned front view isolated on white background. VR is an immersive experience in which your head movements are tracked in 3d world, VR is the future of gaming.	Alexey Boldin/ Shutterstock
FIG06-06a	175	Crowd of anonymous people walking on busy city street	BABAROGA/ Shutterstock
FIG06-07a	177	Collage of a group of people portrait smiling	ESB Professional/ Shutterstock
FIG06-07	177	Group of Diverse Multiethnic Medical People	Rawpixel.com/ Shutterstock

Figure No.	Page Number	Selection Title	Attribution/Credit Line
FIG06-07	177	City hospital building with ambulance and helicopter in flat design. vector.	Sapann Design/ Shutterstock
FIG06-07f	177	An examination room at a doctors office	Rob Byron/123RF
FIG06-14	195	A Drone illustration isolated on white background. Quadcopter icon. Design element for logo, label, emblem, sign. Vector illustration	Liubov Kotliar/123RF
FIG06-14	195	Remote control helicopter toy in white isolated background.	Lutsenko_Oleksandr/ Shutterstock
FIG07-04	210	Desktop computer (monitor, keyboard, mouse) isolated	Dmitry Rukhlenko/ Shutterstock
FIG08-03a	243	Server icon set	Tele52/123RF
FIG08-03c; FIG10-10j	243; 341	Wireless router	Oleksiy Mark/ Shutterstock
FIG08-03; FIG08-08; FIG08-09	243; 254; 257	Colored abstract gear wheels. Raster version	A-R-T/Shutterstock
FIG08-03k; FIG08-05; FIG08-09f	243; 247; 257	Engineering gear	Meilun/Shutterstock
FIG08-03l; FIG08-08d; FIG08-09	243; 254; 257	Rack, of five servers	Fenton Wylam/123RF
FIG08-03n; FIG08-08c; FIG08-09	243; 254; 257	Server rackmount chassis vector graphic illustration isolated	Vadymg/123RF
FIG08-11	262	Servers and communication Internet World	Fenton Wylam/123RF
FIG09-03a	278	Surveillance camera vector	Huston Brady/ Shutterstock
FIG09-03e	278	Fitness tracker isolated on white background	Magraphics/123RF
FIG10-10	341	Computer server icon. Communication and hosting objects series. Vector illustration	Blue Flourishes/ Shutterstock

Figure No.	Page Number	Selection Title	Attribution/Credit Line
FIG10-10	341	An eight-port ethernet network hub. With an uplink port and a DC power-in port	Zern Liew/Shutterstock
FIG10-10c	341	3d illustration of black laptop isolated on white background	dencg/Shutterstock
FIG10-10	341	Wireless router isolated on white background	Sashkin/Shutterstock
FIG10-10	341	Pressure gauge isolated on a white background	Marynchenko Oleksandr/ Shutterstock
FIG11-07a	021	Silhouette factory building with offices and production facilities in perspective	Svetlana Murtazina/123RF
FIG11-07b	021	Lateral red sport car isolated on a white background isolated on a white background: 3D rendering	Cla78/Shutterstock
FIG11-07c	021	Skyscrapers buildings. Towers city business architecture, apartment and office building, urban landscape. Vector illustration in trendy flat style isolated on white background	EgudinKa/Shutterstock
FIG12-09	015	Computer mobility Concept	Aleksanderdn/123RF
FIG12-09d; FIG12-14; FIG12-15; FIG15-17	015; 022; 023; 026	Illustration of smart phone	Radhika Banerjee/ Pearson India Education Services Pvt. Ltd
FIG12-15	023	Illustration of a giant tree on a white background	Iimages/123RF
FIG12-15e	023	Isometric icon representing city building	Tele52/123RF
FIG12-15f	023	Vector isometric icon representing firefighters station building with fire truck and garage	Tele52/Shutterstock
UNPHCov-01	Cover	Smart city and internet of things, wireless communication network, abstract image visual	Krunja/Shutterstock

Chapter 1

Cellular Networks: Concepts and Evolution

Learning Objectives

After studying this chapter, you should be able to:

- Provide an overview of cellular network organization
- Distinguish among four generations of mobile telephony
- Present an overview of 2G, 3G, and 4G systems

Of all the tremendous advances in data communications and telecommunications, perhaps the most revolutionary is the development of cellular networks. Cellular technology is the foundation of mobile wireless communications and supports users in locations that are not easily served by wired networks. Cellular technology is the underlying technology for mobile telephones, personal communications systems, wireless Internet and wireless web applications, and much more.

This chapter begins with a brief discussion of the evolution of cellular networks. This is followed with an overview of basic concepts of cellular networks. The remaining sections provide summaries of the architecture and functionality of the first four generations of cellular networks. This chapter will enable you to better grasp the complexities of fifth-generation networks.

1.1 Evolution of Cellular Networks

Cellular radio is a technique that was developed to increase the capacity available for mobile radio telephone service. Prior to the introduction of cellular communication, mobile radio telephone service was only provided by a high-power transmitter/receiver. A typical system supported about 25 channels with an effective radius of about 80 km. Increasing the capacity of the system required the use of lower-power transmitters with shorter radius as well as numerous transmitters/receivers. This is the principle behind cellular networks.

Since their introduction in the mid-1980s, cellular networks have evolved rapidly. For convenience, industry and standards bodies group the technical advances into *generations*; the fifth generation (5G) is currently being deployed. Each generation has seen improvements in bandwidth, the range of devices supported, the range of applications supported, the number of simultaneous users in a given area, security, and reliability. Figure 1.1 highlights key aspects of the five generations.

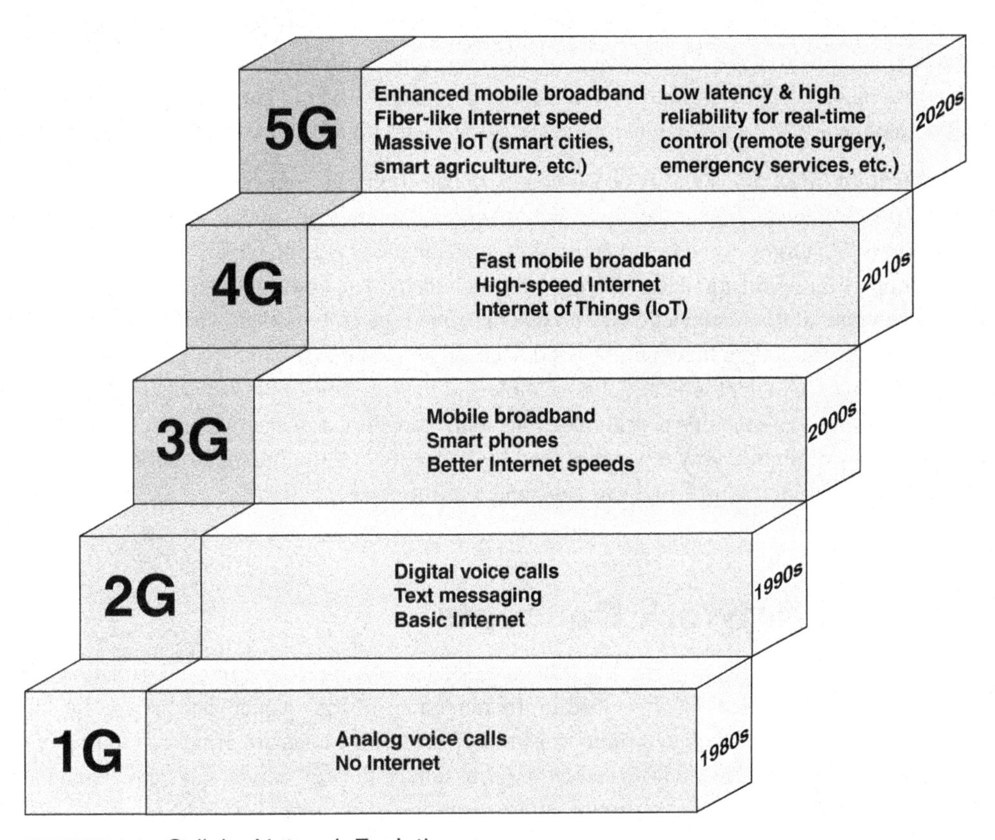

FIGURE 1.1 Cellular Network Evolution

The five generations can be briefly summarized as follows:

- **First generation (1G):** 1G is an analog technology for supporting voice calls. 1G phones had relatively poor battery life and voice quality, were about the size of a brick, and provided little or no security.

- **Second generation (2G):** 2G introduced the use of digital transmission. As with all the other generations, 2G supports voice. It also supports data transmission at modest speeds, so it is suitable only for low-data-rate applications, such as text messaging.
- **Third generation (3G):** Cellular networks as a versatile technology appear with 3G. 3G supports much higher data rates and provides greater security and reliability than 2G. 3G enables the use of smartphones for Internet-related applications such as web browsing, emailing, video downloading, and picture sharing.
- **Fourth generation (4G):** 4G is a significant advance over 3G in almost all aspects. Its purpose is to provide broadband speeds, high quality, and high capacity to users while improving security and lowering the cost of voice and data services, multimedia, and Internet access. Applications include improved mobile web access, IP telephony, gaming services, high-definition mobile TV, video conferencing, 3D television, and cloud computing.
- **Fifth generation (5G):** 5G is a truly revolutionary expansion of the capability of cellular networks. 5G provides much higher data rates, higher connection density, and a far broader range of applications. 5G targets three broad areas of usage. Enhanced mobile broadband (eMBB) provides support for applications demanding very high data rates, equivalent to what can be achieved with optical fiber connections. Massive machine type communications (mMTC) provides the ability to support a huge number of devices in a given geographic area, such as a large Internet of Things (IoT) deployment. Ultra-reliable and low-latency communications (URLLC) is a form of machine-to-machine communications that enables delay-sensitive and mission-critical services that require very low end-to-end delay, such as tactile Internet, remote control of medical or industrial robots, driverless cars, and real-time traffic control.

1.2 Cellular Network Concepts

This section provides an overview of basic cellular network concepts, thus providing a foundation for the survey of the first four generations described in the remainder of the chapter. The section begins with a discussion of the way in which cell radio transmitter/receiver systems are organized to provide the required network coverage. This is followed by a discussion of the key network system elements. Finally, the section provides a general description of the operation of cellular networks.

Cellular Organization

A cellular network uses multiple low-power transmitters, on the order of 100 W or less. Because the range of such a transmitter is small, an area can be divided into cells, each one served by its own antenna, typically located at the center of the cell.[1] Each cell is allocated a band of frequencies and

---

1. One way to increase the subscriber capacity of a cellular network is to replace the omnidirectional antenna at each base station with directional antennas. For example, a typical cellular coverage area is split into three 120-degree sectors using three sets of directional antennas in a triangular configuration of antennas. The base station can either be located at the center of the original (large) cell or at the corner of an original (large) cell. Chapter 12, "Antennas," discusses cell sectorization.

is served by a **base station**, consisting of transmitter, receiver, and control unit. Adjacent cells are assigned different frequencies to avoid interference or crosstalk. However, cells sufficiently distant from each other can use the same frequency band.

The first design decision to make is the shape of cells to cover an area. A matrix of square cells, as shown in Figure 1.2a, would be the simplest layout to define. However, this geometry is not ideal. If the width of a square cell is d, then a cell has four neighbors at a distance d and four neighbors at a distance $\sqrt{2}\ d$. As a mobile user within a cell moves toward the cell's boundaries, it is best if all of the adjacent antennas are equidistant. This simplifies the task of determining when to switch the user to an adjacent antenna and which antenna to choose. A hexagonal pattern provides for equidistant antennas, as shown in Figure 1.2b. The radius of a hexagon is defined to be the radius of the circle that circumscribes it (equivalently, the distance from the center to each vertex, which is also equal to the length of a side of a hexagon). For a cell radius R, the distance between the cell center and each adjacent cell center is $d = \sqrt{3}\ R$.

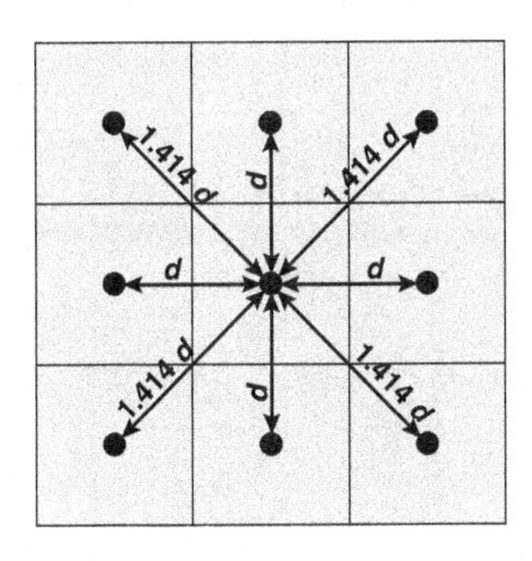

(a) Square pattern

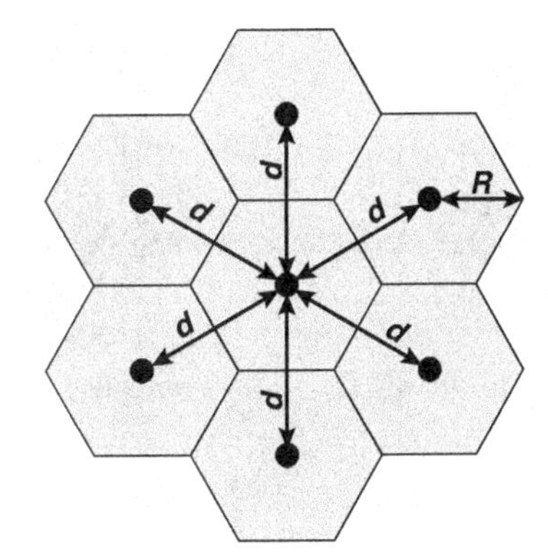

(b) Hexagonal pattern

FIGURE 1.2 Cellular Geometries

In practice, a uniform hexagonal pattern will not precisely map to a coverage area of a base station. Certainly an antenna is not designed to have a hexagonal pattern. Variations from the ideal are also due to topographical limitations such as hills or mountains, local signal propagation conditions such as shadowing from buildings, and practical limitation on siting antennas.

A wireless cellular system limits the opportunity to use the same frequency for different communications because the signals, not being constrained, can interfere with one another even if geographically separated. Systems supporting a large number of communications simultaneously need mechanisms to conserve spectrum.

Frequency Reuse

In a cellular system, each cell has a base transceiver. The transmission power is carefully controlled (to the extent possible in the highly variable mobile communication environment) to allow communication within the cell using a given frequency while limiting the power at that frequency that escapes the cell into adjacent cells. In some cellular architectures, it is not practical to attempt to use the same frequency band in two adjacent cells.[2] In such cases, the design uses the same frequency in other nearby (but not adjacent) cells, thus allowing the frequency to be used for multiple simultaneous conversations. Generally, 10 to 50 frequencies are assigned to each cell, depending on the traffic expected.

A key design issue involves determining the minimum separation between two cells using the same frequency band so that the two cells do not interfere with each other. Various patterns of frequency reuse are possible. Figure 1.3 shows some examples. If the pattern consists of N cells and each cell is assigned the same number of frequencies, each cell can have K/N frequencies, where K is the total number of frequencies allotted to the system. For one first-generation system, $K = 395$ and $N = 7$ is the smallest pattern that can provide sufficient isolation between two uses of the same frequency. This implies that there can be at most $395/7 \approx 57$ frequencies per cell, on average.

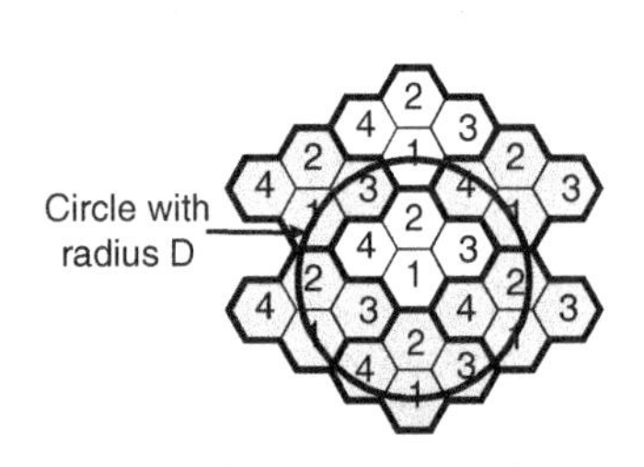

(a) Frequency reuse pattern for $N = 4$

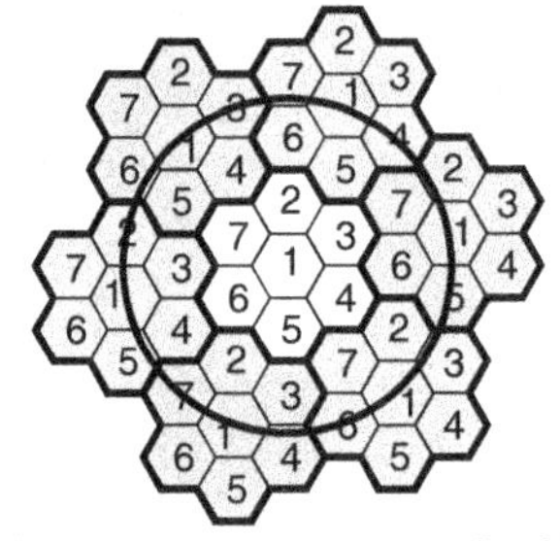

(b) Frequency reuse pattern for $N = 7$

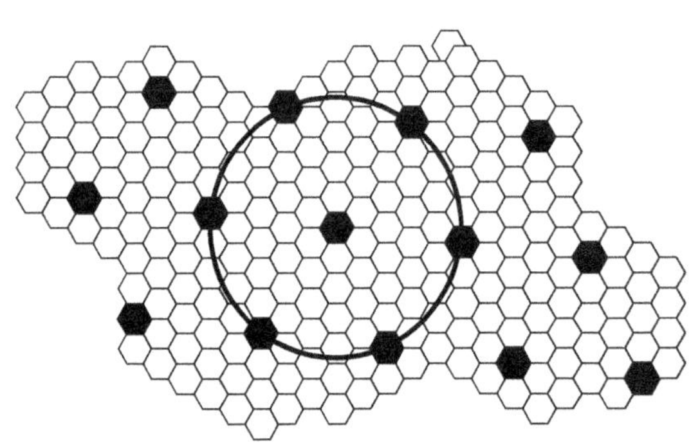

(c) Black cells indicate the frequency reuse for $N = 19$

FIGURE 1.3 Frequency Reuse Patterns

In characterizing frequency reuse, the following parameters are commonly used:

- D = minimum distance between centers of cells that use the same band of frequencies (called *cochannels*)

2. Exceptions include code-division multiple access (CDMA) systems and fourth-generation inter-cell interference coordination and coordinated multipoint transmission systems, described subsequently.

- R = radius of a cell
- d = distance between centers of adjacent cells ($d = \sqrt{3}\,R$)
- N = number of cells in a repetitious pattern (with each cell in the pattern using a unique band of frequencies), termed the **reuse factor**

In a hexagonal cell pattern, only the following values of N are possible:

$$N = I^2 + J^2 + (I \times J) \qquad I, J = 0, 1, 2, 3, \ldots$$

Hence, possible values of N are 1, 3, 4, 7, 9, 12, 13, 16, 19, 21, and so on. The following relationship holds:

$$\frac{D}{R} = \sqrt{3N}$$

This can also be expressed as $D/d = \sqrt{N}$.

Increasing Capacity Through Network Densification

In time, as more customers use a cellular system, traffic may build up so that there are not enough frequency bands assigned to a cell to handle calls. A number of approaches have been used to cope with this situation, including the following:

- **Addition of new channels:** Typically, when a system is set up in a region, not all of the channels are used, and growth and expansion can be managed in an orderly fashion by adding new channels.
- **Frequency borrowing:** In the simplest case, frequencies are taken from adjacent cells by congested cells. The frequencies can also be assigned to cells dynamically.
- **Cell splitting:** In practice, the distribution of traffic and topographic features is not uniform, and this presents opportunities for capacity increase. Cells in areas of high usage can be split into smaller cells. Generally, the original cells are about 6.5 to 13 km in size. The smaller cells can themselves be split. Also, special small cells can be deployed in areas of high traffic demand. (See the subsequent discussion of small cells such as picocells and femtocells.) To use a smaller cell, the power level used must be reduced to keep the signal within the cell. Also, as the mobile units move, they pass from cell to cell, which requires that the call be transferred from one base transceiver to another. This process is called a **handover**.[3] As the cells get smaller, these handovers occur much more frequently. Figure 1.4 indicates schematically how cells can be divided to provide more capacity. A radius reduction by a factor of F reduces the coverage area and increases the required number of base stations by a factor of F^2.

---

3. The term *handoff* is used in U.S. cellular standards documents. ITU documents use the term *handover*, and both terms appear in the technical literature. The meanings are the same.

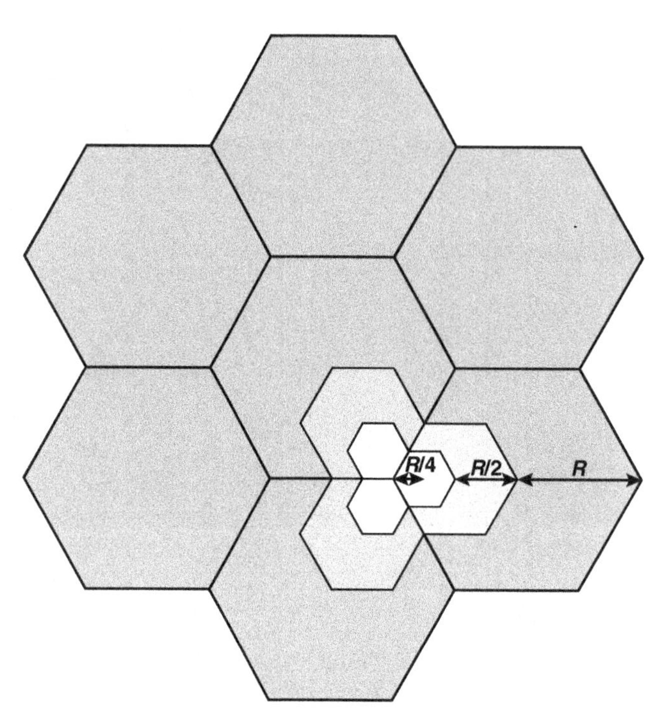

FIGURE 1.4 Cell Splitting with a Cell Reduction Factor of $F = 2$

- **Cell sectoring:** With cell sectoring, a cell is divided into a number of wedge-shaped sectors, each with its own set of channels—typically three sectors per cell. Each sector is assigned a separate subset of the cell's channels, and directional antennas at the base station are used to focus on each sector. This can be seen in the triangular shape of typical cellular antenna configurations, where the antennas mounted on each side of the triangle are directed toward their respective one of the three sectors.

- **Small cells, or micro cells:** As cells become smaller, antennas move from the tops of tall buildings or hills, to the tops of small buildings or the sides of large buildings, and finally to lamp posts, where they form **picocells**. Each decrease in cell size is accompanied by a reduction in the radiated power levels from the base stations and the mobile units. Picocells are useful in city streets in congested areas, along highways, and inside large public buildings. If placed inside buildings, these are called **femtocells**, and they might be open to all users or only to authorized users (e.g., only those who work in the building). If a femtocell is for only a restricted set of users, this is called a **closed subscriber group**. This process of increasing capacity by using small cells is called **network densification**. The large outdoor cells called **macro cells** are intended to support high-mobility users. There are a variety of frequency use strategies for sharing frequencies but avoiding interference problems between small cells and macrocells, such as having separate frequencies for macrocells and small cells or dynamic spectrum assignment between them. In the case of dynamic assignment, **self-organizing networks** of base stations make quick cooperative decisions for channel assignment as needs require.

Ultimately, the capacity of a cellular network depends on how often the same frequencies—or subcarriers, in the case of orthogonal frequency-division multiple access (OFDMA)—can be reused for different mobile devices. Regardless of their location, two mobile devices can be assigned the same frequency if their interference is tolerable. Thus, interference and not location is the limiting factor. If interference can be addressed directly, then the channel reuse patterns in Figure 1.3 might not even be required. For example, if two mobile devices are close to their respective base stations, transmit powers could be greatly reduced for each connection but still provide adequate service. Then the two mobile devicess could use the same frequencies, even in adjacent cells. Modern systems take advantage of these opportunities through techniques such as **inter-cell interference coordination (ICIC)** and **coordinated multipoint transmission (CoMP)**. These techniques perform various functions, such as warning adjacent cells when interference might be significant (e.g., a user is near the boundary between two cells) or performing joint scheduling of frequencies across multiple cells.

Example

Assume a system of 32 cells with a cell radius of 1.6 km, a total of 32 cells, a total frequency bandwidth that supports 336 traffic channels, and a reuse factor of $N = 7$. If there are 32 total cells, what geographic area is covered, how many channels are there per cell, and what is the total number of concurrent calls that can be handled? Repeat for a cell radius of 0.8 km and 128 cells.

Figure 1.5a shows an approximately rectangular pattern. The area of a hexagon of radius R is $1.5\,R^2\sqrt{3}$. A hexagon of radius 1.6 km has an area of 6.65 km^2, and the total area covered is $6.65 \times 32 = 213$ km^2. For $N = 7$, the number of channels per cell is $336/7 = 48$, for a total channel capacity (total number of calls that can be handled) of $48 \times 32 = 1536$ channels. For the layout in Figure 1.5b, the area covered is $1.66 \times 128 = 213$ km^2. The number of channels per cell is $336/7 = 48$, for a total channel capacity of $48 \times 128 = 6144$ channels. A reduction in cell radius by a factor of ½ thus increases the channel capacity by a factor of 4.

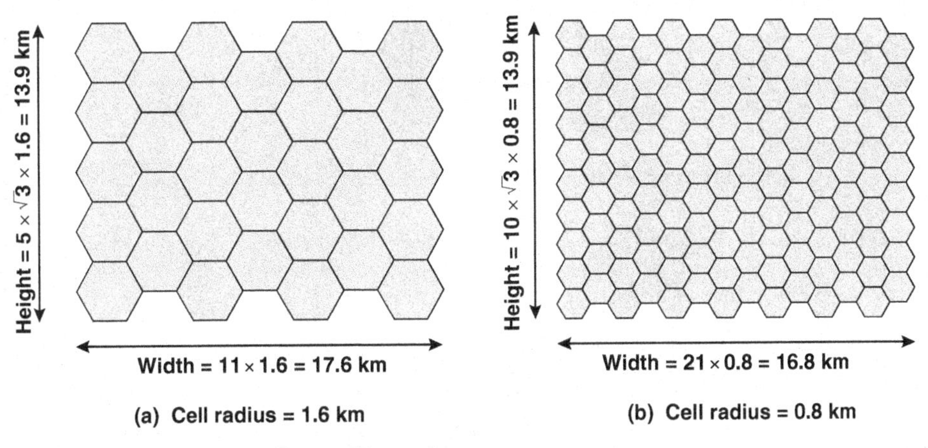

FIGURE 1.5 Frequency Reuse Example

Network System Elements

Figure 1.6a shows the principal elements of a cellular system. The key elements depicted in the figure are:

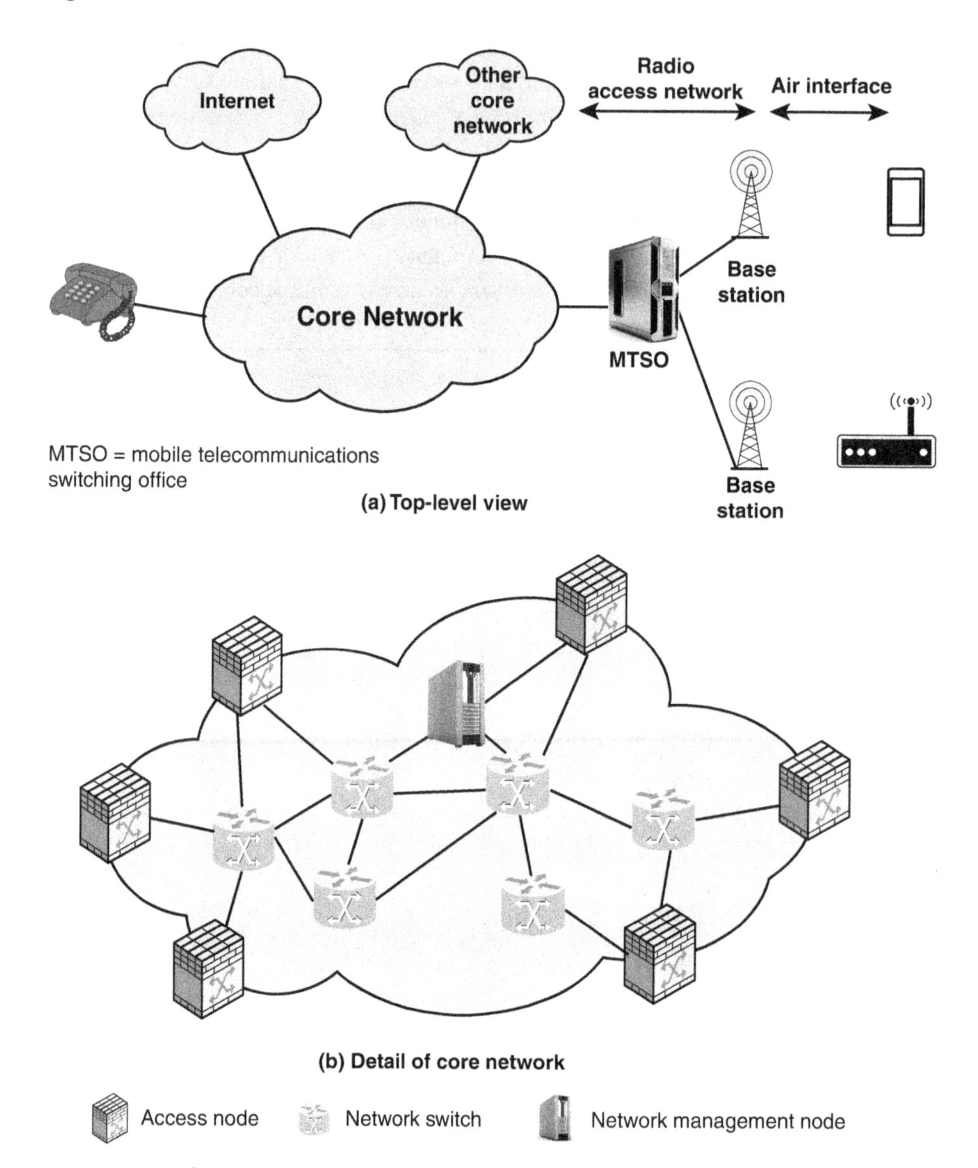

FIGURE 1.6 Simplified Depiction of a Cellular Network System

- **Base station:** A network element in a radio access network responsible for radio transmission and reception in one or more cells to or from the user equipment. A base station can have an integrated antenna or can be connected to an antenna by feeder cables. The base station interfaces the user terminal (through an air interface) to a radio access network infrastructure.

- **Air interface:** Wireless interface between user equipment and the base station, also called a **radio interface**. The air interface specifies the method for transmitting information over the air between base stations and mobile units, including protocols, frequency, channel bandwidth, and the modulation scheme.
- **Mobile telecommunications switching office (MTSO):** Used by a cellular service provider for originating and terminating functions for calls to or from end user customers of the cellular provider. Also known as **mobile switching center** (**MSC**).
- **Radio access network (RAN):** The network that connects radio base stations to the core network. The RAN provides and maintains radio-specific functions, which may be unique to a given radio access technology, that allow users to access the core network. RAN components include base stations and antennas, MTSOs, and other management and transmission elements.
- **Core network:** A central network that provides networking services to attached distribution and access networks.

A base station (BS) provides the radio coverage for a cell. The BS includes an antenna, a controller, and a number of transceivers for communicating on the channels assigned to that cell. The controller is used to handle the call process between the mobile unit and the rest of the network. At any time, a number of mobile units may be active and moving about within a cell, communicating with the BS. Each BS is connected to a mobile telecommunications switching office (MTSO), with one MTSO serving multiple BSs. Typically, the link between an MTSO and a BS is a wire line, although wireless links are also used. An MTSO connects calls between mobile units. The MTSO is also connected to the public telephone or telecommunications network and can make a connection between a fixed subscriber to the public network and a mobile subscriber to the cellular network. The mobile is also given access to the Internet and to subscribers served by other core networks. The MTSO assigns the voice channel to each call, performs handoffs (discussed subsequently), and monitors the call for billing information.

Figure 1.6b indicates in general terms the architecture of the core network. Access nodes provide an interface to the radio access network and other elements that can interface with the core network. These access nodes provide the entry point to a packet-switched network based on Internet Protocol (IP) technology, which includes switches and various network management servers.

Operation of Cellular Systems

The use of a cellular system is fully automated and requires no action on the part of the user other than placing or answering a call or setting up a data connection. Two types of channels are available between the mobile unit and the BS: control channels and traffic channels. **Control channels** are used to exchange information having to do with setting up and maintaining connections and with establishing a relationship between a mobile unit and the nearest BS. **Traffic channels** carry a voice or data connection between users. Figure 1.7 illustrates the steps in a typical voice call between two mobile users within an area controlled by a single MTSO. The steps are as follows:

Step 1. **Mobile unit initialization:** When the mobile unit is turned on, it scans and selects the strongest setup control channel used for this system (see Figure 1.7a). Cells with different frequency bands repetitively broadcast on different setup channels. The receiver selects the strongest setup channel and monitors that channel. The effect of this procedure is that the mobile unit has automatically selected the BS antenna of the cell within which it will operate.[4] Then a handshake takes place between the mobile unit and the MTSO controlling this cell, through the BS in this cell. The handshake is used to identify the user and register its location. As long as the mobile unit is on, this scanning procedure is repeated periodically to account for the motion of the unit. If the unit enters a new cell, then a new BS is selected. In addition, the mobile unit is monitoring for pages, discussed subsequently.

Step 2. **Mobile-originated call:** A mobile unit originates a call by sending the number of the called unit on the preselected setup channel (see Figure 1.7b). The receiver at the mobile unit first checks that the setup channel is idle by examining information in the forward (from the BS) channel. When an idle is detected, the mobile unit may transmit on the corresponding reverse (to BS) channel. The BS sends the request to the MTSO.

Step 3. **Paging:** The MTSO attempts to complete the connection to the called unit. The MTSO sends a paging message to certain BSs to find the called mobile unit, depending on the called mobile unit number and the latest information on the unit's whereabouts (see Figure 1.7c). The MTSO does not always know the location of every mobile if certain mobiles have been in idle modes. Each BS transmits the paging signal on its own assigned paging channel.

Step 4. **Call accepted:** The called mobile unit recognizes its number on the paging channel being monitored and responds to that BS, which sends the response to the MTSO. The MTSO sets up a circuit between the calling and called BSs. At the same time, the MTSO selects an available traffic channel within each BS's cell and notifies each BS, which in turn notifies its mobile unit (see Figure 1.7d). The two mobile units tune to their respective assigned channels.

Step 5. **Ongoing call:** While the connection is maintained, the two mobile units exchange voice or data signals, going through their respective BSs and the MTSO (see Figure 1.7e).

Step 6. **Handoff:** If a mobile unit moves out of range of one cell and into the range of another during a connection, the traffic channel has to change to one assigned to the BS in the new cell (see Figure 1.7f). The system makes this change without either interrupting the call or alerting the user.

---

4. Usually, but not always, the antenna and therefore the base station selected is the closest one to the mobile unit. However, because of propagation anomalies, this is not always the case.

Other functions performed by the system but not illustrated in Figure 1.7 include:

- **Call termination:** When one of the two users hangs up, the MTSO is informed, and the traffic channels at the two BSs are released.

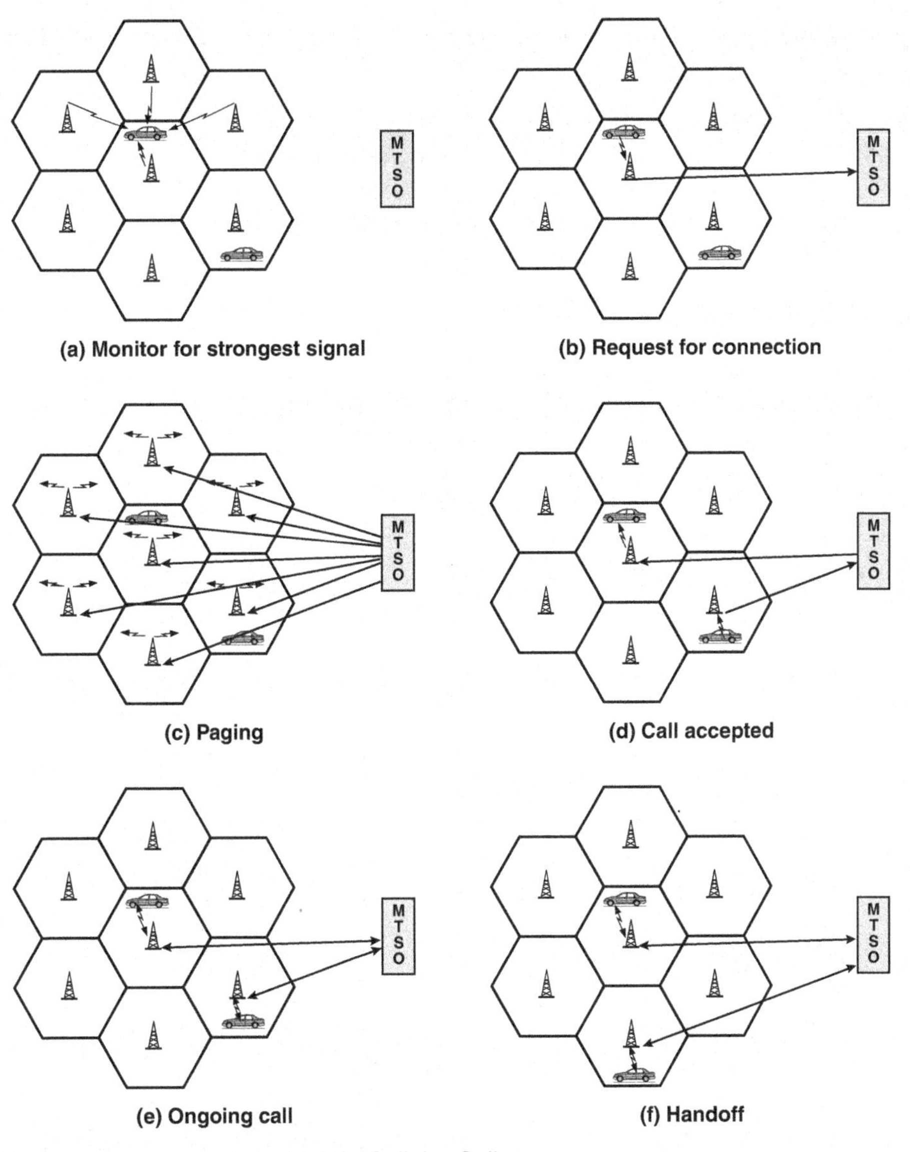

FIGURE 1.7 Example of Mobile Cellular Call

- **Call drop:** During a connection, because of interference or weak signal spots in certain areas, if the BS cannot maintain the minimum required signal strength for a certain period of time, the traffic channel to the user is dropped, and the MTSO is informed.
- **Calls to/from fixed and remote mobile subscriber:** The MTSO connects to the public switched telephone network. Thus, the MTSO can set up a connection between a mobile user in its area and a fixed subscriber via the telephone network. Further, the MTSO can connect to a remote MTSO via the telephone network or via dedicated lines and set up a connection between a mobile user in its area and a remote mobile user.
- **Emergency call prioritization:** If a user identifies the call as an emergency call, calls that may experience blocking due to a busy will get priority, which implies that another existing call may be dropped

1.3 First Generation (1G)

The original cellular telephone networks provided analog traffic channels; these are now referred to as first-generation (or 1G) systems. For 1G, numerous incompatible standard specifications exist. Later generations saw a gradual consolidation to fewer and fewer standards, culminating in 5G, which is a single global specification and standard. The most widely used 1G schemes were the following:

- Nordic Mobile Telephone (NMT), used in Nordic countries, Switzerland, the Netherlands, Eastern Europe, and Russia
- Advanced Mobile Phone System (AMPS), used in the United States and most other Western Hemisphere countries and also in Australia and many Asian countries
- TACS (Total Access Communications System) in the United Kingdom and some Middle East countries
- C-NETZ in West Germany, Portugal, and South Africa
- Radiocom 2000 in France
- TMA in Spain
- Radio Telephone Mobile System (RTMS) in Italy
- Multiple systems in Japan, including TZ-801, TZ-802, and TZ-803 developed by NTT (Nippon Telegraph and Telephone Corporation) and Japan Total Access Communications System (JTACS) operated by Daini Denden Planning, Inc. (DDI)

As an example, this section presents a brief overview of AMPS, which remained in use until about 2010.

In North America, two 25-MHz bands were allocated to AMPS: one for transmission from the base station to the mobile unit (869–894 MHz) and the other for transmission from the mobile to the base station (824–849 MHz). Each of these bands was split in two to encourage competition (i.e., so that in each market two operators could be accommodated). An operator was allocated only 12.5 MHz in each direction for its system. The channels were spaced 30 kHz apart, which allowed a total of 416 channels per operator. Twenty-one channels were allocated for control, leaving 395 to carry calls. The control channels were data channels operating at 10 kbps.

The conversation channels carried the conversations in analog using frequency modulation (FM). Simple frequency-division multiple access (FDMA) was used to provide multiple access. FDMA for cellular systems can be described as follows: Each cell is allocated a total of $2M$ channels of bandwidth δ Hz each. Half the channels (the reverse channels) are used for transmission from the mobile unit to the base station: $f_c, f_c + \delta, f_c + 2\delta, \ldots, f_c + (M - 1)\delta$, where f_c is the center frequency of the lowest-frequency channel. The other half of the channels (the forward channels) are used for transmission from the base station to the mobile unit: $f_c + \Delta, f_c + \delta + \Delta, f_c + 2\delta + \Delta, \ldots, f_c + (M - 1)\delta + \Delta$, where Δ is the spacing between the reverse and forward channels. When a connection is set up for a mobile user, the user is assigned two channels, at f and $f + \Delta$, for full-duplex communication. This arrangement is quite wasteful because much of the time, one or both of the channels are idle.

Each AMPS service included 21 full-duplex 30-kHz control channels, consisting of 21 reverse control channels (RCCs) from subscriber to base station, and 21 forward channels from base station to subscriber. These channels transmitted digital data using frequency-shift keying (FSK).[5] In essence, the two binary values were represented by two different frequencies near the carrier frequency.

This number of channels was inadequate for most major markets, and it became necessary to find some way either to use less bandwidth per conversation or to reuse frequencies. Both approaches were taken in the various approaches to 1G telephony. AMPS used frequency reuse.

1.4 Second Generation (2G)

First-generation cellular networks, such as AMPS, quickly became highly popular, threatening to swamp available capacity. Second-generation (2G) systems were developed to provide higher-quality signals, higher data rates for support of digital services, and greater capacity. Key differences between 1G and 2G networks include:

- **Digital traffic channels:** The most notable difference between the two generations is that 1G systems are almost purely analog, whereas 2G systems are digital. In particular, 1G systems are designed to support voice channels using FM; digital traffic is supported only by the use of a

5. FSK is a simple form of modulation, described in Chapter 13, "Air Interface Physical Layer."

modem that converts the digital data into analog form. 2G systems provide digital traffic channels. These systems readily support digital data; voice traffic is first encoded in digital form before transmission.

- **Encryption:** Because all of the user traffic, as well as control traffic, is digitized in 2G systems, it is a relatively simple matter to encrypt all of the traffic to prevent eavesdropping. All 2G systems provide this capability, whereas 1G systems send user traffic in the clear, providing no security.
- **Error detection and correction:** The digital traffic stream of 2G systems also lends itself to the use of error detection and correction techniques, such as those discussed in Chapter 14, "Air Interface Channel Coding." The result can be very clear voice reception.
- **Channel access:** In 1G systems, each cell supports a number of channels. At any given time, a channel is allocated to only one user. 2G systems also provide multiple channels per cell, but each channel is dynamically shared by a number of users using time-division multiple access (TDMA) or code-division multiple access (CDMA). TDMA is described next; CDMA is defined in the discussion of 3G systems.

Beginning around 1990, a number of different second-generation systems were deployed. The most widely used of them was the Global System for Mobile Communications (GSM), which is still in use today. This section next looks at an underlying technology for GSM: time-division multiple access (TDMA). The remainder of the section examines some details of GSM.

Time-Division Multiple Access

With TDMA for cellular systems, as with FDMA, each cell is allocated a number of channels—half reverse channels and half forward channels. For full-duplex communication, a mobile unit is assigned capacity on matching reverse and forward channels. In addition, each physical channel is further subdivided into a number of logical channels. Transmission is in the form of a repetitive sequence of frames, each of which is divided into a number of time slots. Each slot position across the sequence of frames forms a separate logical channel.

Figure 1.8 illustrates the difference between FDMA and TDMA. With FDMA, each user communicates with the base station on its own narrow frequency band. For TDMA, the users share a wider frequency band and take turns communicating with the base station. Because of the use of a wider frequency band with TDMA, the two configurations shown give each station that same data rate whether using FDMA or TDMA.

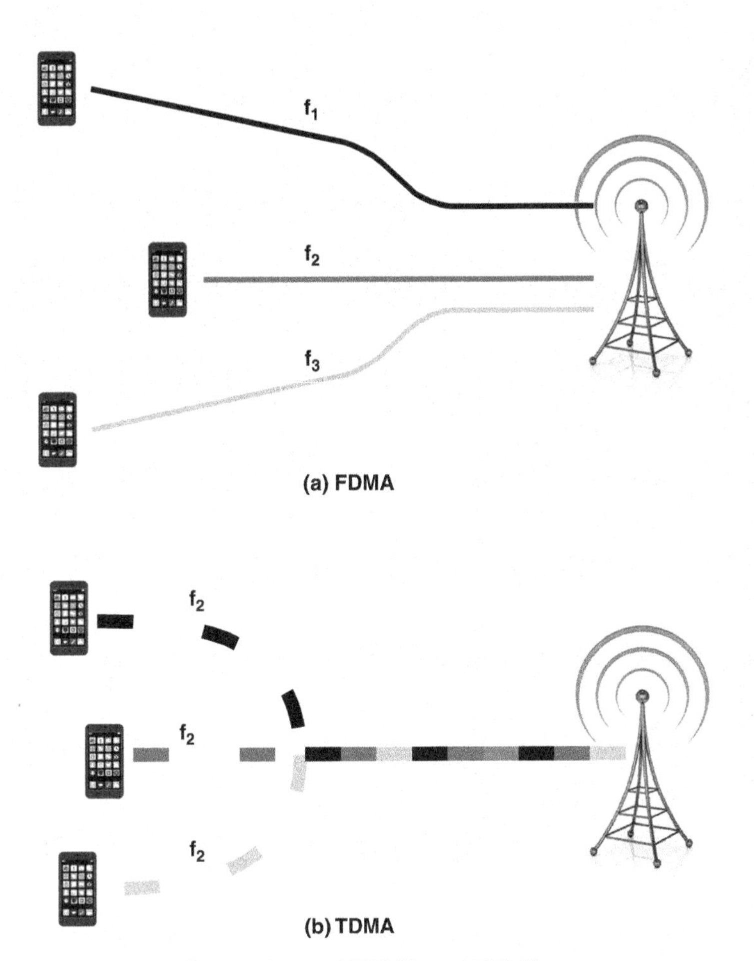

FIGURE 1.8 Comparison of FDMA and TDMA

GSM Architecture

Before the Global System for Mobile Communications (GSM) was developed, the countries of Europe used a number of incompatible first-generation cellular phone technologies. GSM was developed to provide a common second-generation technology for Europe so that the same subscriber units could be used throughout the continent. The technology was extremely successful. GSM first appeared in 1990 in Europe. Similar systems were implemented in North and South America, Asia, North Africa, the Middle East, and Australia.

GSM Network Architecture

Figure 1.9 shows the key functional elements in the GSM system. The boundaries at Um, Abis, and A refer to interfaces between functional elements that are standardized in the GSM documents. Thus, it is possible to buy equipment from different vendors with the expectation that the various devices will successfully interoperate.

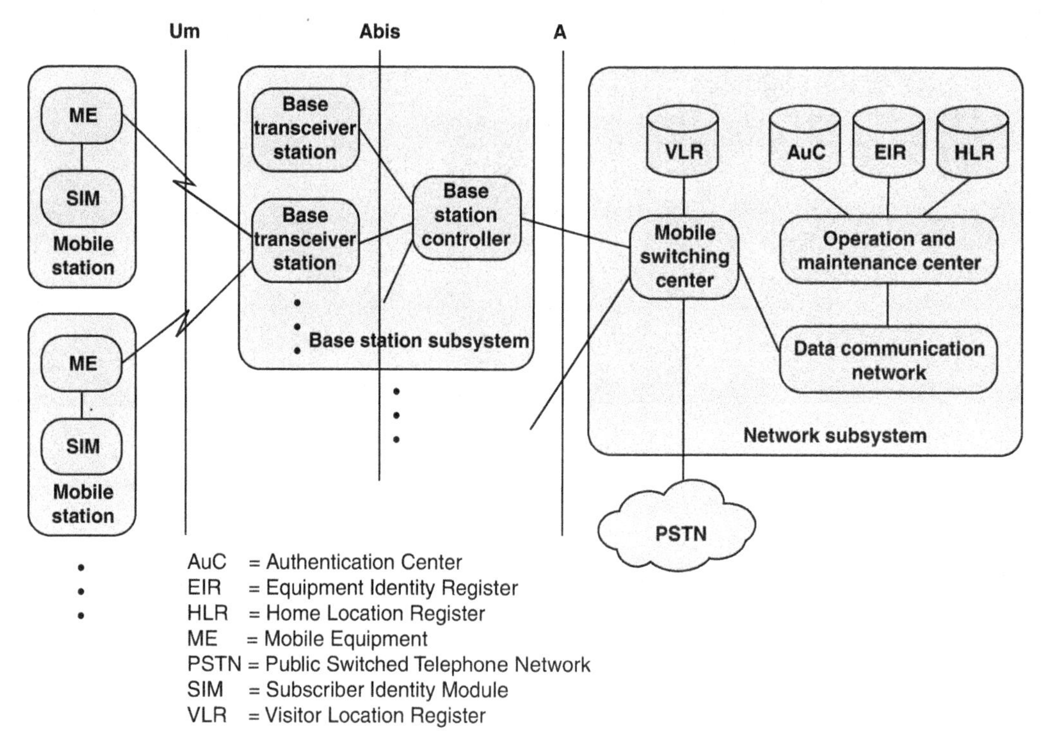

FIGURE 1.9 2G Architecture: GSM

Mobile Station

A mobile station (MS) communicates across the Um interface, also known as the air interface, with a base station transceiver in the same cell in which the mobile unit is located. A mobile station consists of two components: the mobile equipment (ME) and the subscriber identity module (SIM) The ME is the physical terminal, such as a cell phone, which includes the radio transceiver, digital signal processors, and a slot for the SIM. The SIM is a portable device in the form of a smart card or plug-in module that stores the subscriber's identification number, the networks the subscriber is authorized to use, encryption keys, and other information specific to the subscriber. The GSM subscriber units are totally generic until a SIM is inserted. A benefit of having a separate removable SIM is that the user can get a new phone while keeping the same subscription and phone number. In addition, a user who is traveling can buy a local subscription (SIM card) to avoid roaming costs, thus using the same phone with a different SIM card.

Base Station Subsystem

A base station subsystem (BSS) consists of a base station controller and one or more base transceiver stations. Each base transceiver station (BTS) defines a single cell; it includes a radio antenna, a radio transceiver, and a link to a base station controller. A GSM cell can have a radius of between 100 m and 35 km, depending on the environment. A base station controller (BSC) may be collocated with a BTS or may control multiple BTS units and hence multiple cells. The BSC reserves radio frequencies, manages the handoff of a mobile unit from one cell to another within the BSS, and controls paging.

Network Subsystem

The network subsystem (NS) provides the link between the cellular network and the public switched telecommunications networks. The NS controls handoffs between cells in different BSSs, authenticates users and validates their accounts, and includes functions for enabling worldwide roaming of mobile users. The central element of the NS is the mobile switching center (MSC). It is supported by four databases that it controls:

- **Home location register (HLR) database:** The HLR stores information, both permanent and temporary, about each of the subscribers that "belongs" to it (i.e., for which the subscriber has its telephone number associated with the switching center).
- **Visitor location register (VLR) database:** One important, temporary piece of information is the location of the subscriber. The VLR maintains information about subscribers that are currently physically located in the region covered by the switching center. It records whether or not the subscriber is active and other parameters associated with the subscriber. For a call coming to the subscriber, the system uses the telephone number associated with the subscriber to identify the home switching center of the subscriber. This switching center can find in its HLR the switching center in which the subscriber is currently physically located. For a call coming from the subscriber, the VLR is used to initiate the call. Even if the subscriber is in the area covered by its home switching center, it is also represented in the switching center's VLR, for consistency.
- **Authentication center (AuC) database:** This database is used for authentication activities of the system; for example, it holds the authentication and encryption keys for all the subscribers in both the home and visitor location registers. The AuC controls access to user data, and it is also used for authentication when a subscriber joins a network. GSM transmission is encrypted, so it is private. A stream cipher, A5, is used to encrypt transmissions from subscriber to base transceiver. However, the conversation is in the clear in the landline network. Another key is used for authentication.
- **Equipment identity register (EIR) database:** The EIR keeps track of the type of equipment that exists at the mobile station. It also plays a role in security (e.g., blocking calls from stolen mobile stations and preventing use of the network by stations that have not been approved).

Radio Link Aspects

The GSM spectral allocation is 25 MHz for base transmission (935–960 MHz) and 25 MHz for mobile transmission (890–915 MHz). Other GSM bands have also been defined outside Europe. Users access the network using a combination of frequency-division multiple access (FDMA) and time-division multiple access (TDMA). There are radio-frequency carriers every 200 kHz, which provide for 125 full-duplex channels. The channels are modulated at a data rate of 270.833 kbps. As with AMPS, there are two types of channels: traffic channels and control channels.

GPRS and EDGE

Phase 2 of GSM introduced the General Packet Radio Service (GPRS), which provides a packet-switching capability to GSM. Previously, sending data traffic required opening a voice connection, sending data, and closing the connection. GPRS allows users to open a persistent data connection. It also establishes a system architecture for carrying the data traffic. GPRS has different error control coding schemes, and the scheme with the highest throughput (no error control coding, just protocol overheads) produces 21.4 kbps from the 22.8 kbps gross data rate. GPRS can combine up to 8 GSM connections; therefore, overall throughputs of up to 171.2 kbps can be achieved.

The next iteration of GSM included Enhanced Data Rates for GSM Evolution (EDGE). EDGE introduced more efficient modulation schemes. This increased the gross max data rates per channel, depending on channel conditions, up to $22.8 \times 3 = 68.4$ kbps (including overhead from the protocol headers). Using all eight channels in a 200 kHz carrier, gross data transmission rates up to 547.2 kbps became possible. Actual throughput can be up to 513.6 kbps. A later release of EDGE, 3GPP Release 7, added even higher-order modulation and coding schemes that adapt to channel conditions. Downlink data rates over 750 kbps and uplink data rates over 600 kbps can be achieved in excellent channel conditions.

1.5 Third Generation (3G)

With the coming of 3G, international standards activity for cellular networks was consolidated within the International Telecommunication Union (ITU). Chapter 2, "5G Standards and Specifications," discusses ITU's role in the development of cellular network standards in some detail. Briefly, International Mobile Telecommunications (IMT) is the generic term used by the ITU community to designate broadband mobile systems. ITU has promulgated standards for 3G, 4G, and 5G systems under the umbrella terms IMT-2000, IMT- Advanced, and IMT-2020, respectively.

The objective of the third-generation (3G) of wireless communication is to provide fairly high-speed wireless communications to support multimedia, data, and video in addition to voice. The IMT-2000 initiative defines ITU's view of third-generation capabilities as:

- Voice quality comparable to that of the public switched telephone network
- 144 kbps data rate available to users in high-speed motor vehicles over large areas

- 384 kbps available to pedestrians standing or moving slowly over small areas
- Support (to be phased in) for 2.048 Mbps for office use
- Symmetrical and asymmetrical data transmission rates
- Support for both packet-switched and circuit-switched data services
- An adaptive interface to the Internet to reflect efficiently the common asymmetry between inbound and outbound traffic
- More efficient use of the available spectrum in general
- Support for a wide variety of mobile equipment
- Flexibility to allow the introduction of new services and technologies

Code-Division Multiple Access

The dominant technology for 3G systems is code-division multiple access (CDMA). CDMA builds on the use of spread spectrum. Spread spectrum is a technique in which the information in a signal is spread over a wider bandwidth than necessary by using a spreading code. CDMA is a multiple-access technique in which multiple logical channels have access to the same frequency band. The logical channels encode bits in such a way that there is zero or little cross-correlation, so that such signals are distinguishable even when they share the same frequency bands and the same time intervals. 5G does not employ CDMA, and so this book does not pursue this topic.

3G Architecture

Although various 3G air interfaces were standardized, the overall architecture, at a high level, is the same for all of these approaches. Figure 1.10 shows this architecture, termed the Universal Mobile Telecommunications System (UMTS), which is an upgrade to the 2G GSM.

UMTS comprises three elements (compare with GSM architecture, shown in Figure 1.9):

- **User equipment:** This corresponds to the mobile station portion of Figure 1.9. However, UMTS documents use the more general term *user equipment* because many of the devices that use the network are digital devices with no voice capability.
- **Radio access network (RAN):** The RAN has two functional elements. One of them corresponds to the RAN in 2G architectures, consisting of base stations and controllers that provide access for GSM radio connections. The other is UTRAN, described subsequently.
- **Universal Mobile Telecommunications System (UMTS) core network:** This is the equivalent of the GSM network subsystem. It has two functional domains. One of them uses

circuit-switched technology to provide for voice connections to the public switched telephone network (PSTN) and to GSM systems. The other domain introduces packet-switched technology to provide efficient service for data traffic.

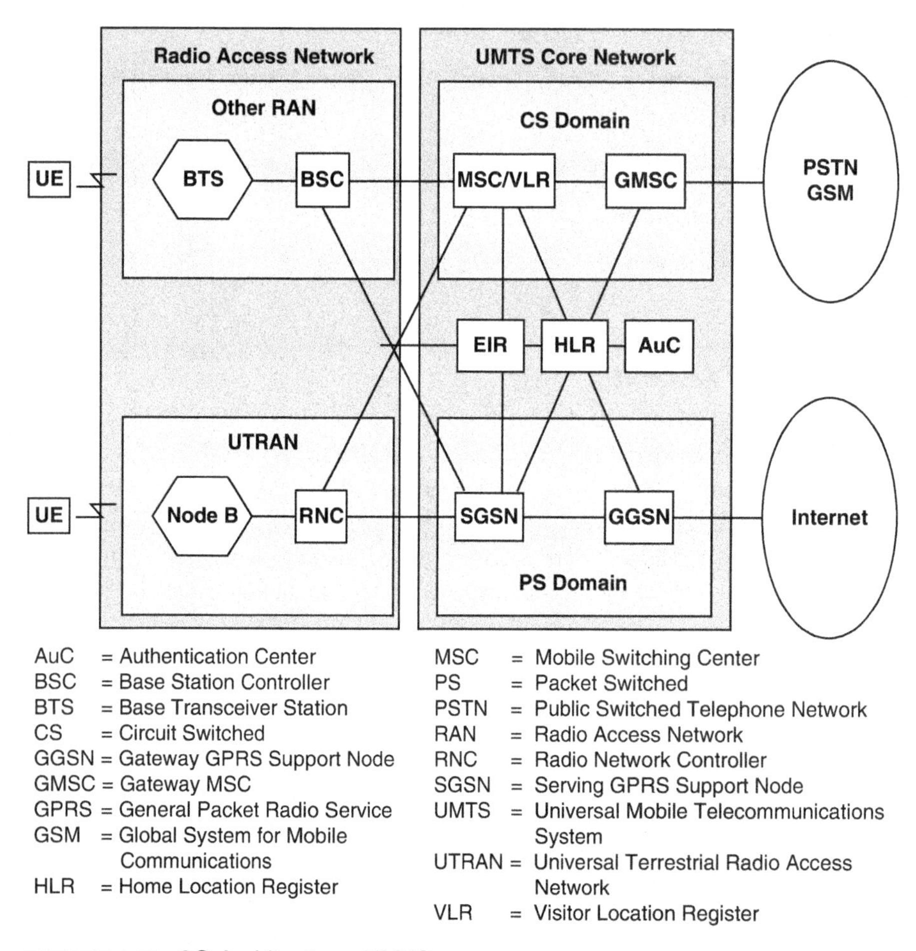

FIGURE 1.10 3G Architecture: UMTS

UTRAN

Universal Terrestrial Radio Access Network (UTRAN) is part of the evolution to packet-switched technology to support efficient digital data communications. The essential elements of the UTRAN are the Node B and the radio network controller (RNC). A Node B is equivalent to a traditional base transceiver station (BTS) but is designed to support high-speed data traffic, using a CDMA-based air interface. Also, a single BTS serves a single cell, whereas a single Node B can use sector antennas to serve multiple sectors. The RNC is equivalent to a base station controller (BSC) but designed to support multiple Node Bs.

UMTS Core Network

The UMTS core network consists of a number of functional elements, some in the circuit-switched (CS) domain, some in the packet-switched (PS) domain, and some shared by both domains. Functional elements in the CS domain are primarily based on the GSM network entities and carry data in a circuit-switched manner (i.e., a permanent channel for the duration of the call). These elements are:

- **Mobile switching center (MSC):** Manages circuit-switched calls. The MSC also includes a visitor location register (VLR), which controls mobile stations roaming in the MSC area. When a mobile station (MS) enters a new location area, it starts a registration procedure. An MSC in charge of that area notices this registration and transfers to a visitor location register the identity of the location area where the MS is situated.
- **Gateway MSC (GMSC):** The interface to the PTSN and GSM networks.

The elements in the PS domain are:

- **Serving GPRS support node (SGSN):** This element was first introduced in 2G GSM systems to support packet-switched functionality. It provides the following functions:
  - **Mobility management (MM):** When user equipment (UE) attaches to the PS domain, the SGSN generates MM information based on the mobile's current location.
  - **Session management:** The SGSN manages the data sessions, providing the required quality of service. It also manages the PDP (Packet Data Protocol) contexts, which are logical connections over which the data packets are sent.
  - **Billing:** The SGSN monitors the flow of data to generate billing information.
- **Gateway GPRS support node (GGSN):** The GGSN is the central element within the PS domain. It handles interworking between the UMTS packet-switched network and external packet-switched networks, thus functioning as a router. In operation, when the GGSN receives data addressed to a specific user, it checks whether the user is active and then forwards the data to the SGSN serving the particular UE.

The elements that interconnect the CS and PS domains are:

- **Home location register (HLR):** This database contains administrative information about each subscriber, along with that individual's last known location. In this way, the UMTS network is able to route calls to the relevant RNC/Node B. When a user switches on his UE, it registers with the network, and from this, it is possible to determine which Node B it communicates with so that incoming calls can be routed appropriately. Even when the UE is not active (but switched on), it re-registers periodically to ensure that the network (HLR) is aware of its latest position with the current or last-known location on the network.

- **Equipment identity register (EIR):** This entity decides whether given user equipment may be allowed on the network. Each piece of user equipment has a number known as the International Mobile Equipment Identity (IMEI). This number is installed in the equipment and is checked by the network during registration.
- **Authentication center (AuC):** The AuC is a protected database that contains the secret key also contained in the user's UMTS SIM (USIM) card. The USIM card includes a microprocessor that provides enhanced capability compared to a SIM card.

1.6 Fourth Generation (4G)

The evolution of smartphones and cellular networks ushered in a new generation of capabilities and standards, collectively called 4G. 4G systems provide broadband Internet access for a variety of mobile devices, including laptops, smartphones, and tablets. 4G networks support mobile web access and high-bandwidth applications such as high-definition mobile TV, mobile video conferencing, and gaming services.

These requirements have led to the development of a fourth generation (4G) of mobile wireless technology that is designed to maximize bandwidth and throughput while also maximizing spectral efficiency. The ITU has issued directives for 4G networks. According to the ITU, an IMT-Advanced (or 4G) cellular system must fulfill a number of minimum requirements, including the following:

- Be based on an all-IP packet switched network.
- Support peak data rates of up to approximately 100 Mbps for high-mobility mobile access and up to approximately 1 Gbps for low-mobility access such as local wireless access.
- Dynamically share and use the network resources to support more simultaneous users per cell.
- Support smooth handovers across heterogeneous networks.
- Support high quality of service for next-generation multimedia applications.

In contrast to earlier generations, 4G systems do not support traditional circuit-switched telephony service but provide only IP telephony services. And, as may be observed in Table 1.1, the spread-spectrum radio technologies that characterized 3G systems are replaced in 4G systems by OFDMA multicarrier transmission and frequency-domain equalization schemes. (Part Four, "5G NR Air Interface and Radio Access Network," examines these technologies.)

TABLE 1.1 Cellular Technology Generations

	1G	2G	3G	4G	5G
Approximate deployment date	1980s	1990s	2000s	2010s	2020s
Theoretical download speed	2 kbps	384 kbps	2 Mbps	1 Gbps	> 1 Gbps
Latency	N/A	629 ms	212 ms	60–98 ms	< 1 ms
Air interface	Analog	Digital, TDMA, CDMA	WCDMA, CDMA2000	OFDM, OFDMA, MIMO	AAS, FD-MIMO, OFDMA, SC-FDMA
Core network	Circuit switching	Circuit switching, packet switching	Packet switching, IP	IP	IP
Primary service	Analog phone calls	Digital phone calls, messaging	Phone calls, messaging, data	Full data services	eMBB, mMTC, URLLC

AAS = Active antenna system

CDMA = Code-division multiple access

eMBB = Enhanced mobile broadband

FD-MIMO = Full-dimension MIMO

IP = Internet Protocol

MIMO = Multiple input/multiple output

mMTC = Massive machine type communications

OFDM = Orthogonal frequency division multiplexing

OFDMA = Orthogonal frequency-division multiple access

SC-FDMA = Single-carrier FDMA

TDMA = Time-division multiple access

URLLC = Ultra-reliable and low-latency communications

The dominant standard, **Long Term Evolution (LTE)**, was developed by the Third Generation Partnership Project (3GPP), a consortium of Asian, European, and North American telecommunications standards organizations. LTE uses pure OFDMA on the downlink, but a technique that is based on OFDMA offers enhanced power efficiency for the uplink. All of the major carriers in the United States, including AT&T, Verizon, and T-Mobile, have adopted a version of LTE based on frequency-division duplex (FDD), whereas China Mobile, the world's largest telecommunication carrier, has adopted a version of LTE based on time-division duplex (TDD).

LTE development began in the 3G era, and its initial releases provided 3G or enhanced 3G service. Beginning with 3GPP Release 10, LTE provides a service that fulfills the IMT-Advanced criteria for 4G, known as **LTE-Advanced**. Table 1.2 compares the performance goals of LTE and LTE-Advanced.

TABLE 1.2 Comparison of Performance Requirements for LTE and LTE-Advanced

System Performance		LTE	LTE-Advanced
Peak rate	Downlink	100 Mbps @20 MHz	1 Gbps @100 MHz
	Uplink	50 Mbps @20 MHz	500 Mbps @100 MHz
Control plane delay	Idle to connected	< 100 ms	< 50 ms
	Dormant to active	< 50 ms	< 10 ms
User plane delay		< 5ms	Lower than LTE
Spectral efficiency (peak)	Downlink	5 bps/Hz @2 × 2	30 bps/Hz @8 × 8
	Uplink	2.5 bps/Hz @1 × 2	15 bps/Hz @4 × 4
Mobility		Up to 350 km/h	Up to 350–500 km/h

The following sections provide a brief overview of LTE and LTE-Advanced. Figure 1.11 illustrates the principal elements in an LTE-Advanced network. The two main components are the radio access network, called Evolved Universal Terrestrial Radio Access Network (E-UTRAN), and the core network, called evolved packet core (EPC).

E-UTRAN

The heart of the E-UTRAN is the base station, designated **evolved NodeB (eNodeB)**. In UMTS, the base station is referred to as NodeB. The key differences between the two base station technologies are:

- The NodeB station interface with subscriber stations (referred to as user equipment [UE]) is based on CDMA, whereas the eNodeB air interface is based on OFDMA.
- eNodeB embeds its own control functionality rather than using a radio network controller (RNC), as does a NodeB. This means that the eNodeB now supports radio resource control, admission control, and mobility management, which were originally the responsibility of the RNC. The simpler structure without an RNC results in simpler operation and higher performance.

The X2 interface shown in Figure 1.11 is used for eNodeBs to interact with each other. The architecture is open so that there can be interconnections between different manufacturers. There is a control plane X2-C interface that supports mobility management, handover preparation, status transfer, UE context release, handover cancel, inter-cell interference coordination, and load management. The X2-U interface is the user plane interface used to transport data during X2-initiated handover.

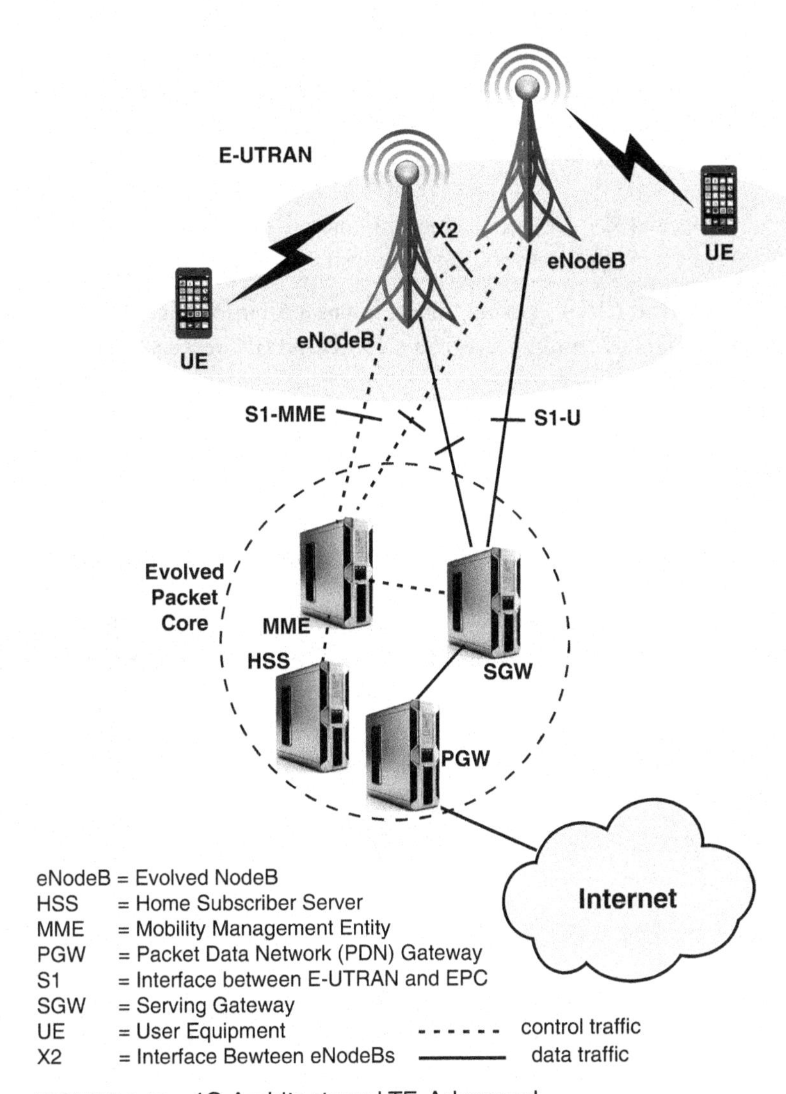

FIGURE 1.11 4G Architecture: LTE-Advanced

Relaying

A key element of an LTE-Advanced cellular network is the use of **relay nodes (RNs)**. As with any other cellular system, an LTE-Advanced base station experiences reduced data rates near the edge of its cell due to lower signal levels and higher interference levels. Rather than use smaller cells, it is more efficient to use small relay nodes, which have a reduced radius of operation compared to an eNodeB, distributed around the periphery of the cell. UE near an RN communicates with the RN, which in turn communicates with the eNodeB.

An RN is not simply a signal repeater. Instead, the RN receives, demodulates, and decodes the data and applies error correction as needed, and then it transmits a new signal to the base station, referred to in this context as a **donor eNodeB**. The RN functions as a base station with respect to its communication with the UE and as UE with respect to its communication with the eNodeB. Note the following:

- The eNodeB → RN transmissions and RN → eNodeB transmissions are carried out in the DL frequency band and UL frequency band, respectively, for FDD systems.
- The eNodeB → RN transmissions and RN → eNodeB transmissions are carried out in the DL subframes of the eNodeB and RN and UL subframes of the eNodeB and RN, respectively, for TDD systems.

RNs can use out-of-band communication using microwave links or inband communication. Inband communication means that the RN–eNodeB interface uses the same carrier frequency as the RN–UE interface. This creates an interference issue that can be described as follows: If the RN receives from the eNodeB and transmits to the UE at the same time, it is both transmitting and receiving on the downlink channel. The RN's transmission will have a much greater signal strength than the DL signal arriving from the eNodeB, making it very difficult to recover the incoming DL signal. The same problem occurs in the uplink direction. To overcome this difficulty, frequency resources are partitioned as follows:

- eNodeB → RN and RN → UE links are time-division multiplexed in a single frequency band, and only one is active at any one time.
- RN → eNodeB and UE → RN links are time-division multiplexed in a single frequency band, and only one is active at any one time.

Femtocells

The industry responded to the increasing data transmission demands from smartphones, tablets, and similar devices by introducing 3G and 4G cellular networks. As demand continues to increase, it becomes increasingly difficult to satisfy the demand, particularly in densely populated areas and remote rural areas. An essential component of the 4G strategy for satisfying demand is the use of femtocells.

A **femtocell** is a low-power, short-range, self-contained base station. Initially used to describe consumer units intended for residential homes, the term has expanded to encompass higher-capacity units for enterprise, rural, and metropolitan areas. Key attributes include IP backhaul, self-optimization, low power consumption, and ease of deployment. Femtocells are by far the most numerous type of small cells.

The term *small cell* is an umbrella term for low-powered radio access nodes that operate in licensed and unlicensed spectrum that have a range of 10 m to several hundred meters indoors or outdoors. These contrast with a typical mobile macrocell, which might have a range of up to several tens of kilometers. Macrocells would best be used for highly mobile users and small cells for low-speed or stationary users. Small cells now outnumber macrocells, and the proportion of small cells in 4G networks has risen dramatically. Deployment of these cells is called **network densification**, and the result is a heterogeneous network of large and small cells called a **HetNet**.

Figure 1.12 shows the typical elements in a network that uses femtocells. The femtocell access point is a small base station, much like a Wi-Fi hotspot base station, placed in a residential, business, or public setting. It operates in the same frequency band and with the same protocols as an ordinary cellular network base station. Thus, a 4G smartphone or tablet can connect wirelessly with a 4G femtocell with no change. The femtocell connects to the Internet, typically over a DSL, fiber, or cable landline. Packetized traffic to and from the femtocell connects to the cellular operator's core packet network via a femtocell gateway.

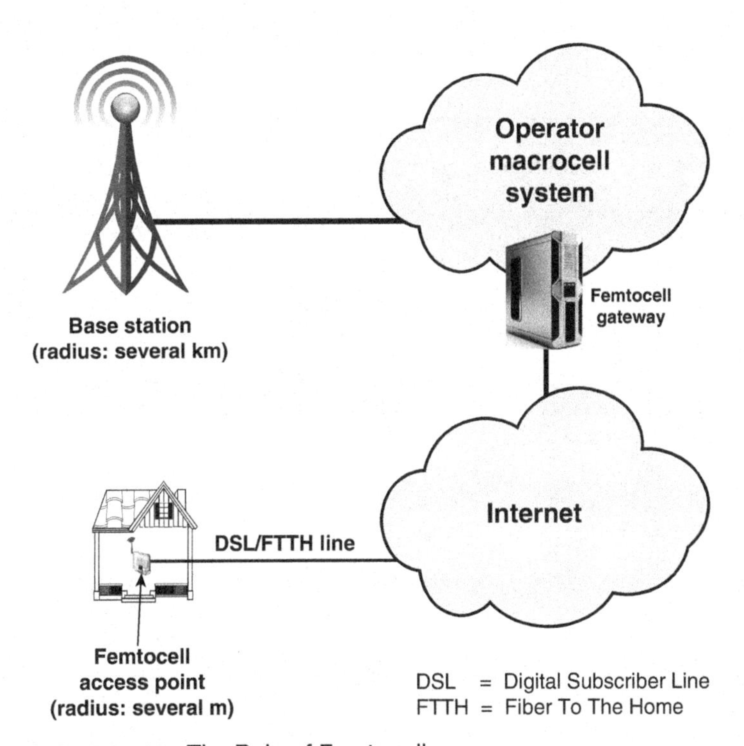

FIGURE 1.12 The Role of Femtocells

LTE-Advanced Transmission Characteristics

LTE-Advanced relies on two key technologies to achieve high data rates and spectral efficiency: orthogonal frequency-division multiplexing (OFDM) and multiple-input/multiple-output (MIMO) antennas. (Both of these technologies are explored in Part Four, "5G NR Air Interface and Radio Access Network")

For the downlink, LTE-Advanced uses OFDMA (orthogonal frequency-division multiple access) and for the uplink SC-FDMA (single-carrier FDMA).

OFDM signals have a high peak-to-average-power ratio (PAPR), requiring a linear power amplifier with overall low efficiency. This is a poor quality for battery-operated handsets. While complex, SC-FDMA has a lower PAPR and is better suited to portable implementation.

FDD and TDD

LTE-Advanced has been defined to accommodate both paired spectrum for frequency-division duplex and unpaired spectrum for time-division duplex operation. Both LTE TDD and LTE FDD are being widely deployed as each form of the LTE standard has advantages and disadvantages. Table 1.3 compares key characteristics of the two approaches.

TABLE 1.3 Characteristics of TDD and FDD for LTE-Advanced

Parameter	LTE-TDD	LTE-FDD
Paired spectrum	Does not require paired spectrum as both transmit and receive occur on the same channel.	Requires paired spectrum with sufficient frequency separation to allow simultaneous transmission and reception.
Hardware cost	Lower cost as no diplexer is needed to isolate the transmitter and receiver. As cost of the UEs is of major importance because of the vast numbers that are produced, this is a key aspect.	Diplexer is needed and cost is higher.
Channel reciprocity	Channel propagation is the same in both directions, which enables transmit and receive to use one set of parameters.	Channel characteristics are different in the two directions as a result of the use of different frequencies.
Uplink (UL)/ downlink (DL) asymmetry	It is possible to dynamically change the UL and DL capacity ratio to match demand.	UL/DL capacity is determined by frequency allocation set out by the regulatory authorities. It is therefore not possible to make dynamic changes to match capacity. Regulatory changes would normally be required
Guard period/ guard band	A guard period is required to ensure that uplink and downlink transmissions do not clash. A large guard period limits capacity. Larger guard periods are normally required if distances are increased to accommodate larger propagation times.	A guard band is required to provide sufficient isolation between the uplink and downlink. A large guard band does not impact capacity.
Discontinuous transmission	Discontinuous transmission is required to allow both uplink and downlink transmissions. This can degrade the performance of the RF power amplifier in the transmitter.	Continuous transmission is required.

Parameter	LTE-TDD	LTE-FDD
Cross-slot interference	Base stations need to be synchronized with respect to the uplink and downlink transmission times. If neighboring base stations use different uplink and downlink assignments and share the same channel, interference may occur between cells.	Not applicable

FDD systems allocate different frequency bands for uplink (UL) and downlink (DL) transmissions. The UL and DL channels are usually grouped into two blocks of contiguous channels (paired spectrum) that are separated by a guard band of a number of vacant radio-frequency (RF) channels for interference avoidance. Figure 1.13a illustrates a typical spectrum allocation in which user *i* is allocated a pair of channels U*i* and D*i* with bandwidths W_U and W_D. The frequency offset used to separate the pair of channels, W_G, should be large enough for the user terminal to avoid self-interference among the links when both links are simultaneously active.

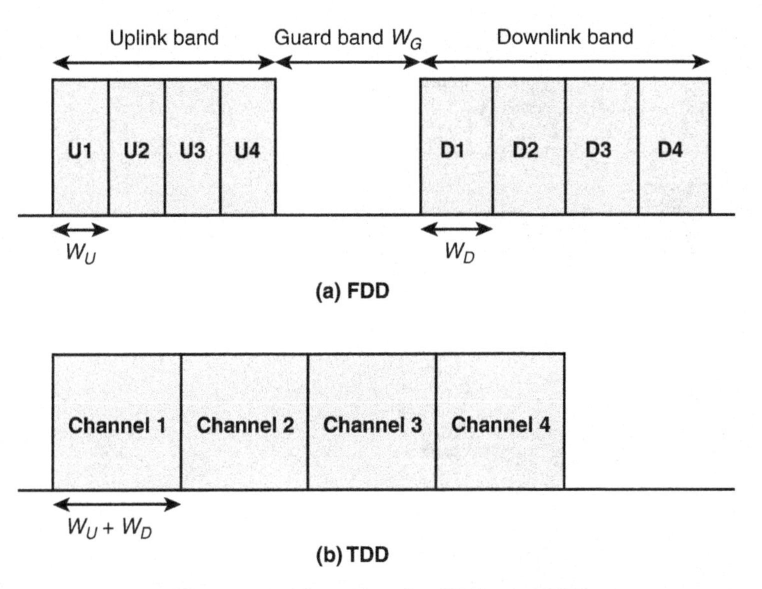

FIGURE 1.13 Spectrum Allocation for FDD and TDD

For TDD, the UL and DL transmissions operate in the same band but alternate in the time domain. Capacity can be allocated more flexibly than with FDD. It is a simple matter of changing the proportion of time devoted to UL and DL within a given channel.

Carrier Aggregation

Carrier aggregation is used in LTE-Advanced in order to increase the bandwidth and thereby increase the bitrates. Because it is important to maintain backward compatibility with LTE, the aggregation is of LTE carriers. Carrier aggregation can be used for both FDD and TDD. Each aggregated carrier is referred to as a component carrier (CC). The component carrier can have a bandwidth of 1.4, 3, 5, 10, 15, or 20 MHz, and a maximum of five component carriers can be aggregated; hence, the maximum aggregated bandwidth is 100 MHz. In FDD, the number of aggregated carriers can be different in the DL and the UL. However, the number of UL component carriers is always equal to or lower than the number of DL component carriers. The individual component carriers can also be of different bandwidths. When TDD is used, the number of CCs and the bandwidth of each CC are the same for the DL and the UL.

Figure 1.14a illustrates how three carriers, each of which is suitable for a 3G station, are aggregated to form a wider bandwidth suitable for a 4G station. As Figure 1.14b suggests, there are three approaches used in LTE-Advanced for aggregation.

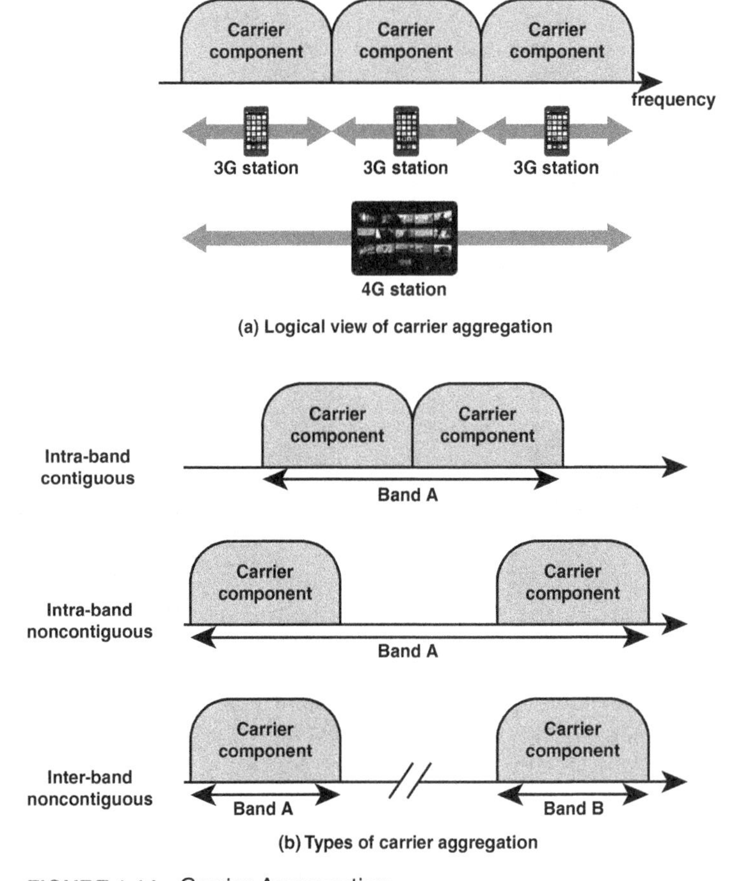

FIGURE 1.14 Carrier Aggregation

The three approaches are as follows:

- **Intra-band contiguous:** This is the easiest form of LTE carrier aggregation to implement. Here, the carriers are adjacent to each other. The aggregated channel can be considered by the terminal as a single enlarged channel from the RF viewpoint. In this instance, only one transceiver is required within the subscriber station. The drawback of this method is the need to have a contiguous spectrum band allocation.
- **Intra-band noncontiguous:** Multiple carrier components (CCs) belonging to the same band are used in a noncontiguous manner. In this approach, the multicarrier signal cannot be treated as a single signal, and therefore multiple transceivers are required. This adds significant complexity, particularly to the UE, where space, power, and cost are prime considerations. This approach is likely to be used in countries where spectrum allocation is noncontiguous within a single band or when the middle carriers are in use by other subscribers.
- **Inter-band noncontiguous:** This form of carrier aggregation uses different bands. It will be of particular use because of the fragmentation of bands—some of which are only 10 MHz wide. For the UE, it requires the use of multiple transceivers within the single item, with the usual impact on cost, performance, and power.

Evolved Packet Core

The operator, or carrier, network that interconnects all of the base stations of the carrier is referred to as the **evolved packet core (EPC)**. Traditionally, the core cellular network was circuit switched, but for 4G, the core is entirely packet switched. It is based on IP and supports voice connections using voice over IP (VoIP).

Figure 1.11 illustrates the essential components of the EPC:

- **Mobility management entity (MME):** The MME deals with control signaling related to mobility and security. The MME is responsible for the tracking and paging of UE in idle mode.
- **Serving gateway (SGW):** The SGW deals with user data transmitted and received by UEs in packet form, using IP. The SGW is the point of interconnect between the radio side and the EPC. As its name indicates, this gateway serves the UE by routing the incoming and outgoing IP packets. It is the anchor point for the intra-LTE mobility (i.e., in the case of handover between eNodeBs). Thus packets can be routed from an eNodeB to an eNodeB in another area via the SGW and can also be routed to external networks such as the Internet (via the PGW).
- **Packet data network gateway (PGW):** The PGW is the point of interconnection between the EPC and external IP networks such as the Internet. The PGW routes packets to and from the external networks. It also performs various functions, such as IP address/IP prefix allocation and policy control and charging.

- **Home subscriber server (HSS):** the HSS maintains a database that contains user-related and subscriber-related information. It also provides support functions in mobility management, call and session setup, user authentication, and access authorization.

Figure 1.11 shows only a single instance of each configuration element. There are, of course, multiple eNodeBs and multiple instances of each of the EPC elements. And there are many-to-many links between eNodeBs and MMEs, between MMEs and SGWs, and between SGWs and PGWs.

Figure 1.11 also shows defined interfaces between the EPC and E-UTRAN. The S1 interfaces involve both the S1-MME and S1-U interfaces. The S1-MME interface is defined for the control plane between the eNodeB and the MME. It has functions to establish, maintain, and release E-UTRAN connections with a given quality of service (QoS). It supports mobility functions for handovers intra-LTE, with other 3GPP technologies, and with CDMA2000 3G systems. The S1-MME also has paging procedures for the EPC to find the location of the UE.

The S1-U interface is for the user plane data transmission to connect with an SGW for each bearer. Multiple S1-U logical interfaces may exist between and eNodeB and the SGW.

1.7 Key Terms and Review Questions

Key Terms

adaptive equalization	femtocells
Advanced Mobile Phone System (AMPS)	flat fading
base station	forward channel
carrier aggregation	forward error correction
cellular network	frequency diversity
code-division multiple access (CDMA)	frequency-division duplex (FDD)
diffraction	frequency reuse
diversity	handover
donor eNodeB	home subscriber server (HSS)
evolved eNodeB (eNodeB)	long-term evolution (LTE)
evolved packet core (EPC)	LTE-Advanced
fading	mobile radio
fast fading	mobility management entity (MME)

packet data network gateway (PGW)	scattering
power control	selective fading
reflection	serving gateway (SGW)
relay node (RN)	slow fading
relaying	space diversity
reuse factor	time-division duplex (TDD)
reverse channel	

Review Questions

1. What geometric shape is used in cellular system design?
2. What is the principle of frequency reuse in the context of a cellular network?
3. List five ways of increasing the capacity of a cellular system.
4. What are the five principal elements of a cellular system?
5. What is the difference between a control channel and a traffic channel?
6. What is a handoff?
7. Explain the paging function of a cellular system.
8. What are the key differences between first- and second-generation cellular systems?
9. What is the difference between TDMA and FDMA?
10. What are some key characteristics that distinguish third-generation cellular systems from second-generation cellular systems?
11. What are NodeB and eNodeB?
12. Briefly explain the X2 interface in LTE-Advanced.

Chapter 2

5G Standards and Specifications

Learning Objectives

After studying this chapter, you should be able to:

- Explain the difference between IMT-2020 and 5G
- Present an overview of the three usage scenarios for IMT-2020
- Discuss the technical performance requirements for IMT-2020
- Describe the roles of ITU-R and ITU-T with respect to 5G
- Understand the concept of network slicing
- Present an overview of the IMT-2020 core network framework
- Explain the role of 3GPP in the development of 5G
- Discuss the detailed requirements defined for 5G by 3GPP

Many of the important developments in information technology and communications, such as the Internet, the Internet of Things (IoT), cloud computing, and virtualization, have been driven in part by international standards. However, in all of these cases, much of the technology was developed and deployed in advance of universally agreed-upon standards. The case of 5G is quite different. Although a reasonably complete set of standards based on fixed specifications is only just coming to fruition, the implementations and deployments that preceded these standards and specifications anticipated their final form. Throughout the 5G ecosystem, which includes device and component manufacturers, cellular network providers, network software providers, and application developers, the work done prior to the introduction of the first set of standards in 2020 closely follows what has ultimately been standardized. There is universal agreement that, going forward, 5G-related implementations will follow the standards.

Because an understanding of 5G depends on an understanding of the process by which the standards are developed and the content of those standards, this chapter provides an overview. The chapter covers the two organizations that are responsible for the development of 5G: the International Telecommunication

Union (ITU) and the 3rd Generation Partnership Project (3GPP). The process of standards development has basically followed this sequence:

1. The ITU has issued and continues to issue standards, called *recommendations*, and other documents, call reports, that define the overall concept for 5G, as well as the technical, performance, and service requirements for 5G.
2. Based on the ITU requirements, as well as requirements generated by national and regional standards organizations and market-based organizations, 3GPP has developed and continues to develop a detailed set of technical specifications for the implementation of 5G.
3. The ITU has translated these specifications into international standards (called recommendations) that dictate how 5G is being implemented.

This process is ongoing, with further refinements and capabilities being added to the requirements and the technical specifications.

This chapter provides a high-level overview of and also goes into some detail about the requirements, objectives, specifications, and standards for 5G.

2.1 ITU-R and IMT-2020

The ITU is a United Nations specialized agency; hence the members of ITU are governments. The U.S. representation is housed in the Department of State. The charter of the ITU says that it "is responsible for studying technical, operating, and tariff questions and issuing Recommendations on them with a view to standardizing telecommunications on a worldwide basis." Its primary objective is to standardize, to the extent necessary, techniques and operations in telecommunications to achieve end-to-end compatibility of international telecommunication connections, regardless of the countries of origin and destinations.

With respect to 5G, the two relevant components of ITU are the ITU Radiocommunication Sector (ITU-R) and the ITU Telecommunication Standardization Sector (ITU-T). In general, ITU-R issues standards related to user requirements and to the air interface between the radio access network (RAN) and user devices. ITU-T issues standards related to the RAN and the core network. This section discusses the role of ITU-R for 5G, and Section 2.2 discusses ITU-T.

ITU-R is responsible for all ITU work in the field of radiocommunication. The main activities of ITU-R are:

- Develop draft ITU-R recommendations on the technical characteristics of, and operational procedures for, radiocommunication services and systems
- Compile handbooks on spectrum management and emerging radiocommunication services and systems

- Ensure optimal, fair, and rational use of the radio-frequency spectrum and satellite-orbit resources and coordinate matters related to radiocommunication services and wireless services

International Mobile Telecommunications

Perhaps the most prominent initiative by ITU-R is the International Mobile Telecommunications (IMT) project. IMT is the generic term used by the ITU community to designate broadband mobile systems. It encompasses IMT-2000, IMT-Advanced, and IMT-2020 collectively, which correspond to 3G, 4G, and 5G, respectively.

ITU-R's role in the development of 5G via IMT-2020 includes developing and adopting the following:

- International regulations on the use of the radio-frequency spectrum, referred to as the Radio Regulations (RR). To take into account the progress of technologies and the changes in spectrum uses, the RR are updated every four years by the ITU World Radiocommunication Conference (WRC). The RR are an international treaty that is binding on the 193 member states of the ITU. They are the basis for the harmonization of IMT spectrum worldwide.
- Global standards for the overall requirements of IMT and for its radio interface (ITU-R recommendations).
- Best practices in the implementation of these standards and regulations (ITU-R reports and handbooks).
- Evaluation criteria and procedures to evaluate technology submissions for IMT-2020, as well as submission templates that proponents must utilize to organize the information that is required in a submission of a candidate technology for evaluation.

Capabilities

A foundational document in the definition of IMT-2020 is ITU-R Recommendation M.2083 (*IMT Vision: Framework and Overall Objectives of the Future Development of IMT for 2020 and Beyond*, published in September 2015). In broad strokes, this document develops a vision of the 5G mobile broadband connected society and future IMT. The two main contributions of this recommendation are a set of target values for key capabilities and a definition of usage scenarios, discussed subsequently.

M.2083 lists the following as the key capabilities for IMT, together with the minimum requirements for each capability:

- **Peak data rate:** This is the maximum achievable data rate under ideal conditions per user/device (in Gbps). The minimum target value is 10 Gbps, with 20 Gbps supported for a number of applications.
- **User-experienced data rate:** This is the achievable data rate that is available ubiquitously across the coverage area to a mobile user/device (in Mbps or Gbps). This rate depends on the type of environment.

A target 100 Mbps is suitable for wide area coverage cases, such as urban and suburban areas. IMT-2020 should be able to achieve up to 1 Gbps for indoor and hotspot users.

- **Latency:** This is the contribution by the radio network to the time from when the source sends a packet to when the destination receives it (in ms). The requirement for IMT-2020 is 1 ms over-the-air latency. This enables the network to support services with very low-latency requirements.
- **Mobility:** This is the maximum speed at which a defined quality of service (QoS) and seamless transfer between radio nodes that may belong to different layers and/or radio access technologies (multilayer/multi-RAT) can be achieved (in km/h). IMT-2020 is expected to enable high mobility up to 500 km/h with acceptable QoS; this is envisioned in particular for high-speed trains.
- **Connection density:** This is the total number of connected and/or accessible devices per unit area (per km^2). The requirement for this parameter is up to $10^6/km^2$ for environments such as a massive IoT deployment.
- **Area traffic capacity:** This is the total traffic throughput served per geographic area (in $Mbps/m^2$). IMT-2020 is expected to support an increase in capacity up to 10 $Mbps/m^2$, which is a factor of 100 increase over IMT-Advanced (4G). This increase is achieved in dense areas by reducing the cell size (e.g., on the order of tens of meters).
- **Energy efficiency:** Energy efficiency has two aspects:
  - On the network side, energy efficiency refers to the quantity of information bits transmitted to/received from users per unit of energy consumption of the radio access network (RAN) (in bits/Joule). The objective is efficient data transmission when there is a substantial load on the network. The energy consumption for the RAN of IMT-2020 should not be greater than for IMT-Advanced, while delivering the enhanced capabilities. The network energy efficiency should therefore be improved by a factor at least as great as the envisaged traffic capacity increase of IMT-2020 relative to IMT-Advanced.
  - On the device side, energy efficiency refers to quantity of information bits per unit of energy consumption of the communication module (in bits/Joule). The objective is low energy consumption when there is no data being sent or received.
- **Spectrum efficiency:** This is average data throughput per unit of spectrum resource and per cell (bps/Hz). The goal is a spectrum efficiency three times higher than that of IMT-Advanced.

The following objectives determined these target values:

- The user experience with IMT-2020 using a mobile device should match, to the extent possible, the experience with fixed networks.[1] The enhancement will be realized by increased peak and

1. The term *fixed network* is not defined in any ITU document. Often the term is limited to wired networks that use physical connections such as coaxial and fiber-optic cable. However, it should also include stationary wireless devices, such as dedicated wireless connections between microwave towers in a wide area network.

user-experienced data rate, enhanced spectrum efficiency, reduced latency, and enhanced mobility support.

- IMT-2020 should support massive machine-to-machine interconnection for a variety of Internet of Things (IoT) environments.
- IMT-2020 should be able to provide these capabilities without undue burden in terms of energy consumption, network equipment cost, and deployment cost to make IMT sustainable and affordable in the future.

The target values were published in 2015, with the admonition that they are presented for purposes of research and development and may be revised in light of future studies and implementation experience.

Figure 2.1, from ITU-R Recommendation M.2083, is a Kiviat graph[2] that compares the capability requirements of IMT-2020 to those of IMT-Advanced. It is clear at a glance that the most substantial required improvements are in the areas of traffic capacity and network energy efficiency.

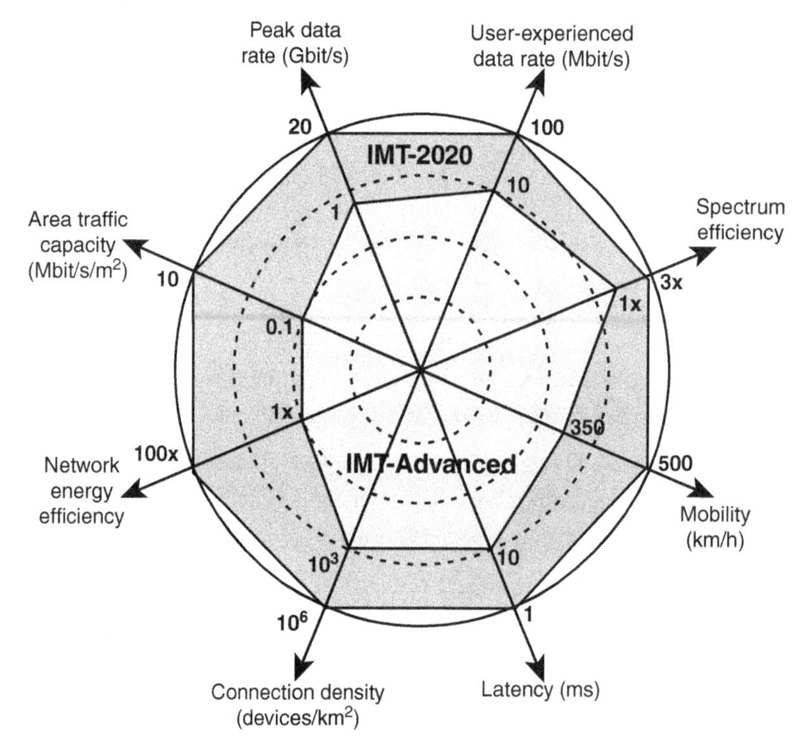

FIGURE 2.1 Enhancement of Key Capabilities from IMT-Advanced to IMT-2020

2. A Kiviat graph provides a pictorial means of comparing systems along multiple variables [MORR74]. The variables are laid out as lines of equal angular intervals within a circle, each line going from the center of the circle to the circumference. A given system is defined by one point on each line. The points are connected to yield a shape that is characteristic of that system.

Usage Scenarios

Two important concepts in M.2083 and related documents are *usage scenario* and *use case*. No ITU document defines these terms, but the following definitions should suffice for this chapter:

- **Usage scenario:** A general description of a way in which an IMT network is used. A usage scenario dictates various performance and technical requirements. A wide but nevertheless constrained variety of use cases are encompassed by a usage scenario.
- **Use case:** A specific application or way of using an IMT network. A general account of a situation or course of actions that uses an IMT network. It is described from the end user perspective and illustrates fundamental characteristics. A use case dictates more specific and refined performance and technical requirements than the corresponding usage scenario.

M.2083 defines three usage scenarios: enhanced mobile broadband, massive machine type communications, and ultra-reliable and low-latency communications. Figure 2.2, from M.2083, indicates the relative importance of the key capabilities for the three usage scenarios.

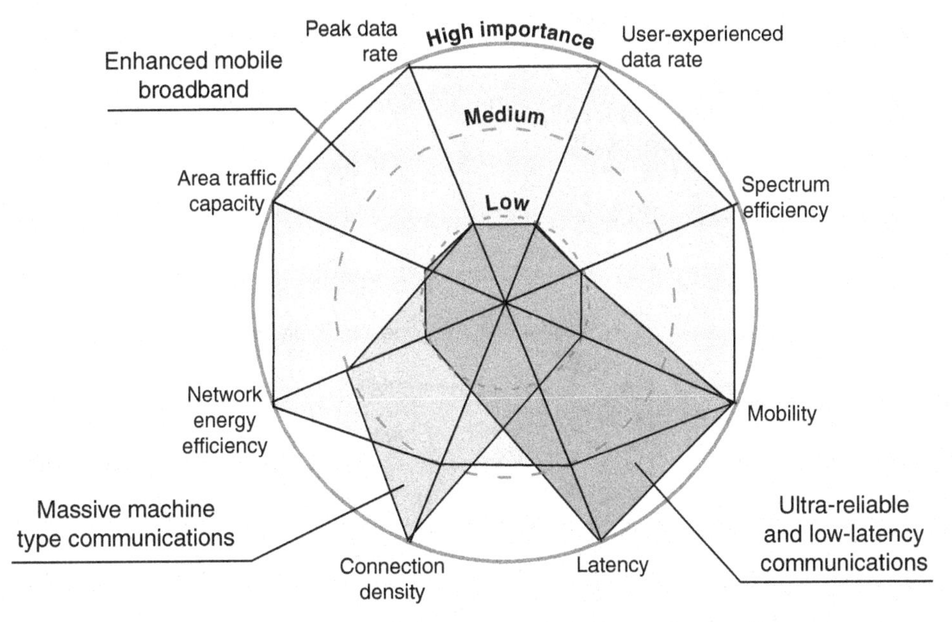

FIGURE 2.2 The Importance of Key Capabilities in Different Usage Scenarios

Enhanced mobile broadband (eMBB) is the feature of 5G provides a significant increase in data rate for a normal mobile Internet user. Enhanced mobile broadband services allow users to experience high-speed and high-quality multimedia services, such as virtual reality, augmented reality (AR), and 4000-pixel horizontal resolution (4K) video, at any time and place. These applications require reasonably low latency and good connection density, with high demand on the other six key capabilities. In addition to the consumption of multimedia content for entertainment purposes, eMBB supports a number of business applications. These include cloud access apps for commuters and other off-site employees, remote workers needing to communicate with the back office, or indeed an entire smart office in which all devices are wirelessly and seamlessly connected. To give an idea of the importance of this usage scenario and its rapid deployment, the Ericsson Mobility Report estimates 220 million 5G subscriptions for eMBB in 2020, rising to 3.5 billion in 2026 [ERIC20].

Massive machine type communications (mMTC) is characterized by a very large number of connected devices typically transmitting a relatively low volume of non-delay-sensitive data. However, machine-to-machine communications involves a range of performance and operational requirements. Devices are required to be low cost and have a very long battery life, such as five years or longer. Ericsson estimates 6 billion cellular IoT connections by 2026 [ERIC20].

Ultra-reliable and low-latency communications (URLLC) is a form of machine-to-machine communications that enables delay-sensitive and mission-critical services that require very low end-to-end delay, such as tactile Internet, remote control of medical or industrial robots, driverless cars, and real-time traffic control.

Figure 2.3 illustrates some examples of applications within the envisioned usage scenarios for IMT-2020 and beyond.

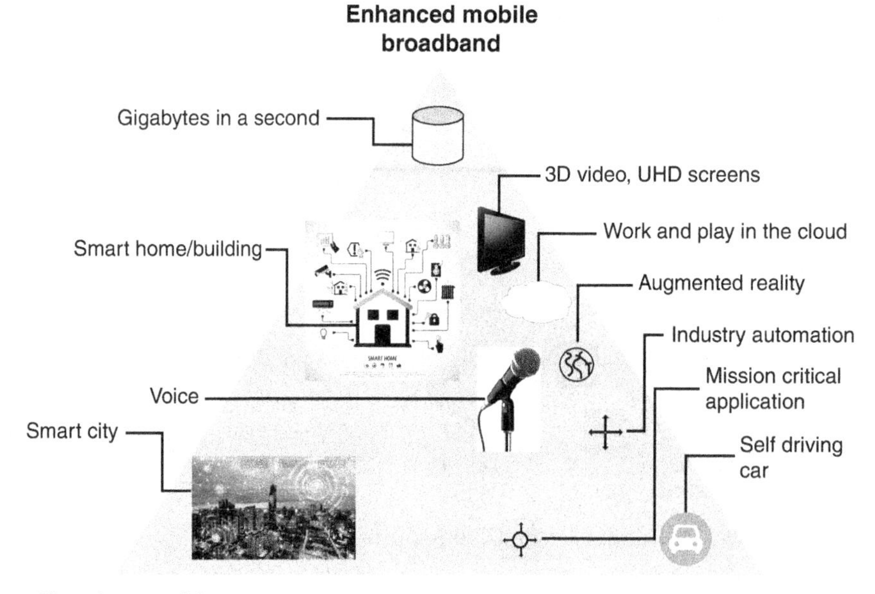

FIGURE 2.3 Usage Scenarios for IMT-2020

Use Cases

ITU-R Report M.2441 (*Emerging Usage of the Terrestrial Component of International Mobile Telecommunication*, published in November 2018) lists the following 16 examples of emerging use cases that can be supported by IMT-2020:

- **Machine-type communication (MTC) (also known as machine-to-machine (M2M) in some jurisdictions):** The prominent use cases in this general category are various deployments of IoT. This category encompasses many industrial, commercial, and government use cases and so overlaps with some of the other categories in this list.
- **Broadband Public Protection and Disaster Relief (PPDR):** Reliable, high-data-rate PPDR networks can greatly enhance emergency response to critical situations. Major incidents such as 9/11 reveal failures due to lack of interoperability. Further, the narrowband nature of typical PPDR networks precludes the use of real-time multimedia services to provide effective PPDR. 5G networks can provide dramatic increases in effectiveness. See, for example, [KUMB17] and ITU-R Report M.2291 (*The Use of International Mobile Telecommunications for Broadband Public Protection and Disaster Relief Applications*, published in November 2016).
- **Transportation applications:** M.2441 lists three types of use cases in this category:
  - **Intelligent transport systems:** Typically classified into three groups: road safety applications, traffic control, and infotainment
  - **Railway and high-speed train communication:** Includes broadband connectivity to the passengers as well as closed-circuit television (CCTV) operations and train control operations
  - **Bus/fleet traffic management:** Includes services that assist drivers to provide comfortable rides by avoiding traffic jams or other obstacles and computer-aided management of crowds during popular events
- **Utilities:** M.2441 lists two examples in this use case category:
  - **Smart grids:** This refers to the application of computer automation and networking to an electricity or natural gas distribution system. Smart grid initiatives seek to improve operations, maintenance, and planning by making sure that each component of the electric grid can both "talk" and "listen." A smart grid requires enormous two-way data flows and complex connectivity.
  - **Water management:** This is a broad area involving all aspects of monitoring and control of water quality, storage, and distribution.
- **Industrial automation:** This is the manufacture of products under the control of computers and programmable controllers. Manufacturing assembly lines as well as standalone machine tools (computer numerical control [CNC] machines) and robotic devices fall into the industrial

automation category. The emergence of ultra-reliable, low-latency wireless communication with 5G presents the opportunity for dramatic improvements in industrial automation processes. With 5G, dispersed IoT sensors, actuators, controllers, and robots, driven by software command and control, can expand the ability to more fully automate an industrial process.

- **Remote control:** Remote control of robots and other actuators in real time is applicable in many contexts, such as construction and maintenance in dangerous areas, repair work in damaged nuclear or chemical plants, and off-shore construction tasks.
- **Surveying and inspection:** Drones, robots, and vehicles that are remotely operated are suitable for applications such as land and sea inspection, where the safety issues arising from the distances covered, adverse weather conditions, and hazardous terrain can be costly to address. Remote operations work well for these types of monitoring applications and are ideal for observing industrial and construction sites in out-of-the way places or large indoor venues and warehouse environments. Video streams and other sensor data are fed back to the operator, enabling appropriate action to be taken. By combining remote inspection with remote manipulation, the level of automation can be raised. For example, a remotely operated robot in a data center can rapidly swap out a malfunctioning server or respond to other types of hardware failures.
- **Healthcare:** Healthcare provider organizations, such as hospitals and clinics, increasingly rely on wireless technology to improve quality and efficiency. The integration of mobile computing, medical sensors, and portable devices supported by 5G networks will dramatically expand the utility of mobile health functionality [MATT16]. M.2441 lists the following sample use case areas: remote surgery, clinical wearables, and mobile health apps.
- **Sustainability/environmental:** The use of millions or even billions of low-cost, low-power sensors and other devices can enhance energy efficiency of various undertakings, such as precision agriculture.
- **Smart city:** A smart city is a municipality that uses information and communication technology (ICT) to increase operational efficiency, share information with the public, and improve both the quality of government services and citizen welfare. Elements include Wi-Fi hotspots, traffic monitoring, bus/subway arrival notifications, streetlights that dim when nobody is nearby, and escalators that operate only when people step on them. A smart city may include drones to monitor traffic and respond to emergencies. Smart cities use a combination of IoT devices, software solutions, user interfaces, and communication networks. 5G enables a comprehensive and integrated urban infrastructure for smart city applications.
- **Wearables:** Wearable devices are devices that can be worn on a person and have the capability to connect and communicate to the network either directly through embedded cellular connectivity or through another device (primarily a smartphone) using Wi-Fi, Bluetooth, or another technology. Wearable devices have a wide variety of applications, including patient monitoring, enhanced emergency responder capabilities, and augmented reality.

- **Smart homes:** A smart home is a highly automated home or other building. A smart home is networked not only for computers and entertainment but also for security, heating, cooling, lighting, and control of appliances, including robotic vacuum cleaners and lawn mowers. Traditionally, wired connections, such as fiber and cable, have provided Internet access for stationary equipment such as that found in homes or other buildings. But to meet growing demand, especially for the work-from-home trend that has been accelerated by Covid-19, service providers are offering fixed wireless access (FWA) connections to cellular networks. An FWA connection provides primary broadband access through wireless wide area mobile network–enabled customer premises equipment (CPE). This includes various form factors of CPE, such as indoor (desktop and window) and outdoor (rooftop and wall-mounted) devices.

- **Agriculture:** Smart farming or smart agriculture refers to the adoption of modern information and communications technology in order to enhance, monitor, automate, or improve agricultural operations and processes. Smart farming solutions provide farmers and the agricultural industry at large with the infrastructure to leverage advanced IoT technologies for tracking, monitoring, automating, and analyzing their agricultural and industrial operations. For example, sensors can collect information such as soil moisture, fertilization, and weather information and transmit it over a cellular wireless network to a cloud-based hub to provide farmers real-time access to information and analysis on their land, crop, livestock, logistics, and machinery. This enables the smart farm to improve its operational performance by analyzing the data collected and acting upon it in ways that increase productivity or streamline operations. 5G enables a massive increase in the use of IoT devices and cloud-based control.

- **Media and entertainment:** Video already dominates mobile traffic, and its share is growing. Ericsson estimates that the video share of mobile traffic in 2020 was 66% and expects it to rise to 77% by 2026 [ERIC20]. Some of this is for video conferencing and other business and government applications, but by far the bulk is for entertainment. Video demand is now being augmented by the increasing use of video for work-from-home applications via FWA. 5G networks are required to provide the performance needed for good user quality of experience.

- **Enhanced personal experiences:** Related to the preceding item, this category refers to two types of personal experience that will be enhanced by 5G. The first is increased use of social media, with an emphasis on streaming video. The second is outdoor activities and games, such as watching 4K movies in a moving car or train and high-definition interactive games on mobile devices.

- **Commercial airspace unmanned aerial systems (UAS) applications:** Small drones equipped with high-resolution cameras can potentially draw in large amounts of data. One potential drone capability involves tracking multiple targets over a wide area, using infrared cameras, heat sensors, and sensors that detect movement.

Evaluation

In addition to defining the minimum technical requirements for ITM-2020, ITU-R is developing recommendations that standardize specific radio interface technologies. ITU-R (*Vocabulary of Terms for International Mobile Telecommunications*, March 2012) defines the radio interface as follows:

> The common boundary between the mobile station and the radio equipment in the network, defined by functional characteristics, common radio (physical) interconnection characteristics, and other characteristics, as appropriate. Note: An interface standard specifies the bi-directional interconnection between both sides of the interface at once. The specification includes the type, quantity and function of the interconnecting means and the type, form and sequencing order of the signals to be interchanged by those means.

ITU-R itself is not developing the technical specifications for radio interface technologies (RITs). Rather, it has used the following process:

1. ITU-R M.2412 defines five test environments (see Table 2.1) that together are representative of the real-world environments for 5G deployment.

TABLE 2.1 Test Environments for IMT-2020

Usage Scenarios	Test Environment	Description
eMBB	Indoor hotspot	An indoor isolated environment at offices and/or in shopping malls based on stationary and pedestrian users with very high user density
	Dense urban	An urban environment with high user density and traffic loads focusing on pedestrian and vehicular users
	Rural	A rural environment with larger and continuous wide area coverage, supporting pedestrian, vehicular, and high-speed vehicular users
mMTC	Urban macro	An urban macro environment targeting continuous coverage focusing on a high number of connected machine type devices
URLLC	Urban macro	An urban macro environment targeting ultra-reliable and low-latency communications

2. ITU-R developed a set of evaluation criteria for assessing the compliance of any proposed RIT with IMT-2020 minimum requirements. These requirements consist of:
    a. The ability to function in one or more of the test environments
    b. The use of approved 5G frequency bands
    c. Technical performance requirements
3. ITU-R invited organizations to submit proposals for RITs or sets of radio interface technologies (SRITs) covering one or more of the test environments listed in Table 2.1.

ITU-R has issued two reports that document the evaluation process: Report M.2411 (*Requirements, Evaluation Criteria and Submission Templates for the Development of IMT-2020*, November 2017) and Report M.2412 (*Guidelines for Evaluation of Radio Interface Technologies for IMT-2020*, October 2017).

Test Environments

It is useful to examine the test environments listed in Table 2.1, as they give an indication of the types of likely 5G deployments. The first test environment is an indoor hotspot with the eMBB usage scenario. Traditionally, a hotspot is defined as an area, often public, such as an airport, coffee shop, or convention center, that is covered with a wireless local area network (WLAN) service, usually Wi-Fi, for providing Internet service. This service is available for the public to use for a nominal charge, for free, or as a premium service. In the context of IMT-2020, an indoor hotspot also provides access to the 5G network. Figure 2.4 illustrates the indoor hotspot-eMBB test environment defined in ITU-R Report M.2412, which consists of one floor of a building. The height of the ceiling is 3 m. The floor has a surface area of 120 m × 50 m, and 12 base station antenna sites are placed in 20-meter spacing. ITU-R uses the term **transmission reception point (TRxP)** to refer to a base station. In essence, a TRxP is an antenna array (with one or more antenna elements) available to the network that is located at a specific geographic location. Figure 2.4 does not explicitly show internal walls; ITU-R provides guidance on developing a stochastic line-of-site (LOS) model for estimating performance.

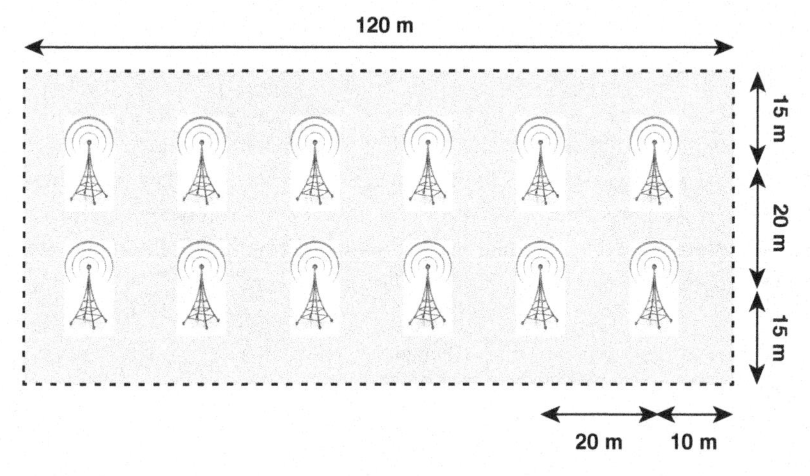

FIGURE 2.4 Indoor Hotspot Site Layout

As the name suggests, rural-eMBB is designed to serve rural areas where user density is much less. Key requirements here are to bring the benefits of high-speed Internet access to rural subscribers and still be able to support massive networks of industrial sensors. Both the urban macro-mMTC and urban macro-URLLC test environments follow the same layout as the rural-eMBB test environment. The differences have to do with cell size, antenna height, and various transmission parameters.

Figure 2.5 illustrates a test environment that consists of TRxPs placed in a regular hexagonal grid. This is an idealized layout; in the real world, topography and other features of the environment may require deviations from this layout.

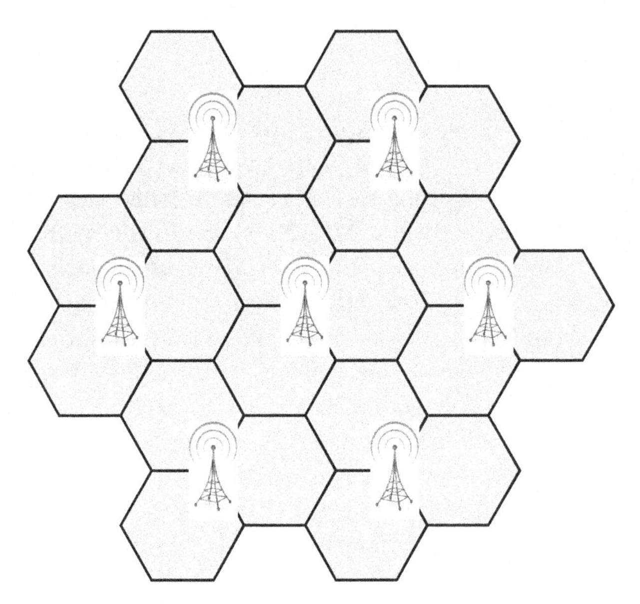

FIGURE 2.5 Hexagonal Site Layout for Rural eMBB, Urban Macro-mMTC, and Urban Macro_URLLC

The dense urban-eMBB test environment addresses the need for high-speed connections for a dense distribution of users. Generally, this requires smaller, lower-range cell towers at higher frequency bands than for other scenarios. The test environment consists of two layers: a macro layer and a micro layer. The macro-layer base stations are placed in a regular grid, as shown in Figure 2.5. For the micro layer, there are three micro sites randomly in each macro cell, as shown in Figure 2.6.

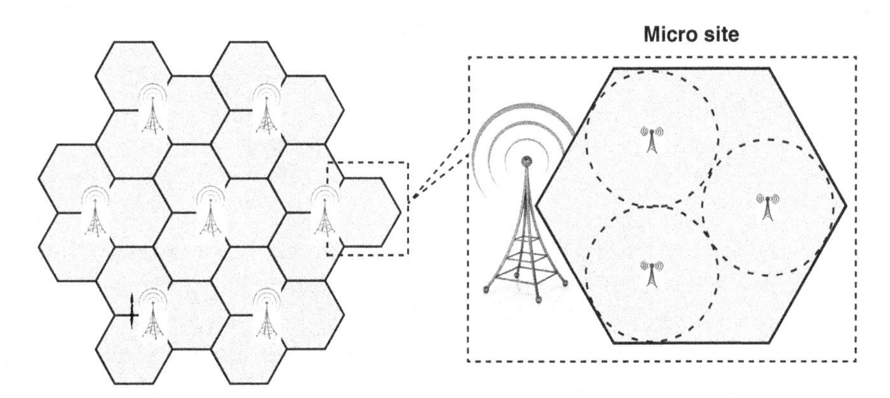

FIGURE 2.6 Layout for Dense Urban eMBB

Frequency Bands

Report M.2411 lists the following frequency bands for use for IMT-2020:

- 450–960 MHz
- 1.427–1.518 GHz
- 1.710–2.025 GHz
- 2.110–2.200 GHz
- 2.300–2.400 GHz
- 2.500–2.690 GHz
- 3.300–3.700 GHz
- 4.800–4.990 GHz

The following bands, which are already allocated to mobile, will also be studied with a view to an IMT-2020 identification:

- 24.25–27.5 GHz
- 37–40.5 GHz
- 42.5–43.5 GHz
- 45.5–47 GHz
- 47.2–50.2 GHz
- 50.4–52.6 GHz
- 66–76 GHz
- 81–86 GHz

The following bands will also be studied, although they are not currently globally allocated to the mobile service:

- 31.8–33.4 GHz
- 40.5–42.5 GHz

The 1G network operated at 850 MHz to 1900 MHz. 2G and 3G added 2100 MHz, and 4G added 600 MHz, 700 MHz, 1.7/2.1 GHz, 2.3 GHz, and 2.5 GHz. 5G networks rely heavily on the use of millimeter waves or bands, as well as using bands employed in previous generations. The term *millimeter wave* (*mmWave*) is somewhat imprecise. Much of the 5G literature uses the term to refer to signals or radio waves with a wavelength from 1 to 10 mm, equivalent to frequencies between 300 GHz and 30 GHz. A looser and also commonly used definition uses the range 1 mm (300 GHz) to 30 mm (10 GHz).

The advantage of mmWave bands is that they can support wider bandwidths and higher data rates. A disadvantage is that they have a significantly shorter effective range for data transfer. Thus, a mmWave system must use smaller cells and a denser population of base stations.

Technical Performance Requirements

Recommendation M.2083 defines eight key capabilities for IMT-2020. This was an initial attempt to characterize the type of performance demands to be addressed by IMT-2020. This list has been expanded and refined into 13 technical performance requirements in Report M.2410 (*Minimum Requirements Related to Technical Performance for IMT-2020 Radio Interface(s)*, November 2017).

The purpose of these performance requirements is to ensure noticeable improvement of user quality of experience (QoE) for legacy 4G services and applications and a high QoE for emerging 5G services and applications. Two terms should be distinguished:

- **Quality of service (QoS):** The measurable end-to-end performance properties of a network service, which can be guaranteed in advance by a service-level agreement between a user and a service provider in order to satisfy specific customer application requirements. These properties may include throughput (bandwidth), transit delay (latency), error rates, priority, security, packet loss, and packet jitter. Chapter 9, "Core Network Functionality, QoS, and Network Slicing," discusses QoS.
- **Quality of experience (QoE):** A subjective measure of performance in a system. QoE relies on human opinion and differs from QoS, which can be precisely measured. A discussion of QoE is beyond the scope of this book. [STAL16] provides extensive coverage of this topic.

In essence, the performance requirements for 5G are QoS measures designed to produce a high QoE. The M.2410 minimum technical performance requirements are as follows:

- **Peak data rate:** The maximum achievable data rate under ideal conditions per user/device (in Gbps). The minimum target values are downlink peak data rate of 20 Gbps and uplink peak data rate of 10 Gbps.
- **Peak spectral efficiency:** The maximum data rate under ideal conditions normalized by channel bandwidth (in bps/Hz). Another way of expressing this is that it is the maximum data rate that can be transmitted over a given bandwidth. The relationship can be expressed as follows:

$$SE_p = \frac{R_p}{W}$$

where R_p is the peak data rate, W is the available bandwidth, and SE_p is the peak spectral efficiency. The minimum for peak spectral efficiencies is downlink of 30 bps/Hz and uplink of 15 bps/Hz.

- **User-experienced data rate:** The achievable data rate that is available ubiquitously across the coverage area to a mobile user/device (in Mbps or Gbps). This rate depends on the type of environment. The target value for dense urban-eMBB is 100 Mbps downlink and 50 Mbps uplink.
- **5th percentile user spectral efficiency:** The 5% point of the cumulative distribution function of the normalized user throughput. The normalized user throughput is defined as the number of correctly received bits—that is, the number of bits contained in the service data units (SDUs) delivered to Layer 3—over a certain period of time divided by the channel bandwidth; it is measured in bps/Hz.
- **Average spectral efficiency:** The average data throughput per unit of spectrum resource and per cell (bps/Hz). The goal is a spectral efficiency three times higher than IMT-Advanced.
- **Area traffic capacity:** The total traffic throughput served per geographic area (in Mbps/m^2). The target downlink value is 10 Mbps/m^2 in the indoor hotspot–eMBB test environment.
- **Latency:** Transmission delays introduced by the network. Report M.2410 considers two types of latency:
  - **User plane latency:** The contribution by the radio network to the time from when the source sends a packet to when the destination receives it (in ms). The minimum requirements for user plane latency are 4 ms for eMBB and 1 ms for URLLC.
  - **Control plane latency:** The transition time from the most "battery-efficient" state (e.g., Idle state) to the start of continuous data transfer (e.g., Active state). The minimum requirement is 20 ms.
- **Connection density:** The total number of connected and/or accessible devices per unit area (per km^2) to fulfill a specific quality of service (QoS). The minimum requirement is 10^6/km^2.
- **Energy efficiency:** In general terms, the relation between useful output and energy consumption. In the context of M.2410, this parameter has two aspects:
  - **Network energy efficiency:** Refers to the quantity of information bits transmitted to/received from users per unit of energy consumption of the radio access network (RAN) (in bits/Joule). The objective is efficient data transmission when there is a substantial load on the network. The energy consumption for the radio access network of IMT-2020 should not be greater than for IMT-Advanced, while delivering the enhanced capabilities. The network energy efficiency should therefore be improved by a factor at least as great as the envisaged traffic capacity increase of IMT-2020 relative to IMT-Advanced.
  - **Device energy efficiency:** Refers to the quantity of information bits per unit of energy consumption of the communication module (in bits/Joule). The objective is low energy consumption when there is no data being sent or received.
- **Reliability:** The probability of successful transmission of a Layer 2/3 packet within a required maximum time, which is the time it takes to deliver a small data packet from the radio protocol Layer 2/3 service data unit (SDU) ingress point to the radio protocol Layer 2/3 SDU egress

point of the radio interface at a certain channel quality. The minimum requirement is $1 - 10^{-5}$ success probability of transmitting a Layer 2 PDU (protocol data unit) of 32 bytes within 1 ms in channel quality of coverage edge for the urban macro–URLLC test environment, assuming small application data (e.g., 20 bytes application data + protocol overhead).

- **Mobility:** The maximum speed at which a defined QoS and seamless transfer between radio nodes that may belong to different layers and/or radio access technologies (multi-layer/-RAT) can be achieved (in km/h). The following classes of mobility are defined:
  - **Stationary:** 0 km/hr
  - **Pedestrian:** 0 km/hr to 10 km/hr
  - **Vehicular:** 10 km/hr to 120 km/hr
  - **High-speed vehicular:** 120 km/hr to 500 km/hr

  This requirement is defined for the purpose of evaluation in the eMBB usage scenario. A mobility class is supported if the traffic channel link data rate on the uplink, normalized by bandwidth, is as shown in Table 2.2. This assumes that the user is moving at the maximum speed in that mobility class in each of the test environments.
- **Mobility interruption time:** The smallest time delay supported by the system, during which the end user device cannot exchange packets with any base stations during transmissions. The mobility interruption time includes the time required to execute any radio access network procedure, radio resource control signaling protocol, or other message exchanges between the mobile station and the radio access network. The required value is 0 ms.
- **Bandwidth:** The maximum aggregated system bandwidth. The minimum requirement is 100 MHz.

Table 2.2 lists the eMBB mobility requirements.

TABLE 2.2 eMBB Mobility Requirements for IMT-2020

Test Environment	Mobility Classes Supported	Normalized Traffic Channel Link Data Rate (bps/Hz)	Mobility (km/h)
Indoor hotspot	Stationary, pedestrian	1.5	16
Dense urban	Stationary, pedestrian, vehicular (up to 30 km/h)	1.12	30
Urban macro	Pedestrian, vehicular, high-speed vehicular	0.8	120
		0.45	500

Evaluation Process

ITU-R has developed evaluation criteria based on the framework and overall objectives of the future development of IMT that support the new capabilities expressed in relevant recommendation(s), taking into account end user requirements and without unnecessary legacy requirements. ITU-R requested proposals for radio interface technologies (RITs) or sets of radio interface technologies (SRITs) covering one or more of the test environments listed in Table 2.1. Each submission had to include either an initial self-evaluation or the proponents' endorsement of an initial evaluation, according to the ITU-R guidelines. ITU-R has registered nine different independent evaluation groups (IEGs) commissioned to verify the performance of candidate proposals for 5G.

Figure 2.7 illustrates the process that ITU-R is using to evaluate candidate schemes. The guidelines for submission are as follows:

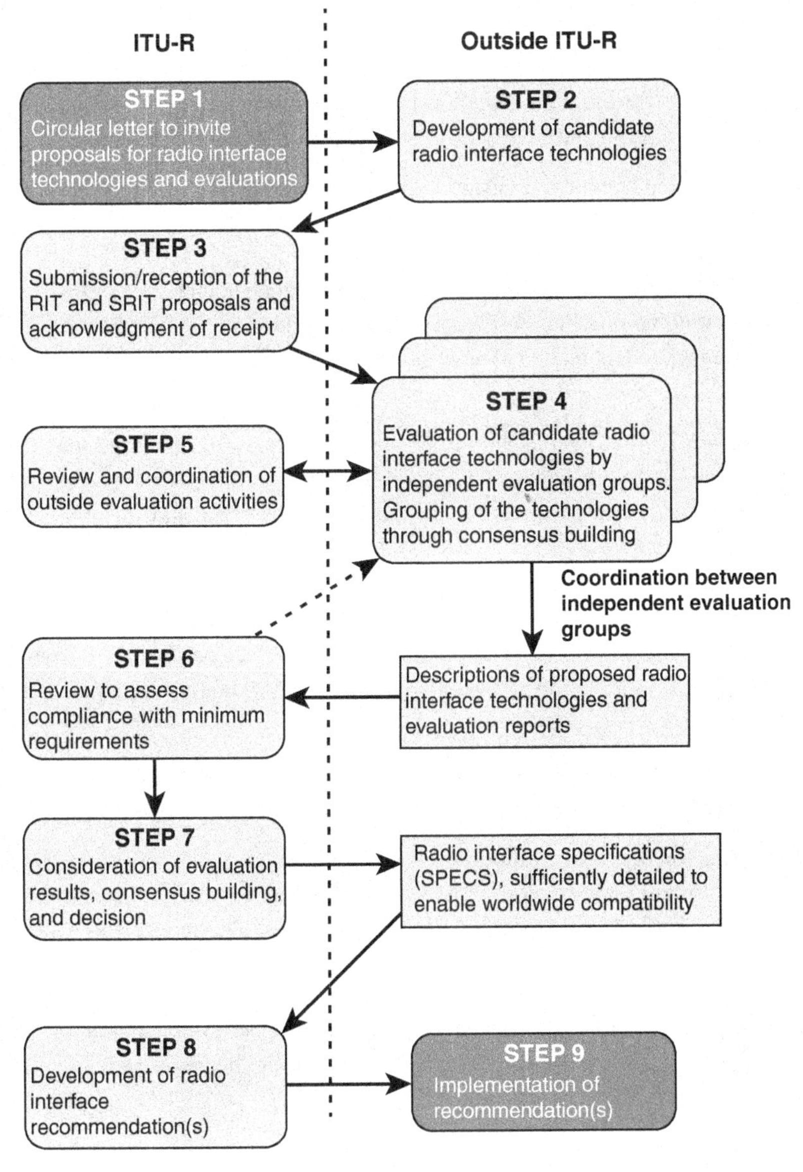

FIGURE 2.7 ITU-R IMT-2020 Radio Interface Technology Evaluation Process

- The submission must satisfy the minimum technical performance requirements (described earlier in this section) and the evaluation criteria (described subsequently) for a given usage scenario.
- An RIT needs to fulfill the minimum requirements for at least three test environments: two test environments under eMBB and one test environment under mMTC or URLLC.
- An SRIT consists of a number of component RITs complementing each other, with each component RIT fulfilling the minimum requirements of at least two test environments and together as an SRIT fulfilling the minimum requirements of at least four test environments comprising the three usage scenarios.

At minimum, each proposal must include information in the form of completed templates. As defined in Report M.2411, there are two categories of templates: description templates and compliance templates.

A description template provides a detailed technical description of the proposed RIT or SRIT. The two templates are:

- **Characteristics template:** This template encompasses all the technical characteristics of the candidate RIT to a level of detail that provides sufficient understanding of the proposed technology in order to enable an independent technical assessment of compliance with the IMT-2020 requirements.
- **Link budget template:** The term *link budget* usually refers to a written description of all the gains and losses by a transmitter during its active state. Here, the template provides a broader description of transmission characteristics, such as bit rate, path loss, error rate, and spectral efficiency.

Figure 2.8 lists the topics covered in these templates. Each of these topics consists of a number of subtopics, enabling a very detailed description. For example, the radio interface functional aspects topic includes multiple-access schemes, modulation schemes, and error control coding schemes and interleaving.

The compliance templates provide the details needed to assess compliance of a candidate RIT or SRIT with the minimum requirements. Figure 2.9 lists the topics included in the three compliance templates.

Report M.2412 provides guidelines for proposal evaluation. The evaluation methodology does not require those submitting proposals to build and deploy a test system. Rather, the evaluation is done based on the proposals, especially the information provided in the templates. The report lists three methods that can be used:

- **Computer simulation:** Simulation involves creating a model of the proposed implementation and then, generally with the use of a random-number generator, executing the model and analyzing the outputs.

- **Analytical approaches:** Analytical approaches involve the use of mathematical models and producing results based on calculations rather than event-by-event simulation.
- **Inspection of the proposal:** This approach simply involves a detailed inspection of the proposal to ensure that it meets the requirements and provides a complete technical description.

Characteristics Template		
Test environments(s) Radio interface functional aspects Channel tracking capabilities Physical channel structure and multiplexing Mobility management (handover) Radio resource management Frame structure Spectrum capabilities and duplex technologies	Support of Advanced antenna capabilites Link adaptation and power control Power classes Scheduler, QoS support and management, data services Radio interface architecture and protocol stack Cell selection Location determination mechanisms Priority access mechanisms	Unicast, multicast, and broadcast Privacy, authorization, encryption, authentication, and legal intercept schemes Frequency planning Interference mitigation within radio interface Synchronization requirements Support for wide range of services Global circulation of terminals Energy efficiency

Link Budget Template		
System configuration Transmitter	Receiver Calculation of available path loss	Range/coverage efficiency calculation

FIGURE 2.8 Description Templates for IMT-2020 Evaluation

Service Template
Service capability requirements

Spectrum Template	
Frequency bands identified for IMT	Higher-frequency range/band(s)

Technical Performance Template	
Peak data rate (Gbps) Peak spectral efficiency (bps/Hz) User experience data rate (Mbps) 5th percentile user spectral efficiency (bps/Hz) Average spectral efficiency (bps/Hz/TRxP) Area traffic capacity (Mbps/m^2) User plane latency (ms) Control plane latency (ms)	Connection density (devices/km^2) Energy efficiency Reliability Mobility classes Mobility traffic channel link data rates (bps/Hz) Mobility interruption time (ms) Bandwidth and scalability

FIGURE 2.9 Compliance Templates for IMT-2020 Evaluation

Six member states and organizations submitted proposals:

- 3rd Generation Partnership Project (3GPP)
- Korea
- China
- The European Telecommunications Standards Institute (ETSI) Technical Committee (TC) Digital Enhanced Cordless Telecommunications (DECTTM)
- Telecommunications Standards Development Society, India (TSDSI)
- Nufront Co., Ltd.

Evaluation of the proposals was received from the following independent evaluation groups:

- 5G India Forum (5GIF)
- 5G Infrastructure Association
- Africa Evaluation Group (AEG)
- ATIS Wireless Technologies and Systems Committee (WTSC) IMT-2020 Evaluation Group
- Beijing National Research Center for Information Science and Technology
- Canadian Evaluation Group (CEG)
- Chinese Evaluation Group (CHEG)
- Chinese Industry and Research Alliance of Telecommunications Evaluation Group
- Telecom Centres of Excellence, India (TCOE)
- Fifth Generation Mobile Communications Promotion Forum (5GMF)
- Trans-Pacific Evaluation Group (TPCEG)
- Telecommunication Technology Association
- Wireless World Research Forum

The outcome of this evaluation process is documented in ITU-R Report M.2483 (*The outcome of the evaluation, consensus building and decision of the IMT-2020 process (Steps 4 to 7), including characteristics of IMT-2020 radio interfaces*. July 2020). Subsequently, ITU-R issued Recommendation M.2150 (*Detailed specifications of the terrestrial radio interfaces of International Mobile Telecommunications-2020*, February 2021). The current version of the Recommendation adopts three radio interface technologies: 3GPP 5G-SRIT, 3GPP 5G-RIT, and 5Gi (India/TSDSI). However, the 5Gi specification is unlikely to achieve widespread adoption outside of India [KAUS21]. Accordingly, the coverage in this book of the air interface standards summarizes the 3GPP specifications.

2.2 ITU-T and IMT-2020

ITU-T fulfills the purposes of the ITU relating to telecommunications standardization by studying technical, operating, and tariff questions and adopting recommendations on them with a view to standardizing telecommunications on a worldwide basis.

With respect to IMT-2020, the role of ITU-T is complementary to that of ITU-R. ITU-R develops and adopts the international regulations on the use of the radio-frequency spectrum. ITU-R also develops and adopts global standards for the overall requirements of IMT and for its radio interface, as well as best practices in the implementation of these standards and regulations. ITU-T specifies requirements for overall non-radio aspects of the IMT-2020 network, especially with respect to network operations and support of service requirements.

Whereas the ITU-R recommendations and reports emphasize the air interface performance characteristics, ITU-T's focus is on increased end-to-end flexibility, taking advantage of software-defined networking (SDN), network functions virtualization (NFV), and cloud computing. ITU-T Recommendation Y.3101 (*Requirements of the IMT-2020 Network*, April 2018) lists the following objectives with respect to IMT-2020:

- Minimized dependency on access network technologies
- Coping with traffic explosion in urban areas
- Easy incorporation of future emerging services
- Provision of a cost-efficient infrastructure
- Expansion of the geographic reach of the network

Requirements

ITU-T Y.3101 defines IMT-2020 requirements for the RAN and core network in two categories: from the service point of view and from the network operation point of view.

Requirements from the Service Point of View

ITU-T uses the same three usage scenarios defined by ITU-R: eMBB, mMTC, and URLCC. For eMBB, ITU-T Y.3101 specifies the following requirements:

- Support capabilities to cope with the explosion in mobile data traffic. An example cited in Y.3101 is moving some network functions to the edge of the core network. Chapter 10, "Multi-Access Edge Computing," examines this approach in detail.
- Be flexible and resilient to support ultra-high-bandwidth services. Examples listed for this requirement include edge computing and network slicing, discussed subsequently.

- Support local offloading in an efficient manner. This refers to the movement of network functionality to the edge of the network in an efficient manner.
- Support diverse mobile fronthaul (MFH) and mobile backhaul (MBH) technologies in order to cope with extreme traffic or connection density. MFH refers to network connections between centralized radio controllers and remote radio units of a base station function. MBH refers to the network path between base station systems and the core network. Chapter 15, "5G Radio Access Network," discusses fronthaul and backhaul.

For mMTC, ITU-T Y.3101 specifies the following requirements:

- Support a massive number of MTC devices in an efficient way.
- Minimize traffic congestion that can be caused by a massive number of MTC devices.
- Support consistent end-to-end (E2E) quality of service (QoS), even in the presence of a large number of concurrent connections.

For URLLC, ITU-T Y.3101 specifies the following requirements:

- Support increased service reliability according to service requirements. An example cited in Y.3101 is the ability to replicate and cache contents in network nodes.
- Provide enhanced service performance by reducing E2E latency according to service requirements.

Requirements from the Network Operation Point of View

ITU-T Y.3101 lists the following requirements from the network operation point of view:

- **Network flexibility and programmability:** This is a major requirement for IMT-2020 networks. The network should be able to support a wide range of devices, users, and applications, with evolving requirements for each. Significant concepts in this regard are network functions virtualization (NFV), separation of user and control planes, and network slicing. NFV is discussed in Chapter 8, "Network Functions Virtualization," and the latter two concepts are discussed later in this section.
- **Fixed mobile convergence:** The goal of this requirement is to enable subscriber access through multiple-access networks in seamless, integrated fashion. This topic is discussed later in this chapter.
- **Enhanced mobility management:** The network should support a wide variety of mobility options.
- **Network capability exposure:** The IMT-2020 network should provide suitable ways (e.g., via APIs) to expose network capabilities and relevant information (e.g., information for connectivity, QoS, and mobility) to third parties. This enables third parties to dynamically customize the network capabilities for diverse use cases within the limits set by the IMT-2020 network operator.

- **Identification and authentication:** There should be a unified approach to user and device identification and authentication mechanisms.
- **Security and personal data protection:** The IMT-2020 network must provide effective mechanisms to preserve security and personal data protection for different types of devices, users, and services, including rapid adaptation to dynamic network changes.
- **Efficient signaling:** There are two aspects to this requirement. The signaling mechanisms should be designed to mitigate risks of control and data traffic bottlenecks. Also, the network should provide lightweight signaling protocols and mechanisms to accommodate limited-resource devices.
- **Quality of service control:** The network should support different QoS levels for different services and applications.
- **Network management:** The network should provide a unified network management framework to support interworking of different providers and management of legacy networks.
- **Charging:** The IMT-2020 network needs to support different charging policies and requirements of network operators and service providers, including third parties that may be involved in a given IMT-2020 network deployment. The charging models to be supported include, but are not limited to, charging based on volume, time, session, and application.
- **Interworking with non-IMT-2020 networks:** IMT-2020 networks should support user-transparent interworking with legacy networks.
- **IMT-2020 network deployment and migration:** The network design should accommodate incremental deployment with migration capabilities for services and related users.

For each of the general requirements listed here, Y.3101 includes a number of specific, more detailed requirements. Figure 2.10 lists these requirements.

Network Slicing

One of the most important requirements for IMT-2020 is network slicing. Indeed, network slicing is essential to the exploitation of the capabilities defined for IMT-2020.

Network slicing permits a physical network to be separated into multiple virtual networks (i.e., logical segments) that can support different radio access networks or several types of services for certain customer segments, greatly reducing network construction costs through more efficient use of communication channels. In essence, network slicing allows the creation of multiple virtual networks atop a shared physical infrastructure. This virtualized network scenario devotes capacity to certain purposes dynamically, according to need. As needs change, so can the devoted resources. Using common resources such as storage and processors, network slicing permits the creation of slices devoted to logical, self-contained, and partitioned network functions. Network slicing supports the creation of virtual networks to provide a given QoS level, such as guaranteed delay, throughput, reliability, and/or priority.

Network Flexibility and Programmability

Programmability of network functions
Separation of control/user planes
Manage network slices
Isolate network slices
Network slice scale-in/scale-out
Network slice API
Associate UEs with network slices
Service-specific security requirements
Network slice selection
Network slice QoS
Network slice context information
Virtualized network function scaling

Fixed Mobile Convergence

Support multiple access networks
Minimize access network technology dependency
Support simultaneous multi-access network connections
Support multi-access coordination

Enhanced Mobility Management

Use context information
Assist choice of most suitable network
Support distributed management
Support consisted user experience

Network Capability Exposure

Expose network capabilities to third-party applications

Identification and Authentication

Support user and device identification
Unified authentication framework
Efficient authentication mechanisms

Security and Personal Data Protection

Confidentiality, integrity, availability
Personal data protection
Differentiated security services

Efficient Signaling

Signaling mechanisms for diverse traffic patterns and communication types
Mitigate control/data traffic bottlenecks
Lightweight signaling

Quality of Service Control

Unified QoS mechanisms
E2E QoS
Finer granularity than legacy networks
User-initiated QoS mechanisms

Network Management

Unified E2E management framework
Life cycle management

Charging

Online and offline charging
Various charging models
Charging data for third parties
Per-network slice charging

Interworking with Non-IMT-2020

Interworking

Deployment and Mitigation

Support incremental deployment
Support migration of services and users

FIGURE 2.10 Requirements from the Network Operation Point of View

ITU-T Y.3112 (*Framework for the Support of Network Slicing in the IMT-2020 Network*, December 2018) defines a *network slice* as a logical network that provides specific network capabilities and network characteristics. This recommendation lays out an overall framework for network slicing, defines high-level requirements, and describes core network functions relevant to network slicing.

Figure 2.11 illustrates the network slicing concept. The requirements of a particular application or user determine the physical and logical network resources needed to provide the desired QoS level. The network slicing function dedicates the appropriate resources to support that QoS level.

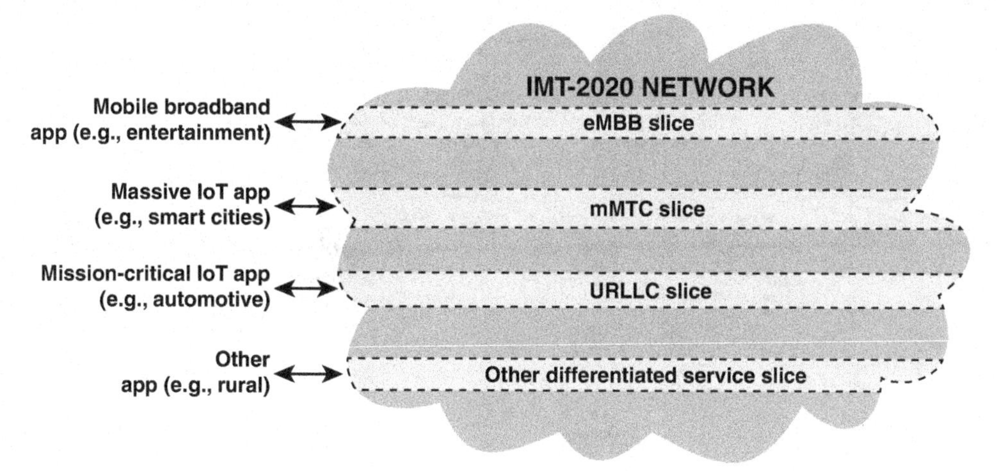

FIGURE 2.11 Network Slicing Concept

Network slicing is examined in detail in Chapter 9, "Core Network Functionality, QoS, and Network Slicing."

Fixed Mobile Convergence

Another important aspect of IMT-2020 is fixed mobile convergence (FMC), which ITU-T defines as the capabilities that provide services and applications to end users, regardless of the fixed or mobile access technologies being used and independently of users' locations. This capability requires a unified core network for the new radio access technologies, as well as existing fixed and wireless networks.

Figure 2.12, from Recommendation Y.3130 (*Requirements of IMT-2020 Fixed Mobile Convergence*, January 2018), depicts the basic concept of FMC. Figure 2.12a shows the use of customer premises equipment (CPE), which may support multiple devices via a LAN at an organization, or a residential gateway, which may support multiple devices on a home system via Wi-Fi or Ethernet connections. In either case, the local device can access either the fixed access network or the mobile access network (the "or" in the figure) or make use of both simultaneously (the "and" in the figure). Thus, IMT-2020 enables transport of traffic on one or the other access networks or transport on both simultaneously, and traffic can be split, combined, and steered according to service requirements and network conditions to optimize the user experience.

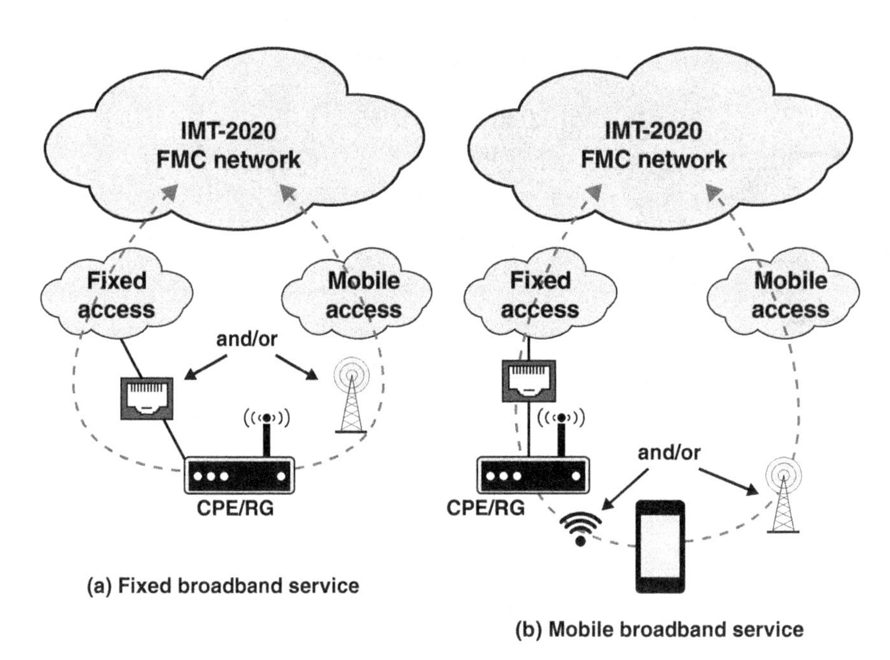

FIGURE 2.12 Broadband Service via Fixed and/or Mobile Access Networks

Figure 2.12b shows that a terminal (e.g., cell phone) of a mobile broadband service can be globally controlled by an IMT-2020 FMC network and get access to data sources (e.g., websites on the Internet) via both fixed and mobile access networks simultaneously or via one of the access technologies at a time.

IMT-2020 Core Network Framework

Recommendation Y.3102 (*Framework of the IMT-2020 Network*, May 2018) provides a framework for overall non-radio aspects of the IMT-2020 network. Figure 2.13, from Recommendation Y.3102, illustrates the interactions between the network functions for providing network service.

The framework delineates three domains: UE, AN, and CN. The user equipment (UE) domain consists of devices that transmit and receive data over the IMT-2020 network. The access network (AN) domain is the wireless connection between the UE and the core network (CN) domain, defined by the ITU-R radio interface recommendations.

The framework diagram in Figure 2.13 also depicts the division between a control plane and a user plane, which cuts across the AN and CN.

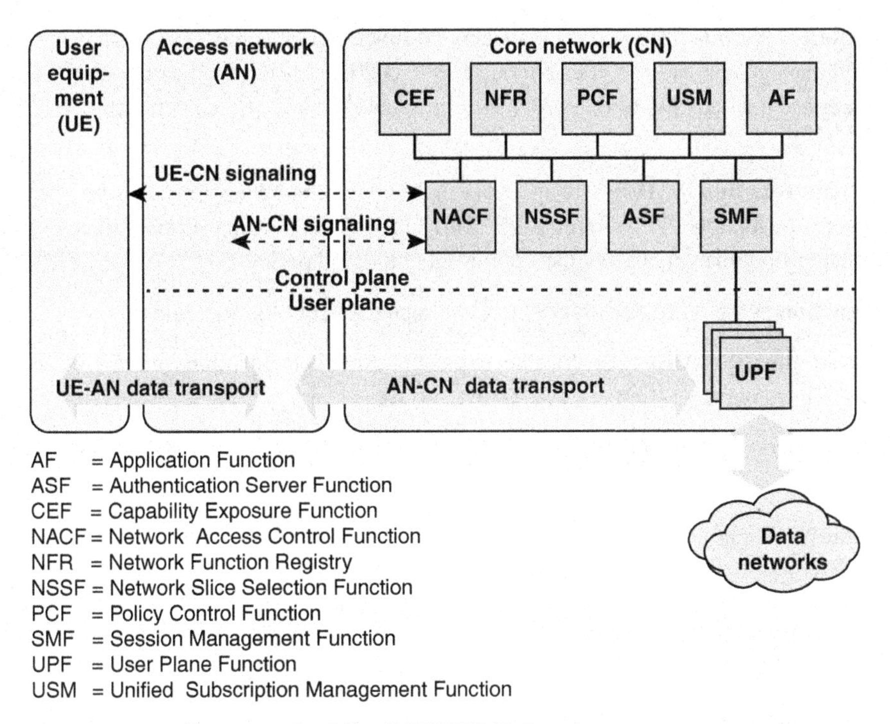

FIGURE 2.13 Framework of the IMT-2020 Network

Control Plane

The control plane performs the call control and connection control functions. For this purpose, a signaling connection between the UE and the CN exchanges signaling messages dealing with managing the signaling connection and with managing the call established for the UE. The control plane functions are requested and managed via control signals that are exchanged between the UE and the AN and between the AN and the CN. Through signaling, the control plane sets up and releases connections, and it may restore a connection in the event of a failure. The control plane also performs other functions in support of call and connection control, such as routing information dissemination.

The core network includes the following functional elements:

- **Network access control function (NACF):** Provides access to the CN services for the AN and for UE. NACF includes:
  - **Registration management:** Enables a UE to register for network access. NACF performs, among other actions, network slice instance selection, UE authentication, authorization of network access and network services, and network access policy control.
  - **Connection management:** Establishes and releases a signaling connection between the UE and the CN.

- **SMF selection:** Determines the session management function that is most appropriate to establish and manage a session. In the context of IMT-2020, a session is an association between a UE and a data network that provides a protocol data unit (PDU) connectivity service.

- **Session management function (SMF):** Sets up and manages one or more sessions that provide connectivity between the local UE and a remote UE. This function deals with user path selection and enforcement of policies, including QoS policy and charging policy.
- **Policy control function (PCF):** Provides for control and management of policy rules.
- **Capability exposure function (CEF):** Enables the exposure of network functions and network slices as a service to third parties.
- **Network function registry (NFR) function:** Assists in the discovery and selection of required network functions.
- **Unified subscription management (USM) function:** Stores and manages UE context and subscription information, including, but not limited to, information on the UE's registration and mobility management, information on network functions that serve the UE, and information on session management. The USM function also provides authentication information of the UE to the ASF.
- **Network slice selection function (NSSF):** When a UE requests registration with the network, the NACF sends a network slice selection request to the NSSF with preferred network slice selection information. The NSSF responds with a message including a list of appropriate network slice instances for the UE.
- **Authentication server function (ASF):** Performs authentication between a UE and the network.
- **Application function (AF):** Interacts with application services that require dynamic policy control. The AF extracts session-related information (e.g., QoS requirements) from application signaling and provides it to the PCF in support of its rule generation.

User Plane

The user plane refers to the set of traffic forwarding components through which traffic flows. Its principal function is to provide transfer of end user information.

The sole functional element in the user plane is the user plane function (UPF). This function includes traffic routing and forwarding, PDU session tunnel management, and QoS enforcement. The PDU session tunnels are used between the AN and UPF(s), as well as between different UPFs as user plane data transport for PDU sessions. The UPF also provides optional functionalities, including packet inspection and collection of UP traffic for lawful intercept. In order to accommodate the diversity of network scenarios, UPF may also provide interworking functionalities among different network segments, such as interworking between the IP-based core network and a non-IP based access network.

Network Services

Y.3102 also lists the primary network services supported by the core network framework. These are:

- **Registration management (RM):** Register or deregister a UE with the IMT-2020 network and establish the user context in the network.
- **Connection management (CM):** Establish and release signaling connection between the UE and NACF.
- **Session management (SM):** Manage PDU sessions, including control of PDU session tunnel establishment, modification, and release.
- **User plane management (UPM):** Forward user traffic, including user traffic rerouting between UPFs due to relocation of the serving UPF and to enforce QoS policies.
- **Handover management (HM):** Handle all aspects related to UE mobility. Mobility management aspects include, but are not limited to, UE reachability management and handover management.

2.3 3GPP

The 3rd Generation Partnership Project (3GPP) was formed in 1998 by a global consortium of regional standard development organizations (SDOs) to develop technology specifications for 3G cellular networks. Because it involved the efforts of the world's leading national standards organizations, 3GPP became the dominant agent in the development of specifications for 3G, then 4G, and now 5G cellular networks.

3GPP began work in 2016 on defining 5G technical specifications for a new radio access technology known as 5G NR (i.e., 5G New Radio) and a next-generation network architecture known as 5G NGN (i.e., 5G NextGen). Unlike with previous generations, there are no longer competing standard bodies working on potential solutions for 5G.

It is important to understand that 3GPP is developing technical specifications, not standards. These specifications are then translated into standards by the seven regional SDOs that form the 3GPP partnership and by ITU to form the IMT-2020 set of recommendations.

Figure 2.14 shows the key players in the 3GPP process and their relationships to one another. Within the 3GPP organization is the Project Co-ordination Group (PCG), which is responsible for the overall time frame and management of technical work to ensure that the 3GPP specifications are produced in a timely manner, as required by the marketplace. Subordinate to the PCG are three Technical Specification Groups (TSGs). Each TSG has the responsibility to prepare, approve, and maintain the specifications within its terms of reference. In addition, it may organize its work in Working Groups (WGs) and liaise with other groups, as appropriate. The TSGs report to the PCG.

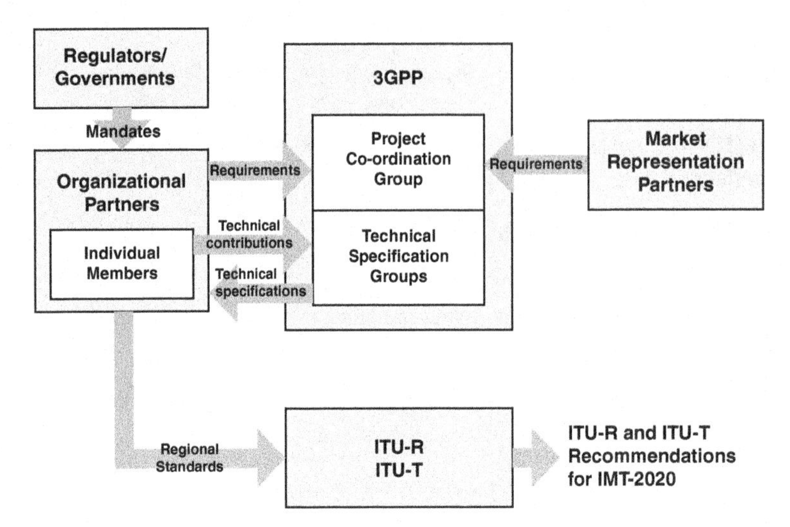

FIGURE 2.14 3GPP Process

Figure 2.15 illustrates the three TSGs.

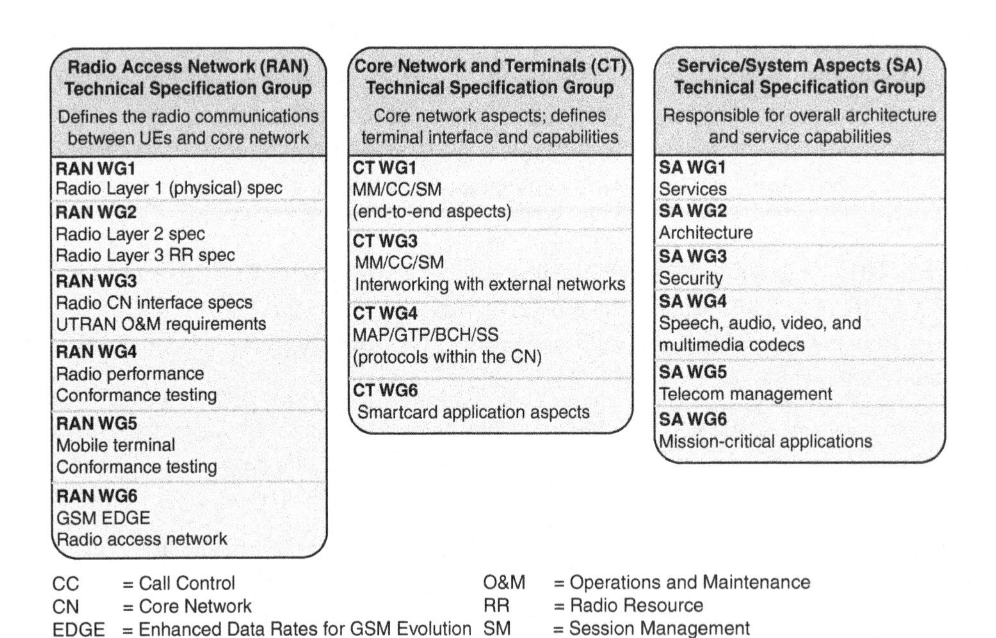

CC = Call Control
CN = Core Network
EDGE = Enhanced Data Rates for GSM Evolution
MM = Mobility Management
O&M = Operations and Maintenance
RR = Radio Resource
SM = Session Management
UTRAN = UMTS (Universal Mobile Telecommunications System) Terrestrial Radio Access Network

FIGURE 2.15 3GPP Technical Specification Groups

The three TSGs are as follows:

- **Radio Access Network (RAN):** Responsible for the definition of the functions, requirements, and interfaces of the radio access network (RAN), including radio performance, physical layer, Layer 2, and Layer 3 RR specifications; specification of the access network interfaces; definition of the operations and maintenance requirements in the RAN; and conformance testing for user equipment and base stations.
- **Core Network and Terminals (CT):** Responsible for specifying terminal interfaces (logical and physical), terminal capabilities (e.g., execution environments), and the core network part of 3GPP systems.
- **Service and Systems Aspects (SA):** Responsible for the overall architecture and service capabilities of systems based on 3GPP specifications and, as such, has a responsibility for cross-TSG coordination.

Key to the 3GPP process are the organizational partners. An organizational partner is a standards organization with a national, regional, or other officially recognized status (in a particular country or region) that has the capability and authority to define, publish, and set standards nationally or regionally. The seven 3GPP organizational partners are:

- Association of Radio Industries and Businesses, Japan (ARIB)
- Alliance for Telecommunications Industry Solutions, United States (ATIS)
- China Communications Standards Association (CCSA)
- European Telecommunications Standards Institute (ETSI)
- Telecommunications Standards Development Society, India (TSDSI)
- Telecommunications Technology Association, Korea (TTA)
- Telecommunication Technology Committee, Japan (TTC)

Individual members are associated with organizational partners. An individual member is a member company that is affiliated with one of the organizational partners. For example, Figure 2.16 shows the individual members affiliated with ATIS.

Finally, there are market representation partners. A market representation partner is an organization invited to participate by the organizational partners to offer market advice to 3GPP and to bring into 3GPP a consensus view of market requirements (e.g., services, features, functionality) falling within the 3GPP scope. The members are:

- 5G Alliance for Connected Industries and Automation (5G-ACIA)
- 5G Automotive Association

- 5G Americas
- 5G Infrastructure Association
- Broadband India Forum
- Cellular Operators Association of India (COAI)
- Cellular Telecommunications and Internet Association (CTIA)
- EMEA (Europe, Middle East, and Africa) Satellite Operators Association (ESOA)
- Global Certification Forum (GCF)
- Global mobile Suppliers Association (GSA)
- GSM Association
- IPv6 Forum
- Next Generation Mobile Networks (NGMN) Alliance
- Public Safety Communication Europe (PSCE) Forum
- Small Cell Forum
- The Critical Communications Association (TCCA)
- Telecommunication Development (TD) Industry Alliance
- Wireless Broadband Alliance

AT&T C Spire Wireless Carnegie Technologies Charter Communications Cisco Systems Cohere Technologies Coherent Logix Comcast Comtech Telecommunications Cybersecurity and Infrastructure Security Agency (CISA)	Department of Defense Department of Transportation DISH Network Ericcson FCC FirstNet Futurewei Technologies, Inc. Google Intel Intelsat InterDigital Communications Johns Hopkins Applied Physics Laboratory	L3 Harris Technologies Lockheed Martin Metaswitch Mobi NextNav NIST Nokia NSI-MI Technologies NTIA Omnispace PC Test Engineering Laboratory Perspecta Labs, Inc. PHY Wireless Pivotal Commware Polaris Wireless	Qualcomm Redline Communications RN-CI Samsung Research America Southern Linc Sprint ST Engineering iDirect Swift Navigation T-Mobile USA TELUS TEOCO Teradyne, Inc. Union Telephone Company

FIGURE 2.16 ATIS 3GPP Individual Members

Figure 2.15 shows, in general terms, the flow of information between the above-mentioned entities. The PCG plans the work of 3GPP based on requirements provided by the organizational partners and the market representation partners. The organizational partners are influenced particularly by their respective national and regional governments and regulators, and the market representation partners generate requirements dictated by the potential market. Individual members provide technical contributions to the TSGs, which ultimately result in technical specifications. These specification are transmitted from the TSGs to the organizational partners, which translate them into national and regional standards. Finally, these standards serve as input to ITU in the development of 5G-related recommendations.

3GPP Releases

3GPP uses a system of parallel releases that provide developers with a stable platform for the implementation of features at a given point and then allow for the addition of new functionality in subsequent releases. Releases are staggered, and work is done on multiple releases in parallel at different stages. When a release is finalized, it means that all new features are functionally frozen and ready for implementation. Furthermore, each 3GPP release is self-contained, meaning that it is possible to build a cellular system based on the set of frozen specifications in that release. As such, releases do not just contain the newly implemented features but instead are introduced in a highly iterative manner that builds upon previous releases.

Table 2.3 provides information on the three releases relating to 5G that were completed or in progress as of early 2021. Release 15 provided an early definition of useful 5G features to enable deployment by 2020. Subsequent releases add progressively more functionality. Release 16 closely resembles the initial set of IMT-2020 recommendations issued by ITU in 2020. Release 17 provides a number of enhancements, especially in the air interface and RAN. As of early 2021, preliminary work on Release 18 was underway, with the timelines not developed.

TABLE 2.3 3GPP Releases for 5G

Release	Status	Functional Freeze	End Date
Release 17	Open	March 18, 2022	June 10, 2022
Release 16	Frozen	March 3, 2020	July 3, 2020
Release 15	Frozen	March 22, 2019	June 6, 2019

Once a release is frozen, the TSGs can add no additional functions to the specifications. However, detailed protocol specifications may not yet be complete. The end date shown in Table 2.3 is indicative only, since for each release, a considerable number of refinements and corrections can be expected for at least two years following this date.

Detailed Requirements

The 3GPP documents include descriptions of 5G requirements that are significantly more detailed than those provided in the ITU documents. As such, they provide an important guide to implementers of 5G networks, components, and systems in terms of the market requirements for 5G success.

Basic Capabilities

3GPP Technical Specification TS 22.261 (*Technical Specification Group Services and System Aspects, Service Requirements for the 5G System, Stage 1 (Release 17)*, December 2020) defines requirements for 34 basic capabilities to be provided by a 5G network. These are listed in Figure 2.17. For each capability, TS 22.261 provides a description and elaborates on the requirements for that capability.

Network slicing	Subscription aspects	Ethernet transport services
Diverse mobility management	Energy efficiency	Non-public networks
Multiple access technologies	Markets requiring minimal service levels	5G LAN-type service
Resource Efficiency	Extreme long-range coverage in low-density areas	Positioning services
Efficient user plane	Multi-network connectivity and service delivery across operators	Cyber-physical control applications in vertical domains
Efficient content delivery	3 GPP access network selection	Messaging aspects
Priority, QoS, and policy control	eV2X aspects	Steering of roaming
Dynamic policy control	NG-RAN sharing	Minimization of service interruption
Connectivity models	Unified access control	UAV aspects
Network capability exposure	QoS monitoring	Video, imaging, and audio for professional applications
Context aware network		Critical medical applications
Self backhaul		
Flexible broadcast/multicast service		

eV@X = Enhanced Vehicle-to-Everything
UAV = Unmanned Aerial Vehicle

FIGURE 2.17 3GPP Basic Capability Requirements

Performance Requirements

TS 22.261 also lists performance requirements that are more detailed and more demanding than those defined in ITU-R Report M.2410. The requirements cover the following categories:

- **High data rates and traffic densities:** Several 5G scenarios require the support of very high data rates or traffic densities, including urban and rural areas, office and home, and special deployments (e.g., massive gatherings, broadcast, residential, high-speed vehicles). Table 2.4 indicates the scenarios and their performance requirements.

TABLE 2.4 Performance Requirements for High-Data-Rate and Traffic-Density Scenarios

Scenario	Experienced Data Rate (DL)	Experienced Data Rate (UL)	Area Traffic Capacity (DL)	Area Traffic Capacity (UL)	Overall User Density	UE Speed
Urban macro	50 Mbps	25 Mbps	100 Gbps/km^2	50 Gbps/km^2	10 000/km^2	Pedestrians and users in vehicles (up to 120 km/h)
Rural macro	50 Mbps	25 Mbps	1 Gbps/km^2	500 Mbps/km^2	100/km^2	Pedestrians and users in vehicles (up to 120 km/h)
Indoor hotspot	1 Gbps	500 Mbps	15 Tbps/km^2	2 Tbps/km^2	250 000/km^2	Pedestrians
Broadband access in a crowd	25 Mbps	50 Mbps	3.75 Tbps/km^2	7.5 Tbps/km^2	500 000/km^2	Pedestrians
Dense urban	300 Mbps	50 Mbps	750 Gbps/km^2	125 Gbps/km^2	25 000/km^2	Pedestrians and users in vehicles (up to 60 km/h)
Broadcast-like services	Maximum 200 Mbps (per TV channel)	N/A or modest (e.g., 500 kbps per user)	N/A	N/A	15 TV channels of 20 Mbps on one carrier	Stationary users, pedestrians, and users in vehicles (up to 500 km/h)
High-speed train	50 Mbps	25 Mbps	15 Gbps/train	7.5 Gbps/train	1 000/train	Users in trains (up to 500 km/h)
High-speed vehicle	50 Mbps	25 Mbps	100 Gbps/km^2	50 Gbps/km^2	4 000/km^2	Users in vehicles (up to 250 km/h)
Airplanes connectivity	15 Mbps	7.5 Mbps	1.2 Gbps/plane	600 Mbps/plane	400/plane	Users in airplanes (up to 1000 km/h)

- **Low latency and high reliability:** Some scenarios require the support of very low latency and very high communications service availability, which implies very high reliability. The overall service latency depends on the delay on the radio interface, transmission within the 5G system, transmission to a server that may be outside the 5G system, and data processing. Some of these factors depend directly on the 5G system itself, whereas for others the impact can be reduced with suitable interconnections between the 5G system and services or servers outside the 5G system, such as to allow local hosting of the services. TS 22.261 provides an overview of potential scenarios and references other technical specifications for specific requirements.

- **High accuracy positioning:** The 5G system needs to provide different 5G positioning services with configurable performance working points (e.g., accuracy, positioning service availability, positioning service latency, energy consumption, update rate, time to first fix) according to the needs of users, operators, and third parties. TS 22.261 lists quantitative requirements for a number of indoor and outdoor scenarios.
- **Key performance indicators (KPIs)[3] for a 5G system with satellite access:** In some contexts, a 5G access network must use at least one satellite link. KPIs defined in TS 22.261 include minimum and maximum UE-to-satellite delay for various earth orbits, as well as maximum propagation delay.
- **High-availability IoT traffic:** This requirement is concerned specifically with medical monitoring but is applicable to other scenarios that require highly reliable machine type communication in both stationary and highly mobile settings.
- **High data rate and low latency:** This requirement defines data and latency requirements for scenarios such as audiovisual interaction, gaming, and virtual reality.
- **KPIs for UE to network relaying in the 5G system:** In several scenarios, it can be beneficial to relay communication between one UE and the network via one or more other UEs. This category includes performance requirements for various scenarios.

2.4 Key Terms and Review Questions

Key Terms

3GPP	energy efficiency
5G	enhanced mobile broadband (eMBB)
5G New Radio (5G NR)	fixed mobile convergence (FMC)
5G NextGen (5G NGN)	International Mobile Telecommunications (IMT)
access network	International Telecommunication Union (ITU)
area traffic capacity	ITU Radiocommunication Sector (ITU-R)
connection density	ITU Telecommunication Standardization Sector (ITU-T)
core network	
customer premises equipment (CPE)	

3. KPIs (key performance indicators) are quantifiable measurements that reflect the critical success factors of a use case.

Kiviat graph	spectrum efficiency
latency	transmission reception point (TRxP)
massive machine type communications (mMTC)	ultra-reliable and low-latency communications (URLLC)
mobility	usage scenario
network slicing	use case
peak data rate	user equipment (UE)
radio access network (RAN)	user-experienced data rate
Radio Regulations (RR)	

Review Questions

1. What are the main activities of ITU-R?
2. What is ITU-R's role in the development of 5G via IMT-2020 documents?
3. Briefly describe eMBB, mMTC, and URLLC.
4. List some examples of emerging use cases that can be supported by IMT-2020.
5. What is ITU-R's role in the development of technical specifications for radio interface technologies for 5G?
6. What is the difference between peak data rate and user-experienced data rate?
7. Describe two types of latency that are relevant to IMT-2020.
8. What is ITU-T's role in the development of 5G via IMT-2020 documents?
9. What are the core network requirements from the network operation point of view defined by ITU-T?
10. Briefly explain network slicing.
11. What is fixed mobile convergence?
12. List the main functional elements of a 5G core network.
13. What is 3GPP's role in the development of 5G technical specifications?
14. List and briefly describe the performance requirements defined by 3GPP.

2.5 References and Documents

References

ERIC20 Ericsson. *Ericsson Mobility Report*. November 2020. https://www.ericsson.com/en/mobility-report

KAUS21 Kaushik, M. "India's own 5G standard could delay its 5G launch." *Business Today*, February 29, 2021.

KUMB17 Kumbhar, A., Koohifar, F., and Güvenç, I. "A Survey on Legacy and Emerging Technologies for Public Safety Communications." *IEEE Communications Surveys and Tutorials*, First Quarter 2017.

MATT16 de Mattos, W., and Gondim, P. "M-Health Solutions Using 5G Networks and M2M Communications." *IT Pro*, May/June 2016.

MORR74 Morris, M. "Kiviat Graphs—Conventions and Figures of Merit." *ACM SIGMETRICS Performance Evaluation Review*, October 1974.

STAL16 Stallings, W. *Foundations of Modern Networking: SDN, NFV, QoE, IoT, and Cloud*. Upper Saddle River, NJ: Pearson Addison-Wesley, 2016.

Documents

3GPP TS 22.261 *Technical Specification Group Services and System Aspects, Service Requirements for the 5G System, Stage 1 (Release 17)*. December 2020.

ITU-R M.1224 *Vocabulary of Terms for International Mobile Telecommunications (IMT)*. March 2012.

ITU-R M.2083 *IMT Vision: Framework and Overall Objectives of the Future Development of IMT for 2020 and Beyond*. September 2015.

ITU-R M.2150 *Detailed specifications of the terrestrial radio interfaces of International Mobile Telecommunications-2020*. February 2021.

ITU-R Report M.2291 *The Use of International Mobile Telecommunications for Broadband Public Protection and Disaster Relief Applications*. November 2016.

ITU-R Report M.2410 *Minimum Requirements Related to Technical Performance for IMT-2020 Radio Interface(s)*. November 2017.

ITU-R Report M.2411 *Requirements, Evaluation Criteria and Submission Templates for the Development of IMT-2020*. November 2017.

ITU-R Report M.2412 *Guidelines for Evaluation of Radio Interface Technologies for IMT-2020*. October 2017.

ITU-R Report M.2441 *Emerging Usage of the Terrestrial Component of International Mobile Telecommunication*. November 2018.

ITU-T Report M.2483 *The outcome of the evaluation, consensus building and decision of the IMT-2020 process (Steps 4 to 7), including characteristics of IMT-2020 radio interfaces.* July 2020.

ITU-T Y.3101 *Requirements of the IMT-2020 Network*. April 2018.

ITU-T Y.3102 *Framework of the IMT-2020 Network*. May 2018.

ITU-T Y.3112 *Framework for the Support of Network Slicing in the IMT-2020 Network.* December 2018.

ITU-T Y.3130 *Requirements of IMT-2020 Fixed Mobile Convergence*. January 2018.

Chapter 3

Overview of 5G Use Cases and Architecture

Learning Objectives

After studying this chapter, you should be able to:

- Present an overview of the types of use cases enabled by 5G
- Discuss the user experience and performance requirements of use cases in various categories
- Present an overview of the NGMN 5G architecture
- Present an overview of the 3GPP 5G core network architecture
- Present an overview of the 3GPP 5G radio access network architecture

This chapter presents a high-level view of the concepts underlying 5G. Section 3.1 provides an overview of the types of use cases enabled by 5G. This section presents three taxonomies or groupings of use cases developed by three different organizations. These different perspectives enable you to develop a solid grasp of the types of use cases that are emerging with 5G and their chief characteristics.

Section 3.2 presents a 5G architecture framework developed by the Next Generation Mobile Networks (NGMN) Alliance. This framework is widely accepted within the telecommunications industry. The framework provides an excellent high-level rendering of the overall 5G system and enables you to understand the overall structure of 5G.

Section 3.3 presents architecture models for the 5G core network and radio access network developed by the 3rd Generation Partnership Project (3GPP), which is the organization developing the technical specifications for 5G. The diagrams presented in this section illustrate the framework being used for the development of detailed technical specifications and enable you to see the context of the separate technical aspects of 5G that are described in subsequent chapters and to see how the various technical elements of 5G fit together.

3.1 5G Use Cases

There is a reciprocal cause–effect relationship between use cases and the capabilities being designed into 5G networks. As users and organizations have come to depend increasingly on wireless and Internet-based applications, these users have come to expect more capability, in terms of data rate and range of features, from wireless networks. These expectations create a demand that justifies the enormous investment required to deploy 5G networks. Conversely, 5G networks are the logical next step in the evolution of cellular networks, and the new capabilities provided by 5G enable users to envision a wide range of new applications, termed *use cases*, in the standards and specifications documents.

This section looks at examples of use cases, or applications, for 5G compiled by three different organizations: ITU Radiocommunication Sector (ITU-R), 5G Americas, and the Next Generation Mobile Networks (NGNM) Alliance. The three compilations provide a good sense of the possibilities enabled by 5G.

ITU-R

ITU-R Report M.2441 describes 16 emerging use cases for IMT-2020, as shown in Figure 3.1. M.2441 provides a description of a number of specific applications encompassed by each use case, as well as a summary of technical capabilities of the IMT systems needed to support each use case.

Machine-type communication	PPDR	Transport applications	Utilities
Industrial automation	Remote Control	Surveying and inspection	Healthcare
Sustainability/ environmental	Smart City	Wearables	Smart Homes
Agriculture	Media and entertainment	Enhanced personal experience	Commercial Airspace UAS

PPDR = Public Protection and Disaster Relief
UAS = Unmanned Aerial Systems

FIGURE 3.1 Emerging 5G Use Cases (ITU-R)

Chapter 2, "5G Standards and Specifications," briefly describes each of the 16 use cases.

5G Americas

5G Americas is an industry trade organization composed of leading telecommunications service providers and manufacturers. The organization's mission is to advocate for and foster the advancement and full capabilities of LTE (Long Term Evolution) wireless technologies and their evolution to 5G throughout the ecosystem's networks, services, applications, and connected devices in the Western Hemisphere. 5G Americas is affiliated with 3GPP as a market representation partner.

The 5G Americas document *5G Services & Use Cases* (November 2017) describes 16 use cases and provides a framework for mapping use cases to 5G capabilities. Figure 3.2, from the 5G Americas document, shows the framework, which has two dimensions. One dimension consists of the three usage scenarios similar to those defined by ITU-R and described in Chapter 2.

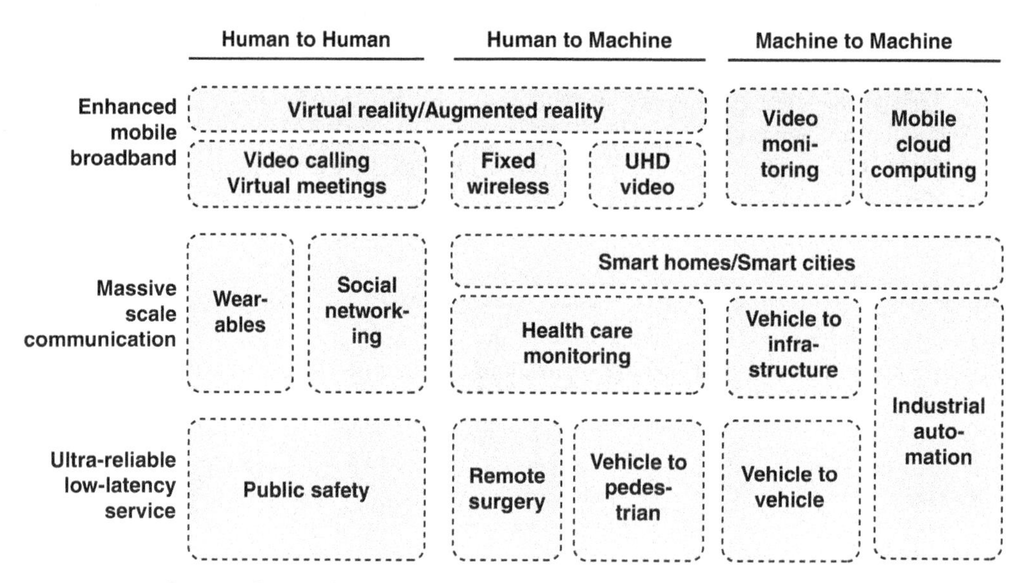

FIGURE 3.2 Some 5G Use Cases Grouped by Type of Interaction and the Range of Performance Requirements (5G Americas)

The three scenarios, based on their performance requirements, are:

- **Enhanced mobile broadband (eMBB):** These use cases generally have requirements for higher data rates and better coverage.
- **Massive scale communication:** These use cases generally have requirements to support a very large number of devices in a small area and, therefore, very high device density.
- **Ultra-reliable low-latency service:** These use cases have very strict requirements on latency and reliability and are also referred to as ultra-reliable and low-latency communications (URLLC).

The other dimension characterizes use cases based on whether they involve human-to-human, human-to-machine, or machine-to-machine communication. As indicated in Figure 3.2, a use case may correspond to more than one type of interaction or more than one usage scenario. This depiction clarifies the requirements for the various use cases.

NGMA Alliance

The Next Generation Mobile Networks (NGNM) Alliance is an association of mobile operators, vendors, manufacturers, and research institutes. It is an open forum whose goal is to ensure that the standards for next-generation network infrastructure, service platforms, and devices meet the requirements of operators and, ultimately, that they will satisfy end user demand and expectations. The NGMN Alliance complements and supports standards organizations by providing a coherent view of what mobile operators require. NGMN is affiliated with 3GPP as a market representation partner.

NGMN's *5G White Paper* (February 2015) defines 24 use cases intended as representative examples of the applications for 5G and meant to highlight the diversity of performance requirements that 5G networks must satisfy. Figure 3.3, from the white paper, illustrates the framework for positioning use cases in a way that clarifies requirements. At a high level, use cases are grouped into eight use case families. These families are roughly similar to the three usage scenarios defined by ITU-R but at a greater granularity. Each family reflects the dominant characteristic of the use cases in that family.

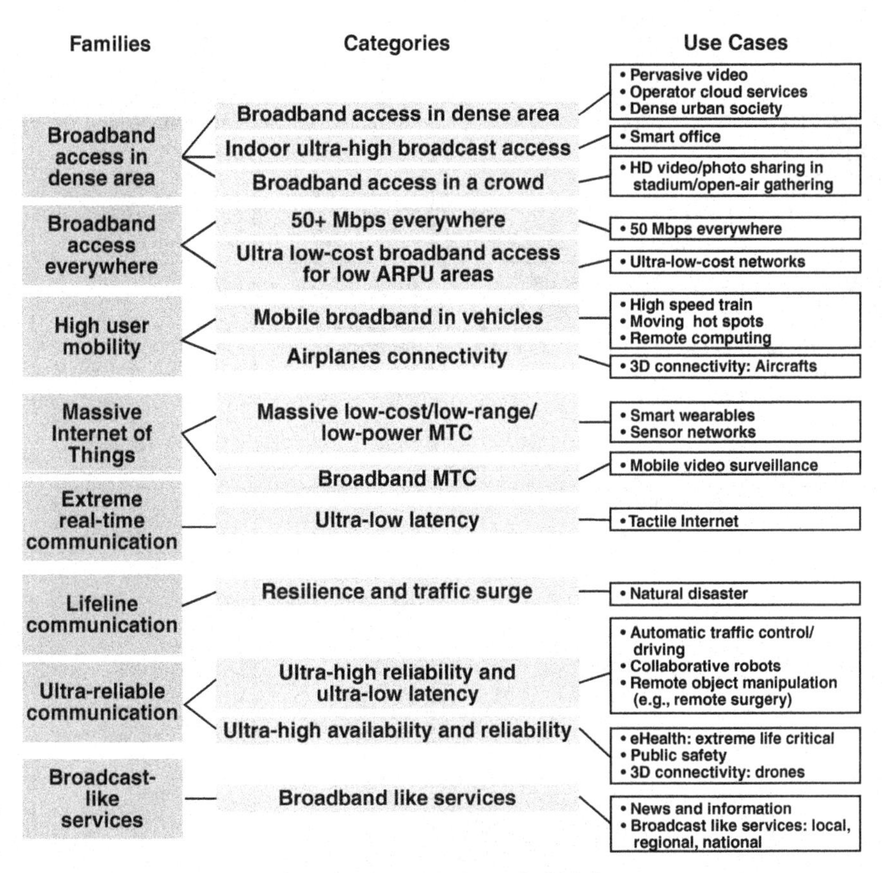

FIGURE 3.3 Use Case Categories Definition (NGNM)

Each family is in turn divided into a number of categories. The categories represent distinct types of demands on the 5G network in terms of user experience requirements and system performance requirements. For each use case category, one set of requirement values is given, which is representative of the extreme use cases(s) in the category. As a result, satisfying the requirements of a category leads to satisfying the requirements of all the use cases in this category. Tables 3.1 and 3.2, from the white paper, summarize the requirements.

TABLE 3.1 User Experience Requirements

Use Case Category	User-Experienced Data Rate	E2E Latency	Mobility
Broadband access in dense areas	DL: 300 Mbps UL: 50 Mbps	10 ms	On demand: 0–100 km/h
Indoor ultra-high broadband access	DL: 1 Gbps UL: 500 Mbps	10 ms	Pedestrian
Broadband access in a crowd	DL: 25 Mbps UL: 50 Mbps	10 ms	Pedestrian
50+ Mbps everywhere	DL: 50 Mbps UL: 25 Mbps	10 ms	0–120 km/h
Ultra-low-cost broadband access for low average revenue per user (ARPU) areas	DL: 10 Mbps UL: 10 Mbps	50 ms	On demand: 0–50 km/h
Mobile broadband in vehicles (cars, trains)	DL: 50 Mbps UL: 25 Mbps	10 ms	On demand: Up to 500 km/h
Airplanes connectivity	DL: 15 Mbps per user UL: 7.5 Mbps per user	10 ms	Up to 1000 km/h
Massive low-cost/long-range/low-power machine-type communication (MTC)	Low (typically 1–100 kbps)	Seconds to hours	On demand: 0–500 km/h
Broadband MTC	See the requirements for the broadband access in dense areas and 50+ Mbps everywhere categories		
Ultra-low latency	DL: 50 Mbps UL: 25 Mbps	< 1 ms	Pedestrian
Resilience and traffic surge	DL: 0.1–1 Mbps UL: 0.1–1 Mbps	Regular communication: not critical	0–120 km/h
Ultra-high reliability and ultra-low latency	DL: 50 kbps–10 Mbps UL: a few bps–10 Mbps	1 ms	On demand: 0–500 km/h
Ultra-high availability and reliability	DL: 10 Mbps UL: 10 Mbps	10 ms	On demand: 0–500 km/h
Broadcast-like services	DL: Up to 200 Mbps UL: Modest (e.g., 500 kbps)	< 100 ms	On demand: 0–500 km/h

TABLE 3.2 System Performance Requirements

Use Case Category	Connection Density	Traffic Density
Broadband access in dense areas	200–2500/km^2	DL: 750 Gbps/km^2 UL: 125 Gbps/km^2
Indoor ultra-high broadband access	75,000/km^2 (75/1000 m^2 office)	DL: 15 Tbps/km^2 (15 Gbps/1000 m^2) UL: 2 Tbps/km^2 (2 Gbps/1000 m^2)
Broadband access in a crowd	150,000/km^2 (30,000/stadium)	DL: 3.75 Tbps/km^2 (0.75 Tbps/stadium) UL: 7.5 Tbps/km^2 (1.5 Tbps/stadium)
50+ Mbps everywhere	400/km^2 suburban 100/km^2 rural	DL: 20 Gbps/km^2 suburban UL: 10 Gbps/km^2 suburban DL: 5 Gbps/km^2 rural UL: 2.5 Gbps/km^2 rural
Ultra-low-cost broadband access for low ARPU areas	16/km^2	16 Mbps/km^2
Mobile broadband in vehicles (cars, trains)	2000/km^2 (500 active users per train × 4 trains or 1 active user per car × 2000 cars)	DL: 100 Gbps/km^2 (25 Gbps per train, 50 Mbps per car) UL: 50 Gbps/km^2 (12.5 Gbps per train, 25 Mbps per car)
Airplanes connectivity	80 per plane 60 airplanes per 18,000 km^2	DL: 1.2 Gbps/plane UL: 600 Mbps/plane
Massive low-cost/long-range/low-power MTC	Up to 200,000/km^2	Not critical
Broadband MTC	See the requirements for the broadband access in dense areas and 50+ Mbps everywhere categories	
Ultra-low latency	Not critical	Potentially high
Resilience and traffic surge	10,000/km^2	Potentially high
Ultra-high reliability and ultra-low latency	Not critical	Potentially high
Ultra-high availability and reliability	Not critical	Potentially high
Broadcast-like services	Not relevant	Not relevant

3.2 NGMN 5G Architecture Framework

NGMN introduced a framework for the 5G architecture in its 2015 white paper and subsequently elaborated the framework in a later document, *5G End-to-End Architecture Framework* (August 2019). The NGMN framework emphasizes the need for modular network functions that could be deployed and scaled on demand to accommodate various use cases in an agile and cost-efficient manner. The NGMN approach is built on the concept of the softwarization of 5G networks. In essence, **softwarization** is an overall approach for designing, implementing, deploying, managing, and maintaining network equipment and/or network components through software programming.

Four approaches to softwarization are important in 5G networks and reflected in the NGMN model:

- **Software-defined networking (SDN):** An approach to designing, building, and operating large-scale networks based on programming the forwarding decisions in routers and switches via software from a central server. SDN differs from traditional networking, which requires configuration of each device separately and relies on protocols that cannot be altered. Chapter 7, "Software-Defined Networking," covers SDN.
- **Network functions virtualization (NFV):** The virtualization of compute, storage, and network functions by implementing these functions in software and running them on virtual machines. Chapter 8, "Network Functions Virtualization," covers NFV.
- **Edge computing:** A distributed information technology (IT) architecture in which client data is processed at the periphery of the network, as close to the originating source as possible.
- **Cloud-edge computing:** A form of edge computing that offers application developers and service providers cloud computing capabilities as well as an IT service environment at the edge of a network. The aim is to deliver compute, storage, and bandwidth much closer to data inputs and/or end users. Chapter 10, "Multi-Access Edge Computing," covers cloud-edge computing.

Layered Functionality

Figure 3.4 illustrates the NGMN architecture framework. The architecture comprises three layers and an end-to-end (E2E) management and orchestration entity.

The **infrastructure resource layer** consists of the physical resources and system software of a fixed-mobile converged (FMC) network. Figure 3.4 shows in detail the core network portion, which includes these types of devices:

- **Cloud nodes:** These nodes provide cloud services, software, and storage resources. There are likely to be one or more central cloud nodes that provide traditional cloud computing service. In addition, cloud-edge nodes provide low latency and higher-security access to client devices at the edge of the network. All of these nodes include virtualization system software to support virtual machines and containers. NFV enables effective deployment of cloud resources to the appropriate edge node for a given application and given fixed or mobile user. The combination of SDN and NFV enables the movement of edge resources and services to dynamically accommodate mobile users.
- **Networking nodes:** These are IP routers and other types of switches for implementing a physical path through the network for a 5G connection. SDN provides for flexible and dynamic creation and management of these paths.

- **Access nodes:** These provide an interface to radio access networks (RANs), which in turn provide access to mobile user equipment (UE). SDN creates paths that use an access node for one or both ends of a connection involving a wireless device.

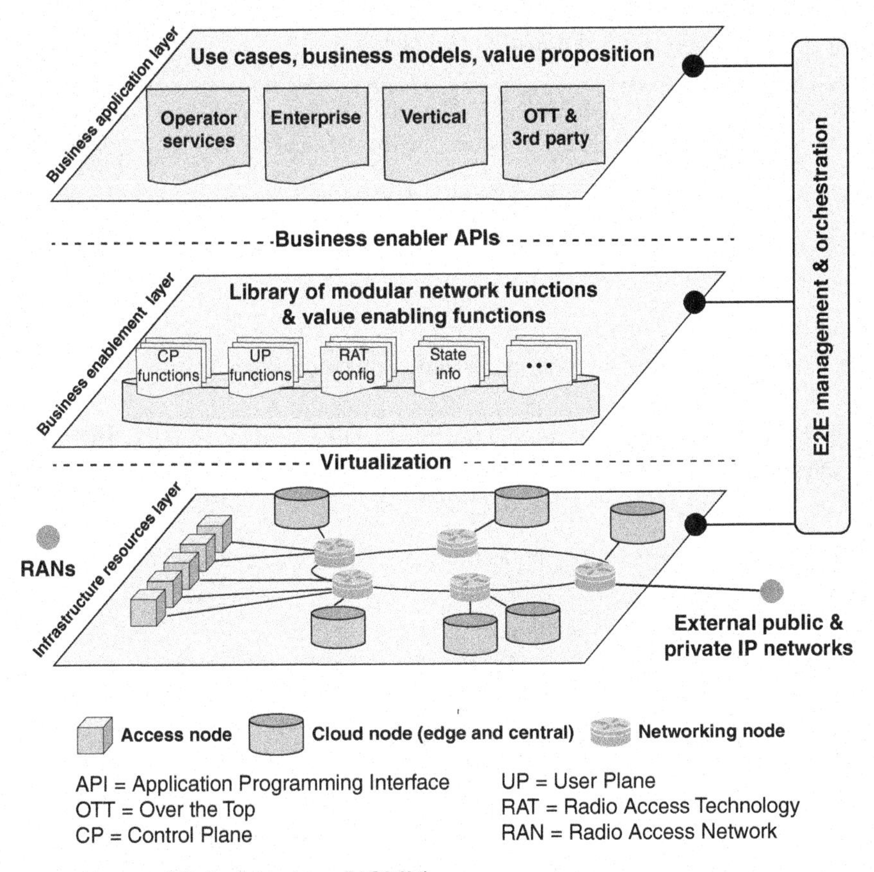

FIGURE 3.4 5G Architecture (NGMN)

Hardware and software resources at the infrastructure resources layer are exposed to higher layers and to the end-to-end management and orchestration entity through relevant application programming interfaces (APIs). Performance and status monitoring as well as configurations are intrinsic parts of such an API.

The **business enablement layer** is a library of all functions required within a converged network in the form of modular architecture building blocks, including functions realized by software modules that can be retrieved from the repository to the desired location, and a set of configuration parameters for certain parts of the network (e.g., radio access). The functions and capabilities are called upon request by the orchestration entity, through relevant APIs. For certain functions, multiple variants might exist. For example, there may be different implementations of the same functionality that have different performance or characteristics. The different levels of performance and capabilities offered could be utilized to differentiate the network functionality much more than in today's networks (e.g., to offer mobility functions such as nomadic mobility, vehicular mobility, or aviation mobility, depending on specific needs). Specific types of components at this layer include:

- **Control plane functions:** These are modules that implement the control signaling functions within the network, as well as control signals associated with SDN and NFV.
- **User plane functions:** These modules deal with the exchange of user data over a connection.
- **Radio access technology (RAT) configuration:** These functions facilitate the configuration of elements in the RAN, including base stations.
- **State information:** State information is split from functions and nodes and managed separately. This includes the state of network connections and RAN radio channels.

With the use of SDN and NFV, 5G networks can support flexible functions and capabilities. The functions at the business enablement layer can tailor connections, configurations, and resource deployment for each use case.

The **business application layer** contains specific applications and services that support the following users of the 5G network:

- **Mobile network operator:** A mobile network operator is a wireless telecommunications organization that provides wireless voice and data communication for its subscribed mobile users. The operator owns or controls the complete telecom infrastructure for hosting and managing mobile communications between the subscribed mobile users with users in the same and external wireless and wired telecom networks, including radio spectrum allocation, wireless network infrastructure, backhaul infrastructure, billing, customer care, provisioning computer systems, and marketing and repair organizations. Mobile network operators are also known as wireless service providers, wireless carriers, and cellular companies.
- **Enterprise:** An enterprise is a business that offers services over the mobile network. These services include applications that run on mobile devices and cloud-based services that enable application portability across multiple devices.
- **Verticals:** An industry vertical is an organization that provides products and/or services targeted to a specific industry, trade, profession, or other group of customers with specialized needs. A vertical might provide a range of products or services useful in the banking industry or healthcare. In contrast, a horizontal provides products or services that address a specific need across multiply industries, such as accounting or billing products and services. Some 5G use cases are realized by standalone private networks managed by the vertical industry itself rather

than the network service provider (NSP). A good example is factory automation. In such cases, the vertical can own and control its own application packages and business application layer.

- **Over-the-top (OTT) and third parties:** OTT or third-party services can be defined as any services provided over the Internet and mobile networks that bypass traditional operators' distribution channel. Cooperation between the mobile network operator and the OTT involves providing quality of service (QoS) and latency attributes in network slices. Examples of OTTs include:
  - **Voice over Internet Protocol (VoIP):** Skype, Viber, etc.
  - **Short Message Service (SMS):** WhatsApp, Kakao Talk, Line, Telegram, etc.
  - **Apps:** Search portals, news portals, banking, weather, shopping, etc.
  - **Cloud services:** Dropbox, Google Drive, Apple iCloud, etc.
  - **Internet television (video streaming):** Netflix, Hulu, YouTube, Amazon Video, etc.

At the interface to the end-to-end management and orchestration entity, users at the business application layer can build dedicated network slices for an application or map an application to existing network slices.

The **E2E management and orchestration** entity is the contact point to translate the use cases and business models into actual network functions and slices. It defines the network slices for a given application scenario, chains the relevant modular network functions, assigns the relevant performance configurations, and finally maps all of this onto the infrastructure resources. It also manages scaling of the capacity of those functions as well as their geographic distribution. In certain business models, it could also possess capabilities to allow for third parties to create and manage their own network slices.

Network Slicing

Chapter 2 introduces the concept of network slicing. Figure 2.11 suggests that slices can divide classes of internal network functions, such as dividing eMBB from mMTC from URLLC. The NGMN framework suggests that slices could effectively partition networks in such a way that different classes of user equipment, utilizing their respective sets of radio access technologies, would perceive quite different infrastructure configurations, even though they would be accessing resources from the same pools.

NGMN defines a network slice as being composed of a collection of 5G network functions and specific RAT settings combined together for the specific use case or business model. Thus, a 5G slice can span all domains of the network—software modules running on cloud nodes, specific configurations of the transport network supporting flexible location of functions, and a dedicated radio configuration or even a specific RAT—as well as configuration of the 5G device.

Figure 3.5 illustrates three use cases highlighted in the NGMN white paper. The blacked-out core network resources represent resources not used to create the network slice.

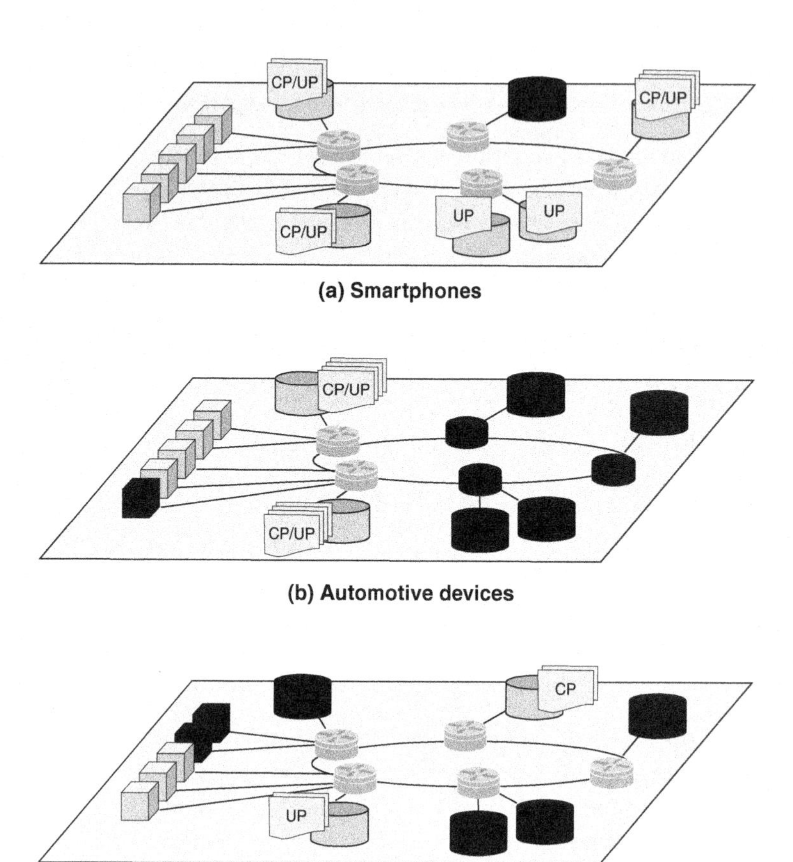

FIGURE 3.5 5G Network Slices Implemented on the Same Infrastructure

Cloud nodes that are part of the slice may include the following:

- Control plane (CP) functions associated with one or more user plane (UP) functions (e.g., a reusable or common framework of control)
- Service- or service category–specific control plane and user plane function pairs (e.g., user-specific multimedia application session)

The first network slice depicted in Figure 3.5 is for a typical smartphone use case. Such a slice might have fully fledged functions distributed across the network. The second network slice in Figure 3.5 indicates the type of support that may be allocated for automobiles in motion. This use case emphasizes the need for security, reliability, and low latency. A configuration to achieve this would limit core network resources to nearby cloud edge nodes, plus the recruitment of sufficient access nodes to support the use case. The final use case illustrated in Figure 3.5 is for a massive IoT deployment, such as a huge number of sensors. The slice can contain just some basic CP and UP functions with, for example, no mobility functions. This slice would only need to engage the access nodes nearest the IoT device deployment.

Thus, as Figures 3.4 and 3.5 suggest, a layered architecture with emphasis on softwarization provides the flexibility needed to fully exploit the 5G infrastructure.

3.3 3GPP 5G Architecture

As discussed in Chapter 2, the 3rd Generation Partnership Project (3GPP) is the organization responsible for developing reports and specifications that define 5G networks. The project covers cellular telecommunications technologies—including the air interface, RAN, core network, and service capabilities—that provide a complete system description for 5G mobile telecommunications. The 3GPP specifications also provide hooks for fixed access to the core network and for interworking with 4G networks and networks outside the 3GPP specifications.

Hundreds of 3GPP reports and specifications related to 5G together describe an extraordinarily complex system. This book provides an overview that encompasses the scope of 5G technology and networks, based on the 3GPP documents. This section provides a "big picture" overview of 5G, summarizing the two key 3GPP documents that together describe the 5G architecture:

- **TS 23.501 (*Technical Specification Group Services and System Aspects; System Architecture for the 5G System (5GS); Stage 2 (Release 16)*, December 2020):** This document describes the core network architecture, together with its services and interfaces.
- **TS 38.300 (*Technical Specification Group Radio Access Network; NR; NR and NG-RAN Overall Description; Stage 2 (Release 16)*, December 2020):** This document describes the radio access network architecture, together with its services and interfaces.

5G Core Network Architecture

The 5G architecture model provides a framework within which detailed specifications can be developed.

Principles

TS 23.501 lists the following as the key principles for the architecture:

- Separate the user plane (UP) functions from the control plane (CP) functions to allow independent scalability, evolution, and flexible deployments (e.g., centralized location or distributed [remote] location).
- Modularize the function design (e.g., to enable flexible and efficient network slicing).
- Wherever applicable, define procedures (i.e., the set of interactions between network functions) as services so that their reuse is possible.

- Enable each network function (NF) and its network function services (NFS) to interact with other NF and its NFS directly or indirectly via a service communication proxy, if required. The architecture does not preclude the use of another intermediate function to help route control plane messages.
- Minimize dependencies between the access network (AN) and the core network (CN). The architecture is defined with a converged core network with a common AN–CN interface that integrates different access types (e.g., 3GPP access and non-3GPP access).
- Support a unified authentication framework.
- Support stateless NFs. A stateless NF separates an NF into a processing module and a data store module containing state information. This design results in a more agile NF in terms of scalability, resilience, and ease of deployment [KABL17].
- Support capability exposure. As mentioned in Chapter 2, this refers to the ability to provide relevant information about network capabilities to third parties.
- Support concurrent access to local and centralized services. To support low-latency services and access to local data networks, UP functions can be deployed close to the access network.
- Support roaming with both home-routed traffic as well as local breakout traffic in the visited public land mobile network (PLMN).

Roaming

The final item in the preceding list references some important concepts. A PLMN, often called a carrier, is a telecommunications network that provides mobile cellular services. A home PLMN for a given mobile phone subscriber is the PLMN that is contracted to provide cellular service to the subscriber. Roaming is the ability for a user to function in a serving network different from the home network, called the visited network. Roaming can occur internationally, in which case mobile subscribers get coverage and at least basic services similar to their domestic package from a network operator in another country. For 5G, any user device should work in any other country. With 4G and earlier, there are multiple technologies in use, and a cell phone from one country may not be able to operate on a network in another country. *National roaming* refers to the ability to move from one mobile operator to another in the same country.

Both the home and visited PLMN use the same network architecture, which defines protocols, services, and interfaces between the home PLMN and the visited PLMN. The 3GPP specifications support two models of operation: home routed and local breakout (LBO). In the home-routed roaming model, the subscriber's data traffic is serviced by the subscriber's home network, which gives the home network operator more control over the user's traffic and is preferable when the relationship between the two operators is not totally trustworthy. In the LBO model, the subscriber's data is serviced by the visited network; this model delivers more efficient routing in terms of bandwidth and latency. In the case of LBO, the home network owner loses control of the customer and has no role in delivering services to that user. The LBO model is used when there is a trusted relationship between the two operators.

An architecture that includes roaming adds complexity but no change to the basic network services and functionality. Therefore, this section focuses on the non-roaming case.

Architecture Diagrams

TS 23.501 contains a number of architecture diagrams from several different points of view and at varying levels of detail. All of the diagrams depict the architecture in terms of a number of interconnected network functions (NFs). An NF is a processing function in a network that has defined functional behavior and interfaces. A network function can be implemented either as a network element on dedicated hardware, as a software instance running on dedicated hardware, or as a virtualized function instantiated on an appropriate platform.

The interconnection between NFs is represented in two ways:

- **Service-based representation:** NFs within the control plane enable other authorized NFs to access their services. This representation illustrates how a set of services is provided/exposed by a given network interface. This interface defines how one network function within the control plane allows other network functions that have been authorized to access its services. This representation also includes point-to-point reference points where necessary.
- **Reference point representation:** This representation uses labeled point-to-point links to show the interaction that exists between two NFs or between an NF and an external functional module or network. The reference point representation is beneficial when showing message sequence charts. It shows the relationships between NFs that are used in the message sequence charts.

There are several advantages to this form of architectural representation. The modular structure provides a framework for developing detailed specifications for each NF. The service-based interfaces and reference points provide a framework for developing detailed specifications of the interaction between NFs in terms of data formats, protocols, and service calls. In addition, the detailed interface specifications promote interoperability between different hardware/software providers. Finally, this type of architecture definition provides a way of ensuring that all 5G functional and service requirements are satisfied.

Service-Based System Architecture

Figure 3.6, based on TS 23.501, depicts the overall non-roaming 5G service-based architecture (SBA) for the core network. The functional components of the control plane of the network are network functions (NFs) that offer their services to any other applicable NFs via a common framework of interfaces accessible to all NFs. Network repository functions (NRFs) allow every NF to discover the services offered by other NFs present in the network; network exposure functions (NEFs) expose capability information and services of the 5G core network NFs to external entities. This model aims to maximize the modularity, reusability, and self-containment of network functions and to foster the ability to grow flexibly while taking advantage of network functions virtualization and software-defined networking.

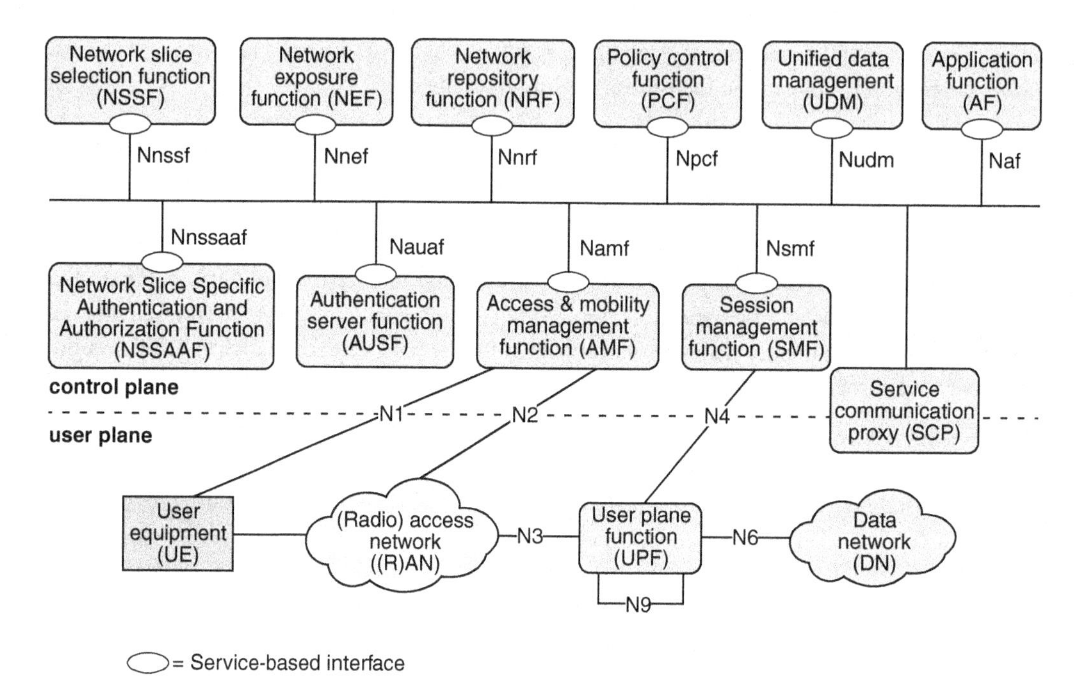

FIGURE 3.6 Non-Roaming 5G System Architecture

The figure includes the following NFs and other modules:

- **Authentication server function (AUSF):** Performs authentication between UE and the network.
- **Access and mobility management function (AMF):** Receives all connection- and session-related information from the user equipment (UE) (N1/N2) but is responsible only for handling connection, registration, reachability, and mobility management tasks. All messages related to session management are forwarded to the session management function (SMF).
- **Network exposure function (NEF):** Provides an interface for outside applications to communicate with the 5G network to obtain network-related information in the following categories:
  - **Monitoring capability:** Allows an external entity to request or subscribe to UE-related events of interest. The monitored events include a UE's roaming status, UE loss of connectivity, UE reachability, and location-related events.
  - **Provisioning capability:** Allows an external entity to provide information about expected UE behavior to the 5G system (e.g., predicted UE movement, communication characteristics).
  - **Policy/charging capability:** Handles QoS and charging policy for the UE, based on a request from an external party.

- **Analytics reporting capability:** Allows an external party to fetch or subscribe/unsubscribe to analytics information generated by the 5G system.

- **Network repository function (NRF):** Allows NFs to register their functionality and to discover the services offered by other NFs present in the network.
- **Network slice selection function (NSSF):** Selects the set of network slice instances to accommodate the service request from a UE. When a UE requests registration with the network, AMF sends a network slice selection request to NSSF with information on the preferred network slice selection. The NSSF responds with a message that includes a list of appropriate network slice instances for the UE.
- **Network slice-specific authentication and authorization (NSSF):** Performs authentication and authorization specific to a slice.
- **Policy control function (PCF):** Provides functionalities for the control and management of policy rules, including rules for QoS enforcement, charging, and traffic routing. PCF enables end-to-end QoS enforcement with QoS parameters (e.g., maximum bit rate, guaranteed bit rate, priority level) at the appropriate granularity (e.g., per UE, per flow, per protocol data unit [PDU] session).
- **Session management function (SMF):** Responsible for PDU session establishment, modification, and release between a UE and a data network. A PDU session, or simply a session, is an association between the UE and a data network that provides a PDU connectivity service. A PDU connectivity service is a service that provides for the exchange of PDUs between a UE and a data network.
- **Unified data management (UDM):** Responsible for access authorization and subscription management. UDM works with the AMF and AUSF as follows: The AMF provides UE authentication, authorization, and mobility management services. The AUSF stores data for authentication of UEs, and the UDM stores UE subscription data.
- **User plane function (UPF):** Handles the user plane path of PDU sessions. This function is described subsequently.
- **Application function (AF):** Provides session-related information to the PCF so that the SMF can ultimately use this information for session management. The AF interacts with application services that require dynamic policy control. The AF extracts session-related information (e.g., QoS requirements) from application signaling and provides it to the PCF in support of its rule generation. An example is the IP multimedia subsystem (IMS), which may interface with the PCRF to request QoS support for VoIP calls.
- **User equipment (UE):** Allows a user access to network services. An example is a mobile phone. For the purpose of 3GPP specifications, the interface between the UE and the network is the radio interface.
- **(Radio) Access Network ((R)AN):** Provides access to a 5G core network. This includes the 5G RAN and other wireless and wired access networks.

- **Data network (DN):** Allows UE to be logically connected by a session. It may be the Internet, a corporate intranet, or an internal services function within the mobile network operator's core (including content distribution networks).
- **Service communication proxy (SCP):** Allows NFs and NFSs to communicate directly or indirectly. The SCP enables multiple NFs to communicate with each other and with user plane entities in a highly distributed multi-access edge compute cloud environment. This provides routing control, resiliency, and observability to the core network.

The ovals on NFs in Figure 3.6 indicate service interfaces that can be accessed by other NFs. Each interface is identified by a label consisting of an uppercase N followed by the abbreviation of the NF in lowercase. For example, the network slice selection function has a service interface labeled Nnssf.

It is informative to compare Figure 3.6 with Figure 2.13, which shows the ITU-T Y.3102 (*Framework of the IMT-2020 Network*, May 2018) representation of the core network, which provides a somewhat different functional breakdown. This can be considered an earlier version of the core network architecture that has been superseded by the current 3GPP architecture.

Reference Point Representation

Figure 3.7, based on TS 23.501, depicts the overall non-roaming 5G architecture using the reference point representation, showing how the NFs interact with each other.

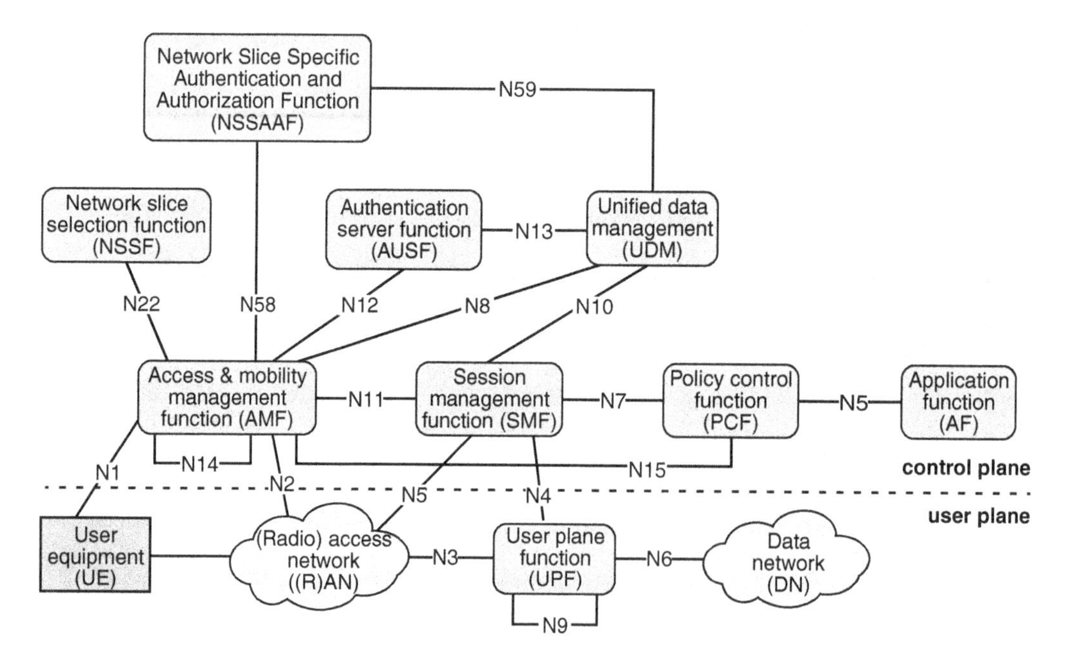

FIGURE 3.7 Non-Roaming 5G System Architecture in Reference Point Representation

Note that there are fewer interconnections depicted in Figure 3.7 than in Figure 3.6. Within the control plane, the interconnections in Figure 3.6 indicate which NFs can access the services of which other NFs. The interconnections of Figure 3.7 indicate which NFs communicate with each other directly, without going through an intermediate NF. The term *directly* does not mean that there is a physical point-to-point link between NFs connected on the diagram. Rather, it means that there is a protocol for the exchange of messages between the connected entities that is not relayed through another NF. Each such link is labeled with a reference point expressed as an uppercase N followed by a number. For example, the logical connection between the session management function and the policy control function is labeled reference point N7.

In Figure 3.7, two reference points loop back to the same function: N9 and N14. The N9 reference point is an interface between two distinct UPFs used for forwarding packets. The N14 reference point is between two AMFs, one acting as a source AMF for a data transfer and the other acting as a destination AMF.

User Plane Function

User plane functions handle the user plane path of PDU sessions. 3GPP specifications support deployments with a single UPF or multiple UPFs for a given PDU session. UPF selection is performed by SMF. UPF functions include:

- Packet routing and forwarding.
- Anchor point for intra-/inter-RAT mobility (when applicable). Anchor points are transit nodes in the network used for forwarding PDUs along a session from a UE to the destination.
- External PDU session point of interconnect to data network.
- Packet inspection (e.g., application detection based on a service data flow [SDF] template). An SDF provides end-to-end packet flow between an end user and an application; this is discussed in Chapter 9, "Core Network Functionality, QoS, and Network Slicing."
- User plane part of policy rule enforcement (e.g., gating, redirection, traffic steering).
- Traffic usage reporting.
- QoS handling for the user plane, such as uplink/downlink rate enforcement.
- Uplink traffic verification (SDFs to QoS flow mapping). A QoS flow is the lowest level of granularity for defining end-to-end QoS policies. A QoS flow may contain multiple SDFs; this is discussed in Chapter 9.
- Transport-level packet marking in the uplink and downlink.
- Downlink packet buffering and downlink data notification triggering.
- Sending and forwarding of one or more end markers to the source NG-RAN node.

Radio Access Network Architecture

Figure 3.8, from TS 38.300, depicts the overall RAN architecture.

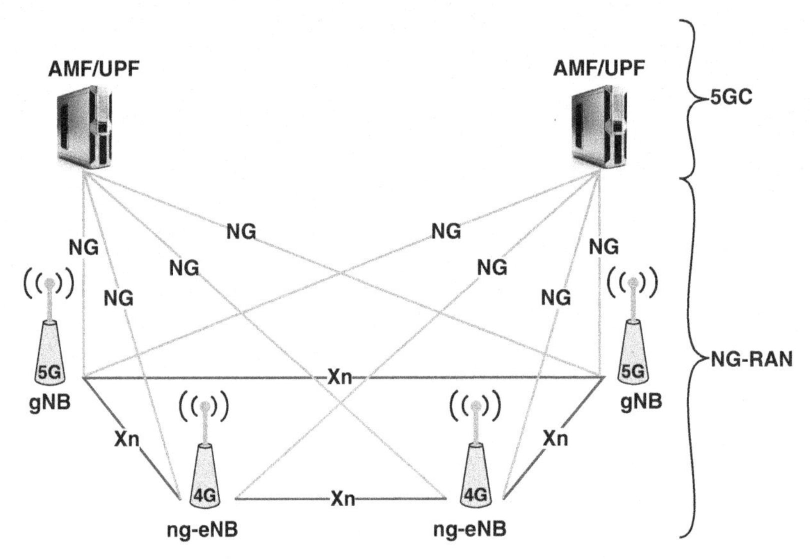

AMF = Access and Mobility Management Function
UPF = User Plane Function
gNB = Provides 5G user plane and control plane protocol terminations toward the UE
ng-eNB = Provides 4G (E-UTRA) user plane and control plane protocol terminations toward the UE
Xn = Interconnects gNBs and ng-eNBs
NG = connects gNBs and ng-eNBs to 5G core network
5GC = Fifth-generation core network
NG-RAN = Fifth-generation (next generation) radio access network

FIGURE 3.8 Overall Radio Access Network Architecture

There are two types of base stations, called NG-RAN nodes:

- **gNB:** Provides 5G user plane and control plane protocol terminations toward the UE.
- **ng-eNB:** Provides 4G (E-UTRA) user plane and control plane protocol terminations toward the UE and connects via the NG interface to the 5G core. This enables 5G networks to support UE that use the 4G air interface. However, the UE must still implement the 5G protocols to interact with the 5G core network.

The gNBs and ng-eNBs are interconnected with each other by means of the Xn interface. The gNBs and ng-eNBs are also connected by means of the NG interfaces to the core network (5GC)—specifically, to the AMF (access and mobility management function) by means of the NG-C interface and to the UPF (user plane function) by means of the NG-U interface.

Figure 3.9, from TS 38.300, shows the major functional elements performed by the RAN, together with functions within the core network that specifically relate to the RAN. The outer shaded boxes depict the logical nodes, and the inner white boxes depict the main functions at each node. TS 38.300 also includes a more comprehensive list of functions for the four logical nodes, and these are discussed in Part Four, "5G NR Air Interface and Radio Access Network."

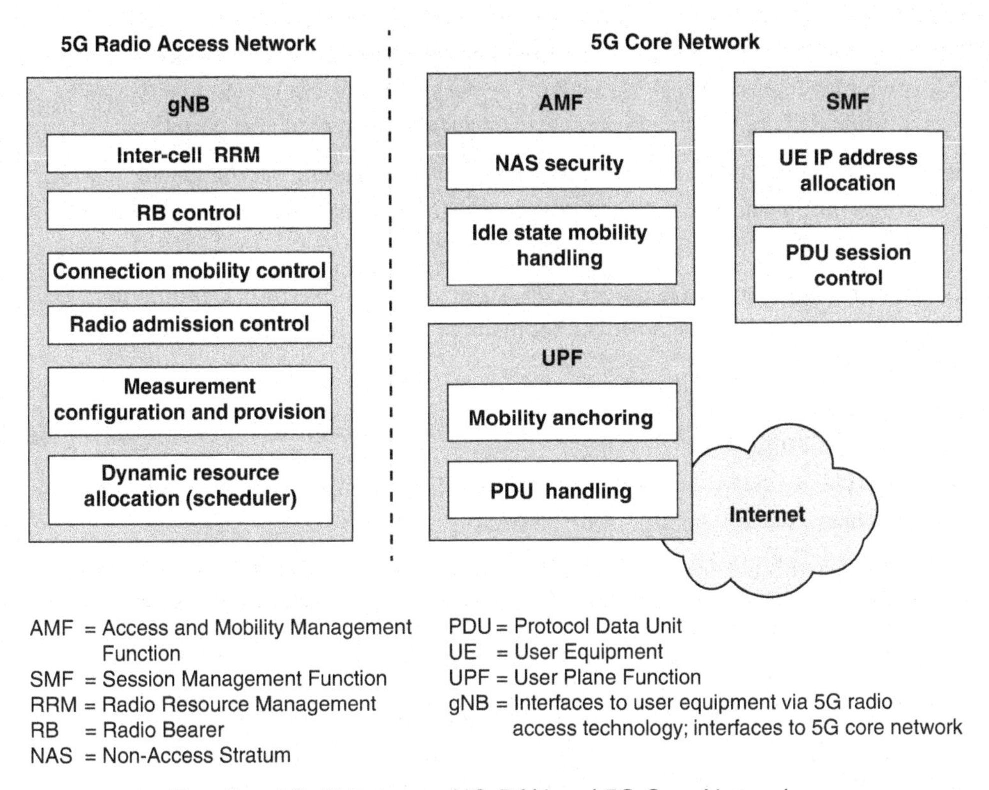

AMF = Access and Mobility Management Function
SMF = Session Management Function
RRM = Radio Resource Management
RB = Radio Bearer
NAS = Non-Access Stratum
PDU = Protocol Data Unit
UE = User Equipment
UPF = User Plane Function
gNB = Interfaces to user equipment via 5G radio access technology; interfaces to 5G core network

FIGURE 3.9 Functional Split Between NG-RAN and 5G Core Network

RAN Functional Areas

Figure 3.9 illustrates the following key functional areas in the NG-RAN:

- **Inter-cell radio resource management:** Allows the UE to detect neighbor cells, query about the best serving cell, and support the network during handover decisions by providing measurement feedback.
- **Radio bearer control (RBC):** Consists of the procedure for configuration (such as security), establishment, and maintenance of the radio bearer (RB) on both the uplink and downlink with different quality of service (QoS). The term *radio bearer* refers to an information transmission path of defined capacity, delay, bit error rate, and other parameters.

- **Connection mobility control (CMC):** Functions both in UE idle mode and connected mode. In idle mode, UE is switched on but does not have an established connection. In connected mode, UE is switched on and has an established connection. In idle mode, CMC performs cell selection and reselection. The connected mode involves handover procedures triggered on the basis of the outcome of CMC algorithms.
- **Radio admission control (RAC):** Decides whether a new radio bearer admission request is admitted or rejected. The objective is to optimize radio resource usage while maintaining the QoS of existing user connections. Note that RAC decides on admission or rejection for a new radio bearer, while RBC takes care of bearer maintenance and bearer release operations.
- **Measurement configuration and provision:** Consists of provisioning the configuration of the UE for radio resource management procedures such as cell selection and reselection and for requesting measurement reports to improve scheduling.
- **Dynamic resource allocation (scheduler):** Consists of scheduling RF resources according to their availability on the uplink and downlink for multiple pieces of UE, according to the QoS profiles of a radio bearer.

Access and Mobility Management

On the core network side, the NG-RAN nodes interact with three functions: the access and mobility management, session management, and user plane functions.

The AMF provides UE authentication, authorization, and mobility management services. The two main functions shown in Figure 3.9 for AMF are NAS security and idle state mobility handling.

The non-access stratum (NAS) is the highest protocol layer of the control plane between UE and the access and mobility management function (AMF) in the core network. The main functions of the protocols that are part of the NAS are the support of mobility of the UE and the support of session management procedures to establish and maintain IP connectivity between the UE and user plane function (UPF). It is used to maintain continuous communications with the UE as it moves. In contrast, the access stratum is responsible for carrying information just over the wireless portion of a connection. NAS security involves IP header compression, encryption, and integrity protection of data based on the NAS security keys derived during the registration and authentication procedure.

Idle state mobility handling deals with cell selection and reselection while the UE is in idle mode, as well as reachability determination.

Session Management Function

The two main functions depicted in Figure 3.9 for SMF are UE IP address allocation and PDU session control.

UE IP address allocation assigns an IP address to the UE at the time of session establishment. This ensures the ability to route data packets within the 5G system and also supports data reception and forwarding to outside networks and provides interconnectivity to external packet data networks (PDNs).

In cooperation with the UPF, the SMF establishes, maintains, and releases a PDU session for user data transfer, which is defined as an association between the UE and a data network that provides PDU connectivity.

User Plane Function

The two main functions depicted in Figure 3.9 for UPF are UE IP mobility anchoring and PDU handling.

UE mobility handling deals with ensuring that there is no data loss when there is a connection transfer due to handover that involves changing anchor points.

Once a session is established, the UPF has a responsibility for PDU handling. This includes the basic functions of packet routing, forwarding, and QoS handling.

Session Establishment

TS 23.502 (*Technical Specification Group Services and System Aspects; Procedures for the 5G System (5GS); Stage 2 (Release 16),* December 2020) defines the session establishment process. Figure 3.10 provides a much simplified view of the interaction between the various network components during session establishment. This section does not examine this process in detail. More detail is provided in Chapter 9.

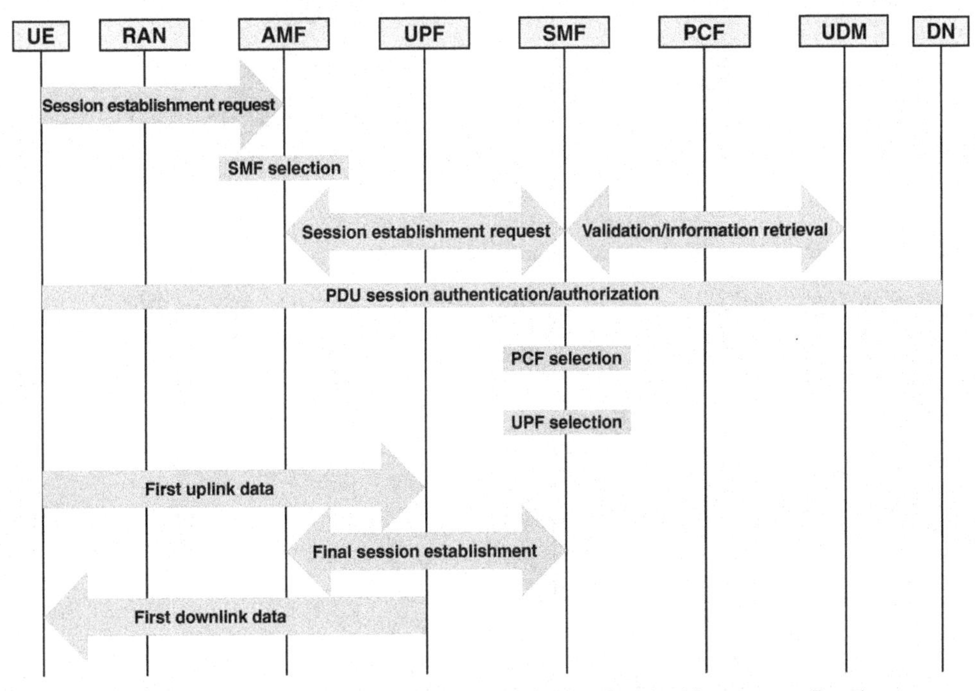

FIGURE 3.10 UE-Requested PDU Session Establishment

Session establishment begins with a request from the UE over the RAN, which is directed to the AMF. An SMF is selected to manage the PDU session. SMF utilizes UDM in the process of creating a session and performing authentication and authorization. SMF selects a PCF for the session. SMF selects a UPF to handle data plane PDU forwarding in both directions. SMF establishes a session with the DN. After a few more exchanges, the UE is able to communicate over a session with the DN.

3.4 Key Terms and Review Questions

Key Terms

3GPP
5G Americas
access and mobility management function (AMF)
cellular company
cloud-edge computing
critical communications
edge computing
enhanced mobile broadband (eMBB)
gNB
ITU Radiocommunication Sector (ITU-R)
massive Internet of Things (MIoT)
mobile network operator
network functions virtualization (NFV)
network slicing
Next Generation Mobile Networks (NGMN) Alliance
ng-eNB
over the top (OTT)
radio access network (RAN)
radio bearer
reference point
reference point representation
roaming
service-based architecture
session management function (SMF)
software-defined networking (SDN)
softwarization
ultra-reliable and low-latency communications (URLLC)
use case
user equipment (UE)
user plane function (UPF)
vertical
wireless carrier
wireless service provider

Review Questions

1. List the emerging 5G use cases defined by ITU-R.
2. Explain the use of two dimensions in characterizing 5G uses cases by 5G Americas.
3. Explain the concepts of families and categories used by NGMN to classify 5G use cases.
4. What is softwarization?
5. What are the main approaches to softwarization?
6. Describe the three layers of the NGMN architecture framework.
7. Define the three types of nodes that are used in the NGMN core network model.
8. In the NGMN model, who are the key 5G users?
9. What is the relationship between network slicing and the types of nodes in the core network?
10. Explain the concept of roaming.
11. What is the difference between a service-based representation and a reference point representation?
12. Summarize the major functions encompassed by the user plane function.
13. What are the two primary types of nodes in a radio access network?
14. List and briefly define the RAN functional areas.
15. Which core network functions interact directly with the RAN?

3.5 References and Documents

References

IVES18 Ives, S. "Where Is Video Monitoring Now?" *Security Systems News*, July 25, 2018.

KABL17 Kablan, M., et al. "Stateless Network Functions: Breaking the Tight Coupling of State and Processing." *Proceedings of the 14th USENIX Symposium on Networked Systems Design and Implementation (NSDI '17)*, March 2017.

Documents

3GPP TS 23.501 *Technical Specification Group Services and System Aspects, System Architecture for the 5G System (5GS), Stage 2 (Release 16)*. December 2020.

3GPP TS 23.502 *Technical Specification Group Services and System Aspects, Procedures for the 5G System (5GS), Stage 2 (Release 16)*. December 2020.

3GPP TS 38.300 *Technical Specification Group Radio Access Network, NR, NR and NG-RAN Overall Description, Stage 2 (Release 16)*. December 2020.

5G Americas *5G Services & Use Cases*. November 2017.

ITU-R Report M.2441 *Emerging Usage of the Terrestrial Component of International Mobile Telecommunication*. November 2018.

ITU-T Y.3102 *Framework of the IMT-2020 Network*. May 2018.

NGMN15 *NGMN 5G White Paper*. February 2015.

NGMN19 *5G End-to-End Architecture Framework*. August 2019.

Chapter 4

Enhanced Mobile Broadband

Learning Objectives

After studying this chapter, you should be able to:

- Describe the three deployment scenarios defined for eMBB
- Summarize the performance requirements for eMBB
- Present an overview of the smart office use case in the indoor hotspot deployment scenario
- Present an overview of the dense urban information society use case in the dense urban deployment scenario
- Present an overview of the radiocommunication systems between train and trackside use case in the rural eMBB deployment scenario

ITU-R (International Telecommunication Union Radiocommunication Sector) M.2083 (*IMT [International Mobile Telecommunications] Vision: Framework and Overall Objectives of the Future Development of IMT for 2020 and Beyond*, September 2015) characterizes enhanced mobile broadband (eMBB) as follows:

> Enhanced Mobile Broadband addresses the human-centric use cases for access to multi-media content, services and data. The demand for mobile broadband will continue to increase, leading to enhanced Mobile Broadband. The enhanced Mobile Broadband usage scenario will come with new application areas and requirements in addition to existing Mobile Broadband applications for improved performance and an increasingly seamless user experience. This usage scenario covers a range of cases, including wide-area coverage and hotspot, which have different requirements. For the hotspot case, i.e. for an area with high user density, very high traffic capacity is needed, while the requirement for mobility is low and user data rate is higher than that of wide area coverage. For the wide area coverage case, seamless coverage and medium to

high mobility are desired, with much improved user data rate compared to existing data rates. However the data rate requirement may be relaxed compared to hotspot.

Of the three usage scenarios defined in M.2083 and described in Chapter 2, "5G Standards and Specifications," eMBB is the only general-purpose case; it is also the scenario that is most familiar to current 4G users. In essence, eMBB is an enhanced version of 4G, providing improved performance and an increasingly seamless user experience.

A white paper from the Next Generation Mobile Networks (NGMN) Alliance (*5G Prospects: Key Capabilities to Unlock Digital Opportunities*, July 2016) lists the following as the key drivers of eMBB:

- **Traffic demand:** It has been estimated that over 80% of cellular Internet traffic is consumed indoors [COMM15]. Users are demanding higher throughputs (e.g., to meet various video needs, such as HD video streaming).
- **Operator competition:** Operators are endeavoring to provide more competitive, attractive data plans, and service offerings are being eroded by over-the-top (OTT) alternatives. Competition is mainly in two directions: to offer cheaper rates, often focusing on particular customer segments as pursued by many mobile virtual network operators (MVNOs), or to offer added value, exemplified by larger data caps, faster data rates, and bundling with, for example, zero-rated videos. Zero-rating is the practice of exempting an app from a user's monthly data plans; an application that is zero-rated does not count against a user's data cap, while all other applications do count against the cap.
- **Incentives to improve attractiveness of countries, cities, and premises:** Many governments and municipalities see eMBB availability as a key to future productivity and economic growth (enriching the lives of citizens, attracting more tourists, and facilitating businesses) and are leading initiatives to improve broadband environments. Similarly, building owners have incentives to invest in connectivity provisioning to sustain property value. The same applies to any kind of space, including airports, hotels, shopping malls, coffee shops, public transport, entertainment venues (e.g., sports, music), and even cars and airplanes. The cost for connectivity provisioning is often paid by indirect sources, making the cost invisible to the end user.

To provide a better understanding of the nature and purpose of eMBB, this chapter examines several perspectives for defining eMBB. Section 4.1 describes the three deployment scenarios envisioned for eMBB. Section 4.2 summarizes the performance characteristics of eMBB. Sections 4.3 through 4.5 present examples of use cases that eMBB is designed to support in the three different deployment scenarios.

4.1 eMBB Deployment Scenarios

ITU-R Report M.2410 (*Minimum Requirements Related to Technical Performance for IMT-2020 Radio Interface(s)*, November 2017) lists three deployment options that characterize the scope of eMBB and that are used for purposes of evaluation of candidate specifications: indoor hotspot, dense urban, and rural.

Indoor Hotspot

ITU-R Report M.2412 (*Guidelines for Evaluation of Radio Interface Technologies for IMT-2020*, October 2017) defines *indoor hotspot* as "an indoor isolated environment at offices and/or in shopping malls based on stationary and pedestrian users with very high user density." This deployment scenario focuses on small coverage per site/TRxP (transmission and reception point) and high user throughput or user density in buildings. The key characteristics of this deployment scenario are high capacity, high user density, and consistent user experience indoors.

5G capabilities should enable a seamless interface for users moving into and out of the indoor zone, without the necessity of joining a Wi-Fi network for indoor use. Types of demand include frequent upload and download of data from a company's servers and real-time video meetings with local as well as remote participants.

One of the main challenges for supporting 5G use cases in the indoor environment is a consequence of the use of much higher-frequency bands for 5G than are used for 4G and earlier generations. These higher bands lead to greater link losses. For example, outdoor signals on the C band are subject to an 8- to 13-dB link loss when penetrating through one concrete wall. The signals on the higher millimeter wave (mmWave) band experience difficulty in penetrating through a wall as the link loss exceeds 60 dB (see Figure 4.1, from [GMSA17]). It is a considerable challenge for outdoor 5G macro signals to cover indoor areas, and a dedicated 5G network consisting of interconnected base stations is required for indoor environments.

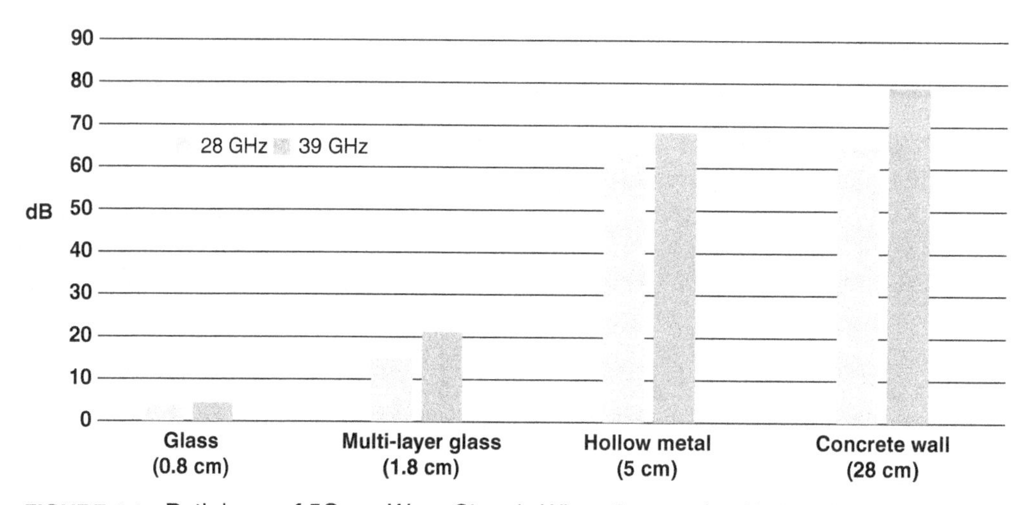

FIGURE 4.1 Path Loss of 5G mmWave Signals When Penetrating Various Walls

Dense Urban

M.2412 defines *dense urban* as "an urban environment with high user density and traffic loads focusing on pedestrian and vehicular users." The dense urban microcellular deployment scenario focuses on macro TRxPs (transmission and reception points) with or without micro TRxPs and high user densities and traffic loads in city centers and dense urban areas. The key characteristics of this deployment scenario are high traffic loads, outdoor coverage, and outdoor-to-indoor coverage.

The dense urban environment for 5G is characterized by the use of a dense collection of small cells to supplement macro cells for two reasons [KULK14]:

- The concentrated collection of stationary, pedestrian, and vehicular users, with 5G use cases, generates a tremendous traffic load.
- 5G mmWave networks are predominantly noise limited. As a result, only small cell sizes can be supported.

Rural

M.2412 defines *rural-eMBB* as "a rural environment with larger and continuous wide area coverage, supporting pedestrian, vehicular and high speed vehicular users." The rural deployment scenario focuses on larger and continuous coverage. The key characteristics of this scenario are continuous wide area coverage supporting high-speed vehicles. This scenario uses macro TRxPs, and is noise-limited and/or interference-limited.

The rural deployment also supports last-mile service to residences and other subscribers to provide telephone and Internet access. Many homes may be near a fiber connection, but the deployment of the last mile of the cabling can be very expensive and not necessarily cost-effective. Adding new subscribers, or households, may be very expensive if new cables need to be installed. It may also require the operator to support two distinct systems, each with its own subscription management, for wired and wireless subscribers. To address this problem, delivering the last mile wirelessly may be a viable option. Such solutions where the last mile is delivered wirelessly are known as WLL (wireless local loop). WLL is an example of fixed wireless access (FWA).

4.2 eMBB Performance Characteristics

Figure 4.2, based on a figure in ITU-R M.2083, indicates in relative terms the importance of eight key performance characteristics for eMBB. Note that latency and connection density are depicted as being of somewhat less importance than the other characteristics.

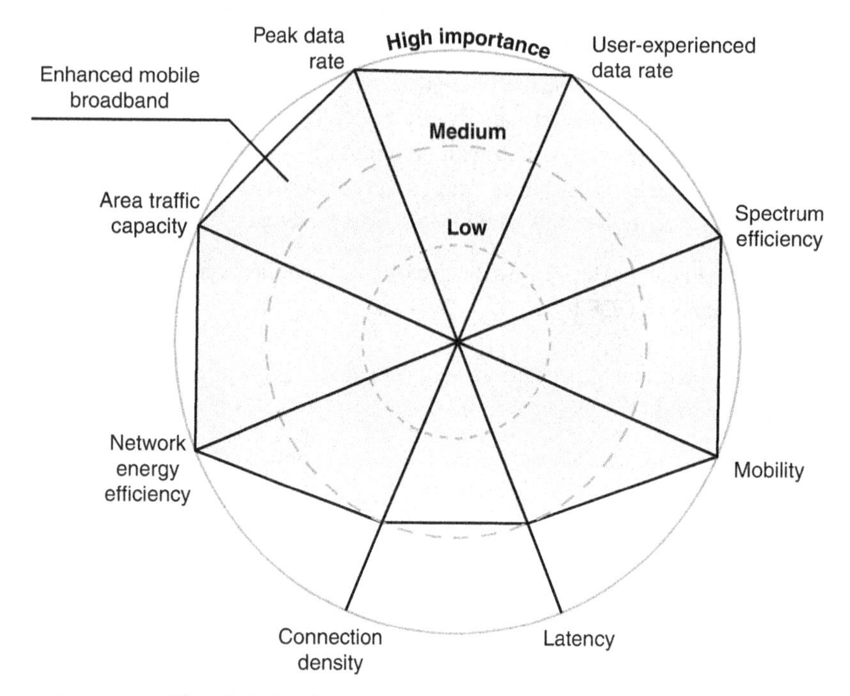

FIGURE 4.2 The Relative Importance of Key Capabilities in the eMBB Usage Scenario

ITU-R Report M.2410 (*Minimum Requirements Related to Technical Performance for IMT-2020 Radio Interface(s)*, November 2017) lists the performance requirements for the three usage scenarios eMBB, massive machine type communications (mMTC), and ultra-reliable and low-latency communications (URLCC). The bulk of the requirements address eMBB, as this is the scenario that is dominating early 5G deployment. The requirements fit into four categories: data rate, spectral efficiency, latency, and mobility. The values for these attributes determine the types of applications that can be supported on a network. The requirements set forth in M.2410 anticipate the needs of applications that require much higher data rates and lower latencies than are supported by 4G.

Data Rate Requirements

Table 4.1 lists requirements related to data rates for three parameters:

- **Peak data rate:** Maximum achievable data rate under ideal conditions for a single user/device (in Gbps), if it is allocated all user data transmission resources.
- **User-experienced data rate:** Achievable data rate that is available ubiquitously[1] across the coverage area to a mobile user/device (in Mbps or Gbps). The required user-experienced data

1. The term *ubiquitous* is related to the considered target coverage area and is not intended to relate to an entire region or country.

rate should be available in at least 95% of the locations (including at the cell edge) for at least 95% of the time within the considered environment. The user-experienced data rate is the minimum data rate required to achieve a sufficient quality of experience.

- **Area traffic capacity:** Total traffic throughput served per geographic area (in Mbps/m^2).

TABLE 4.1 eMBB Data Rate Requirements

Parameter	Downlink Value	Uplink Value	4G Downlink Requirement
Peak data rate	20 Gbps	10 Gbps	1 Gbps
User-experienced data rate (dense urban test environment)	100 Mbps	50 Mbps	10 Mbps
Area traffic capacity (indoor hotspot test environment)	10 Mbps/m^2	—	0.1 Mbps/m^2

The final column of Table 4.1 shows the data rate requirements set by ITU-R for the downlink on 4G networks (IMT-Advanced). The substantial increase in data rates enables 5G networks to provide a wide range of high-traffic services with high user quality of experience (QoE).

Spectral Efficiency Requirements

Table 4.2 lists the requirements related to spectral efficiency, also called *bandwidth efficiency*. In general, the greater the spectral efficiency, the higher the data rate that can be achieved in a given bandwidth. The table lists three parameters:

- **Peak spectral efficiency:** The maximum data rate under ideal conditions, normalized by channel bandwidth (in bps/Hz).
- **5th percentile user spectral efficiency:** The 5% point of the cumulative distribution function of the normalized user throughput. The normalized user throughput is defined as the number of correctly received bits—that is, the number of bits contained in the service data units (SDUs) delivered to Layer 3 over a certain period of time, divided by the channel bandwidth; it is measured in bps/Hz.
- **Average spectral efficiency:** The average data throughput per unit of spectrum resource and per cell (bps/Hz).

TABLE 4.2 eMBB Spectral Efficiency Requirements

Parameter	Test Environment	Downlink Value (bps/Hz)	Uplink Value (bps/Hz)
Peak spectral efficiency	All	30	15
5th percentile user spectral efficiency	Indoor hotspot	0.3	0.21
	Dense urban	0.225	0.15
	Rural	0.12	0.045

Parameter	Test Environment	Downlink Value (bps/Hz)	Uplink Value (bps/Hz)
Average spectral efficiency	Indoor hotspot	9	6.75
	Dense urban	7.8	5.4
	Rural	3.3	1.6

Spectral efficiency is a good, quantifiable, and comparable metric characterizing the performance of a wireless network.

The only spectral efficiency parameter that is defined for both IMT-2020 and IMT-Advanced is peak spectral efficiency. ITU-R M.2134 lists the following requirements for IMT-Advanced: downlink peak spectral efficiency of 15 bps/Hz and uplink peak spectral efficiency of 6.75 bps/Hz. These efficiency levels are quite high, and the requirements for IMT-2020 are even higher.

Latency Requirements

Table 4.3 lists the requirements related to latency.

TABLE 4.3 eMBB Latency Requirements

Parameter	Value	4G Requirement
User plane latency	4 ms	10 ms
Control plane latency	20 ms	100 ms

The table lists two parameters:

- **User plane latency:** This is the contribution by the radio network to the time from which the source sends a packet to when the destination receives it (in ms).
- **Control plane latency:** This is the transition time from a most "battery efficient" state (e.g., Idle state) to the start of continuous data transfer (e.g., Active state).

The final column of Table 4.3 shows the latency requirements set by ITU-R for the downlink on 4G networks (IMT-Advanced). As with data rate, the requirements for 5G are more stringent than for 4G.

Mobility Requirements

Mobility refers to a system's ability to provide seamless service experience to users who are moving. More precisely, mobility is the maximum speed at which a defined quality of service (QoS) and seamless transfer between radio nodes—which may belong to different layers and/or radio access technologies (multilayer/multi-RAT)—can be achieved (in km/h). The speed refers to the rate of motion of user equipment (UE). The following classes of mobility are defined:

- **Stationary:** 0 km/h
- **Pedestrian:** 0–10 km/h
- **Vehicular:** 10–120 km/h
- **High-speed vehicular:** 120–500 km/h

The mobility requirements are defined for the purpose of evaluation in the eMBB usage scenario. A mobility class is supported if the traffic channel link data rate on the uplink, normalized by bandwidth, is as shown in Table 2.2 in Chapter 2. This assumes that the user is moving at the maximum speed in that mobility class in each of the test environments.

Table 4.4 lists the requirements related to mobility in four categories. For indoor hotspots, the requirements address stationary UE and UE used by someone who is walking, perhaps at a very brisk pace. The dense urban requirements also include vehicular traffic but assume that the maximum speed in a dense urban area does not exceed 30 km/hr. Finally, the rural environment is broken up into two categories. One category encompasses pedestrian and ordinary vehicular traffic. The other category addresses the high speeds of trains and aircraft moving through a rural area. Table 4.4 also provides a comparison with corresponding 4G requirements.

TABLE 4.4 eMBB Mobility Requirements

Test Environment	Indoor Hotspot	Dense Urban	Rural	
Mobility classes supported	Stationary, pedestrian	Stationary, pedestrian, vehicular	Pedestrian, vehicular	High-speed vehicular
Traffic channel link data rate (bps/Hz)	1.5	1.12	0.8	0.45
4G traffic channel link data rate (bps/Hz)	1.0	0.75	0.55	0.25
Mobility (km/h)	10	30	120	500
4G mobility (km/h)	10	30	120	350

System Requirements

The requirements detailed in Tables 4.1 through 4.4 are QoS parameters intended to provide acceptable and enhanced QoE for 5G users. Providers must translate these into system requirements that will support the QoS requirements. In general terms, the system requirements are divided into two categories: one related to the air interface and one related to the network architecture and functionality.

Supporting the data rate and spectral efficiency requirements (refer to Tables 4.1 and 4.2) primarily involves design aspects of the air interface, or radio interface. These include antenna design, signal coding techniques, user plane protocol design, and spectrum allocation. Part Four, "5G NR Air Interface and Radio Access Network," examines these issues in detail.

Supporting the latency and mobility requirements (refer to Tables 4.3 and 4.4) primarily involves design aspects of the radio access network and the core network, especially the latter. One particular design strategy stands out: the use of edge computing to move as much processing as possible to the edge of the network, thus closest to the UE. Also important is the use of virtualization techniques that support dynamic allocation of resources to meet user needs. Part Three, "5G NextGen Core Network," examines these issues in detail.

4.3 Smart Office: An Indoor Hotspot Use Case

The installation of 5G networks in the office environment can enable dramatic changes in the capabilities that businesses can exploit. The following are examples of features now in use or that may soon be in use in 5G-enabled workplaces:

- Facial recognition can be used for entrance security. The employee need not carry an identification tag or use some sort of token to gain entrance.
- A 5G virtual desktop infrastructure enables a worker to connect a mobile device on a docking pad to a cloud computing system.
- Workers can convene remote conferences and talk to each other's avatars in cyberspace.
- Security systems can use high-definition video to monitor in greater detail and expand the ability to scan for security threats.
- Workers have faster access to a broader selection of apps.
- 5G enables real-time collaboration between people and things, possibly including augmented reality features.
- Real-time video interaction allows capabilities such as real-time troubleshooting and ad hoc meetings.
- Synchronization of local data with the cloud occurs almost instantaneously, further enhancing collaboration.
- Sensors or facial recognition can tell if a person is in the building and where he or she is at any given moment.

In essence, the smart office use case is characterized by heavy data use, with a particular reliance on high-definition video, in an indoor environment with low mobility requirements. This is a use case scenario where hundreds of users require ultra-high bandwidth to serve intense bandwidth applications. To some extent, Wi-Fi supports these capabilities, but with the increasing demands for high traffic volume, high density of users, and seamless integration of local and wide area communications, a unified 5G solution has inherent advantages over a mixed Wi-Fi/cellular environment.

Table 4.5 indicates the values of key performance indicators (KPIs) for a representative smart office. The second column is based on data in the NGMN white paper. Compared to the ITU-R performance requirements for eMBB, these values are somewhat more demanding. However, with a dense array of mmWave base stations in the office environment, such performance values are achievable.

TABLE 4.5 Indoor Hotspot Key Performance Indicators

KPI	Smart Office (NGMN)	Virtual Reality Office (METIS)
DL user-experienced data rate	1 Gbps Average load: 0.2 Gbps/user	At least 1 Gbps with 95% location availability and 5 Gbps with 20% location availability
UL user-experienced data rate	500 Mbps Average load: 27 Mbps/user	Same as DL
Connection density	75,000/km^2 (0.75/10m^2)	200,000/km^2 (2/10m^2)
Traffic density	DL: 15 Tbps/km^2 UL: 2 Tbps/km^2	100 Tbps/km^2 in DL and UL
Mobility	Pedestrian	Static or low mobility nomadic (less than 6 km/h)
Availability	User-experienced data rate should be available in at least 95% of the locations (including at the cell-edge) for at least 95% of the time	95% for 1 Gbps and 20% for 5 Gbps
Reliability	95%	99% during working hours
Latency	10 ms end-to-end	10 ms round-trip time

The final column relates to a virtual reality office use case defined by the EU project METIS (Mobile and Wireless Communications Enablers for the Twenty-Twenty Information Society). The case study is reported in METIS Deliverable D1.1 (*Scenarios, Requirements and KPIs for 5G Mobile and Wireless System*, April 2013). This case study assumes a company working with 3D telepresence and virtual reality. The work involves interaction with high-resolution 3D scenes and is typically performed in teams of some 5 to 10 individuals simultaneously interacting with a scene. Some of the team members are sited within the building; others are working remotely from other office buildings. Each scene may include the virtual representation of the team members or computer-generated characters and items. The high-resolution quality of the scene provides an as-if-you-were-here feeling. Since each team member may affect the scene, all must continuously update the scene by streaming data to the others. In order to provide the real-time interaction, the work is supported by bidirectional streams with very high data rates and low latencies. The KPIs for this use case are more stringent than those of the NGMN smart office use case. Even so, they should be achievable with an indoor eMBB 5G deployment.

4.4 Dense Urban Information Society: A Dense Urban Use Case

This use case, presented in METIS Deliverable D1.1, refers to the connectivity and data rates required for users of high-volume services at any place and at any time in a dense urban environment. This includes both user interaction with cloud services and data- and device-centric services.

5G enables enhanced cloud services beyond the traditional services of web browsing, file download, and social media. Enhanced services include high-definition video streaming and video sharing. Enhanced device-centric services include augmented reality with information fetched from sensors, smartphones, wirelessly connected cameras, and other sources.

The main features of this use case are as follows:

- High traffic loads
- Low mobility
- High data rate
- Outdoor coverage
- Outdoor-to-indoor coverage
- Supports both low and high frequencies
- Interference limited
- High user density

This use case presents two unique challenges:

- Users expect the same quality of experience in any context, including at their workplace, enjoying leisure activities such as shopping, or being on the move on foot or in a vehicle.
- Users in urban environments tend to dynamically cluster. Examples include people waiting at a traffic light or bus stop and conference room meetings at the workplace. These clusters lead to sudden peaks of geographically concentrated mobile broadband demand.

Table 4.6, based on METIS Deliverable D1.1, indicates KPIs for this use case. These are reasonably in line with the requirements from ITU-R Report M.2410.

TABLE 4.6 Dense Urban Information Society Key Performance Indicators (METIS)

KPI	Value
User-experienced throughput	DL: 300 Mbps UL: 60 Mbps
Traffic volume density	700 Gbps/km^2 (0.7 Gbps/m^2)
Latency	Web browsing: < 0.5 s for download of an average-size web page Video streaming: < 0.5 s Augmented reality processed in the cloud and locally: < 2–5 ms Device-to-device feedback: ≤ 1 ms
User/device density	200,000/km^2 (2/10 m^2)
User mobility	Most of the users and devices have velocities up to 3 km/h and, in some cases, up to 50 km/h

4.5 Radiocommunication Systems Between Train and Trackside: A Rural eMBB Use Case

A prominent use case for eMBB is a radiocommunication system between train and trackside (RTTS). RTTS is a particularly challenging use case. Some of the requirements include [ROTH19]:

- Railways require connectivity at speeds up to 500 km/hr, thus involving numerous rapid handovers.
- Trains often travel in cuttings and tunnels that typically have poor radio-frequency (RF) coverage.
- Trains require very high availability, reaching or exceeding 99.999%, due to the need to control driverless trains.
- To improve security and safety, real-time passenger surveillance and front-looking obstacle detection cameras add the requirement of high uplink throughput capacity.

This use case has received considerable attention from both ITU-R and 3rd Generation Partnership Project (3GPP), which have produced the following documents:

- **ITU-R Report M.2395 (*Introduction to Railway Communication Systems*, November 2016):** Focuses on a case study of measurement results of radiocommunication characteristics between train and ground stations in the mmWave frequency ranges for a number of railway deployment scenarios. The purpose is to assess, among other factors, the impacts of future broadband transmission and high mobility of more than 300 km/h in mmWave frequency ranges on current and future railway radiocommunication systems.
- **ITU-R Report M.2418 (*Description of Railway Radiocommunication Systems Between Train and Trackside (RSTT)*, November 2017):** Addresses the architecture, applications, technologies, and operational scenarios of RSTT for all types of trains (high-speed trains, passenger trains, freight trains, and metro trains).
- **ITU-R Report M.2442 (*Current and Future Usage of Railway Radiocommunication Systems Between Train and Trackside*, November 2018):** Provides a comprehensive survey of spectrum usage and requirements of railways globally.
- **3GPP TR 22.889 (*Study on Future Railway Mobile Communication System; Stage 1*, December 2019):** Analyzes Future Railway Mobile Communication System (FRMCS) use cases, system principles of FRMCS, and interworking between legacy systems and FRMCS in order to derive potential requirements.
- **3GPP TS 22.289 *Technical Specification Group Services and System Aspects; Mobile Communication System for Railways; Stage 1 (Release 17).* December 2019**. Provides rail communication service requirements for 5G system.

Elements of RSTT

Figure 4.3 shows a conceptual diagram of the broadband wireless transmission system between moving trains and the backbone network using mmWave transmission.

FIGURE 4.3 Conceptual View of RSTT

The key elements are:

- **Onboard transceiver:** The transceiver on a train that communicates with the 5G network includes radio equipment installed in the train as well as handsets (e.g., mobile terminals for automatic train control).
- **Trackside antenna:** A base station is positioned near the track. The distance between base stations depends on the frequency of the carrier. For higher frequencies, the distances must be less.
- **Backhaul network:** An intermediate network connects the trackside antennas to the core network.

Table 4.7, derived from 3GPP TR 38.913 (*Study on Scenarios and Requirements for Next Generation Access Technologies*, June 2018), shows antenna placement for two likely carrier frequencies for RSTT.

TABLE 4.7 Base Station Antenna Placement for RSTT

	Carrier Frequency	
	4 GHz	**30 GHz**
Distance between antennas along track	1732 m	580 m
Distance between antenna and track	100 m	5 m

Applications of RSTT

M.2418 lists the following as the main application areas of RSTT:

- Train radio
- Train positioning information

- Train remote
- Train surveillance

Train Radio

Train radio is used for communication between the train and the track side for signaling and traffic management to promote safe train operation. Train radio provides mobile interconnection to landline and mobile-to-mobile voice communication and also serves as the data transmission channel within various bearer services. For voice communication, train radio provides call functions (point-to-point/group/emergency/conference) with specialized modes of operation (e.g., location-dependent addressing, call priorities, late entry, preemption). The following applications fall into the train radio category:

- **Voice/dispatch:** RSTT enables a variety of dispatching modes using voice communication from a dispatch center to trains. These include point-to-point calls to a specific train, public emergency voice calls, broadcast voice calls, and multiparty voice calls.
- **Maintenance:** This application provides voice communication (point-to-point call, point-to-multi-point call, or group call) and data communication for maintenance services in railway infrastructure.
- **Train control:** Communications-Based Train Control (CBTC) is a railway signaling system that makes use of the telecommunications between the train and track equipment for traffic management and infrastructure control. It determines the movement of trains, including controlling the distance between trains traveling on the same line. By means of the CBTC systems, the exact position of a train is known more accurately than with the traditional signaling systems. This results in a more efficient and safe way to manage the railway traffic. CBTC can support automatic train protection (ATP) functions, as well as automatic train operation (ATO) and automatic train supervision (ATS) functions. Train control services usually generate small amounts of traffic, but each traffic packet must be transferred with low latency and high reliability at any time.
- **Emergency:** This category encompasses automatic and operator detection of emergency situations on a track, with signals alerting a central train controller and other moving trains approaching the emergency location.
- **Train information:** Information transmitted by RSTT includes the following categories:
  - Railway transportation information for the train operators, such as train operating status, mobile ticketing, and check-in services
  - Railway transportation information for passengers, such as travel information

Train Positioning Information

RSTT systems gather all kind of train positioning information (e.g., exact locations of all units on trackside) relevant to train operations. This includes line- and location-oriented information. Knowing the locations of all trains and other vehicles on the tracks is essential for railway traffic control, passenger safety, and security of train operations. Techniques for gathering train position information include devices on or near the tracks that can transmit information and radar systems.

Train Remote

Train remote refers to the ability for a ground-based operator to remotely control the movements of a train. This feature is typically used for shunting operations in depots.

Train Surveillance

Train surveillance systems enable the capture and transmission of video of the public and trackside areas, driver cabs, passenger compartments, platforms, and device monitoring. Train surveillance contributes to analysis of the railway environment, improvement of maintenance services, and gathering of information on infrastructure. A set of cameras at specific locations (front, interior, rear view) is used in low to high resolution and at low and high frame rates, depending on the event. Data may be either stored on board/locally or streamed (e.g., real-time video) to control centers via dedicated radiocommunication system.

Broadband Connectivity for Passengers

ITU-R Reports M.2395, M.2418, and M.2442 focus on the railway and train control functions of RTTS. But the advent of 5G also enables eMBB-quality Internet access for train passengers. ITU-R Report M.2441 lists this as one of the use cases for IMT-2020. The objective is to provide broadband Internet access for uses such as streaming video and high-speed interactive applications.

Thus, the key characteristics of this use case are consistent passenger user experience and critical train communication reliability with very high mobility. Figure 4.4 illustrates these requirements.

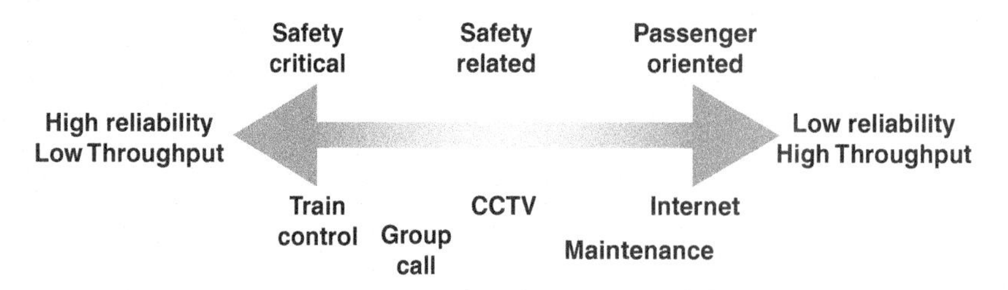

FIGURE 4.4 Railway Reliability and Throughput Requirements

Linear Cell Architecture

To support 5G communication on trains, a linear cell layout is used, as illustrated in Figure 4.5. In a typical deployment, multiple antennas using optical fiber go through an optical router to a central control unit and also link to the Internet via the 5G core network (not shown) [KANN18].

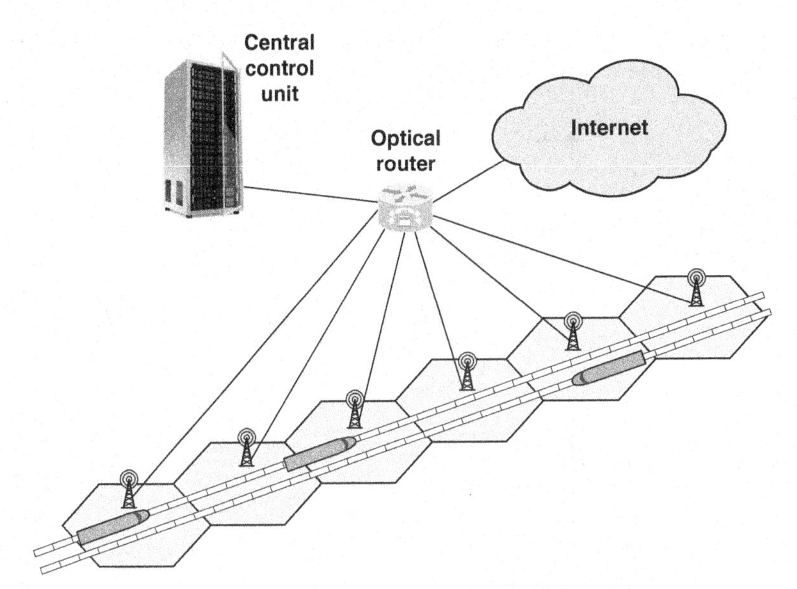

FIGURE 4.5 Conceptual Diagram of RSTT with Optical Network

A variation on this configuration defines a linear cell as comprising multiple contiguous antennas, with the same signal and same frequency, connecting to a base station. For example, a linear cell may contain three antennas distributed linearly along the track. Optical feeders are used to connect between the trackside antennas and the base stations. The linear cells with frequency 1 and 2 are alternately repeated. Using this concept helps avoid frequent handovers that cause throughput reduction. Furthermore, the spectral utilization is efficient because if the linear cell is long enough, only two frequencies are needed to prevent inter-cell interference.

The linear layout of cells, as opposed to a two-dimensional array of cells in other contexts, fortunately makes it practical to meet the stringent mobility requirement of railway travel. It is easy to predict successive locations of a train based on the train location and speed information available through the train control center. Thus, an RSTT system has time to prepare for the handover from one antenna to the next.

4.6 Key Terms and Review Questions

Key Terms

area traffic capacity	radiocommunication systems between train and trackside (RSTT)
backhaul network	rural deployment scenario
control plane latency	smart office
dense urban deployment scenario	spectral efficiency
dense urban information society	stationary
enhanced mobile broadband (eMBB)	trackside antenna
high-speed vehicular	train positioning information
indoor hotspot deployment scenario	train radio
latency	train remote
linear cell architecture	train surveillance
mobility	user-experienced data rate
onboard transceiver	user plane latency
peak data rate	vehicular
pedestrian	

Review Questions

1. What are the key drivers for the deployment of eMBB?
2. What challenge is presented in the use of high-frequency bands in the indoor hotspot deployment scenario?
3. Why does the dense urban environment require the use of a dense collection of small cells to supplement macro cells?
4. Summarize the eMBB data rate requirements.
5. Summarize the eMBB spectral efficiency requirements.
6. Summarize the eMBB latency requirements.
7. Summarize the eMBB mobility requirements.
8. What are the key performance indicators for a smart office indoor hotspot use case?
9. What are the key performance indicators for a dense urban information society use case?

10. List some key requirements for supporting RSTT.

11. List and briefly describe four main application areas of RSTT.

4.7 References and Documents

References

COMM15 CommScope. "Indoor Wireless in Mobile Society: Research Reveals Gap Between Expectations of Wireless Consumers and Those Who Design and Manage Buildings." *Commscope Press Release*. November 19, 2015. https://www.commscope.com/NewsCenter/PressReleases/Indoor-Wireless-in-Mobile-Society-Research-Reveals-Gap-Between-Expectations-Of-Wireless-Consumers-and-Those-Who-Design-and-Manage-Buildings/

GMSA17 Global Mobile Suppliers Association. *5G-Oriented Indoor Digitalization Solution White Paper.* November 2017.

KANN18 Kann, A., Dat, P., and Kawanishi, T. "Millimeter-Wave Radio-over-Fiber Network for Linear Cell Systems." *Journal of Lightwave Technology*, January 2018.

KULK14 Kulkarni, M., Singh, S., and Andrews, J. "Coverage and Rate Trends in Dense Urban mmWave Cellular Networks." *2014 IEEE Global Communications Conference*, December 2014.

ROTH19 Rothbaum, D. "5G for the Future Railway Mobile Communication System." *ITU News Magazine*, No. 4, 2019.

Documents

3GPP TR 22.889 *Study on Future Railway Mobile Communication System, Stage 1.* December 2019.

3GPP TR 23.796 *Study on Application Architecture for the Future Railway Mobile Communication System, Stage 2.* March 2019.

3GPP TR 38.913 *Study on Scenarios and Requirements for Next Generation Access Technologies.* June 2018.

3GPP TS 22.289 *Technical Specification Group Services and System Aspects, Mobile Communication System for Railways, Stage 1 (Release 17).* December 2019.

ITU-R M.2083 *IMT Vision: Framework and Overall Objectives of the Future Development of IMT for 2020 and Beyond.* September 2015.

ITU-R Report M.2134 *Requirements Related to Technical Performance for IMT-Advanced Radio Interface(s)*. 2008.

ITU-R Report M.2395 *Introduction to Railway Communication Systems*. November 2016.

ITU-R Report M.2410 *Minimum Requirements Related to Technical Performance for IMT-2020 Radio Interface(s)*. November 2017.

ITU-R Report M.2412 *Guidelines for Evaluation of Radio Interface Technologies for IMT-2020*. October 2017.

ITU-R Report M.2418 *Description of Railway Radiocommunication Systems Between Train and Trackside (RSTT)*. November 2017.

ITU-R Report M.2441 *Emerging Usage of the Terrestrial Component of International Mobile Telecommunication (IMT)*. November 2018.

ITU-R Report M.2442 *Current and Future Usage of Railway Radiocommunication Systems Between Train and Trackside*. November 2018.

METIS Deliverable D1.1 *Scenarios, Requirements and KPIs for 5G Mobile and Wireless System*. April 2013.

NGMN15 *NGMN 5G White Paper*. February 2015.

NGMN16 *5G Prospects: Key Capabilities to Unlock Digital Opportunities*. July 2016.

Chapter 5

Massive Machine Type Communications

Learning Objectives

After studying this chapter, you should be able to:

- Summarize the performance requirements for mMTC
- Explain the scope of the Internet of Things
- List and discuss the five principal components of IoT-enabled things
- Understand the relationship between cloud computing and IoT
- Explain the relationship between mMTC and IoT
- Summarize the five categories of use cases for smart agriculture developed by IoF2020
- Present an overview of the IoF2020 Precision Crop Management use case
- Present an overview of the smart city use case involving citizen identification by biometrics

ITU-R M.2083 (*IMT Vision: Framework and Overall Objectives of the Future Development of IMT for 2020 and Beyond*, September 2015) characterizes massive machine type communications (mMTC) as follows:

> Massive machine type communications: This usage scenario is characterized by a very large number of connected devices typically transmitting a relatively low volume of non-delay-sensitive data. Devices are required to be low cost, and have a very long battery life.

"Examples of use cases in this category are the Internet of Things (IoT), asset tracking, smart agriculture, smart cities, energy monitoring, smart home, remote monitoring."

A white paper from the Next Generation Mobile Networks (NGMN) Alliance (*5G Prospects: Key Capabilities to Unlock Digital Opportunities*, July 2016) lists the following as the key drivers of mMTC:

- **Business precision and efficiency:** Data-oriented optimization of business processes and resource usage can provide benefit in many industries, including in areas such as production, logistics, customer services, energy consumption, and maintenance. Organizations are increasingly using dispersed devices to collect information about the environment and objects in the environment and coupling this with body-of-knowledge databases to increase the precision and efficiency of value delivery while at the same time improving sustainability of the environment.
- **Technology:** Recent advances in technology (e.g., in sensors/actuators, access technology, data storage, cloud computing, big data, and artificial intelligence) are making information and communications technology (ICT) solutions in this domain feasible and affordable. In addition, ICT players are positioning themselves in the emerging IoT value space, promoting both solutions and platforms.

Section 5.1 summarizes the performance requirements for mMTC. Then, Section 5.2 provides an overview of the Internet of Things (IoT), which is closely related to the concept of mMTC. Section 5.3 discusses more specifically the relationship between mMTC and IoT. Section 5.4 looks at the relationship between mMTC, which has been defined by ITU-R, and two concepts introduced by 3GPP: NB-IoT and eMTC. Sections 5.5 and 5.6 look in some detail at two important applications of mMTC: smart agriculture and smart cities.

5.1 mMTC Performance Requirements

Figure 5.1, based on a figure in ITU-R M.2083, indicates in relative terms the importance of eight key performance characteristics for mMTC. The one parameter that is of high importance is connection density. Connection density is expressed as the total number of connected and/or accessible devices fulfilling a specific quality of service (QoS) per unit area (per km^2).

Of medium importance is network energy efficiency. Network energy efficiency refers to the quantity of information bits transmitted to/received from users, per unit of energy consumption of the radio access network (RAN) (in bits/Joule). The objective is efficient data transmission when there is a substantial load on the network.

ITU-R Report M.2410 (*Minimum Requirements Related to Technical Performance for IMT-2020 Radio Interface(s)*, November 2017) lists the performance requirements for three usage scenarios: eMBB (enhanced mobile broadband), mMTC, and URLCC (ultra-reliable and low-latency communications). For mMTC, the only defined performance requirement is for connection density. M.2410 sets the minimum requirement as 10^6 devices per km^2.

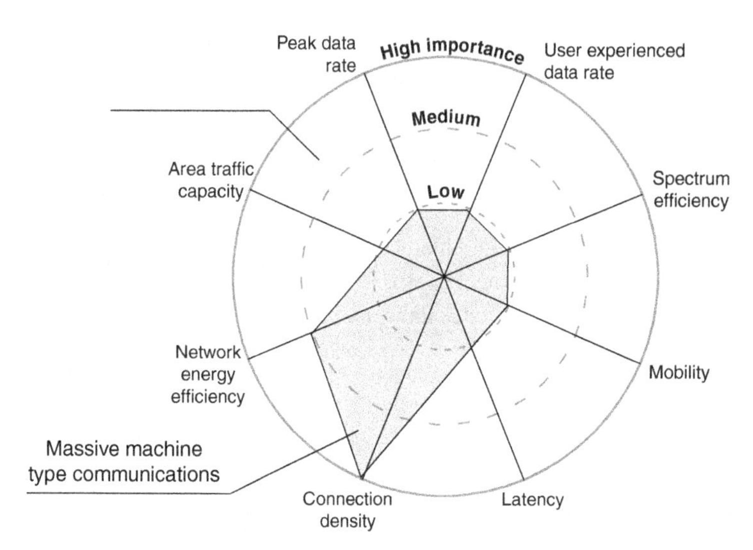

FIGURE 5.1 The Relative Importance of Key Capabilities in the mMTC Scenario

M.2410 is not specific about what QoS or other attributes must be supported at the minimum connection density. It states that connection density should be achieved for a limited bandwidth and number of transmission reception point (TRxPs). The target quality of service (QoS) is to support delivery of a message of a certain size within a certain time and with a certain success probability, as specified in ITU-R Report M.2412 (*Guidelines for Evaluation of Radio Interface Technologies for IMT-2020*, October 2017). M.2412 lays out the following parameters for evaluation:

- Ten pieces of user equipment (UE) per TRxP
- Each UE transmitting 32-byte messages at a rate of one every 2 hours
- QoS defined as successful delivery of 99% of messages within 10 s
- Inter-site distance of 500 m

5.2 The Internet of Things

The term IoT refers to the expanding interconnection of smart devices, ranging from appliances to tiny sensors. A dominant theme is the embedding of short-range mobile transceivers into a wide array of gadgets and everyday items, enabling new forms of communication between people and things and between things themselves. The Internet now supports the interconnection of billions of industrial and personal objects, usually through cloud systems. The objects deliver sensor information, act on their environment, and in some cases modify themselves, enabling overall management of a larger system, such as a factory or city.

The Scope of the Internet of Things

ITU-T Y.2060 (*Overview of the Internet of Things*, June 2012) provides the following definitions that suggest the scope of the IoT:

- **Internet of Things (IoT):** The IoT is a global infrastructure for the information society, enabling advanced services by interconnecting (physical and virtual) things based on existing and evolving interoperable information and communication technologies.
- **Thing:** With regard to the Internet of Things, a thing is an object of the physical world (physical thing) or the information world (virtual thing), which is capable of being identified and integrated into communication networks.
- **Device:** With regard to the Internet of Things, this is a piece of equipment with the mandatory capabilities of communication and the optional capabilities of sensing, actuation, data capture, data storage, and data processing.

Most of the literature views the IoT as involving intercommunicating smart objects. Y.2060 extends this to include virtual things. Physical things exist in the physical world and are capable of being sensed, actuated, and connected. Examples of physical things include the surrounding environment, industrial robots, goods, and electrical equipment. Virtual things exist in the information world and are capable of being stored, processed, and accessed. Examples of virtual things include multimedia content and application software.

Y.2060 characterizes the IoT as adding the dimension "any THING communication" to the information and communication technologies, which already provide "any TIME" and "any PLACE" communication, as shown in Figure 5.2.

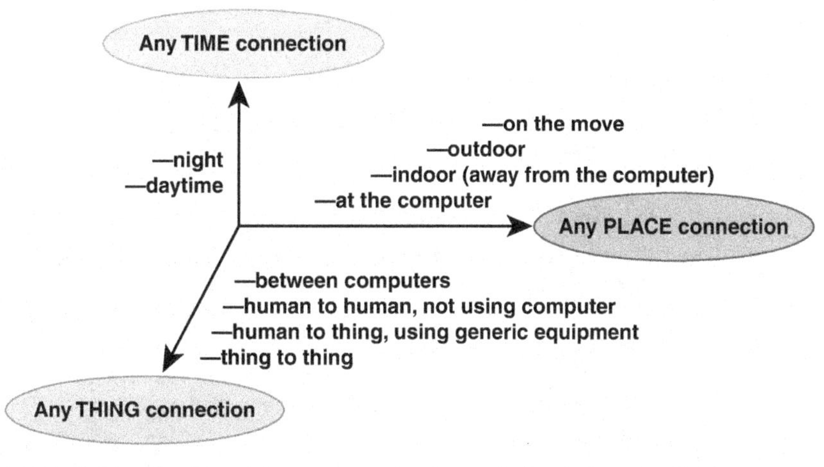

FIGURE 5.2 The New Dimension Introduced in the Internet of Things

[MCEW13] condenses the elements of the IoT into a simple equation:

Physical Objects + Controllers, Sensors, Actuators + Internet = IoT

This equation neatly captures the essence of the Internet of Things. An instance of the IoT consists of:

- A collection of physical objects, each of which contains a microcontroller that provides intelligence
- A sensor that measures some physical parameter and/or an actuator that acts on some physical parameter
- A means of communicating via the Internet or some other network

One item that is not covered in the equation but is referred to in the Y.2060 definition, is a means of identification of an individual thing, usually referred to as a *tag*. Tags are discussed later in this section.

Note that although the phrase *the Internet of Things* is always used in the literature, a more accurate description would be *an Internet of Things*, or *a network of things*. A smart home installation, for example, consists of a number of things in the home that are interconnected via Wi-Fi or Bluetooth with some central controller. In a factory or farm setting, there may be a network of things enabling enterprise applications to interact with the environment and run applications to exploit the network of things. In these examples, it is usually but not invariably the case that remote access over the Internet is available. Whether or not such Internet connection is available, the collection of smart objects at a site, plus any other local compute and storage device, can be characterized as a network or an Internet of Things.

Table 5.1, based on a graphic from Beechem Research [BEEC18], gives an idea of the scope of the IoT.

TABLE 5.1 The Internet of Things

Service Sector	Application Groups	Locations	Examples of Devices
IT and networks	Public	Services, e-commerce, data centers, mobile carriers, fixed carriers, ISPs (Internet service providers)	Servers, storage, PCs, routers, switches, PBXs (private branch exchanges)
	Enterprise	IT/data center, office, private nets	
Security/ public safety	Surveillance	Radar/satellite, military security, unmanned weapons, vehicles, ships, aircraft, gear	Tanks, fighter jets, battlefield communications
	Tracking, Public infrastructure	Human, animal, postal, food/ health, packaging, baggage, water treatment, building environmental, general environmental	Cars, breakdown lane workers, homeland security, fire, environmental monitors
	Emergency services	Equipment and personnel, police, fire, regulatory	Ambulances, public security vehicles

Service Sector	Application Groups	Locations	Examples of Devices
Retail	Specialty	Fuel stations, gaming, bowling, cinema, discos, special events	POS (point of sale) terminals, tags, cash registers, vending machines, signs
	Hospitality	Hotels, restaurants, bars, cafes, clubs	
	Stores	Supermarkets, shopping centers, single site, distribution center	
Transportation	Non-vehicular	Air, rail, marine	Vehicles, lights, ships, planes, signage, tolls
	Vehicles	Consumer, commercial, construction, off-road	
	Transportation systems	Tolls, traffic management, navigation	
Industrial	Distribution	Pipelines, materials handling, conveyance	Pumps, valves, vats, conveyers, pipelines, motors, drives, converting, fabrication, assembly/packing, vessels, tanks
	Converting, discrete	Metals, paper, rubber, plastic, metalworking, electronics assembly, test	
	Fluid/processes	Petrochemical, hydrocarbon, food, beverage	
	Resource automation	Mining, irrigation, agricultural, woodland	
Healthcare and life science	Care	Hospital, ER, mobile PoC, clinic, labs, doctor office	MRIs, implants, surgical equipment, pumps, monitors, telemedicine
	In vivo, home	Implants, home monitoring systems	
	Research	Drug discovery, diagnostics, labs	
Consumer and home	Infrastructure	Wiring, network access, energy management	Digital cameras, power systems, dishwashers, eReaders, desktop computers, washers/dryers, meters, lights, TVs, MP3, games consoles, lighting, alarms
	Awareness and safety	Security/alert, fire safety, environmental safety, elderly, children, power protection	
	Convenience and entertainment	HVAC/climate, lighting, appliances, entertainment	
Energy	Supply/demand	Power generation, transmission and distribution, low voltage, power quality, energy management	Turbines, windmills, UPSs (uninterruptible power supplies), batteries, generators, meters, drills, fuel cells
	Alternative	Solar, wind, co-generation, electrochemical	
	Oil/gas	Rigs, derricks, well heads, pumps, pipelines	
Buildings	Commercial, institutional	Office, education, retail, hospitality, healthcare, airports, stadiums	HVAC, transport, fire and safety, lighting, security, access
	Industrial	Process, clean room, campus	

Things on the Internet of Things

The IoT is primarily driven by **deeply embedded devices**. These devices are low-bandwidth, low-power, low-repetition data capture and low-bandwidth data usage appliances that communicate with each other and with higher-level devices, such as gateway devices that funnel information to cloud systems. Embedded appliances, such as high-resolution video security cameras, video voice over IP (VoIP) phones, and a handful of others, require high-bandwidth streaming capabilities. Yet countless products simply require packets of data to be intermittently delivered.

The user interface for an IoT device may be tightly constrained by limited display size and functionality, or the device might only be controlled via remote means. Thus, it may be difficult for individuals to know what data devices are collecting about them and how the information will be processed after collection.

Components of IoT-Enabled Things

The key components of an IoT-enabled device are shown in Figure 5.3.

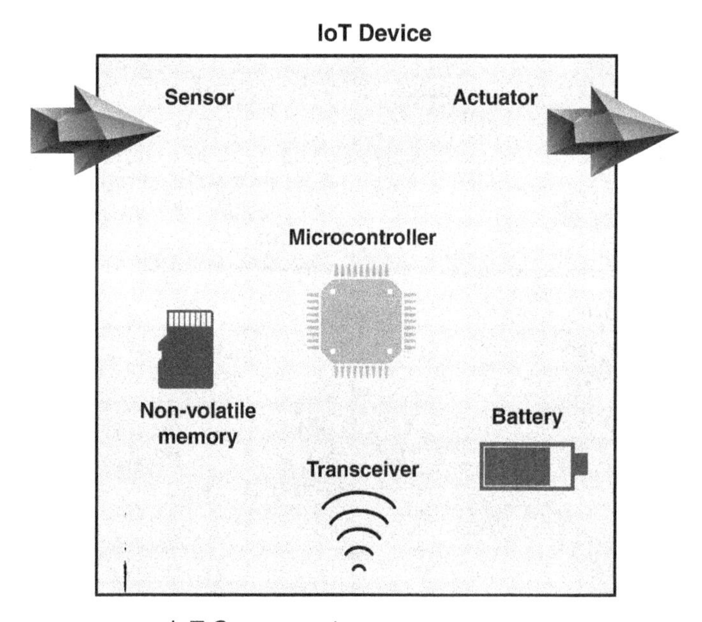

FIGURE 5.3 IoT Components

The key components are as follows:

- **Sensor:** A sensor measures some parameter or set of parameters of a physical, chemical, or biological entity and delivers an electronic signal proportional to the observed characteristic, either in the form of an analog voltage level or a digital signal. In both cases, the sensor output is typically input to a microcontroller or other management element. Examples include temperature measurement, radiographic imaging, optical sensing, and audio sensing.

- **Actuator:** An actuator receives an electronic signal from a controller and responds by interacting with its environment to produce an effect on some parameter or set of parameters of a physical, chemical, or biological entity. Examples include heating coils, cardiac electric shock delivery, electronic door locks, unmanned aerial vehicle operation, servo motors, and robotic arms.
- **Microcontroller:** The "smart" in a smart device is provided by a deeply embedded microcontroller.
- **Transceiver:** A transceiver contains the electronics needed to transmit and receive data. An IoT device typically contains a wireless transceiver that is capable of communication using Wi-Fi, ZigBee, or some other wireless protocol. By means of the transceiver, IoT devices can interconnect with other IoT devices, with the Internet, and with gateway devices to cloud systems.
- **Power supply:** Typically, this is a battery.

An IoT device also typically contains a **radio-frequency identification (RFID)** component. RFID technology, which uses radio waves to identify items, is increasingly becoming an enabling technology for IoT. The main elements of an RFID system are tags and readers. RFID tags are small programmable devices used for object, animal, and human tracking. They come in a variety of shapes, sizes, functionalities, and costs. RFID readers acquire and sometimes rewrite information stored on RFID tags that come within operating range (a few inches up to several feet). Typically, RFID readers communicate with a computer system that records and formats the acquired information for further uses. RFID components on some devices can transmit considerable information about their devices, raising privacy issues.

Constrained Devices

Increasingly, the term *constrained device* is used to refer to the vast majority of IoT devices. In an IoT, a constrained device is a device with limited volatile and nonvolatile memory, limited processing power, a low-data-rate transceiver, and limited electric power. Many devices in the IoT, particularly the smaller, more numerous devices, are resource constrained. As pointed out in [SEGH12], technology improvements following Moore's law continue to make embedded devices cheaper, smaller, and more energy efficient—but not necessarily more powerful. Typical embedded IoT devices are equipped with 8- or 16-bit microcontrollers that possess very little RAM and small storage capacities. A resource-constrained device is likely to be equipped with an IEEE 802.15.4 radio, which enables low-power, low-data-rate wireless personal area networks (WPANs) with data rates of 20–250 kbps and frame sizes of up to 127 octets.

RFC 7228 (*Terminology for Constrained-Node Networks*, May 2014) defines three classes of constrained devices (see Table 5.2).

TABLE 5.2 Classes of Constrained Devices

Class	Data Size (RAM)	Code Size (flash, ROM)
Class 0	<< 10 kbyte	<< 100 kbyte
Class 1	~ 10 kbyte	~ 100 kbyte
Class 2	~ 50 kbyte	~ 250 kbyte

The classes are as follows:

- **Class 0:** These devices are very constrained devices, typically sensors, called *motes*, or *smart dust*. Motes can be implanted or scattered over a region to collect data and pass it on from one device to another to some central collection point. For example, a farmer, a vineyard owner, or an ecologist could equip motes with sensors that detect temperature, humidity, and so on, making each mote a mini weather station. Scattered throughout a field, vineyard, or forest, these motes would allow the tracking of microclimates. Class 0 devices generally cannot be secured or managed comprehensively in the traditional sense. They are most likely to be preconfigured with a very small data set, and they are reconfigured rarely, if at all.
- **Class 1:** These devices are quite constrained in code space and processing capabilities, such that they cannot easily talk to other Internet nodes employing a full protocol stack. However, they are capable enough to use a protocol stack specifically designed for constrained nodes (such as the Constrained Application Protocol [CoAP]) and participate in meaningful conversations without the help of a gateway node.
- **Class 2:** These devices are less constrained and are fundamentally capable of supporting most of the same protocol stacks as used on notebooks or servers. However, they are still very constrained compared to high-end IoT devices. Thus, they require lightweight and energy-efficient protocols and low-transmission traffic.

Class 0 devices are so constrained that a conventional OS is not practical. These devices have a very limited, specialized function or set of functions that can be programmed directly onto the hardware. Class 1 and Class 2 devices are typically less specialized. An OS, with its kernel functions and support libraries, allows software developers to develop applications that make use of OS functionality and can be executed on a variety of devices. However, many embedded operating systems, such as μClinux, consume too many resources and too much power to be usable for these constrained devices. Instead, an OS designed specifically for constrained devices is needed. Such an OS is typically referred to as an IoT OS.

IoT and Cloud Context

To better understand the function of an IoT, it is useful to view it in the context of a complete enterprise network that includes third-party networking and cloud computing elements. Figure 5.4 provides an overview illustration.

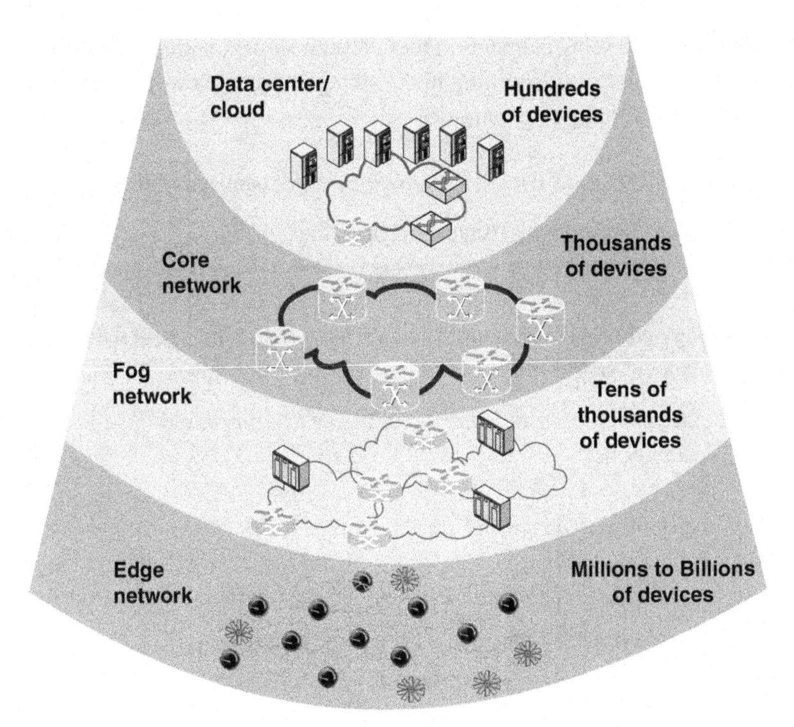

FIGURE 5.4 The IoT/Cloud Context

Edge

At the **edge** of a typical enterprise network is a network of IoT-enabled devices, consisting of sensors and perhaps actuators. These devices may communicate with one another. For example, a cluster of sensors may all transmit their data to one sensor that aggregates the data to be collected by a higher-level entity. At this level, there may also be a number of gateways. A gateway interconnects the IoT-enabled devices with the higher-level communication networks. It performs the necessary translation between the protocols used in the communication networks and those used by devices. It may also perform a basic data aggregation function.[1]

Fog

In many IoT deployments, massive amounts of data may be generated by a distributed network of sensors. For example, offshore oil fields and refineries can generate a terabyte of data per day. An airplane can create multiple terabytes of data per hour. Rather than store all of that data permanently (or at least for a long period) in central storage accessible to IoT applications, it is often desirable to do as much data processing close to the sensors as possible. Thus, the purpose of what is sometimes referred to as the **fog** computing level is to convert network data flows into information that is suitable

1. An edge network is a different concept from edge computing, which involves moving data and applications from a central location to a system closer to the end user.

for storage and higher-level processing. Processing elements at these level may deal with high volumes of data and perform data transformation operations, resulting in the storage of much lower volumes of data. The following are examples of fog computing operations:

- **Evaluation:** Evaluating data for criteria as to whether it should be processed at a higher level
- **Formatting:** Reformatting data for consistent higher-level processing
- **Expanding/decoding:** Handling cryptic data with additional context (e.g., the origin)
- **Distillation/reduction:** Reducing and/or summarizing data to minimize the impact of data and traffic on the network and higher-level processing systems
- **Assessment:** Determining whether data represent a threshold or an alert; this could include redirecting data to additional destinations

Generally, fog computing devices are deployed physically near the edge of the IoT network—that is, near the sensors and other data-generating devices. Thus, some of the basic processing of large volumes of generated data is offloaded and outsourced from IoT application software located at the center.

Fog computing represents an opposite trend in modern networking from cloud computing. With cloud computing, massive, centralized storage and processing resources are made available to distributed customers over cloud networking facilities to a relatively small number of users. With fog computing, massive numbers of individual smart objects are interconnected with fog networking facilities that provide processing and storage resources close to the edge devices in an IoT. Fog computing addresses the challenges raised by the activity of thousands or millions of smart devices, including security, privacy, network capacity constraints, and latency requirements. The term *fog computing* is inspired by the fact that fog tends to hover low to the ground, whereas clouds are high in the sky.

Core

The core network, also referred to as a **backbone network**, connects geographically dispersed fog networks as well as providing access to other networks that are not part of the enterprise network. Typically, the core network uses very high-performance routers, high-capacity transmission lines, and multiple interconnected routers for increased redundancy and capacity. The core network may also connect to high-performance, high-capacity servers, such as large database servers and private cloud facilities. Some of the core routers may be purely internal, providing redundancy and additional capacity without serving as edge routers.

Cloud

The cloud network provides storage and processing capabilities for the massive amounts of aggregated data that originate in IoT-enabled devices at the edge. Cloud servers also host the applications that interact with and manage the IoT devices and that analyze the IoT-generated data.

Table 5.3 compares cloud and fog computing.

TABLE 5.3 Comparison of Cloud and Fog Features

Feature	Cloud	Fog
Location of processing/ storage resources	Center	Edge
Latency	High	Low
Access	Fixed or wireless	Mainly wireless
Support for mobility	Not applicable	Yes
Control	Centralized/hierarchical (full control)	Distributed/hierarchical (partial control)
Service access	Through core	At the edge/on handheld device
Availability	99.99%	Highly volatile/highly redundant
Number of users/devices	Tens/hundreds of millions	Tens of billions
Main content generator	Human	Devices/sensors
Content generation	Central location	Anywhere
Content consumption	End device	Anywhere
Software virtual infrastructure	Central enterprise servers	User devices

5.3 Relationship Between mMTC and the IoT

The mMTC usage scenario defined by ITU-R represents a subset of the total IoT universe. A white paper from Ericsson [ERIC19] lists four segments that comprise IoT, as shown in Figure 5.5.

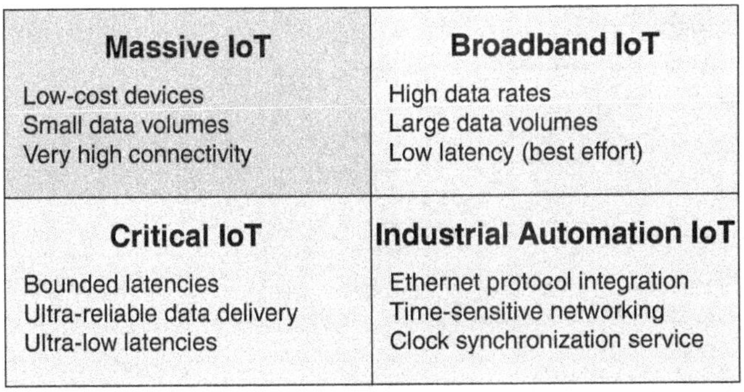

Massive IoT	**Broadband IoT**
Low-cost devices Small data volumes Very high connectivity	High data rates Large data volumes Low latency (best effort)
Critical IoT	**Industrial Automation IoT**
Bounded latencies Ultra-reliable data delivery Ultra-low latencies	Ethernet protocol integration Time-sensitive networking Clock synchronization service

FIGURE 5.5 Cellular IoT Segments

The four segments are as follows:

- **Massive IoT:** Massive IoT is characterized by huge volumes of constrained devices that send and/or receive messages infrequently. The traffic is often tolerant of delay. Examples of use cases include low-cost sensors, meters, wearables, and trackers. Such devices are often deployed in challenging radio conditions, such as the basement of a building. Therefore, they require extended coverage and may rely solely on a battery power supply, which puts extreme requirements on the device's battery life.

- **Broadband IoT:** Broadband IoT is an application eMBB to the IoT environment, providing high data rates and relatively low latencies. Examples of use cases are in the areas of automotive, drones, augmented reality/virtual reality (AR/VR), utilities, manufacturing, and wearables.
- **Critical IoT:** Critical IoT is an application of URLCC to the IoT environment, providing extremely low latencies and ultra-high reliability at a variety of data rates. In contrast to broadband IoT, which achieves low latency on best effort, critical IoT is intended to deliver data within strict latency bounds with required guarantee levels, even in heavily loaded networks. Examples of use cases are in the areas of intelligent transportation systems, smart utilities, remote healthcare, smart manufacturing, and fully immersive AR/VR.
- **Industrial automation IoT:** This segment supports seamless integration of cellular connectivity into the wired industrial infrastructure used for real-time advanced automation. These applications have extremely demanding requirements such as very accurate indoor positioning and time synchronization across devices and networks.

Massive IoT, as defined by Ericsson, is equivalent to mMTC defined by ITU-R. In terms of the number of connections, mMTC is the most rapidly growing segment of IoT. Figure 5.6, based on data in the 2020 Ericsson Mobility Report [ERIC20a], shows the projection that mMTC will begin to dominate in the next few years.

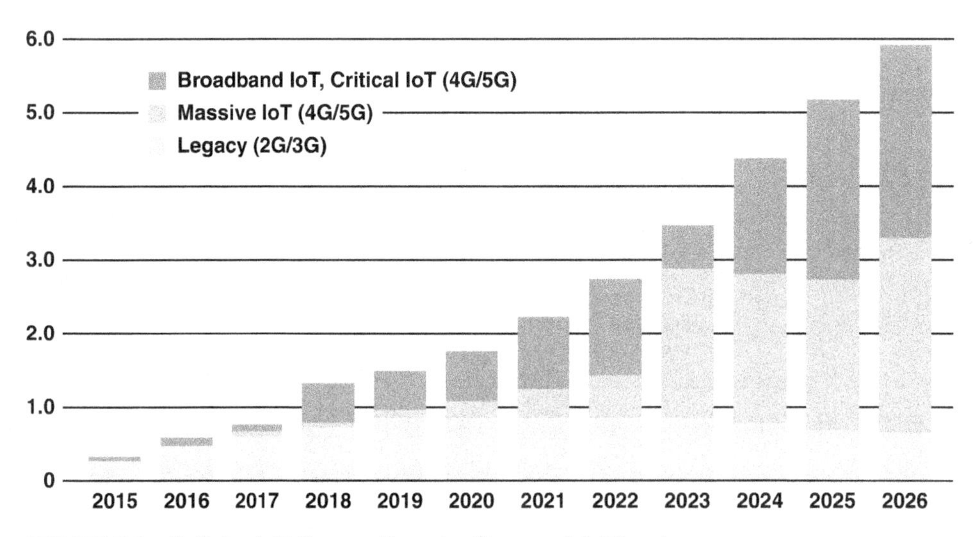

FIGURE 5.6 Cellular IoT Connections by Segment (billions)

Table 5.4, based on a 2020 Ericsson white paper [ERIC20b], indicates likely mMTC use cases supported by 5G.

TABLE 5.4 Industry and Society Applications Enabled by Massive IoT

Application Area	Use Cases
Transport and logistics	■ Fleet management ■ Goods tracking
Agriculture	■ Climate/agriculture monitoring ■ Livestock tracking
Environment	■ Flood monitoring and alerting ■ Environmental monitoring (water, air, noise, etc.)
Industrial	■ Process monitoring and control ■ Maintenance monitoring
Utilities	■ Smart metering ■ Smart grid management
Smart cities	■ Parking sensors ■ Smart bicycles ■ Waste management ■ Smart lighting
Smart buildings	■ Smoke detectors ■ Alarm systems ■ Home automation
Consumers	■ Wearables ■ Child/elderly tracking ■ Medical monitoring

5.4 Relationship Between mMTC and NB-IoT and eMTC

Beginning with Release 13 in 2012, 3GPP has been working on two complementary specifications to support IoT. These two specifications are now called NB-IoT (narrowband IoT) and eMTC (enhanced machine type communications). Both specifications, in their current versions, are designed to support mMTC.

Comparison of NB-IoT and eMTC

Table 5.5 summarizes key parameters of NB-IoT and eMTC.

TABLE 5.5 Comparison of NB-IoT and eMTC

Parameter	NB-IoT	eMTC
Peak data rate	250 kbps	1 Mbps
Carrier bandwidth	200 kHz	1.4 MHz

Parameter	NB-IoT	eMTC
Mobility	Low speed, cell reselection	Low/medium/high speed, cell switching
Voice	Not supported	Supported
Latency	Seconds	Several hundred milliseconds
Duplex mode	Half duplex FDD	Full or half duplex FDD/TDD
Deployment	Standalone, inband, guard band	Inband
Power consumption	Best at low data rates	Best at medium data rates
Penetration indoors	Excellent	Good

The eMTC network is primarily designed for high-data-rate (compared to NB-IoT), low-latency, mobile, and voice-supporting applications. These include industrial handhelds, health monitors, wearables, and high-precision mobile trackers. It is able to leverage existing 4G base stations. Thus, eMTC is optimized for the broadest range of IoT applications.

NB-IoT is optimized for low-data, low-complexity, delay-tolerant, and non-critical applications that need greater connection density, such as utility meters, industrial and environmental sensors, agricultural monitors, and low-precision mobile trackers.

Both technologies support years of battery life and much larger coverage than regular 4G or 5G devices.

Low-Power Wide Area (LPWA)

NB-IoT and eMTC both support LPWA (low-power wide area) networking. This subsection introduces LPWA and some alternative schemes for supporting LPWA.

The term *low-power wide area* refers to wireless networks that have the following characteristics:

- Low power consumption that enables devices to operate for many years on a single charge
- Low device unit cost
- Improved outdoor and indoor coverage compared with existing wide area technologies
- Potential of high connection density
- Secure connectivity and strong authentication
- Optimized data transfer for small, intermittent blocks of data
- Simplified network topology and deployment
- Network scalability for capacity upgrade

The first two items in this list imply support for inexpensive constrained devices.

LPWA solutions use two portions of the frequency spectrum:

- **Unlicensed frequency bands:** These are frequency bands for which the controlling government agency (e.g., the FCC in the United States) allows operation without a license or callsign. They are free and available to all. There are some restrictions on how they can be used, primarily around transmit power limitations.
- **Licensed frequency bands:** In the case of licensed spectrum, an operator purchases the spectrum and is then given exclusive rights to use it; no one else can use it in the geographic area for which the license is granted. Power restrictions are relaxed. A company can also purchase licensed spectrum for private use.

Licensed LPWA solutions use licensed cellular network bands and coexist with other 2G through 5G networks. The advantage of the use of a licensed cellular band is that the LPWA network benefits from all the security and privacy features of mobile networks, such as support for user identity, confidentiality, entity authentication, confidentiality, data integrity, and mobile equipment identification.

Early offerings for supporting wireless IoT were developed in the unlicensed bands, as cellular technology was still evolving. A number of industry groups support various industry standards for unlicensed LPWA. These include:

- **LoRaWAN:** This is an open standard backed by more than 500 member companies. It is deployed in many countries around the world. It can establish long links (up to 18 km in rural settings) with high robustness and low power consumption, even under mobility conditions [SANC18].
- **Sigfox:** Sigfox typically works with one partner—called a Sigfox network operator—per country and has a focus on deploying as much Sigfox LPWA coverage as possible across the globe. To achieve this, it works with (and is also funded by) several companies that have access to towers and other high places where they can put their antennas. Sigfox enables communication using the industrial, scientific, and medical (ISM) radio band, at 868 MHz in Europe and 902 MHz in the United States. It utilizes a wide-reaching signal that passes freely through solid objects, called Ultra Narrowband, and requires little energy.

NB-IoT and eMTC have been and are being developed by 3GPP to operate in licensed bands dedicated to cellular networks. Figure 5.7 indicates how these channels are deployed. The eMTC scheme uses a data channel within a frequency band dedicated to 4G or 5G networks; this is referred to as **inband**. NB-IoT data channels can also be positioned inband, using a much narrower synchronization channel. NB-IoT channels can also fit in the guard bands. NB-IoT channels can also fit in the guard bands NB-IoT can also use a standalone 200 kHz carrier, for example, by re-purposing spectrum currently used by 2G.

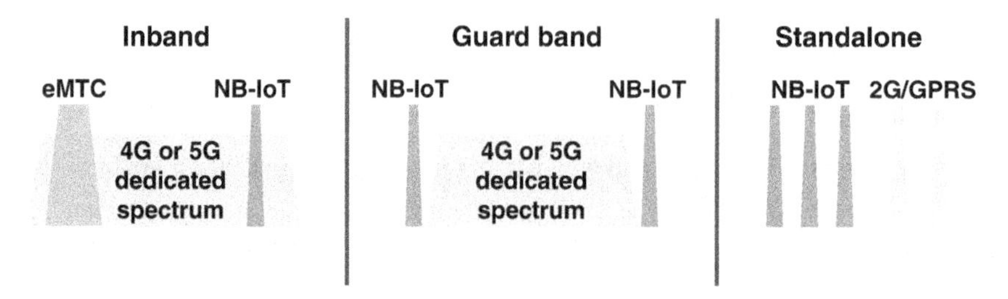

FIGURE 5.7 Spectrum Allocation for NB-IoT and eMTC

5.5 Smart Agriculture

The Food and Agricultural Organization of the United Nations (FAO) estimates that the global population will likely reach 9.8 billion by 2050 and that, in order to feed the world, food production must increase substantially by 2050. A major factor in achieving this growth will be IoT applied to agriculture [FAO17].

The IoT will enable more efficient management and optimized use of farm inputs and resources. Through the IoT, sensors can be deployed where needed, including on the ground, in water, and in vehicles, to collect data on target inputs such as soil moisture and crop health. Specific applications include farm vehicle tracking, livestock monitoring, storage monitoring, environmental conditions monitoring, and other farm operations. This application of IoT to agriculture is often referred to as **smart agriculture**, **smart farming**, or **precision agriculture**.

Model of IoT Deployment

As with any IoT deployment, smart agriculture involves essentially three layers:

- **IoT devices:** These are the sensors and actuators deployed in the agricultural environment.
- **Gateways:** Gateways act as intermediaries between things and the cloud, providing the needed connectivity, security, and manageability. In addition, the gateway may act as a platform for, or provide access to, local edge processing facilities that can react quickly to provide needed services.
- **Cloud-based storage and software:** All the sensor data is ultimately consolidated on a cloud-based system. The cloud-based system can run applications that analyze and manage data from devices and sensors in order to generate services that produce information used in decision making.

Figure 5.8, based on a figure in [OJHA15], illustrates a typical smart agriculture IoT architecture—in this case, a crop field with numerous IoT sensor devices. The sensors contain application-specific software. The sensors collect local environmental data, which might be soil moisture and air temperature and humidity. These data are passed either directly or indirectly via other sensor nodes, to a gateway node that serves as an interface to the rest of the architecture. Communication between nodes and to the gateway uses radio-frequency (RF) links of industrial, scientific, and medical (ISM) radio bands (such as 902–928 MHz and 2.4–2.5 GHz). The gateway node collects data from the deployed sensor nodes and passes this data to a nearby edge computing node using a cellular radio frequency, such as that assigned for 4G or 5G.

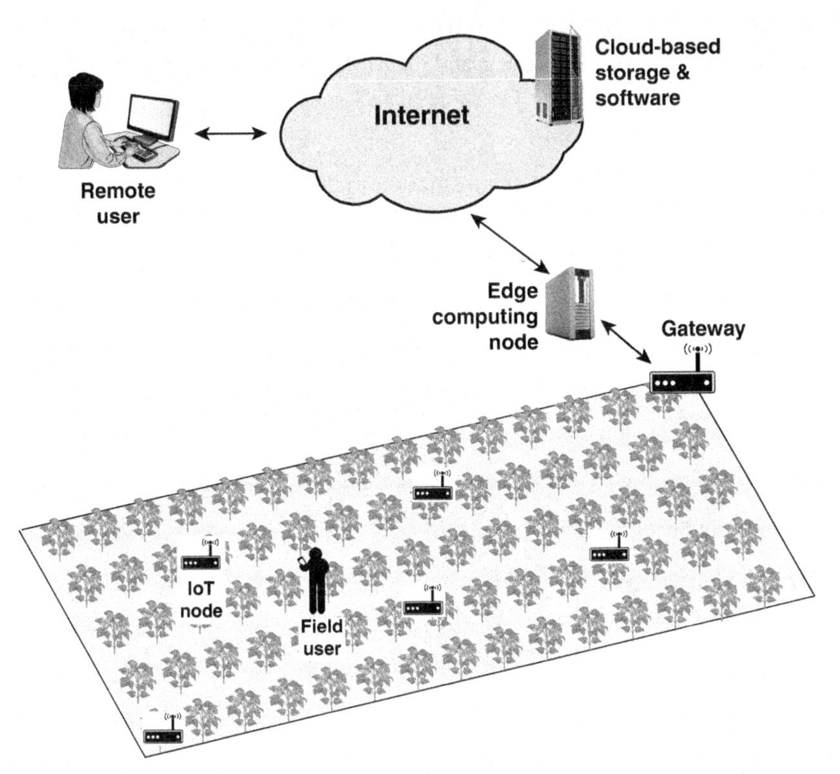

FIGURE 5.8 Agricultural Field with IoT Devices Deployed

The edge computing node can perform some intermediate consolidation of data before passing it to a central system. In addition, if the deployed IoT nodes include actuators, the edge computing node can perform actions in response to sensor data more quickly than if the data needed to be acted upon from a remote central server.

All of the data collected by sensors, plus a history of all actions taken with actuators, are stored in a cloud-based system, which also includes application-specific software. Both remote users and users in or near the agricultural field can access the central system to monitor the state of the field and control the on-field IoT devices.

Use Cases

A comprehensive source of the types of use cases that can be developed for smart agriculture is the European Union's (EU's) project Internet of Food & Farm 2020 (IoF2020) [VERO19]. IoF2020 aims to consolidate Europe's leading position in the IoT technology applied to the agri-food sector. Participants include farmers, food companies, policymakers, technology providers, research institutes, and consumers. The project aims to solve the European food and farming sectors' social challenges, maintain their competitiveness, and increase their sustainability.

As described in the IoF2020 booklet *Internet of Food & Farm*, the project is organized around five agriculture sectors: arable crops, dairy, fruits, vegetables, and meat. A total of 33 use cases, organized into five coherent trials, one per sector, demonstrate the value of IoT solutions for the European food and farming sectors.

This section provides an overview of the sectors and demonstration trials.

Internet of Arable Farming

Arable farming represents the largest agricultural sector in the EU in terms of acreage (60% in 2013) and number of primary production holdings. The Internet of Arable Farming trial seeks to integrate IoT technologies, data acquisition (e.g., soil, crop, climate) using sensors and earth observation systems, crop growth models, and yield gap analysis tools. The objectives for this IoF2020 sector include:

- Increase production for food, feed, bio-based products, and energy with the same or less input.
- Reduce pesticide, fertilizer, and energy use.
- Increase transparency and food safety.
- Halt the loss of soil fertility.
- Prevent the pollution of groundwater.
- Overcome disease/weed resistance.

The arable trial focuses on wheat, soybean, and potato production and processing in Europe's different climate zones. The demonstration trials in this sector are as follows:

- **Within-Field Management Zoning:** This project aims to develop specific IoT devices for acquisition of soil, crop, and climate data in production and storage of key arable and vegetable crops. The goal is to be able to define specific field management zones by developing and linking sensing and actuating devices with external data, mainly in potato production.
- **Precision Crop Management:** This project aims to incorporate IoT devices into a larger computer/network infrastructure that includes decision-making tools and services. Accurate and timely observations of crop status and the growing environment enable precise response to increase productivity. This use case is discussed in detail subsequently.

- **Soya Protein Management:** This use case addresses the current lack of technological innovation in the cultivation and processing of protein plants. Through smart farming technologies, such as decision support systems and better sensor data to optimize machine task operations, this use case aims to reintroduce and increase soybean cultivation in the EU.
- **Farm Machine Interoperability:** Interoperability is the ability of agricultural equipment to exchange and interact or communicate. Lack of interoperability obstructs the adoption of new IoT technologies and inhibits the gain of production efficiency through smart farming methods. This use case aims to integrate different communication standards to unlock the potential of data exchange between field machinery and farm management information systems.
- **Potato Data Processing Exchange:** This use case intends to provide information regarding the location and origin of different batches and types of potatoes. Being able to track produce back to the field regarding food security and quality supports buyers and processors, and it also helps farmers to identify problems and improve their yields in the following years.
- **Data-Driven Potato Production:** The use case is about combining IoT technology with earth observation data in order to help farmers reduce the costs of potato production and improve quality while reducing their environmental footprint.
- **Traceability for Food and Feed Logistics:** This use case develops a fully automated wireless silo detection system which can guarantee that the right bulk contents are delivered to the right silos and that the specifics of that delivery are registered. This will help farmers ensure that they receive and store the right raw materials, and it has the potential of giving consumers more insight in the origin of the food they consume.
- **Solar-Powered Field Sensors:** This use case relies on solar-powered plug-and-play sensors to provide farmers with instant access to data on their soil properties and crop health. These sensors measure parameters such as temperature, humidity, PH value, or nutrients and subsequently store the information in online databases accessible through smartphones.
- **Within-Field Management Zoning Baltics:** This use case develops a remote-sensing deployment to determine which nutritional elements and how much of them a plant is lacking at different stages of its growth. It enables farmers to use a soil map for variable rate planting and weed control. This use case demonstrates the added value of spectral data analysis and IoT technology for precise decision making and optimized crop management in potato and winter wheat production.

Internet of Dairy Farming

This trial aims to demonstrate the use of real-time sensor data (e.g., neck collars and movement sensors for livestock) combined with GPS location data to create value in the dairy chain from "grass to glass"— to more efficiently use resources and produce quality foods and provide a better animal health, welfare, and environment implementation. The trial focuses on feeding and reproduction of cows through early warning systems and real-time data that can be used for remote calibration and validation of sensors.

The use cases in this sector are as follows:

- **Grazing Cow Monitoring:** This use case involves monitoring and managing the outdoor grazing of cows, using GPS tracking in Ultra Narrowband communication networks.
- **Happy Cow:** This use case aims to improve dairy farm productivity by using 3D cow activity sensing and cloud machine learning technologies.
- **Silent Herdsman:** This use case enables herd alert management through a high-node-count distributed sensor network and a cloud-based platform for decision making.
- **Remote Milk Quality:** This use case is focused on remote quality assurance of accurate instruments and analysis and proactive control in the dairy chain.
- **Early Lameness Detection Through Machine Learning:** Lameness is a substantial issue in the dairy industry; it entails pain and discomfort for the cow and results in decreasing fertility and milk yield for the farmer. Current solutions are cost intensive and involve complex equipment. Lameness can be addressed without having to spend a large amount of resources. Leg-mounted sensors and machine learning algorithms can identify lame cattle at an early stage, and the data acquired can be sent directly to the farmer so that treatment of lameness can start immediately.
- **Precision Mineral Supplementation:** This use case involves precision supplementation of dairy cows, using an advanced mineral feeder, cloud-based services, and data integration combined with the identification of cows via electronic ear tags, thereby allowing tailored and individual mineral supplementation. The data gathered is displayed in a web application to provide easy access to monitor the feeding habits of individual cows, which may be early indicators of health problems. This use case's dairy management tool significantly contributes to animal welfare and the resource efficiency of farms.
- **Multi-Sensor Cow Monitoring:** This use case aims to further develop and promulgate a precise and reliable cattle monitoring ecosystem based on the needs of multi-country dairy and beef farmers. The system is made up of a small rumen bolus[2] and collar that monitor various physiological data, and a cloud-based server application to provide accurate information for daily operations. It helps farmers to guard, track, and monitor all assets.

---

2. A rumen bolus is a capsule that is inserted into an animal from three months of age. After application, the bolus comes to rest in the reticulum for the life of the animal. The bolus contains an electronic transponder that can be read electronically and is issued with a matching non-electronic ear tag for manual reading and/or visual confirmation that the bolus has been installed.

Internet of Fruits

This trial aims to demonstrate IoT technology that is integrated throughout the whole supply chain from the field, to logistics, to processing, to the retailer. Sensors in orchards and vineyards (e.g., weather stations, thermal cameras) are connected in the cloud and used for monitoring to provide early warning of pests and diseases and control (e.g., variable rate spraying or selective harvesting). Big data analysis can further optimize all processes in a whole chain. This is intended to result in reduced pre- and post-harvest losses, fewer inputs, higher (fresh) quality, and better traceable products.

The use cases in this sector are as follows:

- **Fresh Table Grapes Chain:** This use case involves the real-time monitoring and control of water supply and crop protection of table grapes and predicting shelf life.
- **Big Wine Optimization:** This use case aims to optimize the cultivation and processing of wine by sensor-actuator networks and big data analysis within a cloud framework.
- **Automated Olive Chain:** This use case includes automated field control, product segmentation, processing, and commercialization of olives and olive oil.
- **Intelligent Fruit Logistics:** This use case involves reusable containers dedicated to the transport of fresh fruit, which are integrated with IT solutions. Equipping the crates with RFID (radio-frequency identification) technology allows for easy identification of objects through the production and distribution cycle.
- **Smart Orchard Spray Application:** This use case focuses on IoT-enabled and highly efficient smart sprayers. The precise and automatic adaptation of sprayers to specific field zones as well as individual plant conditions lowers farmers' costs while mitigating the environmental impacts of cultivation.
- **Beverage Integrity Tracking:** This use case focuses on the distribution of high-quality beverages (specifically alcoholic drinks such as wines and craft beers) that are susceptible to damage during storage and transportation. Many factors (e.g., temperature, humidity, shocks) can negatively affect the quality of beverages during their journey from producer to consumer. The design of this use case involves a cloud-based, plug-and-play application for monitoring and recording detailed information on the stress factors experienced by a beverage during every stage of its distribution chain and assessing the impact of these stress factors on the beverage quality.

Internet of Vegetables

This trial focuses on a combination of environmental control levels: fully controlled indoor growing with an artificial lighting system, semi-controlled greenhouse production, and nonregulated ambient conditions in open-air cultivation of vegetables. It demonstrates the automatic execution of growth recipes by an intelligent combination of sensors that measure crop conditions and control processes (including lighting, climate, irrigation, and logistics) and analysis of big data that is collected through these sensors and advanced visioning systems with location specification. This is intended to result in improved production control and better communication throughout a supply chain (including harvest prediction and consumer information).

The use cases in this sector are as follows:

- **City Farming Leafy Vegetables:** This use case refers to leafy vegetables used in convenience products such as cut lettuce and ready-to-eat salads, where tolerance for dirt, insects, or other unwanted ingredients is almost zero. This use case employs a commercial city farm to demonstrate the smooth integration of IoT technologies into the production of high-quality vegetables in a predictable and reliable manner, leveraging advantages in the production approach such as independence from seasonal influences and absence of plant diseases as well as pesticides.
- **Chain-Integrated Greenhouse Production:** This use case aims to integrate the value chain and quality innovation by using a full sensor-actuator-based system in tomato greenhouses.
- **Added Value Weeding Data:** This use case focuses on boosting the value chain by harvesting weeding data of organic vegetables, leveraging advanced visioning systems.
- **Enhanced Quality Certification System:** This use case has the objective of enhanced trust and simplification of quality certification systems through use of sensors, RFID tags, and intelligent chain analyses.
- **Digital Ecosystem Utilization:** This use case aims to provide more precision in the use of plant protection products, such as insecticides, so that the minimum amount is used only when and where needed. The deployment includes IoT devices in the field, cloud computing, and analytics technologies.

Internet of Meat

This trial aims to demonstrate how the growth of animals can be optimized and how communication in the whole supply chain can be improved using automated monitoring and control of advanced sensor-actuator systems. The data generated by events provides early warning (e.g., on health status) and helps improve the transparency and traceability of meat. This will assure meat quality, reduce mortality, optimize labor, and improve animal health and welfare, leading to reduction in antibiotic use.

The use cases in this sector are as follows:

- **Pig Farm Management:** This use case aims to optimize pig production management through interoperable on-farm sensors and slaughterhouse data.
- **Poultry Chain Management:** This use case aims to optimize production, transport, and processing of poultry meat through automated ambient monitoring and control and data analyses.

- **Meat Transparency and Traceability:** This use case aims to enhance transparency and traceability of meat based on a monitored chain of event data.
- **Decision-Making Optimization in Beef Supply Chain:** The beef supply chain is a complex system, involving crop farms, livestock farms, feedlots, transporters, slaughterhouses, retailers, and consumers. This use case integrates data acquisition throughout the entire supply chain, using a variety of IoT devices, including smart collars and tags for cattle, IoT scales to measure calves' growth rates, and IoT multi-sensor stations for transport and slaughtering conditions (e.g., temperature, dust, noise).
- **Feed Supply Chain Management:** This use case develops an integral feedstock management system to optimize the entire supply chain. Elements of the system include IoT-enabled volumetric sensors in feed silos and a cloud data platform that collects data from silos along with relevant production information from livestock farmers and feed suppliers. Cloud-based algorithms enable optimization in refilling orders, production batches, shipping routes, and raw materials purchases.
- **Interoperable Pig Health Tracking:** This use case aims to provide pig producers better control of the production process through the use of an IoT sensor integrated into a commercial off-the-shelf identification ear tag. The sensor can monitor activity level, heart rate, and blood oxygenation. Changes in physiological parameters of pigs are good indicators of their state of health. This use case thus relies on intensive scrutiny of each animal through IoT sensors, enabling the farmer to swiftly intervene in the event of health risks or diseases.

Precision Crop Management

This section looks in more detail at one of the IoF2020 use cases to provide a better understanding of the application of IoT to farming. This subsection examines the IoF2020 Precision Crop Management use case. This use case demonstration focuses on nitrogen and irrigation in a precision crop–managed wheat field.

The system employs four types of IoT-enabled devices in the wheat field:

- **Soil sensor:** Monitors soil water potential and temperature.
- **Climate sensor:** Monitors air temperature, relative humidity, and solar radiation.
- **Plant sensor:** Monitors crop growth with spectral reflectance and transmittance.
- **RGB camera:** Visually monitors the field or a portion of the field.

Functional View

Figure 5.9, from IoF2020 D3.2 (*The IoF2020 Use Case Architectures and Overview of the Related IoT Systems*, September 29, 2017), provides a functional view of the system. This model is based on the

ITU-T Y.2060 IoT reference model. The model provides an overview of the technical architecture and allows designers to identify suitable technologies and technology providers.

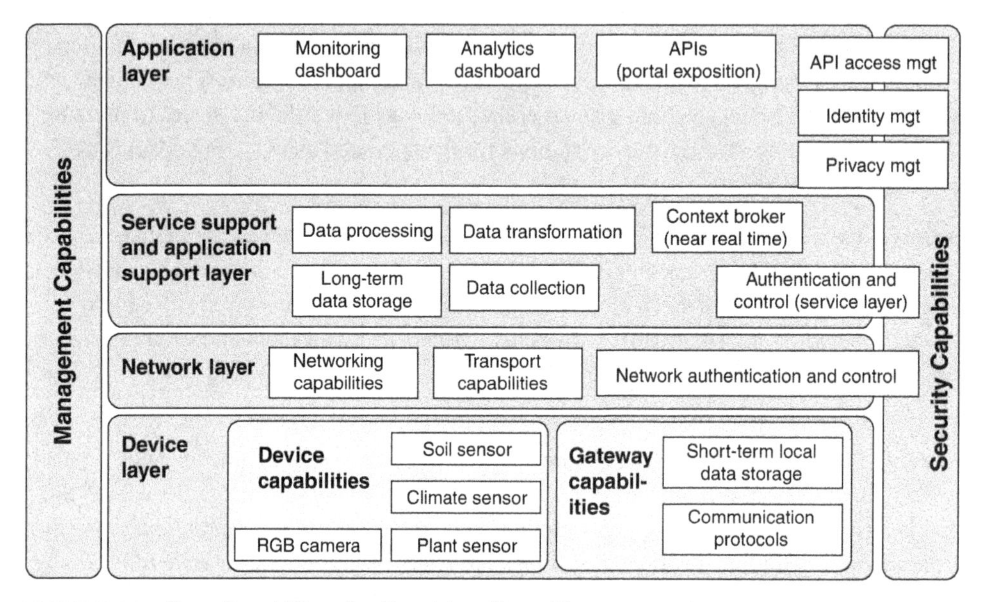

FIGURE 5.9 Functional View for Precision Crop Management

The functional view is structured as follows:

- **Device layer:** This layer is subdivided into device capabilities, which are IoT-enabled devices, and gateway capabilities, which are network capabilities at the gateway level (e.g., multiple interfaces that support protocol conversion). This layer includes the four types of IoT-enabled devices described above, and it also includes a local gateway device that collects the IoT device data and communicates with the remote cloud-based use case platform.
- **Network layer:** This layer performs three basic functions. Networking capabilities refer to the interconnection of devices and gateways. Transport capabilities refer to the transport of IoT service and application specific information as well as IoT-related control and management information. Roughly, these correspond to OSI model network and transport layers. Network authentication and control refers to the network-level security functions that provide secure communications.
- **Service support and application support layer:** This layer provides capabilities that are used by applications. Generic support capabilities can be used by many different applications. Examples include common data processing and database management capabilities. Specific support capabilities cater to the requirements of a specific subset of IoT applications.

- **Application layer:** This layer contains all the applications that interact with IoT devices and analyze device-generated data.

Supplementing the four layers are two components that span all four layers. The **management capabilities layer** covers the traditional network-oriented management functions fault, configuration, accounting, and performance management. Y.2060 lists the following as examples of generic management capabilities:

- **Device management:** Examples include device discovery, authentication, remote device activation and deactivation, configuration, diagnostics, firmware and/or software updating, and device working status management.
- **Local network topology management:** Examples include network configuration management.
- **Traffic and congestion management:** Examples include the detection of network overflow conditions and the implementation of resource reservation for time-critical and/or life-critical data flows.

Specific management capabilities are tailored to specific classes of applications. An example is monitoring management-related signals sent from IoT devices.

The **security capabilities layer** includes generic security capabilities that are independent of applications. Y.2060 lists the following as examples of generic security capabilities:

- **Application layer:** Examples include authorization, authentication, application data confidentiality and integrity protection, privacy protection, security audit, and antivirus.
- **Network layer:** Examples include authorization, authentication, user data and signaling data confidentiality, and signaling integrity protection.
- **Device layer:** Examples include authentication, authorization, device integrity validation, access control, data confidentiality, and integrity protection.

Specific security capabilities relate to specific application requirements, such as IoT security requirements.

The functional view across the four layers depicts the functional components, including equipment, computer hardware, and software. This enables system designers to identify suitable providers of infrastructure or technology that are capable of offering each component.

Business Process View

Another way to characterize the IoF2020 Precision Crop management System is as a business process hierarchy view, as shown in Figure 5.10, from IoF2020 D3.2. This model is based on IEC 62264 (*Enterprise-Control System Integration, Part 1: Models and Terminology Reference Model,*

May 2013). This model provides an overview of business processes and their interrelationships, including the physical product flow of input material to end products, the main objects (things) involved, and the position of business processes in the production control hierarchy, ranging from operational control of physical objects to the enterprise management level.

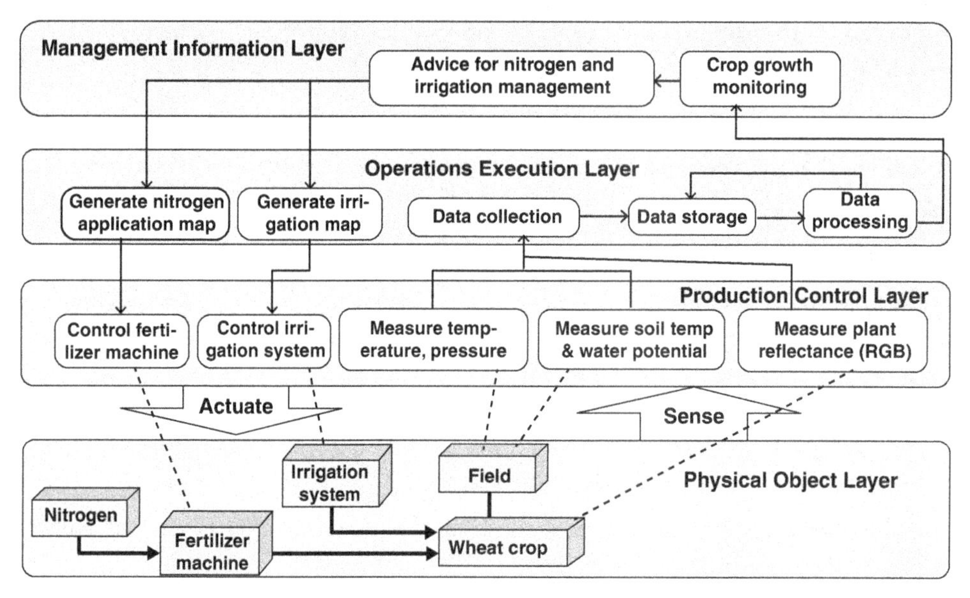

FIGURE 5.10 Business Process Hierarchy View for Precision Crop Management

The model is structured as follows:

- **Physical object layer:** Describes the physical environment. It depicts the elements used to manage the supply of water and nitrogen in the crop, including the machinery. The IoF2020 document does not explicitly show the IoT-enabled devices.
- **Production control layer:** Represents the link between the physical world and the virtual world. So in this part, it can find processes that on one hand measure different kinds of information in the field and on the other hand act on crops (irrigation or fertilization). These processes have a short time horizon (e.g., minutes, seconds, milliseconds).
- **Operations execution layer:** Groups all computational tasks using collected information as well as processes that take into account results of the management information layer. An example is the creation of maps to plan irrigation or apply fertilizer. This layer contains processes related to the definition, control, and performance of tasks with an intermediate time horizon (e.g., days, hours, minutes).
- **Management information layer:** Includes processes that take into account data from the operations layer and apply business rules to obtain several indicators. With some additional analysis,

this results in the development of an overall plan related to the control of an entire enterprise. These processes have the longest time horizon (e.g., months, weeks, days).

Thus, the business process model reflects the real-world operation of the use case and shows how operations in the field are monitored and managed at different time horizons and how the field operations integrate with management processes.

Key Performance Indicators

IoF2020 Deliverable D4.1 (*KPI Catalogue for Each Use Case*, September 22, 2017) defines the key performance indicators (KPIs) for each use case. In general, a KPI is a measurable value that shows the effectiveness in achieving key objectives. KPIs are usually used to evaluate the success in reaching defined targets at various levels. KPIs help to quantify and measure the potential impact of IoF2020 solutions per use case. The quantification is based on predefined variables, measurement procedures, the weight of each variable, and the final calculation.

The KPIs are grouped into three dimensions:

- **Economic:** KPIs in this dimension reflect a potential added value or cost reduction. They address the need to adopt new technologies and methods in order to remain competitive.
- **Environmental:** This dimension refers to the environmental impact of the deployment of the use case.
- **Social:** This dimension relates to the impacts on the social system dealing with human well-being, the fulfilment of human needs, and the equal development of opportunities for all people.

Table 5.6 shows the KPIs for the IoF2020 Precision Crop Management use case.

TABLE 5.6 KPIs for the Precision Crop Management Use Case

Dimension	Category	Indicators
Economic	Productivity increase	Yield increase
	Efficiency improvement	Work time use efficiency Water use reduction
	Cost reduction	Production costs reduction Water cost reduction
	Quality	Quality improvement
Environmental	Lower input	Nitrogen use reduction Nitrogen and water use efficiencies
	Lower emissions and leaching	Nitrogen leaching reduction GHG reduction
Social	Ease of work	Effective time use Stress reduction
	User satisfaction	IoT user satisfaction

Network Performance Requirements

In smart agriculture, the networking demands depend on the specific use case. To provide an idea of what the demands are likely to be, an NGMN white paper (*Perspectives on Vertical Industries and Implications for 5G*, September 2016), estimates performance requirements for two general categories:

- **Data-oriented farming:** This category requires support for massive connectivity for sensors and actuators. The IoF2020 Precision Crop Management use case is an example.
- **Automated farm machinery:** This category requires V2X (vehicle-to-anything) for cooperative farm machinery (tractors, combine harvesters, irrigators, etc.). V2X refers to wireless communications from vehicle to vehicle, vehicle to other devices in the infrastructure, and vehicle to user. V2X is the passing of information from a vehicle to any entity that may affect the vehicle and vice versa.

Table 5.7, which is from the NGMN white paper on vertical industries, summarizes the network capabilities required for these two smart agriculture categories.

TABLE 5.7 Capabilities Required for Relevant Agriculture Use Cases

Use Case Attribute	Data-Oriented Farming	Automated Farm Machinery
Potential technologies	Wi-Fi, LPWA, 4G, 5G	Wi-Fi based, 4G, 5G
Experienced data rate	1 Mbps or less	5 Mbps per link
Latency	Seconds to minutes	10–30 ms end-to-end
Reliability (IP packet delivery within latency bound)	Not critical	Critical
Number of devices	10^4/km^2	~10 devices in vicinity (100 m to 1 km)
Battery	15 years for wireless sensors	Not critical
Coverage	Important	Critical Up to 300 m range
Mobility	Stationary to pedestrian speeds	Up to 50 km/h
Interwork/roaming	Interworking and roaming not needed for fixed sensors	Interworking and roaming needed (as farm machinery can be leased)
Security	Critical (data integrity, privacy)	Critical (authentication, data integrity)
Positioning	Not critical for applications where device location would be known (e.g., fixed devices)	30 cm–1 m

5.6 Smart Cities

ITU-T Y.4900 (*Overview of Key Performance Indicators in Smart Sustainable Cities*, June 2016) defines a **smart sustainable city**, or simply **smart city**, as follows:

> A smart sustainable city is an innovative city that uses information and communication technologies (ICTs) and other means to improve quality of life, efficiency of urban operation and

services and competitiveness, while ensuring that it meets the needs of present and future generations with respect to economic, social, environmental, as well as cultural aspects.

The sustainability of a smart city is based on four main aspects:

- **Economic:** The ability to generate income and employment for the livelihood of the inhabitants
- **Social:** The ability to ensure that the welfare (e.g., safety, health, education) of the citizens can be equally delivered, despite differences in class, race, or gender
- **Environmental:** The ability to protect future quality and reproducibility of natural resources
- **Governance:** The ability to maintain social conditions of stability, democracy, participation, and justice

Smart City Use Cases

Some smart city use cases, such as public protection and disaster relief (PPDR) fit into the URLCC usage scenario and are covered in Chapter 6, "Ultra-Reliable and Low-Latency Communications." This chapter is concerned with use cases that fall into the mMTC usage scenario category.

ITU-T Series Y Supplement 56 (*Supplement on Use Cases of Smart Cities and Communities*, December 2019) lists eight demonstration examples of smart city use cases, in four categories, as shown in Table 5.8. This subsection provides an overview of each use case.

TABLE 5.8 Classification of Smart City Use Cases

Category	Use Cases
Safer city	■ Pedestrian monitoring for decisive disaster response ■ River water-level measurement system using smartphones and augmented reality (AR) ■ Citizens' safety services
Infrastructure	■ Lift monitoring services ■ Infrastructure monitoring
e-Government	■ Citizen identification system using biometric
City operation	■ City operations center ■ Intelligent traffic management system, adaptive traffic control system, CCTV-based real-time public safety system, solid waste management, and integrated platform with command and control center for a smart city

Pedestrian Monitoring for Decisive Disaster Response

This use case involves the installation of surveillance cameras throughout a city that can monitor crowd size and behavior and transmit the gathered information to a central monitoring/management source. The cameras monitor locations that are likely to draw large groups of people, such as near a railroad or subway station or near a school. In the event of a disaster near one of these sites, the system provides real-time information about the size of the crowd at risk. In addition, the pedestrian monitoring system facilitates understanding of the behavior of crowds and the detection of abnormal situations by analyzing images captured by surveillance cameras. In the event of any abnormality, the system automatically provides information or instructions for evacuation from the disaster site or for prevention of accidents.

NEC Corporation has installed such a system in Toshima City, Japan [NEC15].

River Water-Level Measurement System Using Smartphones and Augmented Reality

This is an example of a smart city use case that does not require a large number of IoT devices. Instead, data is collected from the smartphones of a number of municipal employees on a regular basis.

Fujitsu has set up a demonstration system in Manado, Indonesia [FUJI16]. Manado, which faces the sea, has four medium-sized rivers flowing through it and is vulnerable to flash floods and river flooding due to heavy rain, as was the case in January 2014, when large-scale flooding and landslides claimed a number of lives. Victims of flooding are common not only in Manado, but across Indonesia, so there is a need for the monitoring of river water levels, the rapid sharing of information by local government workers when water levels reach flood-warning levels, and provision of rapid and accurate evacuation instructions to residents.

To minimize the damage, the Manado government had been observing rainfall, river water levels, groundwater, and weather data three times a day at 96 locations. Manado local government workers visually monitored and recorded river water levels; however, this method lacked accuracy and immediacy. The government also implemented telemetry systems to monitor river water levels by installing outdoor sensors, but the systems were difficult to use on an ongoing basis due to the high cost of maintaining the equipment.

The solution provided by this use case is for municipal employees to measure water level from a location at a safe distance from the river, using AR technology applications installed in a smartphone. The system includes a number of simple measurement sticks, called AR markers, installed at various points along the rivers. When the observer directs a smartphone camera to an AR marker installed at an observation point, a scale is displayed on the screen, and the observer can perform a simple operation, such as tapping the screen to indicate the water level, and send accurate water levels, on-site photographs, comments on the water levels, and other data to the data center server.

As a result, it has become possible to send accurate water levels at observation points in real time. Also, those responsible for taking measurements can observe water levels from a safe location, and on-site photographs can be stored as evidence.

Citizens' Safety Services

This use case describes interworking between smart city operations center and fire and police stations for the citizens' safety services. The key ingredients of this use case are:

- **Video surveillance cameras:** Surveillance cameras are deployed throughout the city to provide extensive coverage with the minimum number of cameras.
- **Traffic sensors:** IoT-enabled traffic sensors that are able to measure rate and volume of traffic are deployed throughout the city.
- **Smart city operations center:** The operations center connects wirelessly to the cameras and sensors to provide a central source of information.

A demonstration project in Daejeon, South Korea, focuses on fire and police support. The operations center has wired or wireless connections to all police and fire stations. In the case of a fire, rescue, first aid, or disaster situation, an initial alarm identifies the location to which the first responders are going. This alarm may be an automated alarm that comes into the local fire station and is then transmitted to the operations center, or it may be a citizen-reported alarm that goes to the operations center. The operations center provides traffic information to the first responders to enable them to take the best route to the scene. If available, they are also provided with live video that they can view while en route to help them function more effectively on arrival.

A similar scenario unfolds with police response to a crime or disturbance scene. In addition, the surveillance images of crime scenes may help in tracking down offenders.

Lift Monitoring Services

This demonstration use case involves monitoring lifts, or elevators, throughout a city. It is a system developed by Surbana Jurong and deployed in Singapore and other Asian cities. The system consists of a central lift monitoring system (LMS) and IoT-enabled sensors installed in lifts throughout the city. The installation in Singapore monitors more than 26,000 lifts across 10,000 housing units.

The most important aspect of the system is the ability to respond rapidly to an event. A typical sequence is as follows:

1. The lift sensors can detect an emergency alarm (activated by a passenger), stuck lift, out-of-service lift, or fire event.
2. The event is sent to the smart city operations center.
3. The operations center sends messages to on-site staff to prevent further attempts to use the elevator and to engineers or emergency responders, as indicated by the nature of the event.
4. When the event is resolved, such as by completing a repair, an engineer sends a message to the operations center. The IoT device also sends an operational status message.

In addition, the sensor devices capture data on an ongoing basis. This data, using machine learning algorithms, is used to predict future failures, allowing for optimized maintenance and reduced downtime.

Citizen Identification System Using Biometrics

The objective of this use case is to provide a digital identity to the entire population to serve as the basis for accessing social services and interacting with the government at various levels. This system, called Aadhaar, is deployed nationwide in India and currently has over one billion people registered.

Figure 5.11 illustrates the operation of the system. Each individual who is to be included in the database of authorized users must first be **registered** in the system. The person presents his or her information (name, address, and biometric information), and a 12-digit number is given to the person if the person did not register before. For the biometric information, the system collects fingerprints, face images, and iris images. The system digitizes the biometric input and then extracts a set of features that can be stored as a number or set of numbers representing this unique biometric characteristic; this set of numbers is referred to as the user's template. The user is at this point registered in the system, which maintains for the user the 12-digit number and the biometric values.

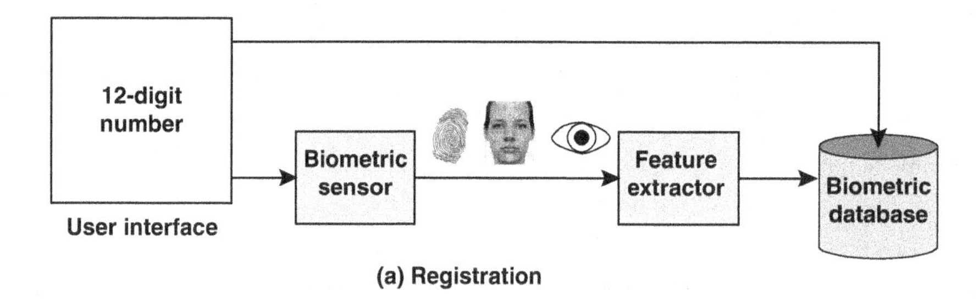

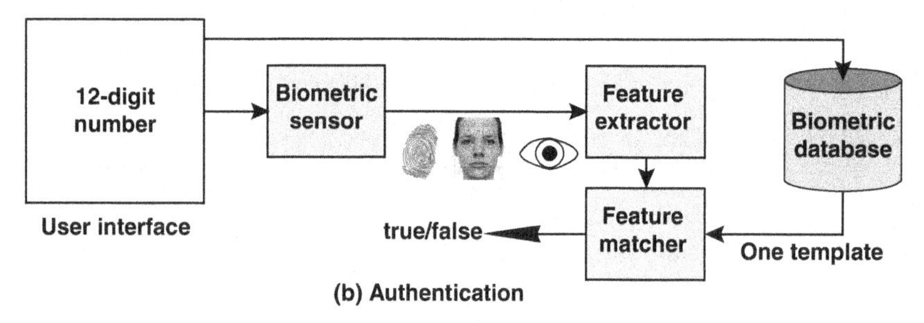

FIGURE 5.11 Citizen Identification System Using Biometrics

When a person wishes to access a service, the biometric sensing device authenticates the person's identity. The system extracts the corresponding biometric information for one or more of the three biometric types and compares that to the template stored for this user. If there is a match, then the system authenticates this user.

These biometric devices fall into two categories:

- **Discrete devices:** These devices are separate pieces of hardware that require connectivity to a host device, such as PC or laptop.
- **Integrated devices:** With this type of device, a sensor is integrated into the device, which may be a smartphone or tablet.

The 12-digit number and biometric authentication is used for fair provision of social services, such as food distribution, government subsidy, and so on. This helps to avoid duplicate provision or fictitious claims of the subsidy. The number is also required when a person opens a new bank account or contracts a new mobile phone. A payment system using fingerprints is also available.

In any large Indian city, there are tens or even hundreds of thousands of Aadhaar devices in use for registration and service access. They form a massive IoT network connected to a central server.

City Operations Center

This use case demonstration project focuses on a central operations center as the key element in a smart city deployment. The operations center is in Bristol, England. The operations center unifies, coordinates, and manages a variety of city functions, including:

- Traffic management and control
- Public transportation management and control
- Closed-circuit TV (CCTV) monitoring for safety and security
- Telecare and telemedicine services
- Alarm monitoring
- Out-of-hours call handling

The operations center connects actors in the public and private sectors both digitally and physically (i.e., within one office), which is fundamental to the delivery of essential services for the city and its citizens. Many of these actors are housed in the same building, allowing for greater contact and coordination and resulting in a better environment for citizens. This includes the city council's Emergency Control Centre, Traffic Control Centre, and Community Safety Control Rooms. Staff from transport providers, such as bus companies, also have a space in the center to work with the city council's traffic management team. Not all of the value is derived from connected technologies. For example, the city's bus drivers also provide timely and valuable information that may then be shared with appropriate teams in the operations center.

Fundamental to the functioning of the operations center is an ICT platform that can enhance service delivery, collaboration, and innovation in the city [TIME20]. The ICT platform is designed to collect and store data from other applications in the city, such as traffic sensors and energy demand sensors, with the aim of optimizing mobility and energy efficiency. The platform is accessible to private businesses and associations (through special contracts and agreements that ensure data protection) so that they can use the platform to collaborate and innovate. They can coordinate operations and transactions across multiple applications and service providers, such as between bus companies to optimize routes and time plans, and to create new public and commercial services, such as programs for users to visualize and manage their energy consumption.

Infrastructure Monitoring

This use case involves using IoT sensor devices to monitor aging infrastructure elements to support automated inspection, diagnosis, confirmation of repair effort, and subsequent status check. The scheme can be applied to bridges, tunnels, and paved roads. This demonstration case was carried out by Fujitsu on bridge decks near Yamaguchi University. The bridge deck is the floor or surface of the bridge, which transfers the weight of vehicles traveling over the bridge to the columns or girders supporting the bridge [FUJI18].

Four different sensor types were developed and deployed:

- **Crack monitoring with camera images:** Cameras positioned on the ground capture images from under the bridge deck. Multiple images from multiple cameras over time are able to detect the existence and growth of cracks.
- **Crack monitoring with displacement meter:** When the floor slab is found to be excessively damaged by regular inspections, maintenance and repair work may not be able to start immediately due to budget and traffic disturbances. In this demonstration experiment, the displacement of the floor slab is measured with a displacement meter attached to a fixed beam, and the damage is monitored. If a threshold is reached, emergency response is required.
- **Crack/strain monitoring with optical fibers:** Optical fibers are fixed to the bottom of the bridge deck. Changes in the amount of light enable detection of cracks and strain distribution.
- **Deformation monitoring with accelerometers:** Multiple accelerometers are installed at the bottom of the slab, and changes in structural performance due to deformation such as peeling or dropping are evaluated from the frequency spectrum or vibration mode unique to the bridge.

Each of these sensor types can provide data transmitted via an IoT network and made available for analysis.

Figure 5.12, from ITU-T Series Y Supplement 56, indicates the IoT architecture for deploying the deformation monitor sensors on a large scale.

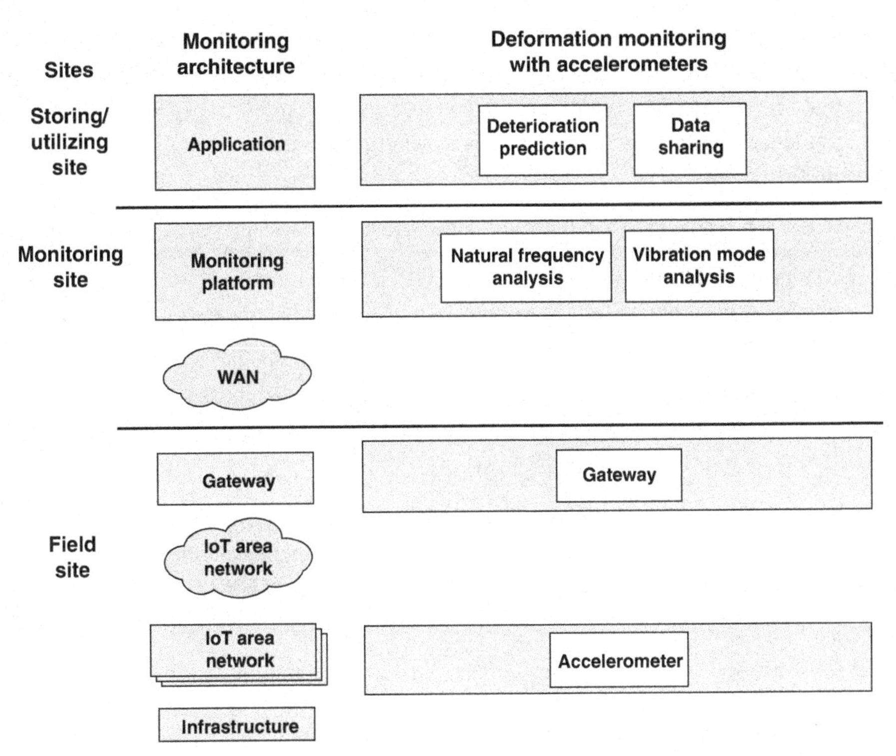

FIGURE 5.12 Architecture for Deformation Monitoring with Accelerometers

The architecture consists of:

- Sensor devices installed in the infrastructure
- A gateway that aggregates data from multiple sensors
- A monitoring platform that monitors the collected data
- Applications that store and use the collected data

Integrated Command and Control Center

This use case relates to the development of a city-wide set of smart city capabilities managed from an integrated command and control center. It demonstrates the evolution toward a truly smart city with many services deployed, coordinated, and integrated. The deployment is in the city of Agra in India.

The key components of this use case are as follows:

- **City communication network:** This involves a combination of optical fiber cable and cellular network capability.

- **Integrated command and control center (ICCC):** The ICCC is the central repository for management and monitoring of all ICT-based smart city components, such as the solid waste management system, the smart street lighting control system, Wi-Fi, smart transport, smart bus stops, CCTV surveillance, digital signages, IoT sensors (environment, etc.), and the public information system (PIS).
- **Data center and disaster recovery:** Cloud-based data storage has backup.
- **City and enterprise geographic information system (GIS):** A comprehensive GIS application is used for planning, management, and governance in the context of the entire functioning of the organization.
- **CCTV-based real-time public safety system:** CCTV cameras are installed at various locations across the city for safety along with public address system and variable message signboard (VaMS), emergency/panic box system, and other services.
- **Intelligent traffic management:** CCTV cameras and traffic violation sensors are installed at various locations across the city for traffic management and enforcement system, such as red light violation detection (RLVD), automatic number plate recognition (ANPR), and speed detection.
- **Environmental sensors:** Smart environmental sensors gather data about pollution, ambient conditions (light, noise, temperature, humidity, and barometric pressure), weather conditions (rain), levels of gases in the city (pollution), and other events on an hourly and subsequently on a daily basis.
- **GIS-/GPS-enabled solid waste management:** A GIS-/GPS-enabled solid waste management system provides end-to-end management and monitoring of garbage collection and processing.
- **Adaptive traffic management system:** A system for control and management of traffic controls the traffic signals on certain stretches of road with sensor-based automation of signals.
- **Integration components:** Present and future systems are integrated with the ICC, including an e-Governance system, smart LED lighting, smart bus stops, a SCADA system, a sewage system, Wi-Fi hotspots, and citizen engagement applications for a smart city.

ICT Architecture for Smart Cities

ITU-T Series Y Supplement 27 (*Smart Sustainable Cities: Setting the Framework for an ICT Architecture*, January 2016) defines an ICT architecture for a smart sustainable city. The architecture defines the technical environment and infrastructure in which all information systems related to the smart city exist.

Figure 5.13, from Supplement 27, shows an architecture consisting of four layers.

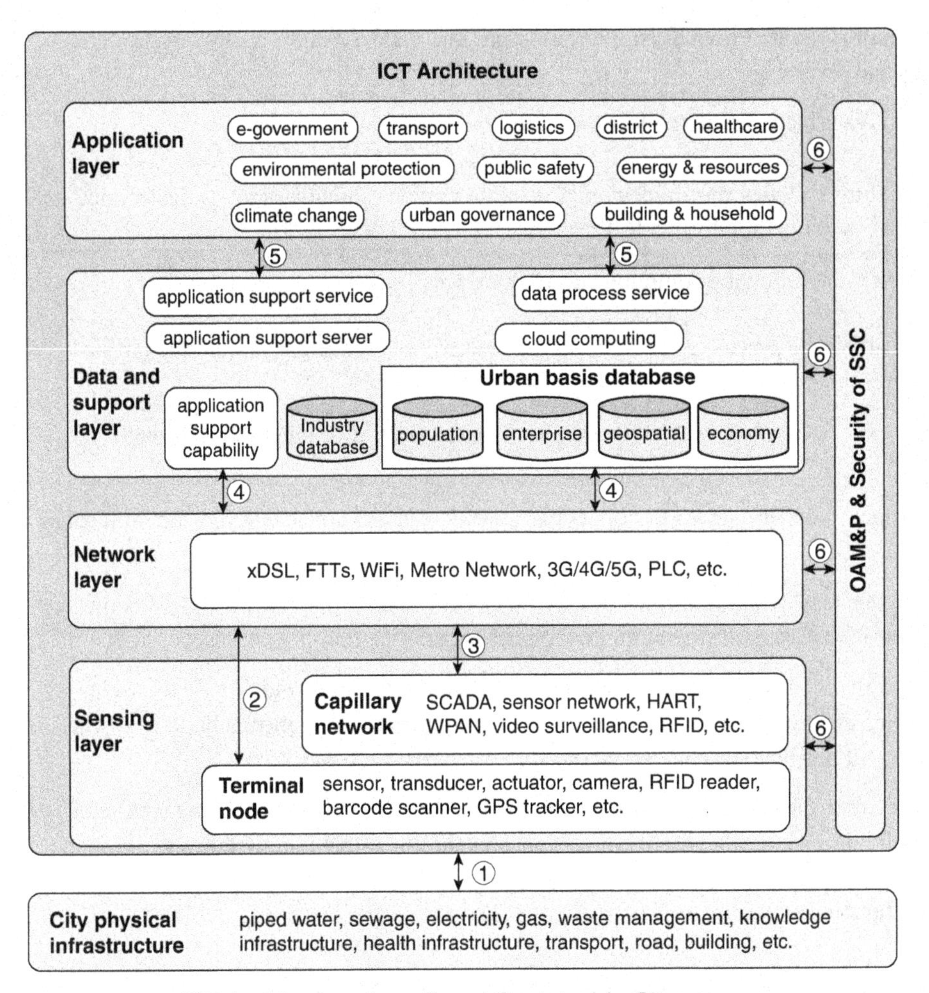

FIGURE 5.13 ICT Architecture for a Smart Sustainable City

The four layers are as follows:

- **Sensing layer:** Includes a collection of IoT-enabled devices with a variety of capabilities, including sensors, actuators, cameras, and RFID readers. Thus, this layer in fact encompasses more than sensing. The other main element of this layer is the capillary network, which is a local network that uses short-range radio-access technologies to provide local connectivity to things and devices.
- **Network layer:** Consists of various networks provided by telecommunication operators, as well as other metro networks provided by city stakeholders and/or an enterprise private communication network.

- **Data and support layer:** Provides support programs and databases for various city-level applications and services.
- **Application layer:** Includes all applications that provide smart city services.

Finally, the architecture includes the framework for the operation, administration, maintenance and provisioning, and security function for the ICT systems.

Figure 5.13 also labels the following communications interfaces:

1. Exchange of information between sensors and actuators in the sensing layer and the physical infrastructure.
2. This interface is between IoT-enabled devices and the network layer, directly or through gateways.
3. For this interface, capillary networks collect the sensing data and connect to the communication networks.
4. This interface enables communications between data centers and lower layers for collecting various information through the communication networks.
5. This interface exists between the data and support layer and the application layer. It enables exchange of information between data centers and/or application support functionalities with corresponding city applications and services.
6. This exists between the operations, administration, management, and provisioning (OAM&P) and security framework and the four layers. It enables the corresponding modules to exchange data flow and control flow and provide operation, administration, maintenance, provisioning, and security functions.

This architectural model is a useful tool for designers of smart city infrastructures.

Network Performance Requirements

For smart cities, the networking demands depend on the specific use case. To provide an idea of what the demands are likely to be, an NGMN white paper (*Perspectives on Vertical Industries and Implications for 5G*, September 2016) estimates performance requirements for three general categories:

- **Real-time video for monitoring and guidance:** This category requires support for massive connectivity of sensors and actuators. The IoF2020 Precision Crop Management use case is an example.
- **Massive connectivity for non-time-critical sensing:** Examples include weather and pollution levels.
- **Massive connectivity for time-critical sensing and feedback:** Examples include detection of natural disasters, smart grid control, and context-aware lighting.

The first two categories fall under the mMTC usage scenario. All of the use cases described in this section are characterized by one or both of these items. The third category falls under the URLCC usage scenario.

Table 5.9, from the NGMN white paper on vertical industries, summarizes the network capabilities required for these three smart city categories.

TABLE 5.9 Capabilities Required for Relevant Smart Cities and Utilities Use Cases

Use Case Attribute	Real-Time Video for Monitoring and Guidance	Massive Connectivity for Non-Time-Critical Sensing	Massive Connectivity for Time-Critical Sensing and Feedback
Potential technologies	5G	Wi-Fi, LPWA, 4G, 5G	4G, 5G
Experienced data rate	4K (2160/60/P): 30~40 Mbps 8K (4320/60/P): 80~100 Mbps HD H.265/HEVC: ~10 Mbps	1 Mbps or less	Uplink: small data rate per sensor but multitude of sensors results in bandwidth demand in backhaul Smart grid: Up to 5 Mbps in downlink and uplink
Latency	100 ms end-to-end	On the order of seconds to minutes	30 ms end-to-end Smart grid: < 5 ms end-to-end for transmission/grid backbone: < 50 ms end-to-end for distribution/grid backhaul < 1 ms for end-to-end access
Reliability (IP packet delivery within latency bound)	Critical	Not critical	Critical for smart grid 99.9%–99.999% for the different domains/applications
Devices	10^3/km^2 activity factor=50%	10^4/km^2	10^4/km^2
Battery	—	15 years for wireless sensors	15 years for wireless sensors
Coverage	Important Including underground, air (up to 3 km above ground), at sea (up to 50 km offshore)	Important	Important
Interwork/ roaming	Interworking and roaming needed for life critical services	Roaming needed as vehicles and aircraft travel abroad	Interworking and roaming needed
Security	Critical (identity, authentication, data integrity, privacy)	Critical (authentication, data integrity, privacy)	Critical (authentication, data integrity, privacy)

Use Case Attribute	Real-Time Video for Monitoring and Guidance	Massive Connectivity for Non-Time-Critical Sensing	Massive Connectivity for Time-Critical Sensing and Feedback
Positioning	Not critical, 1–10 m	Not critical for applications where device location would be known (e.g., fixed devices)	Not critical for applications where device location would be known (e.g., fixed devices)

5.7 Key Terms and Review Questions

Key Terms

actuator	low-power wide area (LPWA)
broadband IoT	massive IoT (MIoT)
cloud	massive machine type communications (mMTC)
constrained devices	microcontroller
core	narrowband IoT (NB-IoT)
critical IoT	precision agriculture
deeply embedded device	Precision Crop Management
edge	radio-frequency identification (RFID)
enhanced machine type communication (eMTC)	sensor
fog	smart agriculture
guard band	smart city
inband	smart farming
industrial automation IoT	smart sustainable city
Internet of Things (IoT)	tag
IoF2020	transceiver
key performance indicator (KPI)	unlicensed frequency band
licensed frequency band	

Review Questions

1. What are the mMTC performance requirements listed in M.2410?
2. Define *Internet of Things*.
3. List and briefly define the principal components of an IoT-enabled thing.
4. Describe the three classes of constrained devices.
5. Explain the relationship between edge, fog, core, and cloud.
6. What is the difference between mMTC and IoT?
7. What is the difference between NB-IoT and eMTC?
8. What is the relationship between mMTC, NB-IoT, eMTC, and LPWA?
9. Provide an overview of the five use case categories defined by IoF2020.
10. List and briefly describe the layers of a business process view model.
11. What are the chief aspects of sustainability for a smart city?
12. What are the key components that might be expected in a smart city integrated command and control center?
13. List and briefly describe the layers of the ICT architecture for smart cities defined by ITU-T.

5.8 References and Documents

References

BEEC18 Beecham Research. *M2M World of Connected Service: The Internet of Things.* 2018. http://www.beechamresearch.com/article.aspx?id=4

ERIC19 Ericsson. *Cellular IoT Evolution for Industry Digitalization.* Ericsson white paper. January 2019.

ERIC20a Ericsson. *Ericsson Mobility Report.* November 2020. https://www.ericsson.com/en/mobility-report

ERIC20b Ericsson. *Cellular Networks for Massive IoT.* Ericsson white paper. January 2020.

FAO17 Food and Agricultural Organization of the United Nations. "The Possibilities of Internet of Things (IoT) for Agriculture." *FAO News*, December 15, 2017. http://www.fao.org/e-agriculture/news/possibilities-internet-things-iot-agriculture

FUJI16 Fujitsu. "Protecting Residents from Serious Flood Damage in Indonesia—River Water-Level Measurement System Using Smartphones and AR Technology." *Fujitsu Journal*, May 16, 2016. https://journal.jp.fujitsu.com/en/2016/05/16/01/

FUJI18 Fujitsu. "Estimating the Degradation State of Old Bridges—Fujitsu Supports Ever-Increasing Bridge Inspection Tasks with AI Technology." *Fujitsu Journal*, March 1, 2018. https://journal.jp.fujitsu.com/en/2018/03/01/01/

MCEW13 McEwen, A., and Cassimally, H. *Designing the Internet of Things*. New York: Wiley, 2013.

NEC15 NEC Corporation. *NEC's Comprehensive Disaster Control System: Toshima City.* 2015. https://www.nec.com/en/case/toshima/index.html

OJHA15 Ojha, T., Misra, S., and Raghuwanshi, N. "Wireless Sensor Networks for Agriculture: The State-of-the-Art in Practice and Future Challenges." *Computers and Electronics in Agriculture*, October 2015.

SANC18 Sanchez-Iborra, R., et al. "Performance Evaluation of LoRa Considering Scenario Conditions." *Sensors*, March 2018.

SEGH12 Seghal, A., et al. "Management of Resource Constrained Devices in the Internet of Things." *IEEE Communications Magazine*, December 2012.

TIME20 Timeus, K., Vinaixa, J., and Pardo-Bosch, F. "Creating Business Models for Smart Cities: A Practical Framework." *Public Management Review*, February 2020.

VERO19 Verouw, C., et al. "Architecture Framework of IoT-Based Food and Farm Systems: A Multiple Case Study." *Computers and Electronics in Agriculture*, October 2019.

Documents

IEC 62264 *Enterprise-Control System Integration, Part 1: Models and Terminology Reference Model.* May 2013.

IoF2020 *Internet of Food & Farm.* IoF2020 booklet. 2020.

IoF2020 *The IoF2020 Use Case Architectures and Overview of the Related IoT Systems. Deliverable D3.2.* September 29, 2017.

IoF2020 *KPI Catalogue for Each Use Case. Deliverable D4.1.* September 22, 2017.

ITU-R Report M.2083 *IMT Vision: Framework and Overall Objectives of the Future Development of IMT for 2020 and Beyond.* September 2015.

ITU-R Report M.2410 *Minimum Requirements Related to Technical Performance for IMT-2020 Radio Interface(s).* November 2017.

ITU-R Report M.2412 *Guidelines for Evaluation of Radio Interface Technologies for IMT-2020.* October 2017.

ITU-T Recommendation Y.2060 *Overview of the Internet of Things*. June 2012.

ITU-T Recommendation Y.4900 *Overview of Key Performance Indicators in Smart Sustainable Cities*. June 2016.

ITU-T Series Y Supplement 27 *Smart Sustainable Cities: Setting the Framework for an ICT Architecture*. January 2016.

ITU-T Series Y Supplement 56 *Supplement on Use Cases of Smart Cities and Communities.* December 2019.

NGMN *5G Prospects: Key Capabilities to Unlock Digital Opportunities*. July 2016.

NGNM *Perspectives on Vertical Industries and Implications for 5G*. September 2016.

RFC 7228 *Terminology for Constrained-Node Networks.* May 2014.

Chapter 6

Ultra-Reliable and Low-Latency Communications

Learning Objectives

After studying this chapter, you should be able to:

- Summarize the performance requirements for URLLC
- Present an overview of URLLC use cases in emerging mission-critical applications
- Present an overview of URLLC applications based on performance requirements
- Understand the need for eMBB, mMTC, and URLLC for Industry 4.0
- Present an overview of the unmanned aircraft system traffic management use case

ITU-R M.2083 (*IMT Vision: Framework and overall objectives of the future development of IMT for 2020 and beyond.* September 2015) characterizes ultra-reliable and low-latency communications (URLCC) as follows:

> Ultra-reliable and low-latency communications: This use case has stringent requirements for capabilities such as throughput, latency, and availability. Some examples include wireless control of industrial manufacturing or production processes, remote medical surgery, distribution automation in a smart grid, transportation safety, etc.

A white paper from the Next Generation Mobile Networks (NGMN) Alliance (*5G Prospects: Key Capabilities to Unlock Digital Opportunities*, July 2016) lists the following as the key drivers of URLCC:

- **Technology:** New technology concepts—such as new radio access technology (RAT), where Layer 1 numerology[1] is expected to scale toward shorter transmit time interval (TTI)[2] to match the wider bandwidths available in the cmW/mmW range, and multi-access edge computing (MEC)[3], as well as advances in device processing power—will enable considerably lower transmission latency. This is creating new value propositions in the URLLC direction.
- **Vertical industry applications:** The automotive industry is undergoing transformation with the advent of autonomous and cooperative driving. Factories are transforming to improve production efficiency through use of cooperative machines that are highly configurable to adapt to quickly changing and individual customer demands. URLLC is a critical component to enable these transformations.

Section 6.1 summarizes the performance requirements for URLLC. Sections 6.2 and 6.3 present two different ways of categorizing URLLC use case: in terms of emerging mission-critical applications and based on performance requirements. Sections 6.4 and 6.5 look in some detail at two important applications of URLLC: Industry 4.0 and unmanned aircraft system traffic management.

6.1 URLLC Performance Requirements

Figure 6.1, based on a figure in ITU-R M.2083, indicates in relative terms the importance of eight key performance characteristics for URLLC. Two parameters are listed as being of high importance: latency and mobility. Reliability, though not mentioned in M.2083, is another vital parameter for URLCC and is included in ITU-R Report M.2410 (*Minimum requirements related to technical performance for IMT-2020 radio interface(s)*, November 2017). This section examines URLLC requirements in the areas of latency, mobility, and reliability. The section concludes by introducing the URLLC requirements defined by NGMN.

1. The term **numerology** refers to the spacing between subcarriers in OFDM (orthogonal frequency-division multiplexing). Chapter 13, "Air Interface Physical Layer," covers OFDM.
2. TTI, which stands for **transmission time interval** or **transmit time interval**, relates to encapsulation of data from higher layers into frames for transmission on the radio link layer. TTI refers to the duration of a transmission on the radio link. The TTI is related to the size of the data blocks passed from the higher network layers to the radio link layer.
3. **Multi-access edge computing (MEC)** refers to harvesting the vast amount of idle computation power and storage space distributed at the network edges to yield sufficient capacities for performing computation-intensive and latency-critical tasks near mobile devices. MEC was originally referred to as **mobile edge computing**, but the term multi-access edge computing is preferred because it emphasizes that some UE may connect through a means other than a cellular radio access network. Chapter 10 covers MEC.

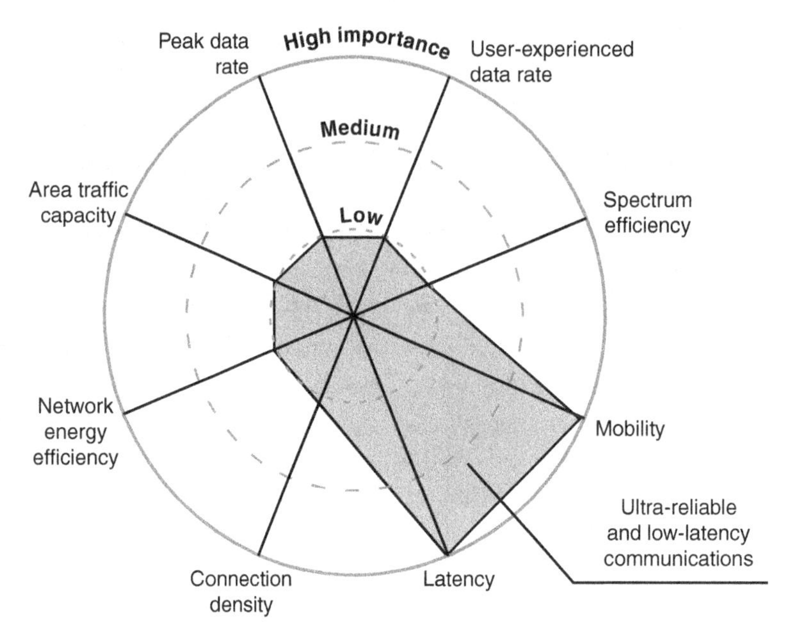

FIGURE 6.1 The Relative Importance of Key Capabilities in the URLLC Usage Scenario

Latency

ITU-R Report M.2410 breaks the latency requirement into two parts:

- **User plane latency:** This is the contribution by the radio network to the time from when the source sends a packet to when the destination receives it (in ms). It is defined as the one-way time it takes to successfully deliver an application layer packet/message from the radio protocol Layer 2/3 service data unit (SDU)[4] ingress point to the radio protocol Layer 2/3 SDU egress point of the radio interface in either uplink or downlink in the network for a given service in unloaded conditions, assuming that the mobile station is in the active state. The minimum requirement (i.e., the maximum allowable value) is 1 ms, assuming unloaded conditions (i.e., a single user) for small IP packets (e.g., 0 byte payload + IP header), for both downlink and uplink.
- **Control plane latency:** This refers to the transition time from a most battery-efficient state (e.g., Idle state) to the start of continuous data transfer (e.g., Active state). The minimum requirement is 20 ms.

4. In a packet, the SDU is data that the protocol transfers between peer protocol entities on behalf of the users of that layer's services. For lower layers, the layer's users are peer protocol entities at a higher layer; for the application layer, the users are application entities outside the scope of the protocol layer model.

User plane latency, however, is only one component that a UE experiences, as illustrated in Figure 6.2. The end-to-end (E2E) latency is generally defined as the time it takes from when a data packet is sent from the transmitting end to when it is received at the receiving entity (e.g., Internet server or other device). The measurement reference is the interface between Layers 2 and 3. It is also referred to as one-trip time (OTT). It includes the user plane latency in one direction, transport network delays, and application processing time. A related measure is round-trip time (RTT), which is the time from when a data packet is sent from a source device until an acknowledgement or response is received from the destination device. Unfortunately, E2E latency is sometimes equated to RTT latency in the literature, even in some 3GPP documents. However, the implication in most standards and specification documents is that E2E latency refers to one-way latency, not round-trip latency.

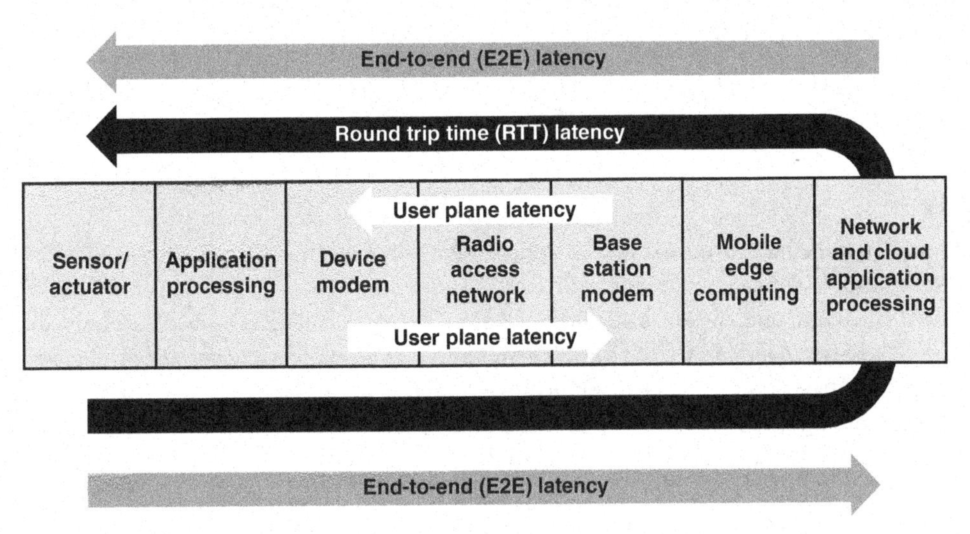

FIGURE 6.2 End-to-End Latency and Round-Trip Time Latency

To reduce the other components of E2E latency besides user plane latency, carriers are moving increasingly to an MEC strategy. Chapter 10, "Multi-Access Edge Computing," explores MEC in detail.

Mobility

Mobility is the maximum UE speed (in km/h) at which a defined quality of service (QoS) can be achieved. Mobility assumes that a seamless transfer between radio nodes—which may belong to different layers and/or radio access technologies (multilayer/multi-RAT)—can be achieved without dropping QoS below a defined threshold. The following classes of mobility are defined:

- **Stationary:** 0 km/h
- **Pedestrian:** 0–10 km/h

- **Vehicular:** 10–120 km/h
- **High-speed vehicular:** 120–500 km/h

M.2410 does not provide a specific measure of QoS. Report ITU-R M.2412 (*Guidelines for Evaluation of Radio Interface Technologies for IMT-2020*, October 2017) defines QoS as successful delivery of 99% of messages within 10 s.

Another aspect of mobility addressed in M.2410 is **mobility interruption time**, which is the shortest time duration supported by the system during which a UE cannot exchange user plane packets with any base station during transitions. This includes the time required to execute any radio access network procedure, radio resource control signaling protocol, or other message exchanges between the mobile station and the radio access network.

The minimum requirement for mobility interruption time is 0 ms. Thus, there should be no interruption of service when a moving UE switches from one base station to another.

Reliability

Reliability, though not mentioned in M.2083, is another vital parameter for URLCC and is included in M.2410. **Reliability** is defined as the probability of successful transmission of a Layer 2/3 packet within a required maximum time, which is the time it takes to deliver a small data packet from the radio protocol Layer 2/3 service data unit (SDU) ingress point to the radio protocol Layer 2/3 SDU egress point of the radio interface at a certain channel quality. The minimum requirement is $1–10^{-5}$ success probability of transmitting a Layer 2 protocol data unit (PDU) of 32 bytes within 1 ms for the urban macro-URLLC test environment.

NGMN Definitions

The 2016 NGMN white paper breaks the URLLC requirements into three broad use case families:

- Ultra-low latency
- Ultra-high reliability and ultra-low latency
- Ultra-high availability and reliability

Table 6.1 shows the performance requirements for these three families.

TABLE 6.1 Performance Requirements for Different Varieties of URLLC

Key Performance Indicator (KPI)	Ultra-Low Latency	Ultra-High Reliability and Ultra-Low Latency	Ultra-High Availability and Reliability
User-experienced data rate	DL: 50 Mbps UL: 25 Mbps	DL: 50 kbps–10 Mbps UL: a few bps–10 Mbps	DL: 10 Mbps UL: 10 Mbps
E2E latency	< 1 ms	1 ms	10 ms

Key Performance Indicator (KPI)	Ultra-Low Latency	Ultra-High Reliability and Ultra-Low Latency	Ultra-High Availability and Reliability
Mobility	Pedestrian	On demand, 0-500 km/h	On demand, 0–500 km/h
Device autonomy	> 3 days	Not critical	> 3 days (standard) Up to several years for some critical MTC services
Connection density	Not critical	Not critical	Not critical
Traffic density	Potentially high	Potentially high	Potentially high

6.2 URLLC Use Cases in Emerging Mission-Critical Applications

A URLLC white paper from 5G Americas (*New Services & Applications with 5G Ultra-Reliable Low-Latency Communications*, November 2018) provides a useful way of understanding the wide variety of URLLC use cases, by focusing on emerging mission-critical applications that have demanding reliability and latency requirements. These are described in this section.

Industrial Automation

The area that has perhaps received the most attention as an application area that requires URLLC support is the smart factory or industrial automation. This application area is typified by extremely demanding KPIs for 5G communication links between sensors, actuators, and controllers. Section 6.4 examines this area in more detail.

Ground Vehicles, Drones, and Robots

This application area refers to remotely controlled mobile devices and robots. Such devices are in common use in factory applications and are also deployed in other contexts, such as smart agriculture. One area of particular interest is unmanned aircraft traffic management; this topic is examined in Section 6.5.

Tactile Interaction

Tactile interaction refers to a level of responsiveness that works at a human scale. For example, remote healthcare or gaming applications may require very low round-trip times to convince human senses that the perceived touch, sight, and sound are lifelike.

These use cases involve interaction between humans and systems, where humans wirelessly control real and virtual objects, and the interaction requires a tactile control signal with audio or visual feedback. Robotic controls and interaction include several scenarios with many applications in manufacturing, remote medical care, and autonomous cars. The tactile interaction requires real-time reactions on the order of a few milliseconds. Remote surgery, discussed later in this chapter, is perhaps the most demanding use case.

Table 6.2 gives typical values of KPIs for tactile Internet applications.

TABLE 6.2 Key Performance Indicators for Tactile Internet Applications

KPI	Value
Traffic volume density	0.03–1 Mbps/m^2 (cell radius 100 m^2)
Experienced user throughput	UL: 0.3–1 Mbps
Latency	User plane latency less than 2 ms
Availability	> 99.999%
Reliability	> 99.999 % for healthcare or remote driving/manipulation 95% for remote gaming or remote augmented reality

Figure 6.3, from an *ITU Technology Watch Report* [ITU14], illustrates a typical latency budget.

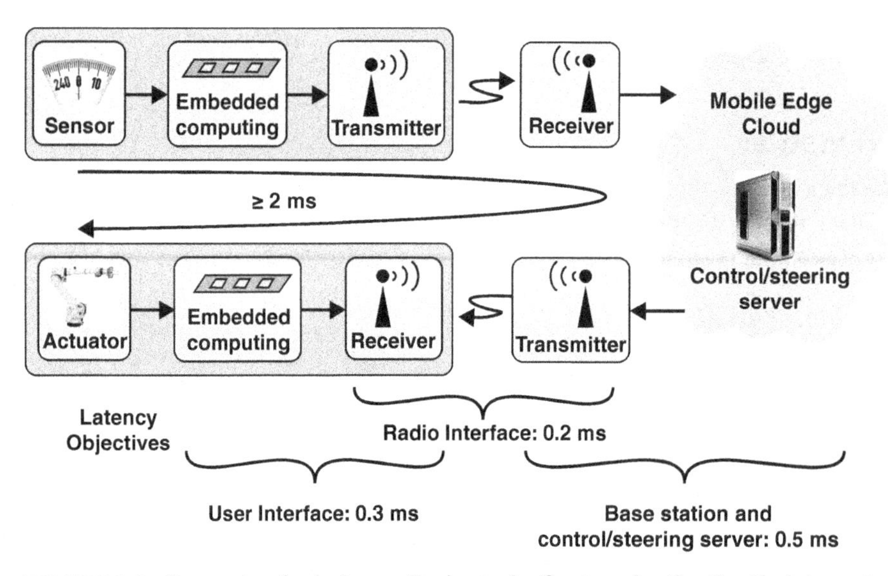

FIGURE 6.3 Example of a Latency Budget of a System for the Tactile Internet

Augmented Reality and Virtual Reality

Augmented reality (AR) and virtual reality (VR) tend to have relatively high data rate requirements. Some specific use cases also have URLLC requirements. An NGMN paper (*Verticals URLLC Use Cases and Requirements*, July 2019) lists three AR/VR examples with URLLC requirements: augmented worker, 360 panoramic VR view video broadcasting, and AR and MR cloud gaming.

Augmented Worker

Augmented work is work that integrates digital technologies into the industrial environment to improve how work is done. Augmented work is appropriate for situations when it is not cost-effective or even possible to fully automate tasks, but it is desirable to augment the capabilities of the human worker. A good example is a task such as equipment repair where access is difficult (e.g., a hazardous environment) or where the expert is physically distant. In such a case, a remote worker can be equipped with an AR headset and some sort of tactile interface for remote control. Sensor information from the remote target location in the form of audio, video, and haptic (tactile) feedback enables the remote operator to control actuators at the target location to achieve the required work. Figure 6.4, from the NGMN paper *Verticals URLLC Use Cases*, illustrates this arrangement.

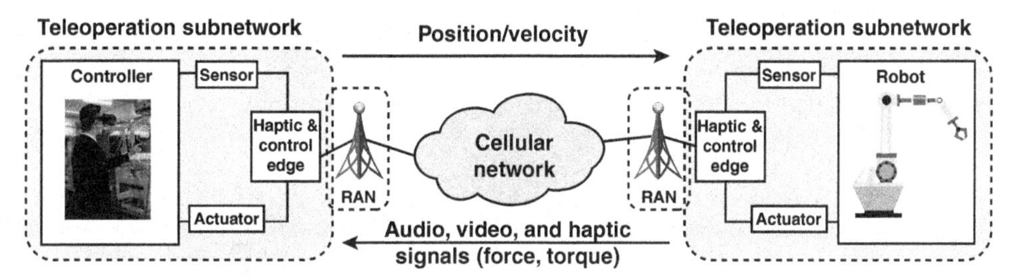

FIGURE 6.4 Augmented Worker

The NGMN document lists the following communications service requirements for this use case:

- **End-to-end latency:** 10 ms
- **End-to-end reliability:** 99.9999%
- **Positioning:** Indoor positioning service with horizontal positioning accuracy better than 1 m, 99% availability, heading < 10 degrees, and latency for positioning estimation < 15 ms for moving UE with speed up to 10 km/h
- **Other requirements:** Application-level requirements:
  - The AR application should have easy access to different context information (e.g., information about the environment, production machinery, the current link state).
  - The (bidirectional) video stream between the AR device and the image processing server should be encrypted and authenticated by the 5G system.
  - Real-time data processing is required.

5G network architecture requirements:

- There is no need for dynamic scalability.
- Mobility at standard values is needed.

- Connectivity will occur frequently.
- Introducing edge computing would be desirable.
- An accurate security mechanism is required.
- A specific and dedicated network slice may be required.

360-Degree Panoramic VR View Video Broadcasting

360-degree videos are video recordings where a view in every direction is recorded at the same time, shot using an omnidirectional camera or a collection of cameras. With 360-degree panoramic VR view video broadcasting, the video is broadcast in real time. Remote users with VR headsets can view the live video feed and, by turning their heads, see the point of view change in real time.

Figure 6.5 illustrates the remote viewing of a stadium via a 5G network. The RTT latency requirement (usually referred to as motion-to-photon [MTP] latency) in VR is commonly targeted at 20 ms [GSMA19].

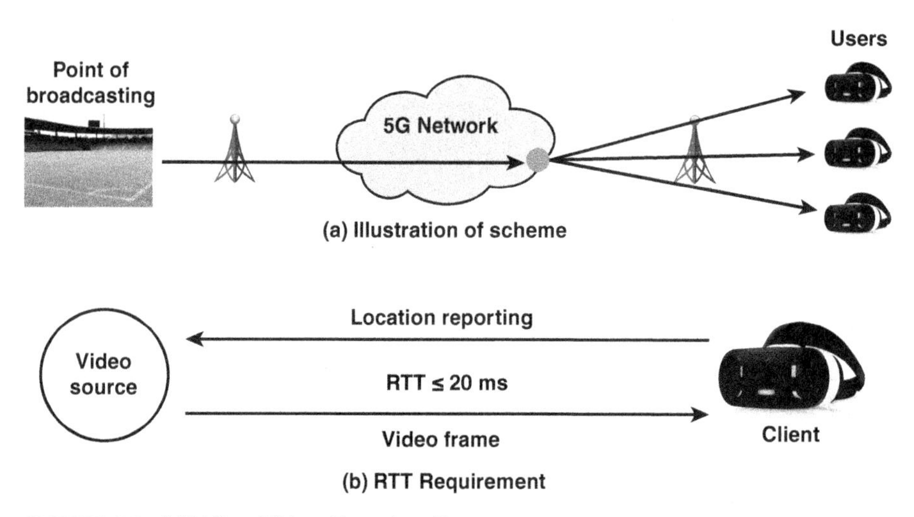

FIGURE 6.5 VR View Video Broadcasting

It is worth expanding on the concept of motion-to-photon latency, as it is a good example of real-world requirements on 5G systems. This latency is the delay between the movement of the user's head and the change of the VR device's display reflecting the user's movement. As the user's head moves, the VR scenery should change to match the movement. Low latency in this case is important for two reasons:

- Low motion-to-photon latency is necessary for user to feel as if they are in the simulated world. As soon as a user's head moves, the VR scenery should match the movement. The more delay (latency) between these two actions, the more unrealistic the VR world seems.

- High motion-to-photon latency leads to a poor virtual reality experience—and possibly motion sickness and nausea. When a user makes a movement wearing the VR headset, the mind expects the screen to be updated correctly to reflect that action. When the screen lags behind the user movement, the user can experience disorientation and motion sickness.

The NGMN document lists the following communications service requirements for this use case:

- **Round trip time:** < 20 ms
- **End-to-end reliability:** 99.999%
- **User-experienced throughput:** 40 Mbps (4K video) to 5 Gbps (12K video, which is equivalent to 4K in TV)
- **Other requirements:**
  - Indoor coverage as the audience may stay at an indoor site
  - Need to support multiple devices (3000/km^2) within an area
  - Support for at least 3 degrees of freedom (3DoF) with 60 frames per second and 6 degrees of freedom (6DoF) with 75 fps

AR and MR Cloud Gaming

A good example of an application in the AR/VR area that requires URLLC performance is AR and mixed reality (MR) cloud gaming, which is real-time game played using a thin client with the bulk of the software on edge servers, as shown in Figure 6.6. This type of online gaming service provides on-demand streaming of games onto computers, consoles, and mobile devices. Thus, the user does not have to upgrade frequently or deal with compatibility issues. Highly interactive games with tight QoS requirements generate the need for low-latency network performance.

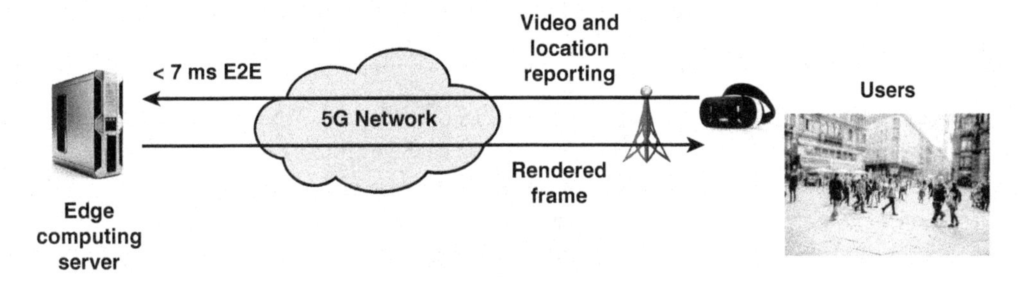

FIGURE 6.6 Cloud Gaming with Edge Computing

The NGMN document lists the following communications service requirements for this use case:

- **Uplink end-to-end latency** < 7 ms
- **End-to-end reliability:** 99.999%

- **User-experienced throughput:** 1 Gbps (uplink and downlink)
- **Need to support multiple devices within an area:** 3000/km^2

Emergencies, Disasters, and Public Safety

Use cases in this category generally require high reliability to enable response to natural disasters and emergencies. Accurate position location and very low latency to enable rapid response are also often critical requirements.

Urgent Healthcare

This category includes applications involving remote diagnosis and treatment. A white paper from 5G Americas (*5G Services & Use Cases*, November 2017) lists the following examples in this category:

- **Remote patient monitoring:** This use case involves remote patient monitoring via communication with devices that measure certain health indicators, such as pulse, blood glucose, blood pressure, and temperature. On an individual basis, the data rate and latency requirements are modest. However, for this use case to become pervasive, 5G is needed to support the massive increase in the number of connections per square meter while still maintaining the requisite QoS.
- **Remote healthcare:** This use case provides for individualized consultation, treatment, and patient monitoring built on a video linkup capability. The video conferencing can be augmented with remote transfer of health-related data in real time. Treatment could also be offered using smart pharmaceutical devices that correctly administer approved dosages of a drug on a schedule specified by the physician or practitioner.
- **Remote surgery:** A demanding use case is remote surgery via control of robotic devices. This application area may be appropriate in ambulances, disaster sites, and remote areas. Important requirements are precise control and very low latency, very high reliability, and tight security.

Figure 6.7, from Deliverable D1.5 of the EU project METIS (Mobile and Wireless Communications Enablers for the Twenty-Twenty Information Society), illustrates the overall interactions of an eHealth system. Cloud-based medical records support both patients and hospital personnel. Wearable IoT devices provide medical telemetry over 5G access to the cloud. Offsite medical personnel access the cloud-based records and software as well as link to the hospital for remote healthcare and surgery. Perhaps the most demanding part of the system, from the 5G point of view, is support for patients being transported by ambulance or even helicopter. High data rates and low latency in this mobility environment call for the type of capability that 5G can provide.

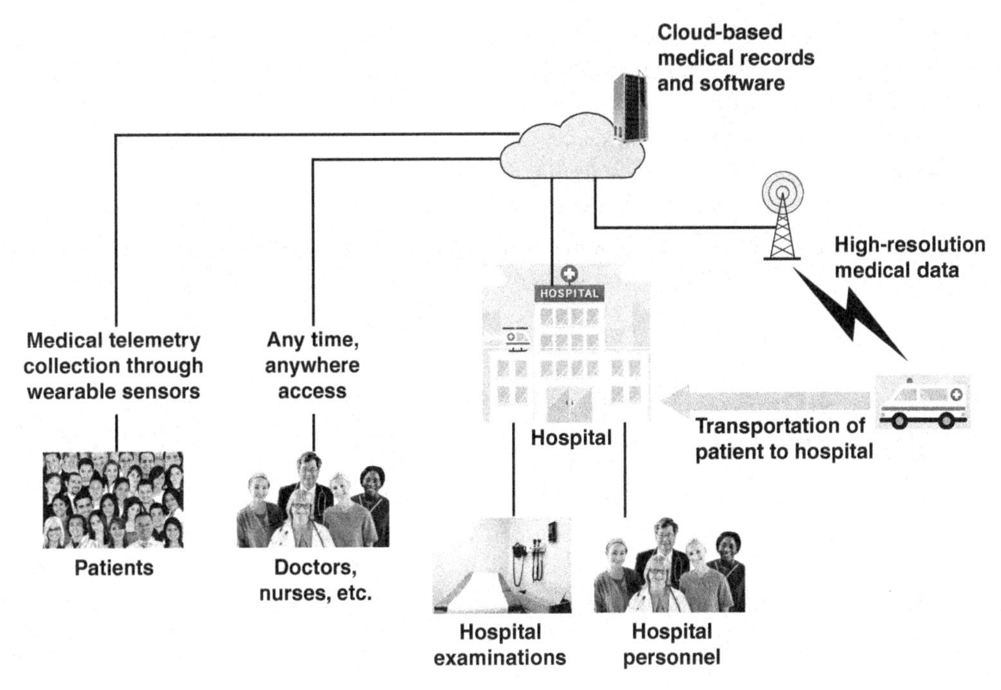

FIGURE 6.7 Use Case Illustration of eHealth

Intelligent Transportation

The NGMN white paper lists the following examples in the intelligent transportation category:

- **Assisted driving:** 5G enables the delivery of advanced driver assistance features that reduce fatal accidents and traffic congestion. These features include real-time maps for navigation, speed warnings, road hazards, vulnerabilities, heads-up display systems, and sensor data sharing. These features enable a vehicle to dynamically change course on the road under certain scenarios and conditions. Vehicle-to-network (V2N) communication is necessary for this use case. Information from the vehicle enables the remote application to perform short-range modeling and recognition of surrounding objects and vehicles as well as mid- to long-range modeling of the surroundings using information on the latest digital maps, traffic signs, traffic signal locations, road construction, and traffic congestion.
- **Autonomous driving:** Fully autonomous driving involves the capability of a vehicle to sense its environment and navigate without human input under all scenarios and conditions. A 5G network with URLLC capability enables a number of necessary features, including using complex algorithms to distinguish between different cars on the road and to identify appropriate navigation paths, given obstacles and the rules of the road, and the exchange of information in real time between thousands of cars connected in the same area.

- **Tele-operated driving:** This use case refers to the use of remote driver assistance in areas where automatic driving is not possible. Tele-operated driving can provide enhanced safety for disabled people, elderly populations, and those in complex traffic situations. Typical application scenarios include disaster areas and unexpected and difficult terrains for manual driving (e.g., in mining and construction). Tele-operated driving requires the wireless network to support V2N communication of video, sound feed information, and diagnostics from the vehicle, along with environmental information, to the remote driver. The network must support transmission of the control commands from the remote driver to the vehicle to maneuver the vehicle in real time.

6.3 URLLC Applications Based on Performance Requirements

3GPP TR 22.862 (*Technical Specification Group Services and System Aspects, Feasibility Study on New Services and Markets Technology Enablers for Critical Communications*, September 2016) addresses the requirements for supporting critical communications in a variety of areas. This report identifies the following use case families in the area of critical communications:

- Higher reliability and lower latency
- Higher reliability, higher availability, and lower latency
- Very low latency
- Higher accuracy positioning
- Higher availability
- Mission-critical services

This categorization provides a convenient way of organizing use cases that require URLLC support but that have somewhat different requirements in the areas of reliability, availability, and latency. Table 6.3 lists examples of use cases in each use case family, and this section provides a brief introduction to each use case family.

TABLE 6.3 URLLC Applications Based on Performance Characteristics

Use Case Families	Use Cases
Higher reliability and lower latency	■ Industrial factory automation ■ Industrial process automation ■ Mission-critical communications ■ Speech, audio, and video in virtual and augmented reality ■ Local UAV collaboration and connectivity
Higher reliability, higher availability, and lower latency	■ Industrial control ■ High reliability, high availability, and high mobility
Very low latency	■ Tactile Internet

Use Case Families	Use Cases
Higher accuracy positioning	■ Outdoor with high-speed movement ■ Low-speed movement (including indoor and outdoor) ■ Low-altitude unmanned aerial vehicle (UAV) in critical condition ■ Massive Internet of Things (MIoT)
Higher availability	■ Secondary connectivity ■ Disaster and emergency response
Mission-critical services	■ Prioritized communications ■ Isolated communications ■ Protected communications ■ Guaranteed communications ■ Optimized communications ■ Supported communications

Higher Reliability and Lower Latency

The use cases in this family typically have modest data rate requirements but need to have messages transmitted reliably and quickly. Use cases include:

- **Industrial factory automation:** This use case may be characterized by large numbers of sensors and actuators, possibly over a large area, that operate in a closed-loop fashion. Sensor data transmitted to controller applications enables the applications to transmit necessary commands to actuators. Section 6.4 examines this use case.
- **Industrial process automation:** This use case may involve a large number of sensors monitoring an industrial process with actuators to control the process and react, under remote command, to events requiring intervention. The demands are similar to those for industrial factory automation.
- **Mission-critical communications:** These services require preferential handling compared to normal telecommunication services (e.g., in support of police or a fire brigade). Examples of mission-critical services include:
  - Industrial control systems (from sensor to actuator, with very low latency for some applications)
  - Mobile healthcare, remote monitoring, diagnosis, and treatment (with high rates and availability)
  - Real-time control of vehicles, road traffic, and accident prevention (including location, vector, context, low round-trip time [RTT])
  - Wide area monitoring and control systems for smart grids
  - Communication of critical information with preferential handling for public safety scenarios

- Multimedia priority service (MPS), providing priority communications to authorized users for national security and emergency preparedness.

- **Speech, audio, and video in virtual and augmented reality:** AR and VR are discussed in Section 6.2.

- **Local UAV collaboration and connectivity:** Unmanned aerial vehicles (UAVs) can collaborate under the control of a single operator to act as a mobile sensor or actuator network. If one or more of the UAVs are beyond the controller's line of sight, communication is maintained via a 5G network. For effective collaboration, the communication path to the remote UAVs needs to be reliable and experience low latency. Figure 6.8 illustrates the communication paths.

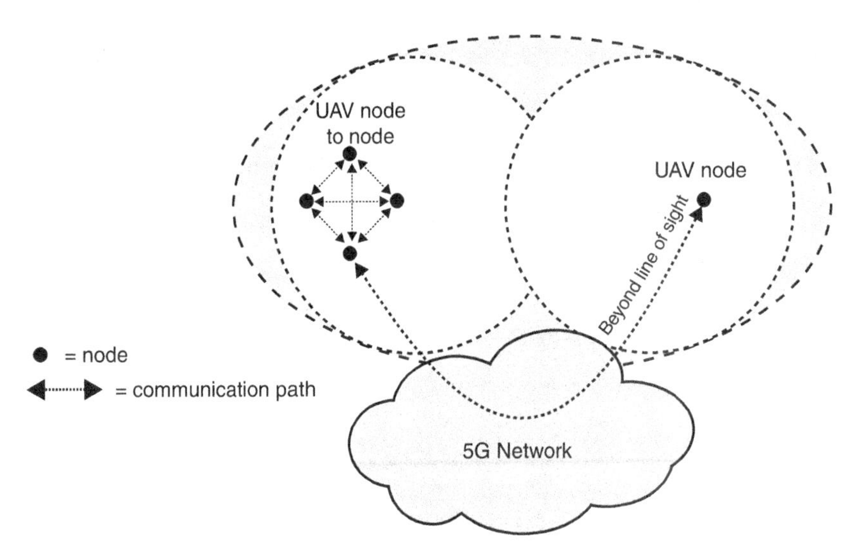

FIGURE 6.8 Depiction of Collaborative Communication Among UAVs

Higher Reliability, Higher Availability, and Lower Latency

In addition to needing high reliability and low latency, the use cases in this family also require that the network and its services be consistently available, with minimum downtime.

One example of a use case in this family is industrial control, which involves integration of hardware, software, and network connectivity to support critical infrastructure. Industrial control systems (ICS) can improve the efficiency of the operation of power plants, water and wastewater plants, transport industries, and other critical infrastructures.

Another general category that falls under this use case family consists of applications that require not only high reliability and availability but also high mobility. A good example in the eHealth area is communication with ambulances. This is discussed in Section 6.2.

Very Low Latency

A prominent example of use cases requiring very low latency is tactile Internet, which is discussed in Section 6.2.

Higher Accuracy Positioning

This use case family is characterized by a high system requirement for positioning accuracy. TR 22.862 lists the following general use cases that fit into this use case family:

- **Outdoor with high-speed movement:** This is a use case of collision avoidance between moving vehicles based on vehicle-to-vehicle exchange of positioning information.
- **Low-speed movement (including indoor and outdoor):** An example in this area is a vehicle feature that enables location of an available parking space in a large parking garage.
- **Low-altitude unmanned aerial vehicle (UAV) in critical condition:** UAVs can be used to monitor emergency situations while first responders are on the way as well as to deliver emergency equipment and supplies.
- **Massive Internet of Things (MIoT):** Some MIoT applications require accurate positioning, such as for monitoring the movement and storage of items in a warehouse.

3GPP TR 22.261 (*Technical Specification Group Services and System Aspects, Service Requirements for the 5G System,* December 2020) lists specific values for positioning KPIs based on service level. 3GPP has designed 5G to have the flexibility to support a range of markets from very high-end markets to very low-end markets. Lower-end markets are characterized by more constraints on power consumption, harsher or more remote locations, and other factors that may reduce the expectations of end users. With this in mind, TR 22.261 provides the following KPI values and ranges for horizontal and vertical positioning:

- **Horizontal accuracy:** 10–0.2 m
- **Vertical accuracy:** 3–0.2 m
- **Positioning service availability:** 85%–99.9%
- **Positioning service latency:** 1 s–10 ms

Higher Availability

The higher availability use case family is concerned with assuring availability of wireless service when the terrestrial network is inadequate, such as during disasters and network outages and when users lose system access. Thus, the 5G network should enable access through satellite links to provide a greater level of service availability.

Another approach to providing higher availability is the use of rapidly deployable networks. All the major cellular providers have the ability to deploy base stations and core network nodes on trucks to provide temporary expansion of coverage.

Mission-Critical Services

With mission-critical services, use cases involve communications requirements that are critical and need a higher priority compared to other communications in the networks, as well as some means of enforcing this priority. Mission-critical communication may need a higher priority over regular traffic because it is used to control devices that have very stringent latency or reliability requirements. Some other types of mission-critical communications may need a higher priority when the network is overloaded; for example, fire brigade personnel may need to have a higher priority than other users at the site of a fire.

TR 22.862 lists the following general categories in this use case family:

- **Prioritized communications:** Some users may need and be authorized to have priority access to network resources very quickly (e.g., emergency communications).
- **Isolated communications:** Networks should be able to isolate prioritized users in terms of resource access to avoid other users negatively impacting higher-priority users.
- **Protected communications:** Networks need to offer different levels of security.
- **Guaranteed communications:** If a network offers different levels of resilience, availability, coverage, and reliability, it is possible to tailor the success rate of the communications to meet the requirements of different users and traffic.
- **Optimized communications:** If the network offers different architectural solutions to deliver services to users, the network can optimize the communications.
- **Supported communications:** When the availability of system resources is limited, either due to very high traffic load or some resources being unavailable, the network must be able to cope with the situation. If the network allows the tailoring of allowed services for different users, the network can operate and still provide users with an acceptable level of service.

6.4 Industry 4.0

Automation has been an essential feature in manufacturing and industrial processes since the beginning of the Industrial Revolution. Advances in automation led to greater efficiency, reliability, and versatility. A host of factors are now leading to a massive transformation, including advances in cellular networks, massive connectivity, cloud computing, big data analytics, and intelligent automation. The following terms are often used in this context:

- **Cyber-physical system (CPS):** A networked, interacting system of digital, analog, physical, and human components. Embedded computers and human operators monitor and control physical processes using feedback loops of sensors and actuators.

- **Industry 4.0:** The fourth generation of the Industrial Revolution, enabled by cyber-physical systems, digitalization, and ubiquitous connectivity provided by 5G and Internet of Things (IoT) technologies.

The term *Industry 4.0* is used to suggest there have been four generations of the Industrial Revolution:

- **First generation:** Mechanization through water and steam power
- **Second generation:** Mass production and assembly lines using electricity
- **Third generation:** Adoption of computers and automation
- **Fourth generation:** Smart and autonomous systems fueled by data and machine learning

Industry 4.0 uses the Internet of Things and cyber-physical systems such as sensors to collect vast amounts of data that can be used by manufacturers and producers to analyze and improve their work.

A white paper from NGMN (*Perspectives on Vertical Industries and Implications for 5G*, September 2016) lists the following as the key drivers of Industry 4.0:

- **Cost-efficiency:** To drive down costs, manufacturers seek to extend the automation of many aspects of the supply chain, apply just-in-time paradigms to the entire value chain, optimize asset utilization, and improve worker productivity. Increasing competitive pressures to drive down operation, maintenance, and downtime costs is also leading the way toward data-aided processes and new paradigms for operation and performance monitoring, predictive maintenance, and asset and inventory management.
- **Personalization and customization:** Manufacturers seek competitive advantage by meeting the demand for personalized and customized products suited to a particular individual and/or environment. In many use cases and contexts, organizations are moving away from traditional paradigms of production that achieve cost-efficiency through economies of scale and mass production. Instead, organizations seek to enable increasing levels of personalization and customization in an economically and environmentally sustainable manner, while addressing safety concerns. Industry 4.0 aims to provide the framework to meet such needs.

Factory Automation Architecture

Industry 4.0 is leading to dramatic changes in factory automation. A white paper from 5G PPP (*5G and the Factories of the Future*, October 2015) classifies innovations in factory automation as follows:

- **Smart factory:** Characterized by extensive use of IoT sensors in manufacturing machines to collect data on their operational status and performance. This extensive sensor network enables automated or human-supervised monitoring for signs that particular parts may fail, enabling preventive maintenance to avoid unplanned downtime on devices. Manufacturers can also analyze trends in the data to try to spot steps in their processes where production slows down or is inefficient in terms of use of materials. IoT networks of actuators can enable more flexible tailored manufacturing processes.

- **Digital factory:** Refers to human-team agile exploitation/analysis of vast amounts of digital information, knowledge management, informed planning, complex simulation, and collaborative product-service engineering support. This includes product life cycle management, modeling, design, and optimization.
- **Virtual factory:** Refers to the use of computers to model, simulate, and optimize the critical operations and entities in a factory plant. The main technologies used in the virtual factory include computer-aided design (CAD), 3D modeling and simulation software, product life cycle management (PLM), virtual reality, high-speed networking, and rapid prototyping. These technologies enable an organization to analyze the manufacturability of a part or product as well as evaluate and validate production processes and machinery and train managers, operators, and technicians on production systems.

Figure 6.9 suggests how the various elements of an automated factory relate to one another in a hierarchical, networked system. The lower portion of the figure depicts the operational technology (OT) domain, which consists of hardware and software that detect or cause a change through the direct monitoring and/or control of physical devices, processes, and events in the enterprise. The upper section of the figure shows the information technology (IT) domain. IT refers to the entire spectrum of technologies for information processing, including software, hardware, communications technologies, and related services. In general, IT does not include embedded technologies.

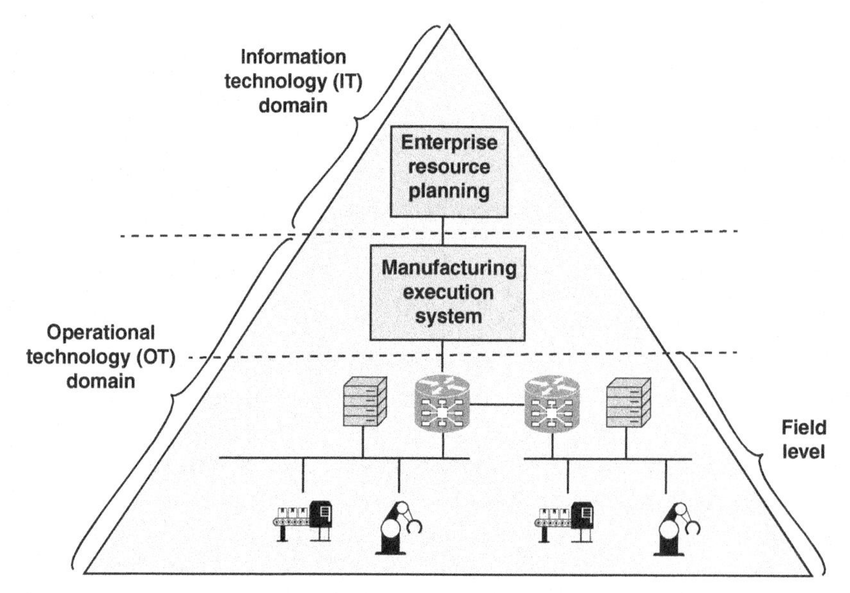

FIGURE 6.9 Hierarchical Network Design Based on the Industrial Automation Pyramid

Figure 6.10, from [SACH19], shows an architecture for an Industry 4.0 automated factory. The architecture employs a local on-premises 5G networking solution. Traditional automated factories rely on wired—typically Ethernet—connectivity. But with thousands or tens of thousands of IoT devices, together with mobile equipment, a high-performance wireless solution is needed. The figure depicts a virtualization of core network (CN) functions, with support for the separation of the control and user plane for flexible CN deployment. The need for local control of a highly dynamic factory dictates local deployment of the control plane. But in some specific applications, a more cost-efficient solution might be to implement some of the control plane functionality from a central location.

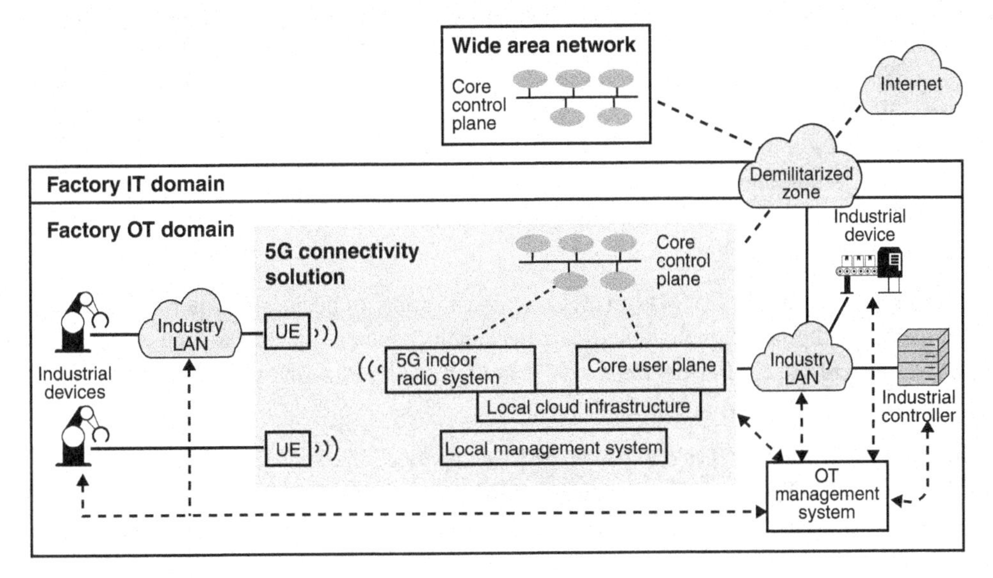

FIGURE 6.10 5G Manufacturing Solution Architecture

Another powerful feature of a 5G solution is the use of network slicing. This allows the provision of dedicated slices both locally and in wide area networks and enhances service differentiation, including isolation of the critical traffic from other service types. It also enables segmentation into security zones, as required for the OT domain.

Application Areas

Factory automation is just one of the application areas encompassed by Industry 4.0. 3GPP TR 22.804 (*Technical Specification Group Services and System Aspects, Study on Communication for Automation in Vertical Domains*, December 2018) lists the following application areas that characterize Industry 4.0:

- **Factory automation:** Factory automation refers to the automated control, monitoring, and optimization of processes and workflows within a factory. This includes aspects like closed-loop control applications (e.g., based on programmable logic or motion controllers), robotics, and aspects of computer-integrated manufacturing. Communication services for factory automation need to fulfill stringent requirements, especially in terms of latency, communication service availability, and determinism. Operation is limited to a relatively small service area, and typically no interaction is required with the public network (e.g., for service continuity and roaming).

- **Process automation:** Process automation refers to the control of production and handling of substances such as chemicals, food, and beverages. Process automation improves the efficiency of production processes, energy consumption, and safety of the facilities. Sensors measuring process values, such as pressures or temperatures, work in a closed loop by means of centralized and decentralized controllers with actuators, such as valves, pumps, and heaters. Process automation also includes monitoring of attributes such as the filling levels of tanks, quality of material, or environmental data, as well as safety warnings or plant shutdowns. A process automation facility may range from a few hundred square meters to many square kilometers or may be distributed over a certain geographic region. Depending on the size, a production plant may have up to tens of thousands of measurement points and actuators. Autarkic (i.e., independent and self-sufficient) device power supply for years is needed in order to stay flexible and to keep the total costs of ownership low.
- **Human–machine interfaces (HMIs):** HMIs include a variety of devices for interaction between people and production facilities. Examples include panels attached to a machine or production line and standard IT devices, such as laptops, tablet PCs, and smartphones. Smart factories increasingly employ AR and VR devices.
- **Production IT:** Production IT encompasses IT-based applications, such as manufacturing execution systems (MESs) as well as enterprise resource planning (ERP) systems. The primary goal of an MES is to monitor and document how raw materials and/or basic components are transformed into finished goods. An ERP system generally provides an integrated and continuously updated view of important business processes. Both types of systems rely on the timely availability of large amounts of data from the production process.
- **Logistics and warehousing:** Logistics and warehousing refers to the organization and control of the flow and storage of materials and goods in the context of industrial production. Intralogistics means logistics on a specific premises (e.g., within a factory), such as ensuring the uninterrupted supply of raw materials on the shop floor level using automated guided vehicles (AGVs), forklifts, and so on. This is in contrast to logistics between different sites, such as for the transport of goods from a supplier to a factory or from a factory to the end customer. Warehousing refers to the storage of materials and goods, such as employing conveyors, cranes, and automated storage and retrieval systems. For practically all logistics use cases, the localization, tracking, and monitoring of assets is of high importance. Communication services for logistics and warehousing need to meet very stringent requirements in terms of latency, communication service availability, and determinism and are limited to a local service area (both indoor and outdoor). Interaction is required with the public network (e.g., for service continuity or roaming).

- **Monitoring and maintenance:** Monitoring and maintenance refers to the monitoring of certain processes and/or assets without an immediate impact on the processes themselves (in contrast to a typical closed-loop control system in factory automation, for example). This particularly includes applications such as condition monitoring and predictive maintenance based on sensor data, but also includes big data analytics for optimizing future parameter sets of a certain process, for instance. For these use cases, the data acquisition process is typically not latency critical, but a large number of sensors may have to be efficiently interconnected, especially since many of these sensors may only be battery driven.

Use Cases

Each of the general application areas listed so far comprises a number of more specific use cases. Figure 6.11, from TR 22.804, maps the application areas to use case categories.

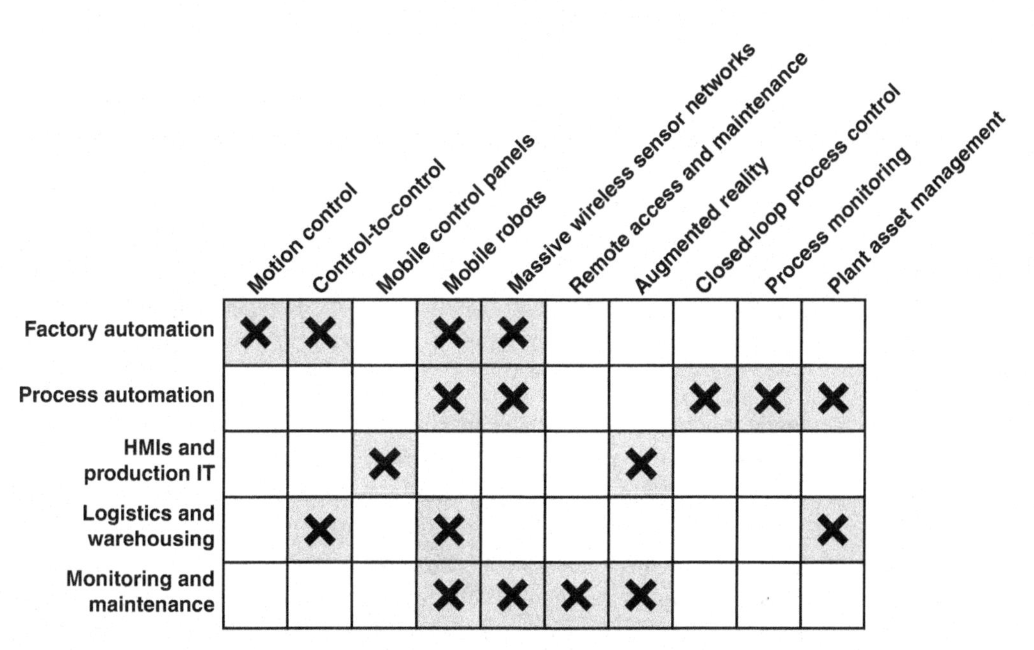

FIGURE 6.11 Areas of Application and Corresponding Use Cases

The use cases include the following:

- **Motion control:** Motion control is a closed-loop feedback mechanism using sensors and actuators to control the movement of a device or piece of equipment. The motion can either be the movement of the entire device or a movement of a component, such as a rotation or clasping/unclasping. A controller sends signals to an actuator, with commands to perform a certain motion. Sensors on the device send back status signals, enabling the controller to continuously adjust the control signals. In general, motion control has the highest requirements in terms of latency and service availability.

- **Control-to-control:** This use case involves communication between two or more controllers of automated equipment. There are two prominent use case categories:
  - One type of use case is a large item of equipment, such as a newspaper printing press, where several controllers are used to control specific functions on the equipment. These controls typically cooperate in real time and have stringent requirements for latency, reliability, and availability.
  - Another type of use case deals with multiple individual machines performing a cooperative task, such as occurs in an assembly line. Again, latency, reliability, and availability requirements are typically stringent.
- **Mobile control panels:** A mobile control panel is a specialized handheld tablet-like device used for visualization, operation, and control of robots, machines, and other industrial devices. Some of these panels include safety-relevant capabilities that enable an operator to quickly halt or modify the action of the controlled device in response to a safety-related event. Due to the criticality of these safety functions, safety control panels currently have mostly a wire-bound connection to the equipment they control. In consequence, there tend to be many such panels for the many machines and production units that typically can be found in a factory. With a URLLC wireless link, it is possible to connect mobile control panels equipped with safety functions wirelessly. This leads to higher usability and allows for flexible and easy reuse of panels for controlling different machines. In general, this use case has very stringent requirements in terms of latency and service availability. The required service area is usually bigger than for motion control.
- **Mobile robots:** A mobile robot is a programmable machine that is able to execute multiple operations, following programmed routes, to perform a variety of tasks. A mobile robot is able to perform activities that assist in work steps and transport of goods, materials, and other objects. Mobile robot systems are characterized by maximum flexibility in mobility relative to the environment, with a certain level of autonomy and perception ability: They can sense and react to their environment. Automated guided vehicles (AGVs) are a subgroup within the mobile robot category. AGVs are driverless vehicles that are steered automatically. They are used to efficiently move goods and materials within a defined area. Mobile robots and AGV systems must frequently interoperate with conveyor assets (e.g., cranes, lifts, conveyors, industrial trucks) and monitoring and control elements (e.g., sensors, actuators). They also need to exchange data for reporting (e.g., inventories, goods movements and throughput, for tracking and monitoring, and for forecasting). Radio-controlled guidance control is necessary to get up-to-date process information to avoid collisions between mobile robots, to assign driving jobs to the mobile robots, and to manage the traffic of mobile robots.
- **Massive wireless sensor networks:** Sensor networks monitor the state or behavior of a particular environment. In the context of a smart factory, wireless sensor networks (WSNs) monitor

processes and equipment, as well as their corresponding parameters. Diverse sensor types—such as microphones, CO_2 sensors, pressure sensors, humidity sensors, and thermometers—provide comprehensive monitoring and coverage. 5G has the potential to take these networks to the next level: massive machine type communications (mMTC) enables massive wireless sensor networks, featuring millions of devices per square kilometer; the size and density of these networks will be far beyond those of today's wireless sensor networks.

- **Remote access and maintenance:** Remote access is the ability to establish contact and communicate with a device from a distant location, and it is often the means for performing remote maintenance. Typical industrial networks are isolated from the Internet and often based on very specific protocols. In Industry 4.0, devices on these industrial networks can be accessed remotely via secure 5G communications, bypassing the Internet.
- **Augmented reality:** In a factory environment, AR can provide factory-floor workers with effective support, such as assistance that allows them to rapidly become familiar with and adept at new tasks and that ensures that they can work in an efficient, productive, and ergonomic manner. Examples of applications appropriate for AR are the following:
  - Monitoring of processes and production flows
  - Delivery of step-by-step instructions for specific tasks, such as for manual assembly
  - Ad hoc support from a remote expert, such as for maintenance or service tasks
- **Closed-loop process control:** For this use case, sensors distributed throughout the production facility continuously measure typical process parameters such as pressure, temperature, flow rate, or pH value. Sensor values are fed back to a controller, which uses these data to determine signals to send to actuators. Examples of actuators are valves, pumps, and heaters/coolers. This use case has very stringent requirements in terms of latency and service availability.
- **Process monitoring:** For this use case, sensors distributed throughout the production facility continuously monitor process or environmental conditions or inventories. Data are transmitted to displays for observation and/or to databases for logging and trend monitoring. The communication service must support high sensor density and provide low latency and high service availability.
- **Plant asset management:** This use case is concerned with maintenance of assets such as pumps, valves, heaters, and instruments. Timely recognition of any degradation and ongoing self-diagnosis are used to support and plan maintenance work. This calls for sensors that provide visibility into process or environmental conditions. Remote software updates modify and enhance components in line with changing conditions and advances in technology.

Performance Requirements

The performance requirements for Industry 4.0 vary widely depending on the application area and use case. As Figure 6.12 indicates, enhanced mobile broadband (eMBB), mMTC, and URLLC 5G capabilities are all required for the internal 5G network used in an industrial environment. For example, eMBB is needed to support high-data-rate use cases, particularly those involving substantial use of video; mMTC supports massive deployments of sensors; and URLLC supports critical use cases such as those involving the use of handheld terminals and control of industrial robots.

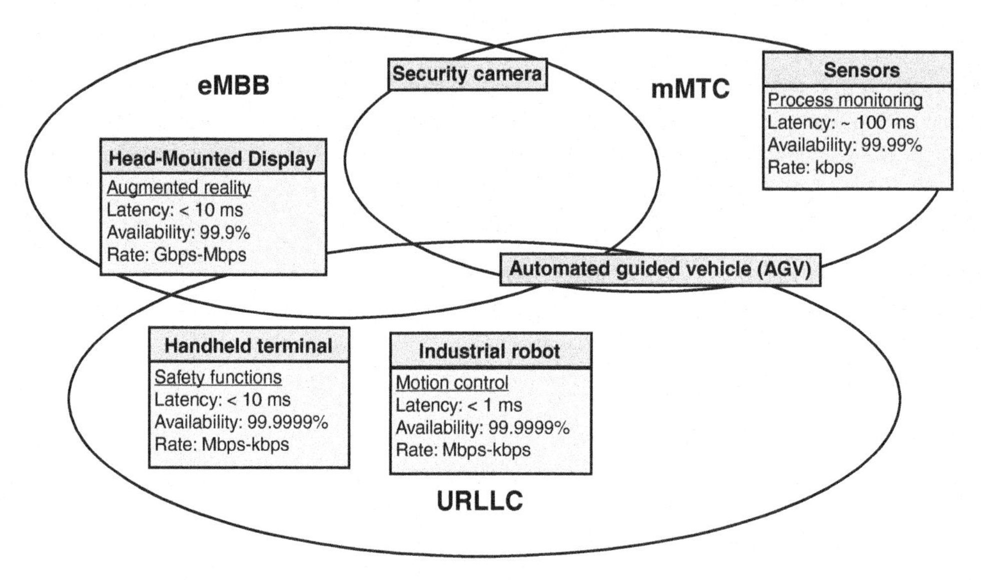

FIGURE 6.12 Wireless Connectivity Requirements

Table 6.4, derived from data in TR 22.804, indicates use cases that require URLLC network support. The term *cycle time* in the table refers to the time allowed for a control system to generate a command, transport it across the network to a sensor or an actuator, and then receive confirmation that the command was successfully delivered. The latency allowable over the network is thus a fraction of the overall time budget. A 2 ms cycle time, for example, may only allow 500 μs for transmission.

TABLE 6.4 Industrial Automation Performance Requirements for 5G

Use Case (High Level)		Availability	Cycle Time	Typical Payload Size	Number of Devices	Typical Service Area
Motion control	Printing machine	> 99.9999%	< 2 ms	20 bytes	> 100	100 m × 100 m × 30 m
	Machine tool	> 99.9999%	< 0.5 ms	50 bytes	~ 20	15 m × 15 m × 3 m
	Packaging machine	> 99.9999%	< 1 ms	40 bytes	~ 50	10 m × 5 m × 3 m

Use Case (High Level)		Availability	Cycle Time	Typical Payload Size	Number of Devices	Typical Service Area
Mobile robots	Cooperative motion control	> 99.9999%	1 ms	40–250 bytes	100	< 1 km^2
	Video-operated remote control	> 99.9999%	10–100 ms	15–150 bytes	100	< 1 km^2
Mobile control panels with safety functions	Assembly robots or milling machines	> 99.9999%	4–8 ms	40–250 bytes	4	10 m × 10 m
	Mobile cranes	> 99.9999%	12 ms	40–250 bytes	2	40 m × 60 m
Process automation		> 99.99%	> 50 ms	Varies	10^4 devices per km^2	

6.5 Unmanned Aircraft System Traffic Management

Air traffic control bodies, such as the U.S. Federal Aviation Administration (FAA) and the European Organisation for the Safety of Air Navigation (EUROCONTROL), make a distinction between the following three terms:

- **Unmanned aerial vehicle (UAV):** An aircraft operated without the possibility of direct human intervention from within or on the aircraft. Equivalent terms are remotely piloted vehicle (RPV), drone, unmanned aircraft (UA), and uncommanded aerial vehicle (UCAV).
- **Autonomous drone:** A type of UAV in which communications management software, instead of a human, coordinates missions and pilots the aircraft.
- **Unmanned aircraft system (UAS):** Refers to the entire system required for UAV operations, consisting of the UAV (including antenna, sensors, software, and power supply), the ground or airborne control system, and communication links and networks.

UASs serve in a variety of use cases, including infrastructure monitoring, precision agriculture, public safety, search and rescue, disaster relief, weather monitoring, and delivery of goods. As these use cases proliferate, they increasingly operate in areas previously used only by general aviation aircraft, helicopters, gliders, balloons, and parachutists, as well as in the airspace around airports and along commercial aircraft routes.

Of particular concern is the operation of UASs in controlled airspace. *Controlled airspace* is an airspace of defined dimensions within which an air traffic control service is provided. The ability to safely incorporate UAS operation in controlled airspace is increasingly difficult due to the total size of the recreational and commercial fleet of small UASs (i.e., those under 55 lb). For example, the FAA projects that in the United States, the combined UAS fleet size will reach over 2.5 million by 2024 [FAA20a]. To meet safety needs, the FAA has been developing the UAS Traffic Management (UTM) system. EUROCONTROL is developing a similar system [EURO18]. The UTM is a good example of a URLLC use case for unmanned aircraft.

UTM Architecture

Figure 6.13, from the [FAA20b], illustrates the overall architecture of the UTM. The architecture identifies, at a high level, the various actors and components, their contextual relationships, and high-level functions and information flows. The vertical dashed line represents the demarcation between FAA and industry responsibilities for the infrastructure, services, and entities that interact as part of UTM.

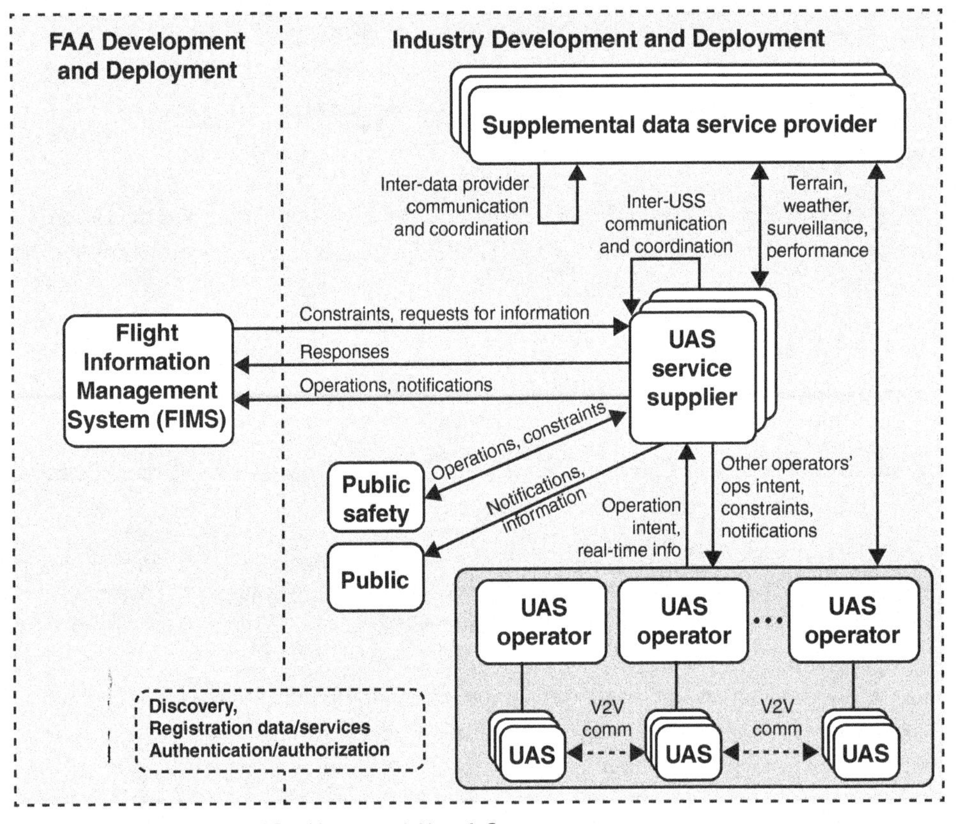

FIGURE 6.13 Unmanned Aircraft System Traffic Management (UTM)

The key elements of the architecture are:

- **Flight Information Management System (FIMS):** FIMS supports data exchange between FAA systems and UTM participants. FIMS enables the exchange of airspace constraint data between the FAA and the ***UAS service supplier*** (USS) network. The FAA also uses this interface as an access point for information on active UTM operations.
- **UAS service supplier (USS):** The role of the USS is to offer support for safe airspace operations. The organizations that provide airspace support may or may not operate UAS themselves. For all intents and purposes, with performance-based airspace operations, these functions are considered separate from the main purpose of UAS operators. USSs share information about their supported operations to promote safety and to ensure that USSs have a consistent view of all UAS operations and thus enable UASs to stay clear of each other.
- **USS network:** This is a collection of USSs connected to each other in a geographic area, exchanging information on behalf of subscribed operators. The USS network shares operational intent data, airspace constraint information, and other relevant details across the network to ensure shared situational awareness for UTM participants.
- **UAS supplemental data service provider (SDSP):** Operators and USSs can access SDSPs for essential or enhanced services, including terrain and obstacle data, specialized weather data, surveillance, and constraint information.
- **Public safety:** Public safety entities, when authorized, can access UTM operations data as a means to ensure safety of the airspace and persons and property on the ground, security of airports and critical infrastructure, and privacy of the general public.
- **Public:** The general public can access data that is determined or required to be publicly available.
- **UAS operator:** This is a human operator of a commercial or recreational UAS.

Table 6.5 lists the key responsibilities of UAS operators, USSs, and the FAA within the UTM architecture.

TABLE 6.5 Allocation of Responsibilities for UTM Actors/Entities

Function		UAS Operator	USS	FAA
Separation	UAS from UAS (VLOS and BVLOS)	✓	S	
	UAS from low-altitude manned aircraft (VLOS and BVLOS)	✓	S	
Hazard/terrain avoidance	Weather avoidance	✓	S	
	Terrain avoidance	✓	S	
	Obstacle avoidance	✓	S	

Function		UAS Operator	USS	FAA
Status	UTM operations status	S	✓	
	Flight information archive	✓	S	
	Flight information status	✓	S	
Advisories	Weather information	✓	S	
	Alerts to affected airspace users of UAS hazard	✓	S	
	Hazard information (e.g., obstacles, terrain)	✓	S	
	UAS-specific hazard information (e.g., powerlines, no-UAS zones)	✓	S	
Planning, intent, and authorization	Operation plan development	✓	S	
	Operation intent sharing (pre-flight)	✓	S	
	Operation intent sharing (in-flight)	✓	S	
	Operation intent negotiation	✓	S	
	Controlled airspace authorization		S	✓
	Control of flight	✓		
	Airspace allocation and constraints definition		S	✓

✓ = Primary Responsibility
S = Support
VLOS = Visual Line of Sight
BVLOS = Beyond Visual Line of Sight
UAS = Unmanned Aircraft System
USS = UAS service supplier

5G Performance Requirements for UTM

3GPP has published a set of network performance requirements for UTM in TS 22.125 (*Technical Specification Group Services and System Aspects, Unmanned Aerial System (UAS) Support in 3GPP*, September 2020). Figure 6.14, from TS 22.125, illustrates the reference model used for defining requirements.

The reference model has the following characteristics:

- A command and control (C2) communications link delivers messages with C2 information for UAV operation from a UAV controller or a UTM.
- A UAS is composed of one UAV controller and one or more UAVs.
- UAVs can communicate via a 5G cellular network.
- A UAV may be controlled by a UAV controller connected via the 5G network. C2 communications may include video, which is required to control the operation of the UAV. C2 messages

may be communicated with the UAV controller, the UTM, or both and may or may not be periodic. UAV control and UTM communications may happen at essentially the same time with different required QoS levels. Any mission-specific communication (e.g., high-definition (HD) video for area surveillance), if required, is additional.

- A UAV may be controlled by a UAV controller that is not connected via the 5G network, using a C2 interface that is not in 3GPP scope.
- The UAS exchanges application data traffic with a UTM. This includes traffic related to UTM services in support of UAS operations, such as UAS identification and tracking, authorization, enforcement, and regulation of UAS operations.

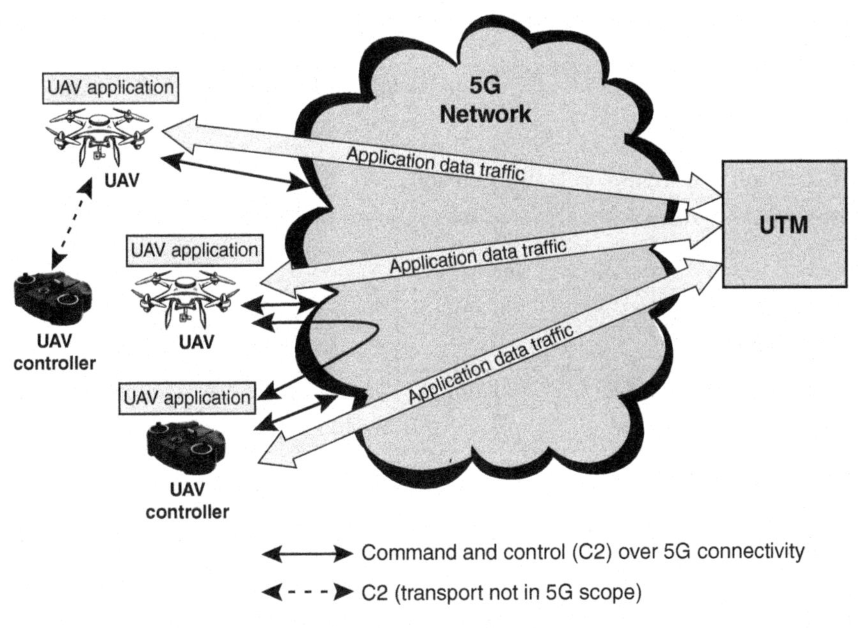

FIGURE 6.14 UAS Model in the 5G Ecosystem

Table 6.6 shows the KPIs defined in TS 22.125 for services provided to the UAV applications. The latency requirements are not as stringent as those found for many other URLLC use cases, but the data rate requirements are substantial.

TABLE 6.6 KPIs for Services Provided to UAV Applications*

Service	Data Rate	E2E Latency
8K video live broadcast	100 Mbps UAV originated	200 ms
	800 kbps UAV terminated	20 ms
Laser mapping, HD video	120 Mbps UAV originated	200 ms
	300 kbps UAV terminated	20 ms

Service	Data Rate	E2E Latency
4×4K AI surveillance	120 Mbps UAV originated	20 ms
	50 Mbps UAV terminated	20 ms
Remote UAV controller via HD video	≥ 25 Mbps UAV originated	100 ms
	300 kbps UAV terminated	20 ms
Real-time video	60 kbps without video UAV originated	100 ms
Video streaming	4 Mbps for 720p video UAV originated 9 Mbps for 1080p video UAV originated	100 ms
Periodic still photos	1 Mbps UAV originated	1 ms

* For 720p video, 720 stands for 720 horizontal scan lines of image display resolution (also known as 720 pixels of vertical resolution). The *p* stands for progressive scan (i.e. non-interlaced). 1080p has a similar meaning.

Table 6.7 shows the KPIs defined in TS 22.125 for C2 communications. The table lists different modes of control and their typical KPIs.

TABLE 6.7 KPIs for Command and Control of UAV Operation

Control Mode	C2 Message	Typical Message Interval	Max UAV Ground Speed	Typical Message Size	E2E Latency	Reliability
Steer to waypoints	UAV terminated	≥ 1 s	300 km/h	100 B	1 s	99.9%
	UAV originated	1 s		84–140 B	1 s	99.9%
Direct stick steering	UAV terminated	40 ms	60 km/h	24 B	40 ms	99.9%
	UAV originated	40 ms		84-140 B	40 ms	99.9%
Automatic flight by UTM	UAV terminated	1 s	300 km/h	≤ 10 kB	5 s	99.9%
	UAV originated	1 s		1.5 kB	5 s	99.9%
Approaching autonomous navigation infrastructure	UAV terminated	500 ms	50 km/h	4 kB	10 s	99%
	UAV originated	500 ms		4 kB	140 ms	99.99%

The modes are:

- **Steer to waypoints:** The control message contains flight waypoints that are sent from the UAV controller or UTM to the UAV. A waypoint is an intermediate point or place on a route or line of travel, a stopping point, or a point at which course is changed. The control mode is used in both direct C2 communication and network-assisted C2 communication.
- **Direct stick steering:** The control message contains direction instructions sent from the UAV controller to the UAV while video traffic is optionally provided as feedback from the UAV to the UAV controller. The control mode is used in both direct C2 communication and network-assisted C2 communication.

- **Automatic flight by UTM:** The control message contains a prescheduled flight plan sent from the UTM to the UAV, which thereafter flies autonomously with periodic position reporting. The control mode is used in UTM-navigated C2 communication.
- **Approaching autonomous navigation infrastructure:** The control message contains waypoints, altitudes, and speeds from the UTM to the UAV. When the UAV is landing or taking off, the UTM coordinates more closely with autonomous navigation infrastructure (e.g., vertiport, package distribution center). The control mode is used in UTM-navigated C2 communication.

In general, Table 6.7 shows fairly stringent latency and mobility requirements and high reliability requirements.

Table 6.8 shows the KPIs defined in TS 23.125 for situations in which video feed from the UAV is used for control. The data rate, latency, and reliability requirements are stringent.

TABLE 6.8 KPIs for Video Used to Aid UAV Control (sent by UAV)

Scenario	Data Rate	E2E Latency	Reliability
VLOS	2 Mbps at 480p video, 30 fps	1 s	99.9%
BVLOS	4 Mbps at 720p video, 30 fps	140 ms	99.99%

VLOS = Visual Line of Sight

BVLOS = Beyond Visual Line of Sight

6.6 Key Terms and Review Questions

Key Terms

assisted driving	factory automation
augmented reality (AR)	Industry 4.0
augmented work	latency
autonomous driving	mission-critical services
autonomous drone	mobile edge computing (MEC)
control plane latency	mobility
cyber-physical system (CPS)	mobility interruption time
digital factory	multi-access edge computing (MEC)
end-to-end (E2E) latency	numerology

one-trip time (OTT) process automation reliability round-trip time (RTT) service data unit (SDU) smart factory tele-operated driving transmit time interval (TTI) UAS Traffic Management (UTM)	ultra-reliable and low-latency communications (URLLC) unmanned aerial vehicle (UAV) unmanned aircraft system (UAS) user plane latency virtual factory virtual reality (VR) waypoint

Review Questions

1. What are the URLLC performance requirements listed in M.2410?
2. What is the difference between user plane latency and control plane latency?
3. Explain the four mobility classes.
4. What are some examples of urgent healthcare use cases?
5. Explain the difference between assisted driving, autonomous driving, and tele-operated driving.
6. What are mission-critical services?
7. What are cyber-physical systems?
8. Summarize the characteristics of the four generations of the Industrial Revolution.
9. Explain the difference between smart factory, digital factory, and virtual factory.
10. What is the difference between IT and OT?
11. What is the difference between factory automation and process automation?
12. Explain the difference between a UAV, an autonomous drone, and a UAS.

6.7 References and Documents

References

EURO18 European Organisation for the Safety of Air Navigation. *UAS ATM Integration: Operational.* November 2018.

FAA20a Federal Aviation Administration. *FAA Aerospace Forecast, Fiscal Years 2020–2040.* March 2020.

FAA20b Federal Aviation Administration. *Unmanned Aircraft System (UAS) Traffic Management (UTM) Concept of Operations.* March 2020.

GSMA19 GSM Association. *Cloud AR/VR Whitepaper.* April 2019. https://www.gsma.com/futurenetworks/wiki/cloud-ar-vr-whitepaper

ITU14 ITU. *The Tactile Internet. ITU Technology Watch Report.* August 2014.

SACH19 Sachs, J., et al. "Boosting Smart Manufacturing with 5G Wireless Connectivity." *Ericsson Technology Review*, February 20, 2019.

Documents

3GPP TR 22.804 *Technical Specification Group Services and System Aspects, Study on Communication for Automation in Vertical Domains.* December 2018.

3GPP TR 22.862 *Technical Specification Group Services and System Aspects, Feasibility Study on New Services and Markets Technology Enablers for Critical Communications.* September 2016.

3GPP TR 22.261 *Technical Specification Group Services and System Aspects, Service Requirements for the 5G System.* December 2020.

3GPP TS 22.125 *Technical Specification Group Services and System Aspects, Unmanned Aerial System (UAS) Support in 3GPP.* September 2020.

5G Americas *5G Services & Use Cases.* November 2017.

5G Americas *New Services & Applications with 5G Ultra-Reliable Low-Latency Communications.* November 2018.

5G PPP *5G and the Factories of the Future.* October 2015.

ITU-R M.2083 *IMT Vision: Framework and Overall Objectives of the Future Development of IMT for 2020 and Beyond*. September 2015.

ITU-R Report M.2410 *Minimum Requirements Related to Technical Performance for IMT-2020 Radio Interface(s)*. November 2017.

ITU-R Report M.2412 *Guidelines for Evaluation of Radio Interface Technologies for IMT-2020*. October 2017.

METIS Deliverable D1.5 *Updated Scenarios, Requirements and KPIs for 5G Mobile and Wireless System with Recommendations for Future Investigations*. April 2015.

NGMN *NGMN 5G White Paper*. February 2015.

NGMN *5G Prospects: Key Capabilities to Unlock Digital Opportunities*. July 2016.

NGMN *Perspectives on Vertical Industries and Implications for 5G*. September 2016.

NGMN *Verticals URLLC Use Cases and Requirements*. July 2019.

Chapter 7

Software-Defined Networking

Learning Objectives

After studying this chapter, you should be able to:

- Make a presentation justifying the position that traditional network architectures are inadequate for modern networking needs
- List and explain the key requirements for an SDN architecture
- Present an overview of an SDN architecture
- Present an overview of the functions of the SDN data plane
- Understand the concept of an OpenFlow logical network device
- List and explain the key functions of the SDN control plane
- Explain the purpose of the southbound, northbound, eastbound, and westbound interfaces
- Present an overview of the SDN application plane architecture
- List and describe six major application areas of interest for SDN

The two technology underpinnings for the 5G NextGen core network are software-defined networking (SDN) and network functions virtualization (NFV). This chapter provides a technical background for SDN, and Chapter 8, "Network Functions Virtualization," covers NFV. Both topics are covered in more detail in [STAL16]. Chapter 9, "Core Network Functionality, QoS, and Network Slicing," discusses the application of SDN and NFV in the 5G core network.

SDN is a network organizing technique that has recently come to prominence. In essence, a software-defined network separates the data and control functions of networking devices, such as routers, packet switches, and LAN switches, with a well-defined application programming interface (API)[1] between

1. An API is a language and message format used by an application program to communicate with the operating system or some other control program, such as a database management system (DBMS) or communications protocol. APIs are implemented by writing function calls that provide links to the required subroutines for execution. An open or standardized API can ensure the portability of the application code and the vendor independence of the called service.

the two. In contrast, in most traditional large enterprise networks, routers and other network devices encompass both data and control functions, making it difficult to adjust the network infrastructure and operation to large-scale addition of end systems, virtual machines, and virtual networks.

7.1 Evolving Network Requirements

Evolving network requirements have led to a demand for a flexible response approach to controlling traffic flows within a network or on the Internet.

One key driving factor is the increasingly widespread use of server virtualization. In essence, server virtualization masks server resources, including the number and identity of individual physical servers, processors, and operating systems, from server users. This makes it possible to partition a single machine into multiple, independent servers and conserve hardware resources. It also makes it possible to quickly migrate a server from one machine to another for load balancing or for dynamic switchover in the case of machine failure. Server virtualization has become a central element in dealing with big data applications and in implementing cloud computing infrastructures. But it creates problems with traditional network architectures. One problem is configuring virtual local area networks (VLANs). Network managers need to make sure the VLAN used by a virtual machine (VM) is assigned to the same switch port as the physical server running the VM. But with the VM being movable, it is necessary to reconfigure the VLAN every time a virtual server is moved. In general, to match the flexibility of server virtualization, a network manager needs to be able to dynamically add, drop, and change network resources and profiles. This is difficult to do with conventional network switches, as the control logic for each switch is collocated with the switching logic.

Another effect of server virtualization is that traffic flows differ substantially from the traffic flows in the traditional client/server model. Typically, there is a considerable amount of traffic among virtual servers, for such purposes as maintaining consistent images of databases and invoking security functions such as access control. These server-to-server flows change in location and intensity over time, demanding a flexible approach to managing network resources.

Another factor leading to the need for rapid response in allocating network resources is the increasing use by employees of mobile devices, such as smartphones, tablets, and notebooks, to access enterprise resources. Network managers must be able to respond to rapidly changing resource, quality of service (QoS), and security requirements.

Prior to the introduction of SDN, network infrastructures could respond to changing requirements for the management of traffic flows, providing differentiated QoS levels and security levels for individual flows; but the process was very time consuming if the enterprise network was large and/or involved network devices from multiple vendors. The network manager had to configure each vendor's equipment separately, and adjust performance and security parameters on a per-session, per-application basis. In a large enterprise, every time a new virtual machine was brought up, it could take hours or even days for network managers to do the necessary reconfiguration [ONF12].

As discussed in this chapter, the SDN architecture and the OpenFlow standard provide an open architecture in which control functionality is separated from the network device and placed in accessible control servers. This enables the underlying infrastructure to be abstracted for applications and network services, which can treat the network as a logical entity.

7.2 The SDN Approach

This section provides an overview of SDN and shows how it is designed to meet evolving network requirements.

Modern Network Requirements

The Open Data Center Alliance (ODCA) provides a good, concise list of requirements for a modern networking approach, which includes the following [ODCA14]:

- **Adaptability:** Networks must adjust and respond dynamically, based on application needs, business policy, and network conditions.
- **Automation:** Policy changes must be automatically propagated so that manual work and errors can be reduced.
- **Maintainability:** Introduction of new features and capabilities (e.g., software upgrades, patches), must be seamless, with minimal disruption of operations.
- **Model management:** Network management software must allow management of the network at a model level rather than implementing conceptual changes by reconfiguring individual network elements.
- **Mobility:** Control functionality must accommodate mobility, including mobile user devices and virtual servers.
- **Integrated security:** Network applications must integrate seamless security as a core service instead of as an add-on solution.
- **On-demand scaling:** Implementations must have the ability to scale up or scale down the network and its services to support on-demand requests.

SDN Architecture

The central concept behind SDN is to enable developers and network managers to have the same type of control over network equipment that they have had over x86 servers. The SDN approach splits the switching function between a data plane and a control plane that are on separate devices (see

Figure 7.1b). The data plane is simply responsible for forwarding packets, and the control plane provides the "intelligence" in designing routes, setting priority and routing policy parameters to meet quality of service (QoS) and quality of experience (QoE) requirements, and coping with the shifting traffic patterns. Open interfaces are defined so that the switching hardware presents a uniform interface, regardless of the internal implementation details. Similarly, open interfaces are defined to enable networking applications to communicate with the SDN controllers.

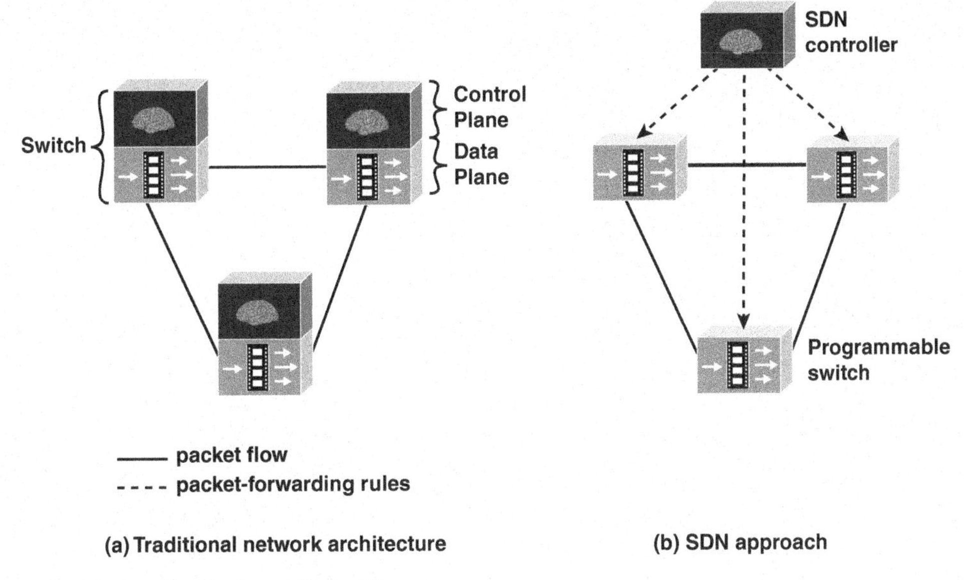

FIGURE 7.1 Control and Data Planes

Figure 7.2 illustrates the SDN architecture. The data plane consists of physical switches and virtual switches, both of which are responsible for forwarding packets. The internal implementation of buffers, priority parameters, and other data structures related to forwarding can be vendor dependent. However, each switch must implement a model, or an abstraction, of packet forwarding that is uniform and open to the SDN controllers. This model is defined in terms of an open API between the control plane and the data plane (i.e., the southbound API). The most prominent example of such an open API is OpenFlow, discussed later in this chapter. The OpenFlow specification defines both a protocol between the control and data planes and an API by which the control plane can invoke the OpenFlow protocol.

It should be noted that SDN is capable of dealing with physical switches for both wireless (e.g., Wi-Fi, cellular) and wired (e.g., Ethernet) transmission links.

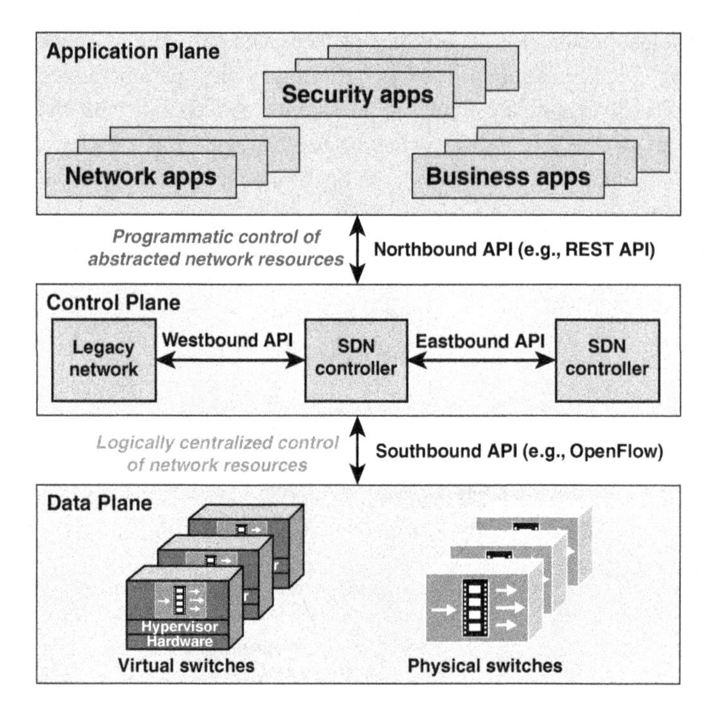

FIGURE 7.2 SDN Architecture

Similarly, SDN controllers can be implemented directly on a server or on a virtual server. OpenFlow or some other open API is used to control the switches in the data plane. In addition, controllers use information about capacity and demand obtained from the networking equipment through which the traffic flows. SDN controllers also expose northbound APIs, which means developers and network managers can deploy a wide range of off-the-shelf and custom-built network applications, many of which were not feasible prior to the advent of SDN. As yet there is no standardized northbound API, and there is no consensus on an open northbound API.

In simple terms, the SDN controller manages the forwarding state of the switches in the software-defined network. This management is done through a vendor-neutral API that allows the controller to address a wide variety of operator requirements without changing any of the lower-level aspects of the network, including topology.

With the decoupling of the control and data planes, SDN enables applications to deal with a single abstracted network device, without concern for the details of how the device operates. Network applications see a single API to the controller. Thus it is possible to quickly create and deploy new applications to orchestrate network traffic flow to meet specific enterprise requirements for performance or security.

There are also horizontal (peer-to-peer) APIs. These APIs are of two types:

- **Eastbound API:** Enables communication and cooperation among groups or federations of controllers.
- **Westbound API:** Provides communication between SDN and non-SDN (or legacy) networks.

At the application plane, a variety of applications interact with SDN controllers. SDN applications are programs that may use an abstract view of the network for their decision-making goals. These applications convey their network requirements and desired network behavior to the SDN controller via a northbound API. Examples of applications are energy-efficient networking, security monitoring, access control, and network management.

Characteristics of Software-Defined Networking

The key characteristics of SDN are as follows:

- The control plane is separated from the data plane. Data plane devices are simple packet-forwarding devices (refer to Figure 7.1).
- The control plane is implemented in a centralized controller or set of coordinated centralized controllers. The SDN controller has a centralized view of the network or networks under its control. The controller is portable software that can run on commodity servers and is capable of programming the forwarding devices based on a centralized view of the network.
- Open interfaces are defined between the devices in the control plane (controllers) and those in the data plane.
- The network is programmable by applications running on top of the SDN controllers. The SDN controllers present an abstract view of network resources to the applications.

7.3 SDN Data Plane

The SDN data plane, referred to as the *resource layer* in ITU-T Y.3300 (*Framework of Software-Defined Networking*, June 2014) and also often referred to as the *infrastructure layer*, is where network forwarding devices perform the transport and processing of data according to decisions made by the SDN control plane. The important characteristic of the network devices in a software-defined network is that these devices perform a simple forwarding function, without embedded software to make autonomous decisions.

Data Plane Functions

Figure 7.3 illustrates the functions performed by the data plane network devices (also called data plane network elements, or switches).

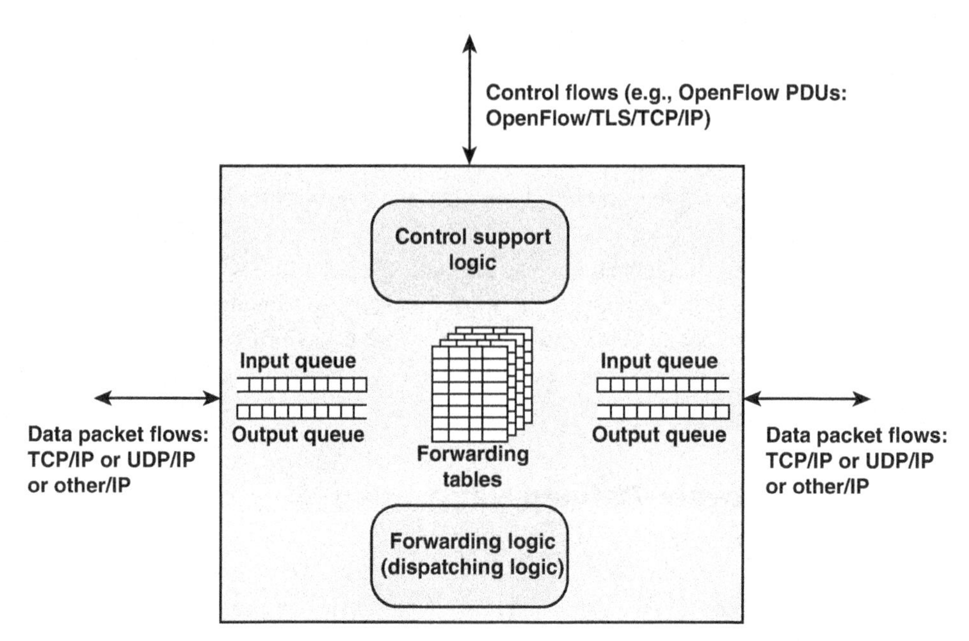

FIGURE 7.3 Data Plane Network Device

The principal functions of the network device are the following:

- **Control support function:** Interacts with the SDN control layer to support programmability via resource-control interfaces. The switch communicates with the controller, and the controller manages the switch via the OpenFlow switch protocol.
- **Data forwarding function:** Accepts incoming data flows from other network devices and end systems and forwards them along the data forwarding paths that have been computed and established according to the rules defined by the SDN applications.

The forwarding rules used by the network device are embodied in forwarding tables that indicate for given categories of packets what the next hop in the route should be. In addition to simple forwarding of a packet, the network device can alter the packet header before forwarding, or discard the packet. As shown, arriving packets may be placed in an input queue, where they await processing by the network device. Forwarded packets are generally placed in an output queue, where they await transmission.

The network device in Figure 7.3 is shown with three I/O ports: one providing control communication with an SDN controller, and two for the input and output of data packets. This is a simple example. A network device may have multiple ports to communicate with multiple SDN controllers and may have more than two I/O ports for packet flows into and out of the device.

Data Plane Protocols

Figure 7.3 suggests the protocols supported by the network device. Data packet flows consist of streams of IP packets. It may be necessary for the forwarding table to define entries based on fields in upper-level protocol headers, such as headers for Transmission Control Protocol (TCP), User Datagram Protocol (UDP), or some other transport or application protocol. The network device examines the IP header and possibly other headers in each packet and makes a forwarding decision.

The other important flow of traffic is via the southbound API, consisting of OpenFlow protocol data units (PDUs) or some similar southbound API protocol traffic.

7.4 OpenFlow

To turn the concept of SDN into practical implementation, two requirements must be met:

- There must be a common logical architecture in all switches, routers, and other network devices to be managed by an SDN controller. This logical architecture may be implemented in different ways on different vendor equipment and in different types of network devices, as long as the SDN controller sees a uniform logical switch functionality.
- A standard secure protocol is needed between the SDN controller and the network device.

Both of these requirements are addressed by OpenFlow, which is both a protocol between SDN controllers and network devices and a specification of the logical structure of the network switch functionality. OpenFlow is defined in the *OpenFlow Switch Specification* [ONF15], published by the Open Networking Foundation (ONF). Three terms are useful to the discussion:

- **OpenFlow switch:** A set of OpenFlow resources that can be managed as a single entity, including a data path and a control channel. OpenFlow switches connect logically to each other via their OpenFlow ports.
- **OpenFlow port:** Where packets enter and exit the OpenFlow pipeline. A packet can be forwarded from one OpenFlow switch to another OpenFlow switch only via an output OpenFlow port on the first switch and an ingress OpenFlow port on the second switch.
- **OpenFlow channel:** Interface between an OpenFlow switch and an OpenFlow controller, used by the controller to manage the switch.

Figure 7.4 illustrates the main elements of an OpenFlow environment, consisting of SDN controllers that include OpenFlow software, OpenFlow switches, and end systems.

Figure 7.5 displays the main components of an OpenFlow switch. An SDN controller communicates with OpenFlow-compatible switches using the OpenFlow protocol running over Transport Layer Security (TLS). On the switch side, the interface is known as an *OpenFlow channel*. Each switch connects to other OpenFlow switches and, possibly, to end user devices that are the sources and destinations of packet flows. These connections are via OpenFlow ports. An OpenFlow port also connects the switch to the SDN controller.

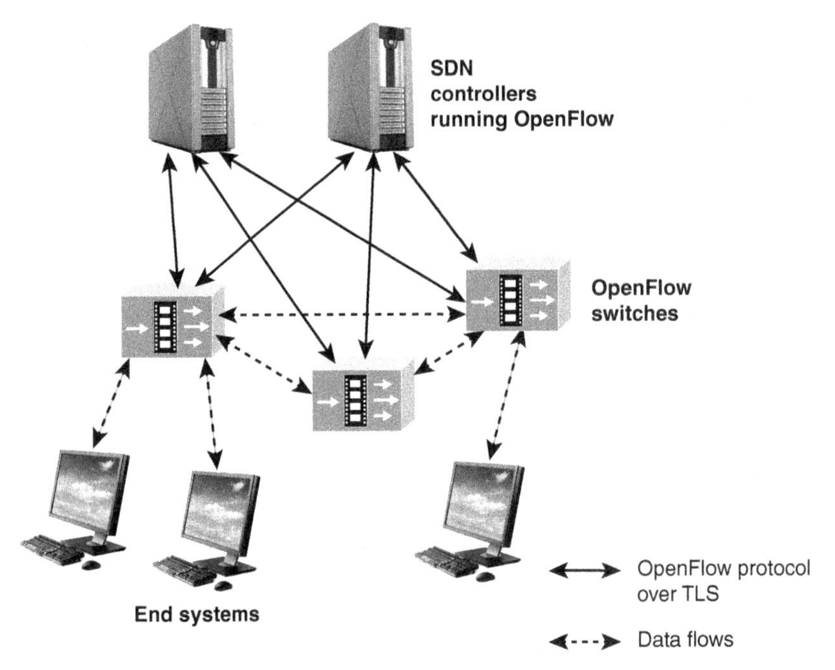

FIGURE 7.4 OpenFlow Switch Context

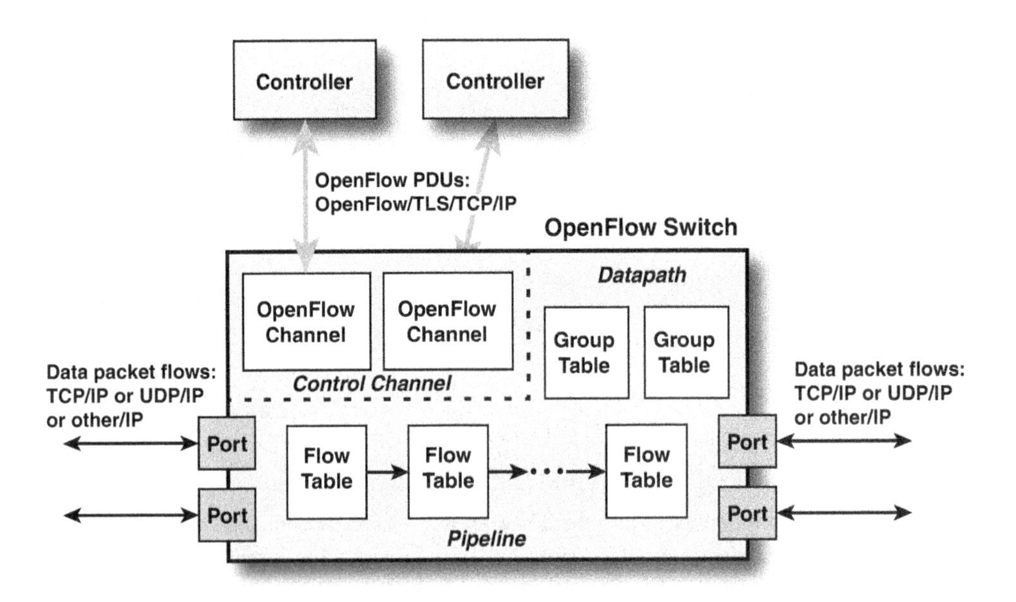

FIGURE 7.5 Main Components of an OpenFlow Switch

OpenFlow defines three types of ports:

- **Physical port:** Corresponds to a hardware interface of the switch. For example, on an Ethernet switch, physical ports map one-to-one to the Ethernet interfaces.
- **Logical port:** Does not correspond directly to a hardware interface of the switch. Logical ports are higher-level abstractions that may be defined in the switch using non-OpenFlow methods (e.g., link aggregation groups, tunnels, loopback interfaces). Logical ports may include packet encapsulation and may map to various physical ports. The processing done by the logical port is implementation dependent and must be transparent to OpenFlow processing. Logical ports must interact with OpenFlow processing in the same manner as OpenFlow physical ports.
- **Reserved port:** Defined by the OpenFlow specification and specifies generic forwarding actions, such as sending to and receiving from the controller, flooding, or forwarding using non-OpenFlow methods, such as "normal" switch processing.

Within each switch, a series of tables are used to manage the flows of packets through the switch. The OpenFlow specification defines three types of tables in the logical switch architecture. A **flow table** matches incoming packets to a particular flow and specifies what functions are to be performed on the packets. There may be multiple flow tables that operate in a pipeline fashion, as explained shortly. A flow table may direct a flow to a **group table**, which may trigger a variety of actions that affect one or more flows. A **meter table** can trigger a variety of performance-related actions on a flow. Using the OpenFlow switch protocol, the controller can add, update, and delete flow entries in tables, both reactively (in response to packets) and proactively.

Before proceeding, it is helpful to define what is meant by the term *flow*. Curiously, this term is not defined in the OpenFlow specification, nor has there been an attempt to define it in virtually all of the literature on OpenFlow. In general terms, a **flow** is a sequence of packets traversing a network that share a set of header field values. For example, a flow could consist of all packets with the same source and destination IP addresses or all packets with the same virtual LAN (VLAN) identifier. We provide a more specific definition later in this chapter.

Flow Table Structure

The basic building block of the logical switch architecture is the flow table. Each packet that enters a switch passes through one of more flow tables. Each flow table consists of a number of rows, called **entries**, consisting of seven components (see Figure 7.6a). The entry components are as follows:

- **Match fields:** These fields are used to select packets that match the values in the fields. The match fields (see Figure 7.6b) identify ingress and egress addresses at various protocol levels. Each of the fields in the match fields component either has a specific value or a wildcard value that matches any value in the corresponding packet header field. A flow table may include a table-miss flow entry, in which case wildcards match all fields (i.e., every field is a match, regardless of value), and the entry has the lowest priority.

- **Priority:** This field shows the relative priority of table entries. It is a 16-bit field with 0 corresponding to the lowest priority. In principle, there could be 216 = 64,000 priority levels.
- **Counters:** This field is updated for matching packets. The OpenFlow specification defines a variety of counters. Table 7.1 lists the required counters that must be supported by an OpenFlow switch.

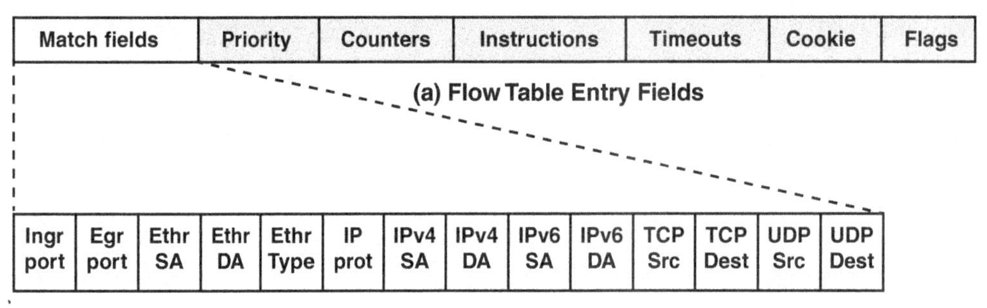

Group identifier	Group type	Counters	Action buckets

(c) Group table entry fields

FIGURE 7.6 OpenFlow Table Entry Formats

TABLE 7.1 Required OpenFlow Counters

Counter	Usage	Bit Length
Reference count (active entries)	Per flow table	32
Duration (seconds)	Per flow entry	32
Received packets	Per port	64
Transmitted packets	Per port	64
Duration (seconds)	Per port	32
Transmit packets	Per queue	64
Duration (seconds)	Per queue	32
Duration (seconds)	Per group	32

- **Instructions:** This field contains instructions to be performed if a match occurs.
- **Timeouts:** This field indicates the maximum amount of idle time before a flow is expired by the switch. Each flow entry has an idle_timeout and a hard_timeout associated with it. A non-zero hard_timeout field causes the flow entry to be removed after the given number of seconds, regardless of how many packets it has matched. A non-zero idle_timeout field causes the flow entry to be removed when it has matched no packets in the given number of seconds.
- **Cookie:** This field contains a 64-bit opaque data value chosen by the controller. It may be used by the controller to filter flow statistics, flow modification, and flow deletion; it is not used when processing packets.
- **Flags:** Flags alter the way flow entries are managed; for example, the flag OFPFF_SEND_FLOW_REM triggers Flow Removed messages (and removes the flow entry) for a flow entry.

It is now possible to offer a more precise definition of the term *flow*. From the point of view of an individual switch, a flow is a sequence of packets that matches a specific entry in a flow table. The definition is packet oriented in the sense that it is a function of the values of header fields of the packets that constitute the flow and not a function of the path the packets follow through the network. A combination of flow entries on multiple switches defines a flow that is bound to a specific path.

The instructions component of a table entry consists of a set of instructions that are executed if the packet matches the entry. Before describing the types of instructions, we need to define the terms *action* and *action set*. **Actions** describe packet forwarding, packet modification, and group table processing operations. The OpenFlow specification includes the following actions:

- **Output:** This action forwards a packet to the specified port. The port could be an output port to another switch or the port to the controller. In the latter case, the packet is encapsulated in a message to the controller.
- **Set-Queue:** This action sets the queue ID for a packet. When the packet is forwarded to a port using the output action, the queue ID determines which queue attached to this port is used for scheduling and forwarding the packet. Forwarding behavior is dictated by the configuration of the queue and is used to provide basic QoS support.
- **Group:** This action processes a packet through the specified group.
- **Push-Tag/Pop-Tag:** This action pushes or pops a tag field for a VLAN or Multiprotocol Label Switching (MPLS) packet.
- **Set-Field:** The various Set-Field actions are identified by their field type and modify the values of the respective header fields in a packet.

- **Change-TTL:** The various Change-TTL actions modify the values of the IPv4 TTL (time to live), IPv6 hop limit, or MPLS TTL in the packet.
- **Drop:** There is no explicit action to represent drops. Instead, packets whose action sets have no output action should be dropped.

An **action set** is a list of actions associated with a packet that are accumulated while the packet is processed by each table and that are executed when the packet exits the processing pipeline.

The types of instructions can be grouped into four categories:

- **Direct packet through pipeline:** The Goto-Table instruction directs a packet to a table farther along in the pipeline. The Meter instruction directs a packet to a specified meter.
- **Perform action on packet:** Actions may be performed on a packet when it is matched to a table entry. The Apply-Actions instruction applies the specified actions immediately, without any change to the action set associated with this packet. This instruction may be used to modify the packet between two tables in the pipeline.
- **Update action set:** The Write-Actions instruction merges specified actions into the current action set for this packet. The Clear-Actions instruction clears all the actions in the action set.
- **Update metadata:** A metadata value can be associated with a packet. It is used to carry information from one table to the next. The Write-Metadata instruction updates an existing metadata value or creates a new value.

Flow Table Pipeline

A switch includes one or more flow tables. If there is more than one flow table, they are organized as a pipeline, with the tables labeled with increasing numbers starting with 0. The use of multiple tables in a pipeline, rather than a single flow table, provides the SDN controller with considerable flexibility.

The OpenFlow specification defines two stages of processing:

- **Ingress processing:** Ingress processing always happens, and it begins with table 0 and uses the identity of the input port. Table 0 may be the only table, in which case the ingress processing is simplified to the processing performed on that single table, and there is no egress processing.
- **Egress processing:** Egress processing is the processing that happens after the determination of the output port. It happens in the context of the output port. This stage is optional. If it occurs, it may involve one or more tables. The separation of the two stages is indicated by the numeric identifier of the first egress table. All tables with a number lower than the first egress table must be used as ingress tables, and no table with a number higher than or equal to the first egress table can be used as an ingress table.

Pipeline processing always starts with ingress processing at the first flow table: The packet must first be matched against flow entries of flow table 0. Other ingress flow tables may be used, depending on the outcome of the match in the first table. If the outcome of ingress processing is to forward the packet to an output port, the OpenFlow switch may perform egress processing in the context of that output port.

When a packet is presented to a table for matching, the input consists of the packet, the identity of the ingress port, the associated metadata value, and the associated action set. For Table 0, the metadata value is blank, and the action set is null. At each table, processing proceeds as follows:

- If there is a match on one or more entries other than the table-miss entry, then the match is defined to be with the highest-priority matching entry. As mentioned in the preceding discussion, the priority is a component of a table entry and is set via OpenFlow; the priority is determined by the user or application invoking OpenFlow. The following steps may then be performed:

  Step 1. Update any counters associated with this entry.

  Step 2. Execute any instructions associated with this entry. This may include updating the action set, updating the metadata value, and performing actions.

  Step 3. The packet is then forwarded to a flow table further down the pipeline, to the group table, to the meter table, or to an output port.

- If there is a match only on a table-miss entry, the table entry may contain instructions, as with any other entry. In practice the table-miss entry specifies one of three actions:
  - Send the packet to the controller. This enables the controller to define a new flow for this and similar packets or decide to drop the packet.
  - Direct the packet to another flow table farther down the pipeline.
  - Drop the packet.
- If there is no match on any entry and there is no table-miss entry, the packet is dropped.

For the final table in the pipeline, forwarding to another flow table is not an option. If and when a packet is finally directed to an output port, the accumulated action set is executed and then the packet is queued for output. Figure 7.7 illustrates the overall ingress pipeline process.

If egress processing is associated with a particular output port, then after a packet is directed to an output port at the completion of the ingress processing, the packet is directed to the first flow table of the egress pipeline. Egress pipeline processing proceeds in the same fashion as for ingress processing, except that there is no group table processing at the end of the egress pipeline.

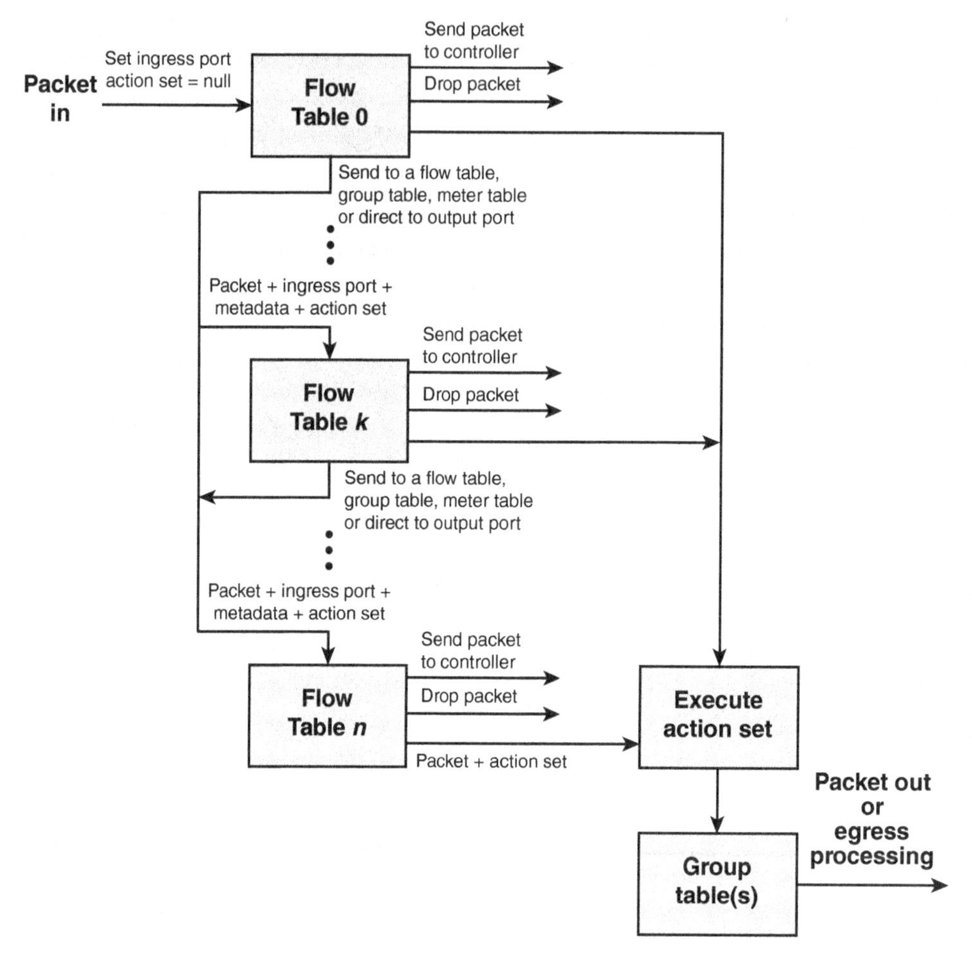

FIGURE 7.7 Packet Flow Through an OpenFlow Switch: Ingress Processing

The Use of Multiple Tables

The use of multiple tables enables the nesting of flows or, put another way, it allows a single flow to be broken down into a number of parallel subflows. For example, an entry in Table 0 defines a flow consisting of packets traversing the network from a specific source IP address to a specific destination IP address. Once a least-cost route between these two endpoints is established, it may make sense for all traffic between these two endpoints to follow that route, and the next hop on that route from this switch can be entered in Table 0. In Table 1, separate entries for this flow can be defined for different transport-layer protocols, such as TCP and UDP. For these subflows, the same output port might be retained so that the subflows all follow the same route. However, TCP includes elaborate congestion control mechanisms not normally found with UDP, so it might be reasonable to handle the TCP and UDP subflows differently in terms of QoS-related parameters. Any of the Table 1 entries could immediately route its respective subflow to the output port, but some or all of the entries may invoke Table 2, further dividing each subflow.

Group Table

In the course of pipeline processing, a flow table may direct a flow of packets to the group table rather than to another flow table. The group table and group actions enable OpenFlow to represent a set of ports as a single entity for forwarding packets. Different types of groups are provided to represent different forwarding abstractions, such as multicasting and broadcasting.

Each group table consists of a number of rows, called group entries, each of which has four components (refer to Figure 7.6c):

- **Group identifier:** This is a 32-bit unsigned integer that uniquely identifies the group. A group is defined as an entry in the group table.
- **Group type:** This component determines the group semantics, as explained later in this chapter.
- **Counters:** This component is updated when packets are processed by a group.
- **Action buckets:** This is an ordered list of action buckets, where each action bucket contains a set of actions to execute and associated parameters.

Each group includes a set of one or more action buckets. Each bucket contains a list of actions. Unlike the action set associated with a flow table entry, which is a list of actions that accumulate while the packet is processed by each flow table, the action list in a bucket is executed when a packet reaches a bucket. The action list is executed in sequence and generally ends with the output action, which forwards the packet to a specified port. The action list may also end with the group action, which sends the packet to another group. This enables the chaining of groups for more complex processing.

OpenFlow Protocol

The OpenFlow protocol describes message exchanges that take place between an OpenFlow controller and an OpenFlow switch. Typically, the protocol is implemented on top of TLS, providing a secure OpenFlow channel.

The OpenFlow protocol enables the controller to perform add, update, and delete actions to the flow entries in the flow tables. It supports three types of messages:

- **Controller-to-switch:** These messages are initiated by the controller and, in some cases, require a response from the switch. This class of messages enables the controller to manage the logical state of the switch, including its configuration and details of flow and group table entries. Also included in this class is the packet-out message, which is sent by the controller to a switch when that switch sends a packet to the controller and the controller decides not to drop the packet but to direct it to a switch output port.

- **Asynchronous:** These types of messages are sent without solicitation from the controller. This class includes various status messages to the controller. Also included is the packet-in message, which may be used by the switch to send a packet to the controller when there is no flow table match.
- **Symmetric:** These messages are sent without solicitation from either the controller or the switch. They are simple yet helpful. Hello messages are typically sent back and forth between the controller and switch when the connection is first established. Echo request and reply messages can be used by either the switch or controller to measure the latency or bandwidth of a controller-switch connection or just verify that the device is up and running. The experimenter message is used to stage features to be built in to future versions of OpenFlow.

In general terms, the OpenFlow protocol provides the SDN controller with three types of information to be used in managing the network:

- **Event-based messages:** Sent by the switch to the controller when a link or port change occurs.
- **Flow statistics:** Generated by the switch based on traffic flow. This information enables the controller to monitor traffic, reconfigure the network as needed, and adjust flow parameters to meet QoS requirements.
- **Encapsulated packets:** Sent by the switch to the controller either because there is an explicit action to send this packet in a flow table entry or because the switch needs information for establishing a new flow.

The OpenFlow protocol enables the controller to manage the logical structure of a switch without regard to the details of how the switch implements the OpenFlow logical architecture.

7.5 SDN Control Plane

The SDN control layer maps application-layer service requests into specific commands and directives to data plane switches and supplies applications with information about data plane topology and activity. The control layer is implemented as a server or cooperating set of servers known as SDN controllers. This section provides an overview of control plane functionality.

Control Plane Functions

Figure 7.8 illustrates the functions performed by SDN controllers.

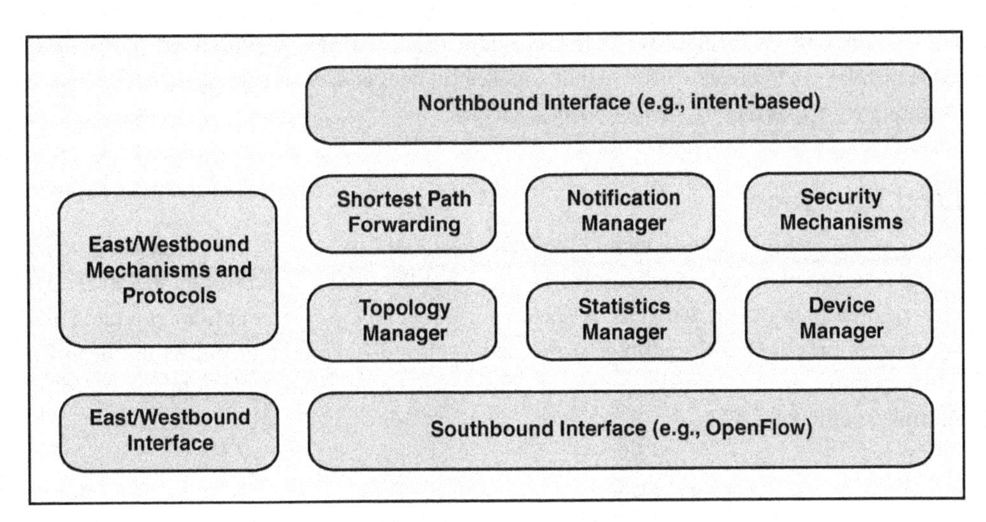

FIGURE 7.8 SDN Control Plane Functions and Interfaces

The figure illustrates the essential functions that any controller should provide, which include the following:

- **Shortest path forwarding:** Uses routing information collected from switches to establish preferred routes.
- **Notification manager:** Receives, processes, and forwards to an application events such as alarm notifications, security alarms, and state changes.
- **Security mechanisms:** Provide isolation and security enforcement between applications and services.
- **Topology manager:** Builds and maintains switch interconnection topology information.
- **Statistics manager:** Collects data on traffic through the switches.
- **Device manager:** Configures switch parameters and attributes and manages flow tables.

The functionality provided by the SDN controller can be viewed as a network operating system (NOS).[2] As with a conventional OS, an NOS provides essential services, common APIs, and an abstraction of lower-layer elements to developers. The functions of an SDN NOS, such as those listed above, enable developers to define network policies and manage networks without concern for the details of the network device characteristics, which may be heterogeneous and dynamic. The northbound interface, discussed subsequently, provides a uniform means for application developers and network managers

2. An NOS is a server-based operating system oriented to computer networking. It may include directory services, network management, network monitoring, network policies, user group management, network security, and other network-related functions.

to access SDN service and perform network management tasks. Further, well-defined northbound interfaces enable developers to create software that is not only independent of data plane details but to a great extent usable with a variety of SDN controller servers.

Southbound Interface

The functionality of the control plane is visible through four interfaces, which are given geographic names [LATI20]. The southbound interface (SBI) provides the logical connection between the SDN controller and the data plane switches, as shown in Figure 7.9. Some controller products and configurations support only a single southbound protocol. A more flexible approach is the use of a southbound abstraction layer that provides a common interface for the control plane functions while supporting multiple southbound APIs.

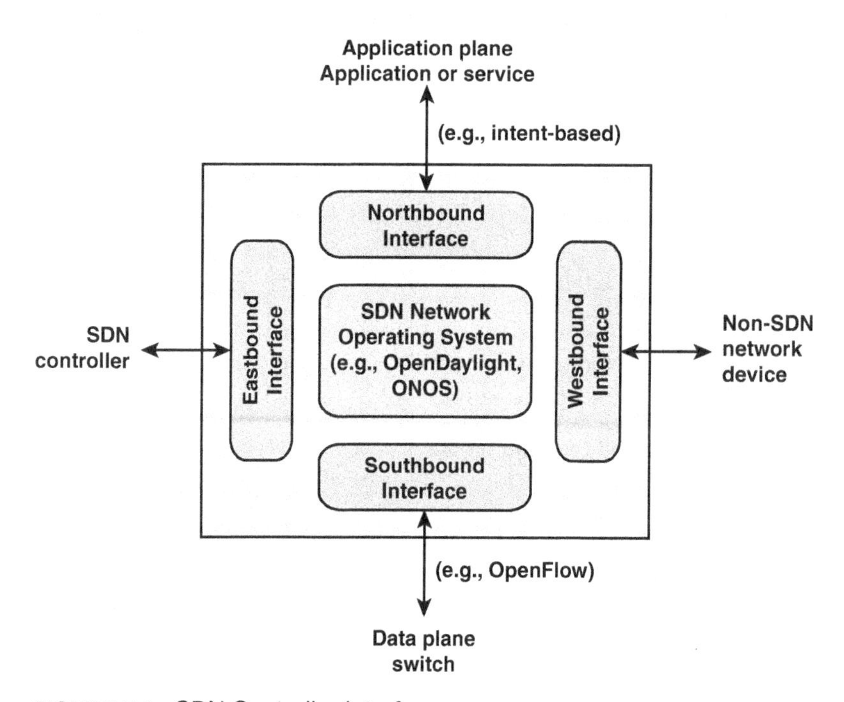

FIGURE 7.9 SDN Controller Interfaces

The most commonly implemented southbound API is OpenFlow.

Northbound Interface

The northbound interface (NBI) enables applications to access control plane functions and services without needing to know the details of the underlying network switches. The northbound interface is typically viewed as a software API rather than as a protocol.

Unlike OpenFlow for the SBI, there is no single API or protocol that different developers/vendors can use for the northbound interface. One reason for this lack of standardization is the variation in applications and their requirements. However, there is consensus on the architectural approach known as intent NBI, and a number of open source SDN packages use it [PHAM16].

There are essentially two approaches for developing an NBI:

- **Prescriptive:** The application specifies or constrains the selection and allocation, virtualization and abstraction, or assembly and concatenation of resources needed to satisfy the request.
- **Nonprescriptive:** This is also referred to as intent NBI. The application describes its requirements in application-oriented language, and the controller becomes an intelligent black box that integrates core network services to construct network applications to serve users' requests.

An important reference on intent NBI is a document issued by the Open Networking Foundation [ONF16], which describes the intent NBI paradigm, its utility and properties, and its potential implementation structure. Figure 7.10, from that document, provides a schematic representation of the architecture.

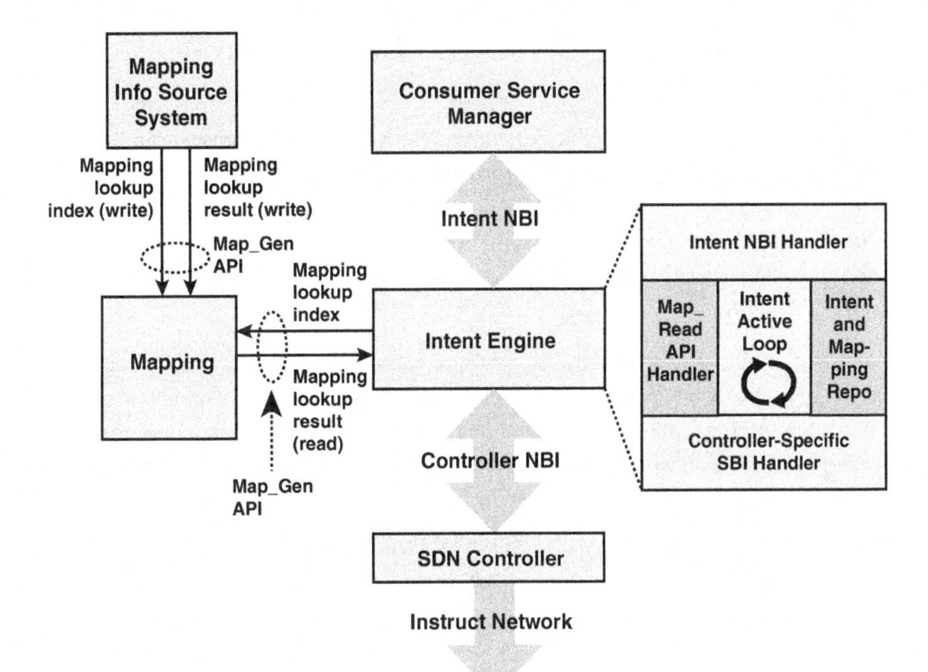

FIGURE 7.10 Schematic Representation of Intent NBI Architecture

The architecture makes use of mappings that can translate intent NBI requests into forms that lower-level entities can understand. Mappings are an information intermediation mechanism that effectively permits consumer and provider systems to communicate in terms that are natural to each. The intent environment is dynamic. The intent system, which is essentially middleware between applications and the SDN controller, continuously evaluates the relationship among a number of elements:

- Existing and new intent requests
- Mappings
- Controlled resource sets and states

The intent engine in Figure 7.10 is a middleware component that has an NBI interface with the controller and an NBI interface with the applications that are managed by a consumer service manager component. The intent engine has five major components:

- **Information repository (repo):** This contains the set of active service intents and mapping lookup values.
- **Map_Read API handler:** This is responsible for reflecting to the repo an up-to-date capture of mapping lookup values.
- **Intent NBI handler:** This is responsible for receiving service intents from the consumer system and reflecting them to the repo (in native form) and for reflecting notifications to the consumer system.
- **Intent active loop:** This element is responsible for continuously evaluating active service intents and mappings from the repo and network information from the SBI handler and taking actions required to instantiate new, or appropriately modify existing, service configurations as a function of detected intent changes (repo), and/or of mapping changes (repo), and/or of network changes (reflected by the SBI handler). Computing appropriate inputs to be reflected to the SBI handler results in the following:
    - Inputs to the SBI handler may adhere to some model-based form, such as specifying one or more pairs of endpoints for piece-wise interconnection, along with parametric constraints on such interconnections. The function of the intent active loop would then include any required and appropriate parsing and translation of intent request terms into such model form. Instructions relayed to the SBI handler would be based on this model and use the specific terms that result from mapping lookups.
    - Where network conditions do not permit instructions that deliver all aspects of specified service intents, appropriate notifications are synthesized to be reflected to the intent NBI handler.
- **Controller-specific SBI handler:** This is the only controller-specific component of the intent engine and has the following functions:
    - It receives inputs from the intent active loop and provides an appropriately modified form of those inputs, as provisioning instructions, to the SDN controller.

- It receives information (e.g., topology information) from the SDN controller and forwards it to the intent active loop in an appropriately modified form.

The figure also depicts a mapping information source system. Conceptually, this system associates intent NBI–specific terms, called *keys* or *indices*, with controller NBI objects and operations.

This architecture has a number of strengths and benefits:

- Intent declarations are declarative and serve to separate consumer and provider system implementations. They are intended to make human and/or machine consumer requests that are forwarded to provider systems as simple as possible. The SDN controller can calculate the optimal result to fulfill the intent request.
- The intent request is independent of the controller platforms and implementations. It only expresses the requirements for the application layer and uses application-related vocabulary and information. One intent request can be implemented on different controllers with various algorithms. This makes applications portable.
- The intent NBI approach may reduce resource allocation conflicts. Because intent NBI requests do not indicate specific resources to be allocated to specific services, the provider can assign resources to services in ways that reduce the possibility of resource contention.

An example of the types of objects that can be referenced in intent declarations is provided in the intent framework ONOS, which is an open source SDN controller. An intent describes an application's request to the ONOS core to alter the network's behavior. At the lowest levels, intents may be described in terms of:

- **Network resources:** A set of object models, such as links, that tie back to the parts of the network affected by an intent.
- **Constraints:** Weights applied to a set of network resources, such as bandwidth, optical frequency, and link type.
- **Criteria:** Packet header fields or patterns that describe a slice of traffic. For example, an intent's TrafficSelector carries criteria as a set of objects that implement the Criterion interface.
- **Instructions:** Actions to apply to a slice of traffic, such as header field modifications or outputting through specific ports. For example, an intent's TrafficTreatment carries instructions as a set of objects that implement the Instruction interface.

Eastbound Interface

Eastbound interfaces are used to import and export information among distributed controllers. The eastbound interface supports the creation of multiple domains. In a large enterprise network, the deployment of a single controller to manage all network devices would be unwieldy or undesirable.

A more likely scenario is that the operator of a large enterprise or carrier network divides the whole network into a number of nonoverlapping SDN domains, as shown in Figure 7.11.

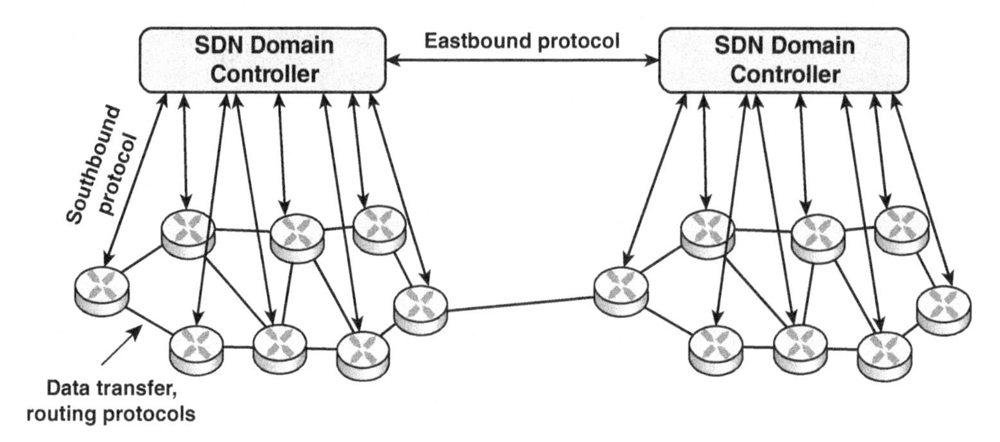

FIGURE 7.11 SDN Domain Structure

Reasons for using SDN domains include the following:

- **Scalability:** The number of devices an SDN controller can feasibly manage is limited. Thus, a reasonably large network may need to deploy multiple SDN controllers.
- **Privacy:** A carrier may choose to implement different privacy policies in different SDN domains. For example, an SDN domain may be dedicated to a set of customers that implement their own highly customized privacy policies, requiring that some networking information in this domain (e.g., network topology) should not be disclosed to an external entity.
- **Incremental deployment:** A carrier's network may consist of portions of legacy and nonlegacy infrastructure. Dividing the network into multiple individually manageable SDN domains allows for flexible incremental deployment.

The existence of multiple domains creates a requirement for individual controllers to communicate with each other via a standardized protocol to exchange routing information. No single approach has gained widespread acceptance, and there have been a wide variety of implementations on both open source and proprietary systems [LATI20].

Westbound Interface

The westbound interface enables communication between an SDN controller and a non-SDN network. These interfaces typically use Border Gateway Protocol (BGP) to bridge the gap between SDN and traditional networks.

7.6 SDN Application Plane

The power of the SDN approach to networking is in the support it provides for network applications to monitor and manage network behavior. The SDN control plane provides the functions and services that facilitate rapid development and deployment of network applications.

While the SDN data and control planes are well defined, there is much less agreement on the nature and scope of the application plane. At minimum, the application plane includes a number of network applications—that is, applications that specifically deal with network management and control. There is no agreed-upon set of such applications or even categories of such applications. Further, the application layer may include general-purpose network abstraction tools and services that might also be viewed as part of the functionality of the control plane.

Application Plane Architecture

The application plane contains applications and services that define, monitor, and control network resources and behavior. These applications interact with the SDN control plane via application control interfaces in order for the SDN control layer to automatically customize the behavior and the properties of network resources. The programming of an SDN application makes use of the abstracted view of network resources provided by the SDN control layer by means of information and data models exposed via the application control interface.

This section provides an overview of application plane functionality, which is illustrated in Figure 7.12.

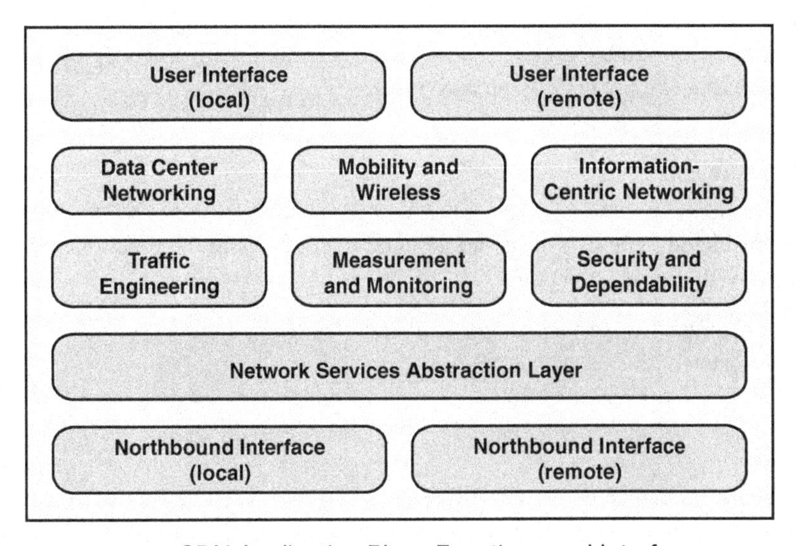

FIGURE 7.12 SDN Application Plane Functions and Interfaces

Northbound Interface

As described in Section 7.5, the northbound interface enables applications to access control plane functions and services without needing to know the details of the underlying network switches. Typically, the northbound interface provides an abstract view of network resources controlled by the software in the SDN control plane.

Figure 7.12 indicates that the northbound interface can be a local or remote interface. For a local interface, the SDN applications are running on the same server as the control plane software (i.e., the controller network operating system). Alternatively, the applications can be run on remote systems, in which case the northbound interface is a protocol or an API that connects the applications to the controller NOS running on a central server. Both architectures are likely to be implemented.

Network Services Abstraction Layer

RFC 7426 (*Software-Defined Networking [SDN]: Layers and Architecture Terminology*, January 2015) from the Internet Engineering Task Force (IETF) defines a network services abstraction layer between the control and application planes and describes it as a layer that provides service abstractions that can be used by applications and services. Several functional concepts are suggested by the placement of this layer in the SDN architecture:

- This layer could provide an abstract view of network resources that hides the details of the underlying data plane devices.
- This layer could provide a generalized view of control plane functionality so that applications could be written that would operate across a range of controller network operating systems. This functionality is similar to that of a hypervisor or virtual machine monitor that decouples applications from the underlying OS and underlying hardware.
- This layer could provide a network virtualization capability that allows different views of the underlying data plane infrastructure.

Arguably, this layer could be considered to be part of the northbound interface, with the functionality incorporated in the control plane or the application plane.

A wide range of schemes have been developed that roughly fall into this layer, although discussing them is beyond the scope of this chapter.

Network Applications

Many network applications could be implemented for a software-defined network. Different published surveys of SDN have come up with different lists and even different general categories of SDN-based network applications. Figure 7.12 includes six categories that encompass the majority of SDN applications. The following sections discuss these six categories.

Traffic Engineering

Traffic engineering involves dynamically analyzing, regulating, and predicting the behavior of data flowing in networks with the aim of performance optimization to meet service-level agreements (SLAs). Traffic engineering involves establishing routing and forwarding policies based on QoS requirements. With SDN, the task of traffic engineering should be considerably simplified compared to in a non-SDN network. SDN offers a uniform global view of heterogeneous equipment and powerful tools for configuring and managing network switches.

Traffic engineering is an area of great activity in the development of SDN applications. [KREU15] lists the following application types that have been implemented as SDN applications:

- On-demand virtual private networks
- Load balancing
- Energy-aware routing
- QoS for broadband access networks
- Scheduling/optimization
- Traffic engineering with minimal overhead
- Dynamic QoS routing for multimedia apps
- Fast recovery through fast-failover groups
- QoS policy management framework
- QoS enforcement
- QoS over heterogeneous networks
- Multiple packet schedulers
- Queue management for QoS enforcement
- Divide and spread forwarding tables

Measurement and Monitoring

The area of measurement and monitoring applications can roughly be divided into two categories: applications that provide new functionality for other networking services and applications that add value to OpenFlow-based SDNs.

An example of the first category is in the area of broadband home connections. If a connection is to an SDN-based network, new functions can be added to the measurement of home network traffic and demand, allowing the system to react to changing conditions. The second category typically involves using different kinds of sampling and estimation techniques to reduce the burden of the control plane in the collection of data plane statistics.

Security and Dependability

Security and dependability applications have one of two goals:

- **Address security concerns related to the use of SDN:** SDN involves a three-layer architecture (application, control, and data layers) and new approaches to distributed control and data encapsulation. All of this introduces the potential for new vectors for attack. Threats can occur at any of the three layers or in the communication between layers. SDN applications are needed to provide for the secure use of SDN.
- **Use the functionality of SDN to improve network security:** Although SDN presents new security challenges for network designers and managers, it also provides a platform for implementing consistent, centrally managed security policies and mechanisms for the network. SDN allows the development of SDN security controllers and SDN security applications that can provision and orchestrate security services and mechanisms.

Data Center Networking

The three areas of SDN applications discussed so far—traffic engineering, measurement and monitoring, and security—relate to a broad range of use cases in many different kinds of networks. The remaining three applications areas in Figure 7.12—data center networking, mobility and wireless, and information-centric networking—have use cases in specific types of networks.

Cloud computing, big data, large enterprise networks, and even, in many cases, smaller enterprise networks depend strongly on highly scalable and efficient data centers. [KREU15] lists the following as key requirements for data centers: high and flexible cross-section bandwidth[3] and low latency, QoS based on application requirements, high levels of resilience, intelligent resource utilization to reduce energy consumption and improve overall efficiency, and agility in provisioning network resources (e.g., by means of network virtualization and orchestration with computing and storage).

With traditional network architectures, many of these requirements are difficult to satisfy because of the complexity and inflexibility of the network. SDN offers the promise of substantial improvement in the ability to rapidly modify data center network configurations, to flexibly respond to user needs, and to ensure efficient operation of the network.

Mobility and Wireless

In addition to all the traditional performance, security, and reliability requirements of wired networks, wireless networks impose a broad range of new requirements and challenges. Mobile users are continuously generating demands for new services with high quality and efficient content delivery, independent of location. Network providers must deal with problems related to managing the available

3. Cross-section bandwidth is the maximum bidirectional data rate that can pass between two parts of a network that is divided into two equal halves.

spectrum, implementing handover mechanisms, performing efficient load balancing, responding to QoS and QoE requirements, and maintaining security.

SDN can provide much-needed tools for a mobile network provider, and in recent years, a number of SDN-based applications for wireless network providers have been designed. Among others, [KREU15] lists the following SDN application areas: seamless mobility through efficient handovers, creation of on-demand virtual access points, load balancing, downlink scheduling, dynamic spectrum usage, enhanced inter-cell interference coordination, per-client and/or per-base station resource block allocations, simplified administration, easy management of heterogeneous network technologies, interoperability between different networks, shared wireless infrastructures, and management of QoS and access control policies.

Information-Centric Networking

Information-centric networking (ICN), also known as content-centric networking, has received significant attention in recent years, mainly driven by the fact that distribution and manipulation of information have become the major functions of the Internet. Unlike in the traditional host-centric networking paradigm, where information is obtained by contacting specified named hosts, ICN is aimed at providing native network primitives for efficient information retrieval by directly naming and operating on information objects.

With ICN, there is a distinction between location and identity, thus decoupling information for its sources. With this approach, information sources can place, and information users can find, information anywhere in the network because the information is named, addressed, and matched independently of its location. In ICN, instead of a source/destination host pair being specified for communication, a piece of information is named. In ICN, after a request is sent, the network is responsible for locating the best source that can provide the desired information. Routing of information requests thus seeks to find the best source for the information, based on a location-independent name.

Deploying ICN on traditional networks is challenging because existing routing equipment needs to be updated or replaced with ICN-enabled routing devices. Further, ICN shifts the delivery model from a host-to-user model to a content-to-user model. This creates a need for clear separation between the task of information demand and supply and the task of forwarding. SDN has the potential to provide the necessary technology for deploying ICN because it provides for programmability of the forwarding elements and a separation of the control and data planes.

A number of projects have proposed using SDN capabilities to implement ICNs. There is no consensus approach to achieving this coupling of SDN and ICN. Suggested approaches include substantial enhancements/modifications to the OpenFlow protocol, developing a mapping of names into IP addresses using a hash function, using the IP option header as a name field, and using an abstraction layer between an OpenFlow switch and an ICN router, so that the layer, the OpenFlow switch, and the ICN router function as a single programmable ICN router.

User Interface

The user interface enables a user to configure parameters in SDN applications and to interact with applications that support user interaction. Again, there are two possible interfaces. A user that is collocated with the SDN application server (which may or may not include the control plane) can use the server's keyboard/display. More typically, a user logs on to the application server over a network or communications facility.

7.7 Key Terms and Review Questions

Key Terms

application programming interface (API)	northbound interface (NBI)
cross-section bandwidth	OpenFlow
eastbound interface	OpenFlow channel
egress processing	OpenFlow port
flow	OpenFlow switch
flow table	physical port
flow table pipeline	prescriptive NBI
group table	reserved port
ingress processing	SDN application plane
intent NBI	SDN control plane
logical port	SDN data plane
meter table	software-defined networking (SDN)
network operating system (NOS)	southbound interface (SBI)
network services abstraction layer	westbound interface
nonprescriptive NBI	

Review Questions

1. List the key requirements for a modern networking approach.
2. How does SDN split the functions of a switch?
3. Describe the elements of an SDN architecture.
4. Explain the function of the northbound, southbound, eastbound, and westbound interfaces.
5. What are the two main data plane functions?
6. Explain the difference between an OpenFlow switch, an OpenFlow port, and an OpenFlow channel.
7. Explain the difference between an OpenFlow physical port, an OpenFlow logical port, and an OpenFlow reserved port.
8. Explain the difference between an OpenFlow flow table, an OpenFlow group table, and an OpenFlow meter table.
9. Define flow.
10. Define action and action set.
11. Describe OpenFlow ingress and egress processing.
12. List and briefly define the essential functions of an SDN controller.
13. Characterize the operation of an intent NBI.
14. List key reasons for using SDN domains.
15. Describe the application plane architecture.
16. What is the purpose of a network services abstraction layer?
17. List and briefly describe six network application areas.

7.8 References and Documents

References

KREU15 Kreutz, D., et al. "Software-Defined Networking: A Comprehensive Survey." *Proceedings of the IEEE*, January 2015.

LATI20 Latif, Z., et al. "A Comprehensive Survey of Interface Protocols for Software Defined Networks." *Journal of Network and Computer Applications*, volume 145, April 15, 2020.

ODCA14 Open Data Center Alliance. *Open Data Center Alliance Master Usage Model: Software-Defined Networking Rev. 2.0.* ODCA white paper. 2014.

ONF12 Open Networking Foundation. *Software-Defined Networking: The New Norm for Networks.* ONF white paper, April 12, 2012.

ONF15 Open Networking Foundation. *OpenFlow Switch Specification Version 1.5.1.* March 26, 2015.

ONF16 Open Networking Foundation. *Intent NBI: Definition and Principles.* ONF-TR-523, October 2016.

PHAM16 Pham, M., and Hoang, D. "SDN Applications: The Intent-Based Northbound Interface Realisation for Extended Applications." *2016 IEEE NetSoft Conference and Workshops*, May 2016.

STAL16 Stallings, W. *Foundations of Modern Networking: SDN, NFV, QoE, IoT, and Cloud.* Upper Saddle River, NJ: Pearson Addison Wesley, 2016.

Documents

RFC 7426 *Software-Defined Networking (SDN): Layers and Architecture Terminology.* January 2015.

ITU-T Y.3300 *Framework of Software-Defined Networking*. June 2014.

Chapter 8

Network Functions Virtualization

Learning Objectives

After studying this chapter, you should be able to:

- Understand the concept of virtual machines
- Explain the difference between type 1 and type 2 hypervisors
- List and explain the key benefits of NFV
- List and explain the key requirements for NFV
- Present an overview of the NFV architecture
- Explain the elements of the NFV infrastructure and their interrelationships
- Understand key design issues related to virtualized network functions
- Discuss the relationship between SDN and NFV

Complementing software defined networking (SDN), is the other key enabling technology for 5G core networks: network functions virtualization (NFV).Virtualization encompasses a variety of technologies for managing computing resources by providing a software translation layer, known as an abstraction layer, between the software and the physical hardware. Virtualization turns physical resources into logical, or virtual, resources. Virtualization enables users, applications, and management software operating above the abstraction layer to manage and use resources without needing to be aware of the physical details of the underlying resources.

The chapter begins by providing some background and motivation for the NFV approach. Then, Section 8.2 examines the fundamental technology underlying NFV: virtual machines. Section 8.3 introduces the basic concepts of NFV. Section 8.4 summarizes the benefits of NFV and the requirements for implementing NFV. Section 8.5 examines the NFV reference architecture. The next two sections look in more detail at the NFV infrastructure and virtualized network functions. Finally, Section 8.8 discusses the relationship between SDN and NFV. Chapter 9, "Core Network Functionality, QoS, and Network Slicing," discusses the application of SDN and NFV in the 5G core network.

NFV is a complex subject. You can get a basic overview in Sections 8.1 through 8.3 and a good grasp of NFV operation by also reading Sections 8.4 and 8.5. Sections 8.6 and 8.7 are for those interested in a more detailed understanding of this topic.

8.1 Background and Motivation for NFV

NFV originated from discussions among major network operators and carriers about how to improve network operations in the high-volume multimedia era. These discussions resulted in the publication of the original ISG NFV white paper, *Network Functions Virtualization: An Introduction, Benefits, Enablers, Challenges & Call for Action*. This white paper indicates that the overall objective of NFV is to leverage standard IT virtualization technology to consolidate many network equipment types onto industry-standard high-volume servers, switches, and storage, which could be located in data centers, network nodes, and end user premises.

The white paper highlights that this new approach was needed because networks included a large and growing variety of proprietary hardware appliances, leading to the following negative consequences:

- New network services often required additional different types of hardware appliances, and finding the space and power to accommodate these boxes was becoming increasingly difficult.
- New hardware means additional capital expenditures.
- Once new types of hardware appliances are acquired, operators are faced with the rarity of skills necessary to design, integrate, and operate increasingly complex hardware-based appliances.
- Hardware-based appliances rapidly reach end of life, requiring much of the procure–design–integrate–deploy cycle to be repeated with little or no revenue benefit.
- As technology and services innovation were accelerating to meet the demands of an increasingly network-centric IT environment, the need for an increasing variety of hardware platforms inhibited the introduction of new revenue-earning network services.

The NFV approach moves away from dependence on a variety of hardware platforms to the use of a small number of standardized platform types, with virtualization techniques used to provide the needed network functionality. In the white paper, the group expresses the belief that the NFV approach is applicable to any data plane packet processing and control plane function in fixed and mobile network infrastructures.

8.2 Virtual Machines and Containers

Traditionally, applications have run directly on an operating system (OS) on a personal computer (PC) or on a server. Each PC or server would run only one OS at a time. Thus, the application vendor had to rewrite parts of its applications for each OS/platform they would run on and support, and this increased time to market for new features/functions, increased the likelihood of defects, increased quality testing efforts, and usually led to increased price. To support multiple operating systems, application vendors needed to create, manage, and support multiple hardware and operating system infrastructures, which was a costly and resource-intensive process. One effective strategy for dealing with this problem, known as **hardware virtualization**, involves the use of software to partition a computer's resources to dedicate them to separate and isolated virtual machines. It enables multiple operating system copies (the same operating system or different ones) to execute on the computer and prevents applications from different VMs from interfering with each other.

A machine with virtualization software can host numerous applications—including those that run on different operating systems—on a single platform. In essence, the host operating system can support a number of **virtual machines (VMs)**, each of which has the characteristics of a particular OS and, in some versions of virtualization, the characteristics of a particular hardware platform.

Hypervisor

The mechanism that enables virtualization is the **hypervisor**, also known as a **virtual machine monitor (VMM)**. This software sits between the hardware and the VMs, acting as a resource broker. Simply put, it allows multiple VMs to safely coexist on a single physical server host and share that host's resources. Figure 8.1 illustrates this type of virtualization in general terms. On top of the hardware platform sits some sort of virtualizing software, which may consist of the host OS plus specialized virtualizing software or simply a software package that includes host OS functions and virtualizing functions, as explained subsequently. The virtualizing software provides abstraction of all physical resources (e.g., processor, memory, network, and storage) and thus enables multiple computing stacks, called virtual machines, to be run on a single physical host.

Each VM includes an OS called the guest OS. This OS may be the same as the host OS or a different one. For example, a guest Windows OS could be run in a VM on top of a Linux host OS. The guest OS, in turn, supports a set of standard library functions and other binary files and applications. From the point of view of the applications and the user, this stack appears as an actual machine, with hardware and an OS; thus the term *virtual machine* is appropriate. In other words, it is the hardware that is being virtualized.

The number of guests that can exist on a single host is measured as a **consolidation ratio**. For example, a host that is supporting 4 VMs is said to have a consolidation ratio of 4 to 1, also written as 4:1. The initial commercially available hypervisors provided consolidation ratios between 4:1 and 12:1, but even at the low end, if a company virtualized all of its servers, it could remove 75% of the servers from its data centers. More importantly, it could remove the cost as well, which often ran into the millions or tens of millions of dollars annually. With fewer physical servers, less power and less cooling were needed. These other reductions led to fewer cables, fewer network switches, and less floor space. Server consolidation became, and continues to be, a tremendously valuable way to solve a costly and wasteful problem. Today, more virtual servers are deployed in the world than physical servers, and virtual server deployment continues to accelerate.

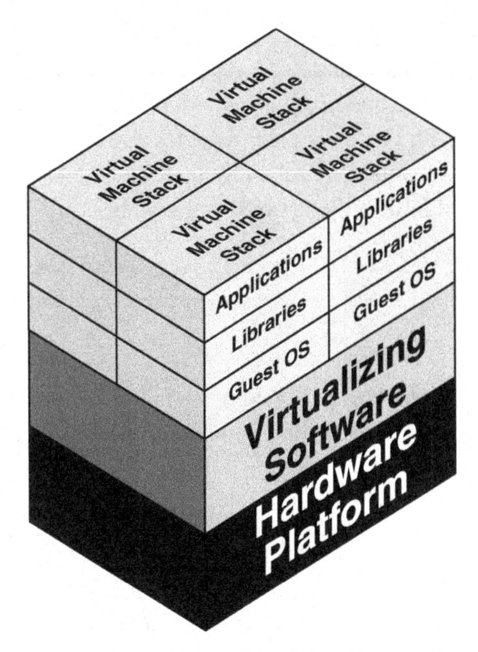

FIGURE 8.1 Virtual Machine Concept

The key reasons that organizations use virtualization are as follows:

- **Legacy hardware:** Applications built for legacy hardware can still be run by virtualizing (emulating) the legacy hardware, making it possible to retire the old hardware.
- **Rapid deployment:** As discussed later in this chapter, whereas it may take weeks or longer to deploy new servers in an infrastructure, it may be possible to deploy a new VM in a matter of minutes. A VM consists of files, and by duplicating those files in a virtual environment, you get a perfect copy of the server.
- **Versatility:** Hardware usage can be optimized by maximizing the number of kinds of applications that a single computer can handle.
- **Consolidation:** A large-capacity or high-speed resource, such as a server, can be used more efficiently by sharing the resources among multiple applications simultaneously.
- **Aggregating:** Virtualization makes it easy to combine multiple resources into one virtual resource, as in the case of storage virtualization.

- **Dynamics:** With the use of virtual machines, hardware resources can be easily allocated in a dynamic fashion. This enhances load balancing, fault tolerance, and the ability to satisfy QoS requirements.
- **Ease of management:** Virtual machines facilitate deployment and testing of software at a faster rate than is possible with bare-metal servers.
- **Increased availability:** Virtual machine hosts are clustered together to form pools of compute resources. Multiple VMs are hosted on each of these servers, and in the event of a physical server failure, the VMs on the failed host can be quickly and automatically restarted on another host in the cluster. Compared with providing this type of availability for a physical server, virtual environments can provide higher availability at significantly less cost and with less complexity.

Architectural Approaches

Virtualization is a form of abstraction. Much as an OS abstracts the disk I/O commands from a user through the use of program layers and interfaces, virtualization abstracts the physical hardware from the VMs it supports. The hypervisor is the software that provides this abstraction. It acts as a proxy for the guests (VMs) as they request and consume resources of the physical host.

A VM is a software construct that mimics the characteristics of a physical server. It is configured with some number of processors, some amount of RAM, storage resources, and connectivity through the network ports. Once that VM is created, it can be powered on like a physical server, loaded with an operating system and software solutions, and utilized in the manner of a physical server. Unlike a physical server, this virtual server only sees the resources it has been configured with, not all of the resources of the physical host. This isolation allows a host machine to run many VMs, each of them running the same or different copies of an operating system, sharing RAM and storage and network bandwidth without problems. An operating system in a VM accesses the resource that is presented to it by the hypervisor. The hypervisor facilitates the translation and I/O from the VM to the physical server devices and back again to the correct VM. In this way, certain privileged instructions that a "native" operating system would be executing on its host hardware are trapped and run by the hypervisor as a proxy for the VM. This creates some performance degradation in the virtualization process, but over time, both hardware and software improvements have minimalized this overhead.

A VM instance is defined in files. A typical VM can consist of just a few files. There is a configuration file that describes the attributes of the VM. It contains the server definition, how many virtual processors (vCPUs) are allocated to this VM, how much RAM is allocated, which I/O devices the VM has access to, and how many network interface cards (NICs) are in the virtual server. It also describes the storage that the VM can access. Often that storage is presented as virtual disks that exist as additional files in the physical file system. When a VM is powered on, or instantiated, additional files are created for logging, for memory paging, and for other functions. Because a VM consists of files, certain functions in a virtual environment are much simpler and quicker than in a physical environment. Because VMs are already files, copying them produces not only a backup of the data but also a copy of the entire server, including the operating system, applications, the state of the applications running on the VM, and the hardware configuration.

Hypervisor Functions

The principal functions performed by a hypervisor are as follows:

- **Execution management of VMs:** This includes scheduling VMs for execution, virtual memory management to ensure VM isolation from other VMs, and context switching between various processor states. It also includes isolation of VMs to prevent conflicts in resource usage and emulation of timer and interrupt mechanisms.
- **Devices emulation and access control:** This includes emulating all network and storage (block) devices that different native drivers in VMs are expecting, mediating access to physical devices by different VMs.
- **Execution of privileged operations by hypervisor for guest VMs:** Certain operations invoked by guest OSs, instead of being executed directly by the host hardware, may have to be executed on its behalf by the hypervisor because of their privileged nature.
- **Management of VMs (also called VM life cycle management):** This includes configuring guest VMs and controlling VM states (e.g., Start, Pause, Stop).
- **Administration of hypervisor platform and hypervisor software:** This involves setting parameters for user interactions with the hypervisor host as well as hypervisor software.

Types of Hypervisor

There are two types of hypervisors, distinguished by whether there is another operating system between the hypervisor and the host. A **type 1 hypervisor**, as shown in Figure 8.2a, is loaded as a thin software layer directly into a physical server, much as an operating system is loaded. Once it is installed and configured, usually within a matter of minutes, the server is capable of supporting VMs as guests. In mature environments, where virtualization hosts are clustered together for increased availability and load balancing, a hypervisor can be staged on a new host, the new host can be joined to an existing cluster, and VMs can be moved to the new host without any interruption of service. The idea that the hypervisor is loaded onto the "bare metal" of a server is usually a difficult concept for people to understand. They are more comfortable with a solution that works as a traditional application, with program code loaded on top of a Microsoft Windows or UNIX/Linux operating system environment. This is exactly how a **type 2 hypervisor** is deployed (see Figure 8.2b).

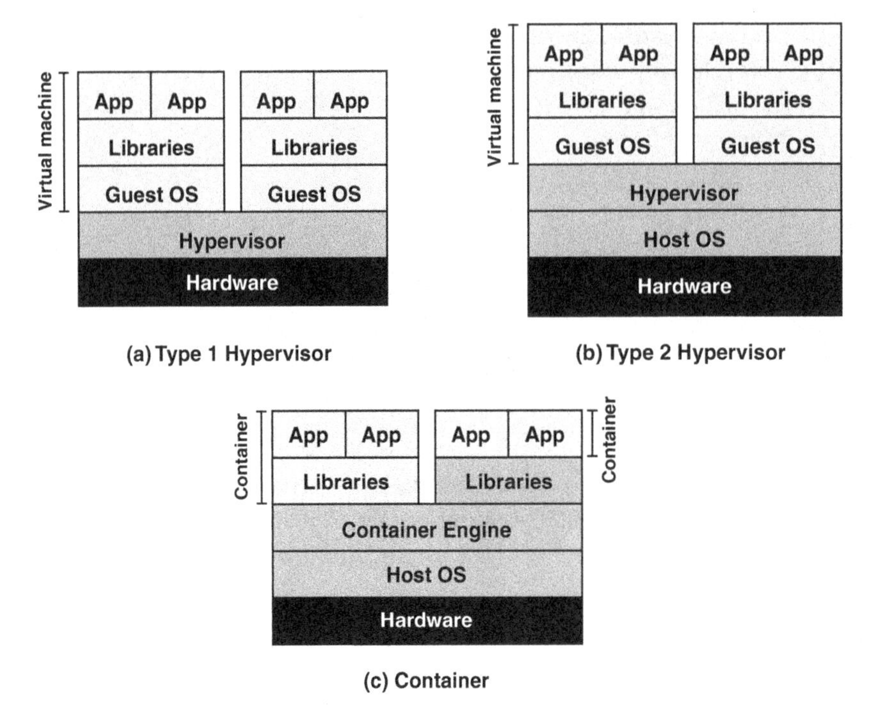

FIGURE 8.2 Comparison of Virtual Machines and Containers

Key differences between the two hypervisor types are as follows:

- Typically, type 1 hypervisors perform better than type 2 hypervisors. Because a type 1 hypervisor doesn't compete for resources with an OS, there are more resources available on the host, and, by extension, more virtual machines can be hosted on a virtualization server using a type 1 hypervisor.
- Type 1 hypervisors are considered to be more secure than type 2 hypervisors. Virtual machines on a type 1 hypervisor make resource requests that are handled external to that guest, and they cannot affect other VMs or the hypervisor they are supported by. This is not necessarily true for VMs on a type 2 hypervisor, and a malicious guest could potentially affect more than itself.
- Type 2 hypervisors allow a user to take advantage of virtualization without needing to dedicate a server to only that function. Developers who need to run multiple environments as part of their process, in addition to taking advantage of the personal productive workspace that a PC OS provides, can do both with a type 2 hypervisor installed as an application on their Linux, Windows, or macOS desktop. The virtual machines that are created and used can be migrated or copied from one hypervisor environment to another, reducing deployment time and increasing the accuracy of what is deployed—ultimately reducing the time to market of a project.

Container Virtualization

An alternative approach to virtualization is known as **container virtualization**. In this approach, software known as a **virtualization container** runs on top of the host OS kernel and provides an isolated execution environment for applications. Unlike hypervisor-based VMs, containers do not aim to emulate physical servers. Instead, all containerized applications on a host share a common OS kernel. This eliminates the resources needed to run a separate OS for each application and can greatly reduce overhead. Container virtualization is likely to be a substantial and perhaps dominant deployment technology for 5G [TOZZ20].

Figure 8.2 compares container and hypervisor software stacks. For containers, only a small container engine is required as support for the containers. The container engine sets up each container as an isolated instance by requesting dedicated resources from the OS for each container. Each container app then directly uses the resources of the host OS. Although the details differ from one container product to another, the following are typical tasks performed by a container engine:

- Maintain a lightweight runtime environment and toolchain that manages containers, images, and builds.
- Create a process for the container.
- Manage file system mount points.
- Request resources from the kernel, such as memory, I/O devices, and IP addresses.

Because all the containers on one machine execute on the same kernel, thus sharing most of the base OS, a configuration with containers is much smaller and lighter weight than a hypervisor/guest OS virtual machine arrangement. Accordingly, an OS can have many containers running on top of it, whereas it can support only a limited number of hypervisors and guest OSs.

Containers have two noteworthy characteristics:

- There is no need for a guest OS in the container environment. Therefore, containers are lighter weight and have less overhead compared to virtual machines.
- Container management software simplifies the process of container creation and management.

Because they are lighter weight, containers are an attractive alternative to virtual machines. An additional attractive feature of containers is that they provide application portability. Containerized applications can be quickly moved from one system to another.

These container benefits do not mean that containers are always a preferred alternative to virtual machines, as the following considerations show:

- Container applications are only portable across systems that support the same OS kernel with the same virtualization support features, which typically means Linux. Thus, a containerized Windows application would only run on Windows machines.

- A virtual machine may require a unique kernel setup that is not applicable to other VMs on the host; this requirement is addressed by the use of the guest OS.
- VM virtualization functions at the border of hardware and an OS. It's able to provide strong performance isolation and security guarantees with the narrowed interface between VMs and hypervisors. Containerization, which sits in between the OS and applications, incurs lower overhead but may potentially introduce greater security vulnerabilities.

8.3 NFV Concepts

NFV builds on standard virtual machine technologies, extending their use into the networking domain. This is a significant departure from traditional approaches to the design, deployment, and management of networking services. NFV decouples network functions, such as Network Address Translation (NAT), firewalling, intrusion detection, Domain Name System (DNS), and caching, from proprietary hardware appliances so they can run in software on virtual machines.

Virtual machine technology, as discussed in Section 8.2, enables migration of dedicated application and database servers to commercial off-the-shelf (COTS) x86 servers. The same technology can be applied to network-based devices, including:

- **Network function devices:** For example, switches, routers, network access points, and deep packet inspectors[1]
- **Network-related compute devices:** For example, firewalls, intrusion detection systems, and network management systems
- **Network-attached storage:** For example, file and database servers attached to the network

In traditional networks, all network elements are enclosed boxes, and hardware cannot be shared. Each device requires additional hardware for increased capacity, but this hardware is idle when the system is running below capacity. With NFV, however, network elements are independent applications that are flexibly deployed on a unified platform comprising standard servers, storage devices, and switches. In this way, software and hardware are decoupled, and capacity for each application is increased or decreased by adding or reducing virtual resources, as shown in Figure 8.3.

1. A deep packet inspector analyzes network traffic to discover the type of application that sent the data. In order to prioritize traffic or filter out unwanted data, deep packet inspection (DPI) can differentiate data, such as video, audio, chat, voice over IP (VoIP), email, and web data. By inspecting the packets all the way up to the application layer, DPI can be used to analyze anything within the packet that is not encrypted. For example, it can determine that packets contain the contents of a web page and also identify which website the page is from. The increasing use of encryption limits the utility of DPI.

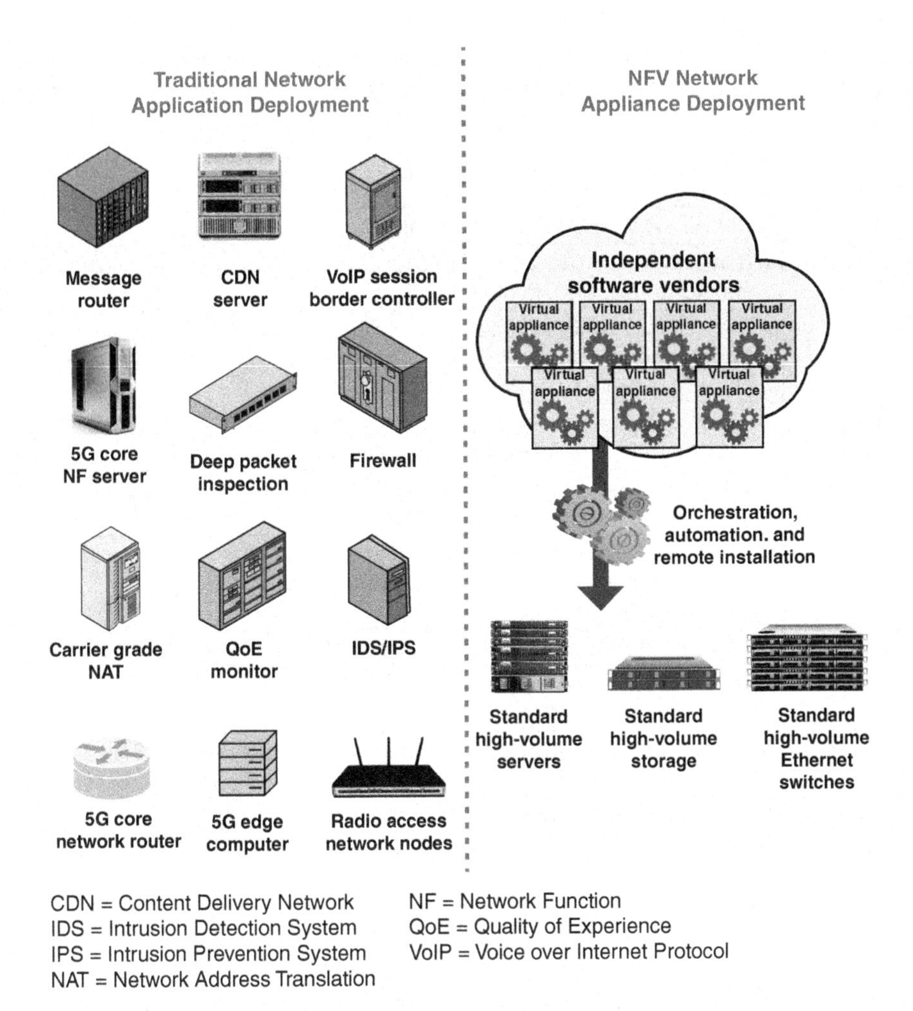

CDN = Content Delivery Network
IDS = Intrusion Detection System
IPS = Intrusion Prevention System
NAT = Network Address Translation
NF = Network Function
QoE = Quality of Experience
VoIP = Voice over Internet Protocol

FIGURE 8.3 Vision for Network Functions Virtualization

By broad consensus, the Network Functions Virtualization Industry Standards Group (ISG NFV), created as part of the European Telecommunications Standards Institute (ETSI), has the lead and indeed virtually the sole role in creating NFV standards. ISG NFV was established in 2012 by seven major telecommunications network operators. Its membership has since grown to include network equipment vendors, network technology companies, other IT companies, and service providers such as cloud service providers.

As 3GPP began developing specifications for 5G, ETSI began working closely with 3GPP to design new features to support 5G networks [ETSI19]. These new features focus on support for network slicing, resource management, and orchestration.

Table 8.1 provides definitions for a number of terms that are used in the ISG NFV documents and the NFV literature in general.

TABLE 8.1 NFV Terminology

Term	Definition
Compute domain	A domain within the NFVI that includes servers and storage.
Infrastructure network domain	A domain within the NFVI that includes all networking that interconnects compute/storage infrastructure.
Network function (NF)	A functional block within a network infrastructure that has well-defined external interfaces and well-defined functional behavior. Typically, this is a physical network node or other physical appliance.
Network functions virtualization (NFV)	The principle of separating network functions from the hardware they run on by using virtual hardware abstraction.
Network functions virtualization infrastructure (NFVI)	The totality of all hardware and software components that build up the environment in which VNFs are deployed.
NFVI-Node	A physical device or devices deployed and managed as a single entity that provides the NFVI functions required to support the execution environment for VNFs.
NFVI-PoP	An N-PoP where a network function is or could be deployed as a VNF.
Network forwarding path	An ordered list of connection points forming a chain of NFs, along with policies associated with the list.
Network point of presence (N-PoP)	A location where a network function is implemented as either a PNF or a VNF.
Network service	A composition of network functions that is defined by its functional and behavioral specification.
NFV infrastructure (NFVI)	The totality of all hardware and software components that build up the environment in which VNFs are deployed. The NFVI can span several locations (i.e., multiple N-PoPs). The network providing connectivity between these locations is regarded as part of the NFVI.
Physical network function (PNF)	An implementation of an NF via a tightly coupled software and hardware system. This is typically a proprietary system.
Virtual machine (VM)	A virtualized computation environment that behaves very much like a physical computer/server.
Virtual network	A topological component used to affect routing of specific characteristic information. The virtual network is bounded by its set of permissible network interfaces. In the NFVI architecture, a virtual network routes information among the network interfaces of VM instances and physical network interfaces, providing the necessary connectivity.
Virtualized network function (VNF)	An implementation of an NF that can be deployed on an NFVI.
VNF forwarding graph (VNF FG)	A graph of logical links connecting VNF nodes for the purpose of describing traffic flow between these network functions.
VNF set	A collection of VNFs with unspecified connectivity between them.

Simple Example of the Use of NFV

In this section we consider a simple example from the ISG NFV specification, *Network Functions Virtualisation (NFV), Architectural Framework* (ETSI GS NFV 002). Figure 8.4a shows a physical realization of a network service. At the top level, the network service consists of endpoints connected by a forwarding graph of network functional blocks, called network functions (NFs). Examples of NFs are firewalls, load balancers, and wireless network access points. In the architectural framework, NFs are viewed as distinct physical nodes. The endpoints are outside the scope of the NFV specifications and include all customer-owned devices. So, in the figure, endpoint A could be a smartphone, and endpoint B a content delivery network (CDN) server.

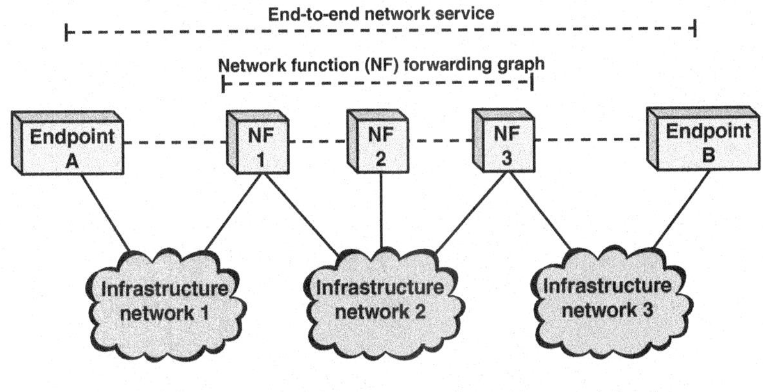

(a) Graph representation of an end-to-end network service

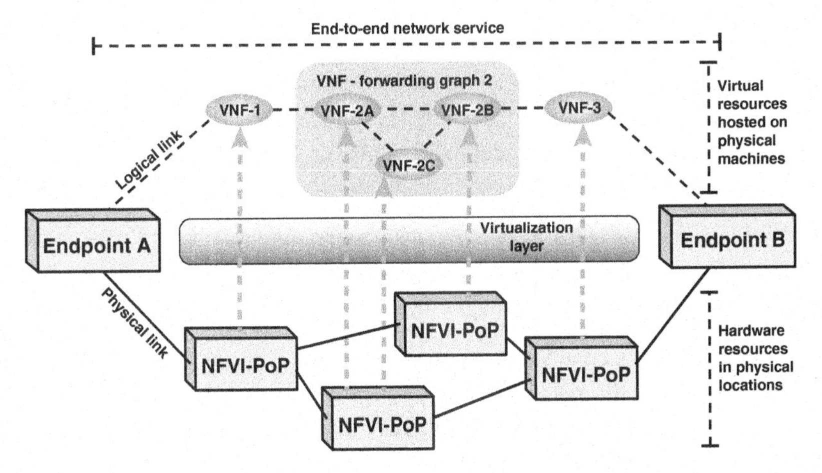

(b) Example of an end-to-end network service with VNFs and nested forwarding graphs

FIGURE 8.4 A Simple NFV Configuration Example

Figure 8.4a is a logical representation that highlights the network functions that are relevant to the service provider and customer. The interconnections among the NFs and endpoints are depicted by dashed lines, which represent logical links. These logical links are supported by physical paths through infrastructure networks (wired or wireless).

Figure 8.4b shows a virtualized network service configuration that could be implemented on a physical configuration corresponding to the logical depiction of Figure 8.4a. Virtualized network function 1 (VNF-1) provides network access for endpoint A, and VNF-3 provides network access for B. The figure also depicts the case of a nested VNF forwarding graph (VNF-FG-2) constructed from other VNFs (i.e., VNF-2A, VNF-2B, and VNF-2C). All of these VNFs run as virtual machines on physical machines, called points of presence (PoPs). This configuration illustrates several important points. First, VNF-FG-2 consists of three VNFs, even though ultimately all the traffic transiting VNF-FG-2 is between VNF-1 and VNF-3. The reason for this is that three separate and distinct network functions are being performed. For example, it may be that some traffic flows need to be subjected to a traffic policing or shaping function, which could be performed by VNF-2C. So, some flows would be routed through VNF-2C, and others would bypass this network function.

A second observation is that two of the virtual machines in VNF-FG-2 are hosted on the same physical machine. Because the two virtual machines perform different functions, they need to be distinct at the virtual resource level but can be supported by the same physical machine. But this is not required, and a network management function may at some point decide to migrate one of the virtual machines to another physical machine for performance reasons. This movement is transparent at the virtual resource level.

NFV Principles

As suggested in Figure 8.4, the VNFs are the building blocks used to create end-to-end network services. Three key NFV principles are involved in creating practical network services:

- **Service chaining:** VNFs are modular, and each VNF provides limited functionality on its own. For a given traffic flow within a given application, the service provider steers the flow through multiple VNFs to achieve the desired network functionality. This is referred to as service chaining.
- **Management and orchestration:** This involves deploying and managing the life cycle of VNF instances. Examples of functions are VNF instance creation, VNF service chaining, monitoring, relocation, shutdown, and billing. MANO also manages the NFV infrastructure elements.
- **Distributed architecture:** A VNF may be made up of one or more VNF components (VNFCs), each of which implements a subset of the VNF's functionality. Each VNFC may be deployed in one or multiple instances. These instances may be deployed on separate, distributed hosts in order to provide scalability and redundancy.

High-Level NFV Framework

Figure 8.5 shows a high-level view of the NFV framework defined in ISG NFV ETSI GS NFV 002. This framework supports the implementation of network functions as software-only NVFs. We use this figure to provide an overview of the NFV architecture, which is examined in more detail subsequently.

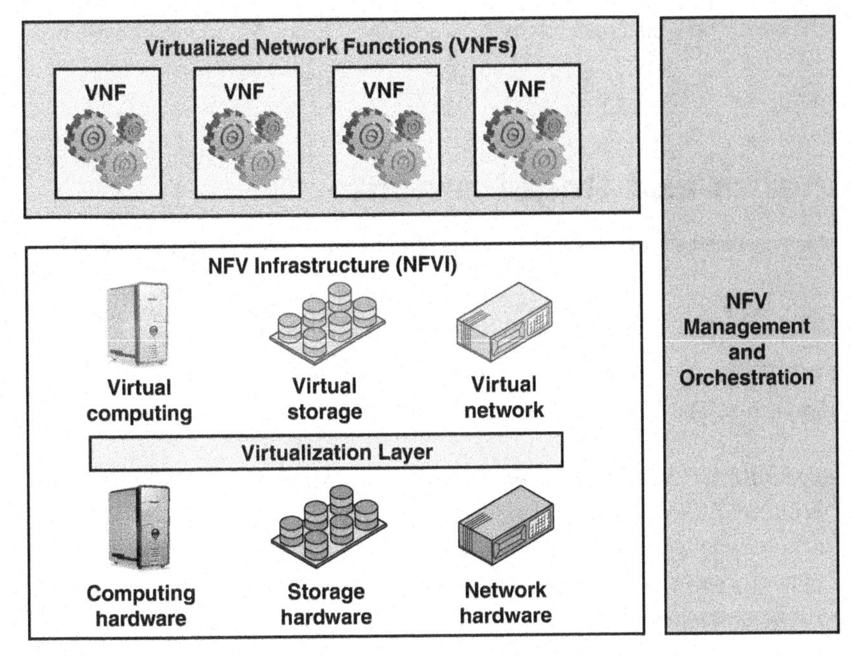

FIGURE 8.5 High-Level NFV Framework

The NFV framework consists of three domains of operation:

- **Virtualized network functions (VNFs):** A collection of VNFs, implemented in software, run over the NFVI.
- **NFV infrastructure (NFVI):** The NFVI performs a virtualization function on the three main categories of devices in the network service environment: computer devices, storage devices, and network devices.
- **NFV management and orchestration:** This encompasses the orchestration and life cycle management of physical and/or software resources that support the infrastructure virtualization and the life cycle management of VNFs. NFV management and orchestration focuses on all virtualization-specific management tasks necessary in the NFV framework.

The ISG NFV Architectural Framework specification ETSI GS NFV 002 specifies that in the deployment, operation, management, and orchestration of VNFs, two types of relationships between VNFs are supported:

- **VNF forwarding graph (VNF-FG):** Covers the case where network connectivity between VNFs is specified, such as a chain of VNFs on the path to a web server tier (e.g., firewall, network address translator, load balancer).

- **VNF Set:** Covers the case where the connectivity between VNFs is not specified, such as a web server pool.

8.4 NFV Benefits and Requirements

This section summarizes key benefits of NFV and requirements for successful implementation.

NFV Benefits

If NFV is implemented efficiently and effectively, it can provide a number of benefits compared to traditional networking approaches. The following are the most important potential benefits:

- **Reduced capital expenditure (CapEx):** A CapEx is a business expense incurred to create future benefits. A CapEx is incurred when a business spends money either to buy fixed assets or to add to the value of an existing asset with a useful life that extends beyond the tax year. NFV reduces CapEx by using commodity servers and switches, consolidating equipment, exploiting economies of scale, and supporting pay-as-you grow models to eliminate wasteful overprovisioning.
- **Reduced operational expenditure (OpEx):** An OpEx is a business expense incurred in the course of ordinary business, such as maintenance and operation of equipment. NFV reduces OpEx in a number of ways. As with CapEx savings, the use of commodity hardware and the consolidation of equipment reduce ongoing costs. Less training of staff is needed. There are fewer hardware boxes to maintain. Overall network management and control expenses are reduced. CapEx and OpEx are key drivers for NFV.
- **The ability to innovate and roll out services quickly:** NFV reduces the time to deploy new networking services to support changing business requirements, seize new market opportunities, and improve return on investment for new services. It also reduces the risks associated with rolling out new services, allowing providers to easily test and evolve services to determine what best meets the needs of customers.
- **Ease of interoperability:** Standardized and open interfaces improve interoperability.
- **Use of a single platform for different applications, users, and tenants:** Network operators can share resources across services and across different customer bases.
- **Agility and flexibility:** It is possible to quickly scale services up or down to address changing demands.
- **Targeted service introduction:** Service introduction can be based on geography or customer sets. Services can be rapidly scaled up/down as required.
- **Wide variety of ecosystems and openness:** NFV opens the virtual appliance market to pure software entrants, small players, and academia, encouraging more innovation to bring new services and new revenue streams quickly at much lower risk.

NFV Requirements

To deliver the potential benefits, NFV must be designed and implemented to meet a number of requirements and technical challenges, including the following list from an ISG NFV 2012 white paper:

- **Portability/interoperability:** It is possible to load and execute VNFs provided by different vendors on a variety of standardized hardware platforms. The challenge is to define a unified interface that clearly decouples the software instances from the underlying hardware, as represented by virtual machines and their hypervisors.
- **Performance trade-off:** Since the NFV approach is based on industry standard hardware (i.e., avoiding any proprietary hardware, such as acceleration engines), a probable decrease in performance has to be taken into account. The challenge is to keep the performance degradation as small as possible by using appropriate hypervisors and modern software technologies so that the effects on latency, throughput, and processing overhead are minimized.
- **Migration and coexistence with respect to legacy equipment:** The NFV architecture must support a migration path from today's proprietary physical network appliance-based solutions to more open standards–based virtual network appliance solutions. In other words, NFV must work in a hybrid network composed of classical physical network appliances and virtual network appliances. Virtual appliances must therefore use existing northbound interfaces (for management and control) and interwork with physical appliances implementing the same functions.
- **Management and orchestration:** A consistent management and orchestration architecture is required. NFV presents an opportunity, through the flexibility afforded by software network appliances operating in an open and standardized infrastructure, to rapidly align management and orchestration northbound interfaces to well-defined standards and abstract specifications.
- **Automation:** NFV scales only if all of the functions can be automated. Process automation is paramount to success.
- **Security and resilience:** The security, resilience, and availability of their networks should not be impaired when VNFs are introduced.
- **Network stability:** The stability of the network is not impacted when managing and orchestrating a large number of virtual appliances between different hardware vendors and hypervisors. Stability is particularly important when any of the following occur:
  - Virtual functions are relocated from one hardware platform to another.
  - Reconfiguration is required; for example, because of hardware or software failures.
  - Response to cyber-attack

- **Simplicity:** It is important to ensure that virtualized network platforms will be simpler to operate than those that exist today. A significant focus for network operators is simplification of the plethora of complex network platforms and support systems that have evolved over decades of network technology evolution, while maintaining continuity to support important revenue generating services.
- **Integration:** Network operators need to be able to mix and match servers from different vendors, hypervisors from different vendors, and virtual appliances from different vendors without incurring significant integration costs and avoiding lock-in. The ecosystem must offer integration services and maintenance and third-party support; it must be possible to resolve integration issues between several parties. The ecosystem requires mechanisms to validate new NFV products.

8.5 NFV Reference Architecture

Whereas Figure 8.5 provides a high-level view of the NFV framework, Figure 8.6 shows the ISG NFV reference architectural framework.

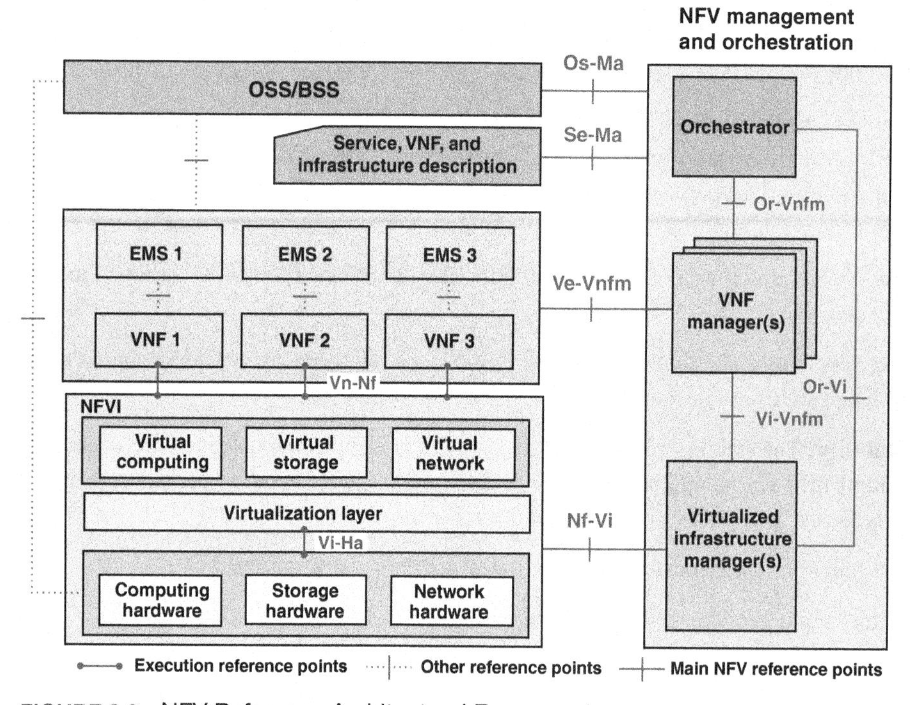

FIGURE 8.6 NFV Reference Architectural Framework

We can view this architecture as consisting of four major blocks:

- **NFV infrastructure (NFVI):** The NFVI comprises the hardware and software resources that create the environment in which VNFs are deployed. The NFVI virtualizes physical computing, storage, and networking and places them into resource pools.
- **VNF/EMS:** A collection of VNFs is implemented in software to run on virtual computing, storage, and networking resources, together with a collection of element management systems (EMSs) that manage the VNFs.
- **NFV management and orchestration (NFV-MANO):** This framework for the management and orchestration of all resources in the NFV environment includes computing, networking, storage, and virtual machine (VM) resources.
- **OSS/BSS:** Operational and business support systems are implemented by the VNF service provider.

It also useful to view the architecture as consisting of three layers. The NFVI together with the virtualized infrastructure manager provide and manage the virtual resource environment and its underlying physical resources. The VNF layer provides software implementation of network functions, together with element management systems and one or more VNF managers. Finally, there is a management, orchestration, and control layer consisting of OSS/BSS and the NFV orchestrator.

NFV Management and Orchestration

The NFV management and orchestration facility includes the following functional blocks:

- **NFV orchestrator:** Responsible for installing and configuring new network services (NS) and virtualized network function (VNF) packages, NS life cycle management, global resource management, and validation and authorization of NVFI resource requests.
- **VNF manager:** Oversees life cycle management of VNF instances.
- **Virtualized infrastructure manager:** Controls and manages the interaction of a VNF with computing, storage, and network resources under its authority, as well as their virtualization.

Reference Points

Figure 8.6 shows a number of reference points that constitute interfaces between functional blocks. The main (named) reference points and execution reference points are shown by solid lines and are in the scope of NFV. These are potential targets for standardization. The dashed line reference points are available in present deployments but might need extensions for handling network function virtualization. The dotted reference points are not a focus of NFV at present.

The main reference points include the following considerations:

- **Vi-Ha:** This reference point marks interfaces to the physical hardware. A well-defined interface specification enables operators to share physical resources for different purposes, reassign resources for different purposes, evolve software and hardware independently, and obtain software and hardware components from different vendors.
- **Vn-Nf:** These interfaces are APIs used by VNFs to execute on the virtual infrastructure. Application developers, whether migrating existing network functions or developing new VNFs, require a consistent interface that provides functionality and the ability to specify performance, reliability, and scalability requirements.
- **Nf-Vi:** These reference points mark interfaces between the NFVI and the virtualized infrastructure manager (VIM). This interface can facilitate specification of the capabilities that the NFVI provides for the VIM. The VIM must be able to manage all the NFVI virtual resources, including allocation, monitoring of system utilization, and fault management.
- **Or-Vnfm:** This reference point is used for sending configuration information to the VNF manager and collecting state information of the VNFs necessary for network service life cycle management.
- **Vi-Vnfm:** This reference point is used for resource allocation requests by the VNF manager and the exchange of resource configuration and state information.
- **Or-Vi:** This reference point is used for resource allocation requests by the NFV orchestrator and the exchange of resource configuration and state information.
- **Os-Ma:** This reference point is used for interaction between the orchestrator and the OSS/BSS systems.
- **Ve-Vnfm:** This reference point is used for requests for VNF life cycle management and exchange of configuration and state information.
- **Se-Ma:** This reference point is the interface between the orchestrator and a data set that provides information regarding the VNF deployment template, VNF forwarding graph, service-related information, and NFV infrastructure information models.

8.6 NFV Infrastructure

The heart of the NFV architecture is a collection of resources and functions known as the NFV infrastructure (NFVI). The NFVI encompasses three domains, as shown in Figure 8.7.

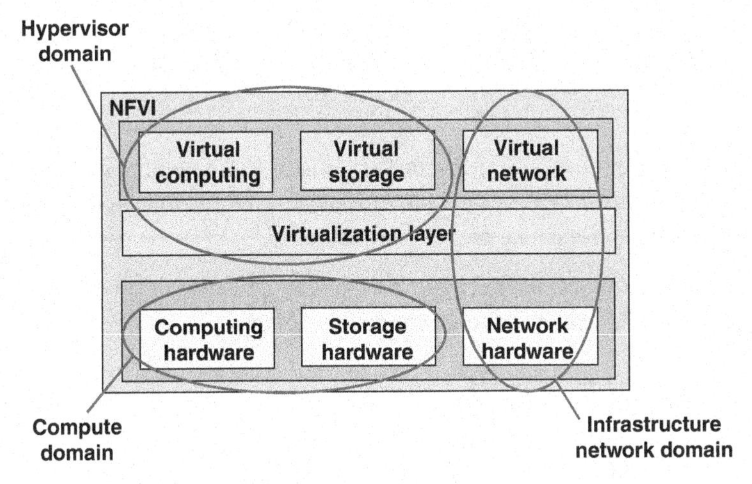

FIGURE 8.7 NFV Domains

The three domains are as follows:

- **Compute domain:** Provides COTS high-volume servers and storage.
- **Hypervisor domain:** Mediates the resources of the compute domain to the virtual machines of the software appliances, providing abstraction of the hardware.
- **Infrastructure network domain:** Comprises all the generic high-volume switches interconnected into a network that can be configured to supply infrastructure network services.

Container Interface

An important concept in NFVI is the **container interface**. Unfortunately, the ETSI documents use the term *container* in a unique sense. The NFV infrastructure document states that *container interface* should not be confused with *container* as used in the context of container virtualization as an alternative to full virtual machines. Further, the infrastructure document states that some VNFs may be designed for hypervisor virtualization, and other VNFs may be designed for container virtualization. This section examines the container interface concept.

The ETSI documents make a distinction between a functional block interface and a container interface, as follows:

- **Functional block interface:** An interface between two blocks of software that perform separate (perhaps identical) functions. The interface allows communication between the two blocks. The two functional blocks may or may not be on the same physical host.

- **Container interface:** An execution environment on a host system within which a functional block executes. The functional block is on the same physical host as the container that provides the container interface.

The concept of container interface is important because, in discussing VMs and VNFs within the NFV architecture, and how these functional blocks interact, it is easy to lose sight of the fact that all of these virtualized functions must execute on actual physical hosts.

Figure 8.8 relates container and functional block interfaces to the domain structure of NFVI. The ETSI NFVI 2012 white paper makes the following points concerning this figure:

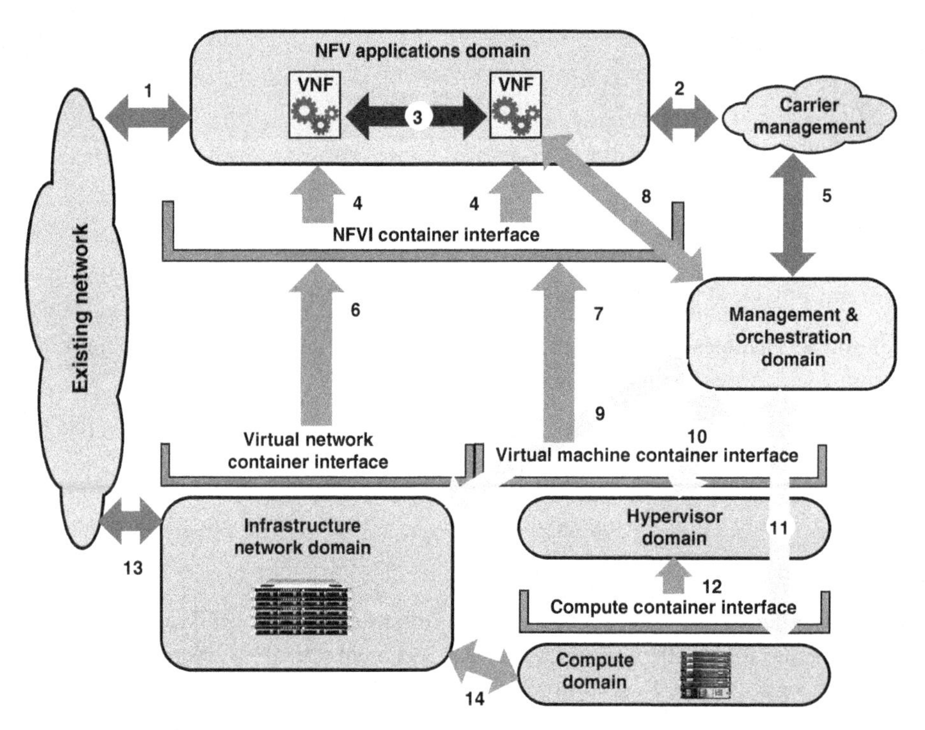

FIGURE 8.8 General Domain Architecture and Associated Interfaces

- The architecture of the VNFs is separated from the architecture hosting the VNFs (i.e., the NFVI).
- The architecture of the VNFs may be divided into a number of domains, with consequences for the NFVI and vice versa.
- Given the current technology and industrial structure, compute (including storage), hypervisors, and infrastructure networking are already largely separate domains and are maintained as separate domains within the NFVI.

- Management and orchestration tends to be sufficiently distinct from the NFVI as to warrant being defined as its own domain; however, the boundary between the two is often only loosely defined with functions such as element management functions in an area of overlap.
- The interface between the VNF domains and the NFVI is a container interface and not a functional block interface.
- The management and orchestration functions are also likely to be hosted in the NFVI (as virtual machines) and therefore are also likely to sit on a container interface.

Figure 8.8 gives insight into the deployment of NFV. The user view of a network of interconnected VNFs is of a virtualized network in which the physical and lower-level logical details are transparent. But the VNFs and logical links between VNFs are hosted on an NFVI container, which in turn is hosted on VMs and VM containers running on physical hosts. Thus, if we view the VNF architecture as having three layers (physical resource, virtualization, and application), all three layers are present on a single physical host. Of course, functionality may be distributed across multiple computer and switch hosts, but all application software ultimately is running on the same physical host as the virtualization software.

Table 8.2 describes the interfaces labeled in Figure 8.8. Interfaces 4, 6, 7, and 12 are container interfaces, so that components on both sides of the interface are executing on the same host. Interfaces 3, 8, 9, 10, 11, and 14 are functional block interfaces and, in most cases, the functional blocks on the two sides of the interface execute on different hosts. However, in some cases, some of the management and orchestration software may be hosted on a system that also hosts other NFVI components. Figure 8.8 also shows interfaces 1, 2, 5, and 13, which are interfaces to existing networks that have not implemented NFV. The NFV documents anticipate that typically NFV will be introduced over time into an enterprise facility so that interaction with non-NFV network is necessary.

TABLE 8.2 Inter-domain Interfaces Arising from the Domain Architecture

Interface Type	Interface Number	Description
NFVI container interfaces	4	Primary interface provided by the infrastructure to host VNFs. The applications may be distributed, and the infrastructure provides virtual connectivity that interconnects the distributed components of an application.
VNF interconnect interfaces	3	Interfaces between VNFs. The specification of these interfaces does not include, and is transparent to, the way the infrastructure provides the connectivity service between the hosted functional blocks.
VNF management and orchestration interface	8	Interface that allows the VNFs to request different resources of the infrastructure (e.g., request new infrastructure connectivity services, allocate more compute resources, or activate/deactivate other virtual machine components of the application).

Interface Type	Interface Number	Description
Infrastructure container interfaces	6	Virtual network container interface: Interface to the connectivity services provided by the infrastructure. This container interface makes the infrastructure appear to NFV applications as instances of these connectivity services.
	7	Virtual machine container interface: Primary hosting interface on which VNF virtual machines run.
	12	Compute container interface: Primary hosting interface on which the hypervisor runs.
Infrastructure interconnect interfaces	9	Management and orchestration interface with the infrastructure network domain.
	10	Management and orchestration interface with the hypervisor domain.
	11	Management and orchestration interface with the compute domain.
	14	Network interconnect between the compute equipment and the infrastructure network equipment.
Legacy interconnect interfaces	1	Interface between the VNF and the existing network. This is likely to be higher layers of protocol only as all protocols provided by the infrastructure are transparent to the VNFs.
	2	Management of VNFs by existing management systems.
	5	Management of NFV infrastructure by existing management systems.
	13	Interface between the infrastructure network and the existing network. This is likely to be lower layers of protocol only as all protocols provided by VNFs are transparent to the infrastructure.

Deployment of NFVI Containers

A single compute or network host can host multiple virtual machines (VMs), each of which can host a single VNF. The single VNF hosted on a VM is referred to as a VNF component (VNFC). A network function may be virtualized by a single VNFC, or multiple VNFCs may be combined to form a single VNF. Figure 8.9a shows the organization of VNFCs on a single compute node. The compute container interface hosts a hypervisor, which in turn can host multiple VMs, each hosting a VNFC.

When a VNF is composed of multiple VNFCs, it is not necessary for all of the VNFCs to execute in the same host. As shown in Figure 8.9b, the VNFCs can be distributed across multiple compute nodes interconnected by network hosts forming the infrastructure network domain.

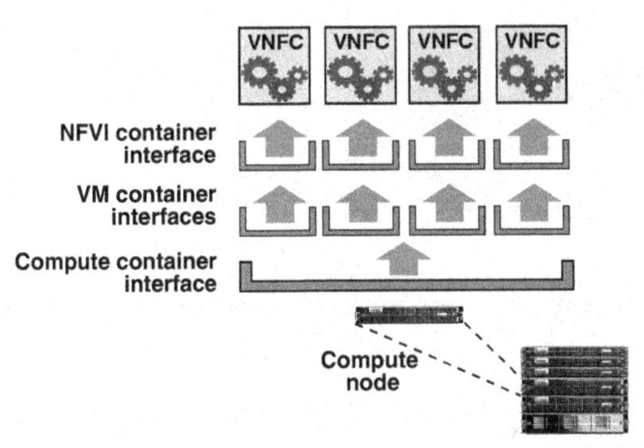

(a) A single compute platform supporting multiple VNFCs

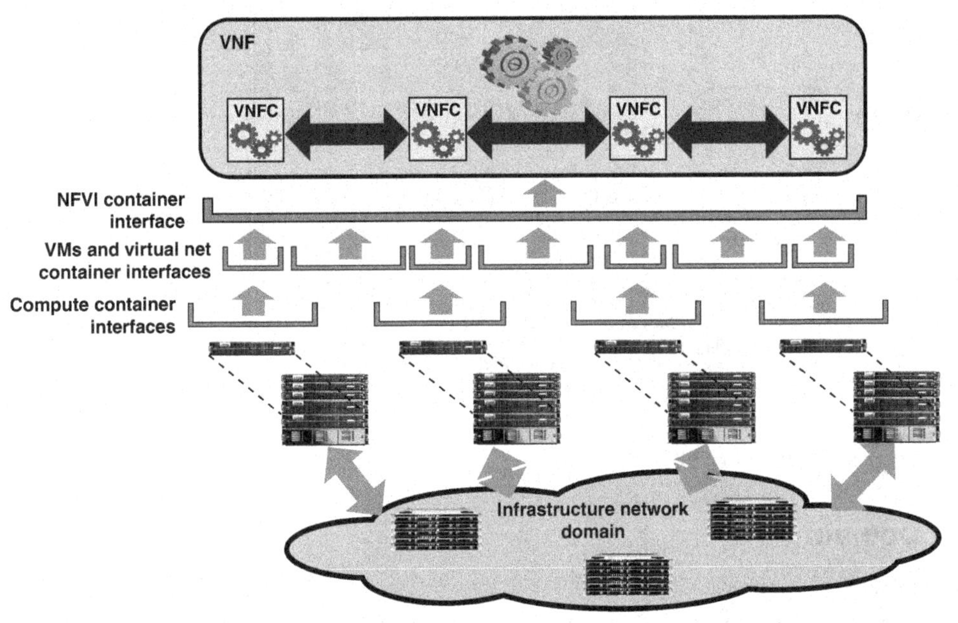

(b) A composed, distributed VNF hosted across multiple compute platforms

FIGURE 8.9 Deployment of NFVI Containers

Logical Structure of NFVI Domains

The ISG NFV specification, *Network Functions Virtualisation (NFV), Infrastructure Overview*, lays out the logical structure of the NFVI domains and their interconnections. The specifics of the actual implementation of the elements of this architecture will evolve in both open source and proprietary implementation efforts. The NFVI domain logical structure provides a framework for such development and identifies the interfaces between the main components, as shown in Figure 8.10.

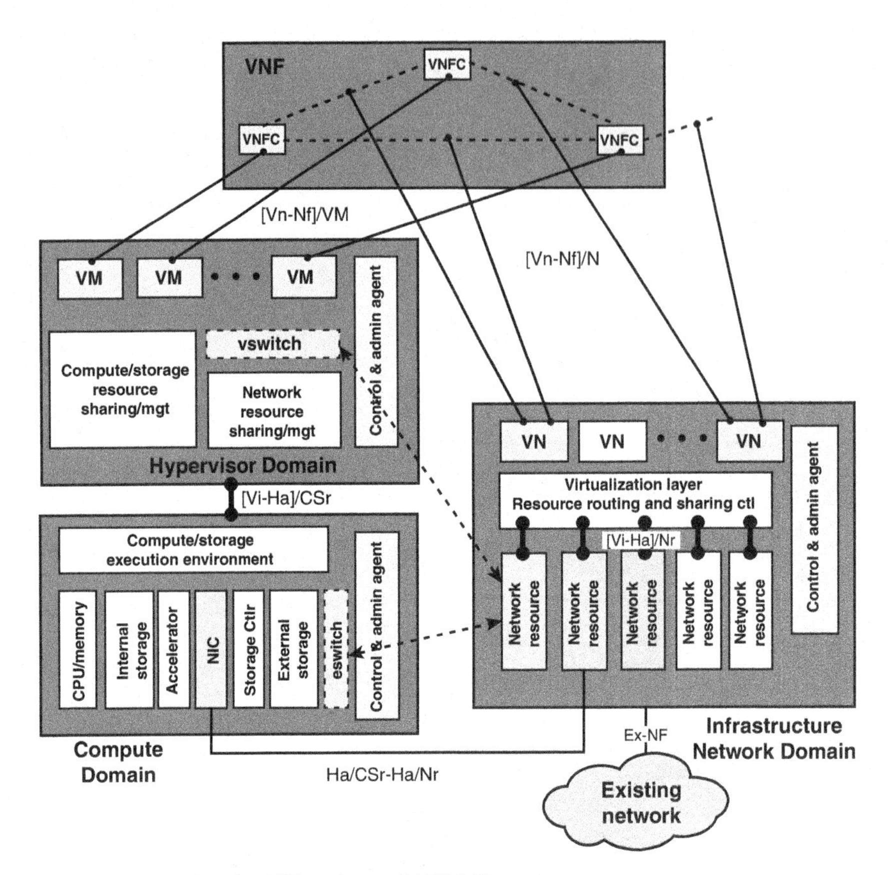

FIGURE 8.10 Logical Structure of NFVI Domains

Compute Domain

The principal elements in a typical compute domain may include the following:

- **CPU/memory:** A multicore processor, with main memory, that executes the code of the VNFC.
- **Internal storage:** Nonvolatile storage housed in the same physical structure as the processor, such as flash memory.
- **Accelerator:** Accelerator functions for security, networking, and packet processing.
- **External storage with storage controller:** Access to secondary memory devices.
- **Network interface card (NIC):** An adapter circuit board installed in a computer to provide a physical connection to a network. It provides the physical interconnection with the infrastructure network domain, which is labeled Ha/CSr-Ha/Nr and corresponds to interface 14 in Figure 8.8.
- **Control and admin agent:** An agent that connects to the virtualized infrastructure manager (VIM); refer to Figure 8.6.

- **Eswitch:** Server embedded switch. The eswitch function, described below, is implemented in the compute domain. However, functionally it forms an integral part of the infrastructure network domain.

- **Compute/storage execution environment:** The execution environment presented to the hypervisor software by the server or storage device ([VI-Ha]/CSr, interface 12 in Figure 8.8).

Eswitch

To understand the functionality of the eswitch, we first note that, broadly speaking, VNFs deal with two different kinds of workloads: control plane and data plane workloads. Control plane workloads are concerned with signaling and control plane protocols such as BGP. Typically, these workloads are more processor intensive than I/O intensive, and they do not place a significant burden on the I/O system. Data plane workloads are concerned with the routing, switching, relaying, or processing of network traffic payloads. Such workloads can require high I/O throughput.

In a virtualized environment such as NFV, all VNF network traffic would go through a virtual switch in the hypervisor domain, which invokes a layer of software between virtualized VNF software and host networking hardware. This can create a significant performance penalty. The purpose of the eswitch is to bypass the virtualization software and provide the VNF with a direct memory access (DMA) path to the NIC. The eswitch approach accelerates packet processing without any processor overhead.

NFVI Implementation Using Compute Domain Nodes

As suggested by Figure 8.9, a VNF consists of one or more logically connected VNFCs. The VNFCs run as software on hypervisor domain containers that in turn run on hardware in the compute domain. Although virtual links and networks are defined through the infrastructure network domain, the actual implementation of network functions at the VNF level consists of software on compute domain nodes.

The term *node* is used frequently in the ISG NFV specifications. The ISG NFVI specification GS NFV-INF 001 defines an **NFVI-Node** as collection of physical devices deployed and managed as a single entity, providing the NFVI functions required to support the execution environment for VNFs. NFVI nodes are in the compute domain and encompass the following types of compute domain nodes:

- **Compute node:** A functional entity that is capable of executing a generic computational instruction set (with each instruction being fully atomic and deterministic) in such a way that the execution cycle time is on the order of tens of nanoseconds, regardless of what specific state is required for cycle execution. In practical terms, this defines a compute node in terms of memory access time. A distributed system cannot meet this requirement as the time taken to access state stored in remote memory cannot meet this requirement.

- **Gateway node:** A single identifiable, addressable, and manageable element within an NFVI-Node that implements gateway functions. Gateway functions provide the interconnection between NFVI-PoPs and the transport networks. They also connect virtual networks to existing network components. A gateway may process packets going between different networks, such as removing headers and adding headers. A gateway may operate at the transport level, dealing with IP and data link packets, or at the application level.

- **Storage node:** A single identifiable, addressable, and manageable element within an NFVI-Node that provides storage resources using compute, storage, and networking functions. Storage may be physically implemented in a variety of ways. It could, for example, be implemented as a component within a compute node. An alternative approach is to implement storage nodes independent of the compute nodes as physical nodes within the NFVI-Node. An example of such a storage node might be a physical device accessible via a remote storage technology, such as Network File System (NFS) and Fibre Channel.
- **Network node:** A single identifiable, addressable, and manageable element within an NFVI-Node that provides networking (switching/routing) resources using compute, storage, and network forwarding functions.

A compute domain within an NFVI node is often deployed as a number of interconnected physical devices. Physical compute domain nodes may include a number of physical resources, such as a multicore processor, memory subsystems, and NICs. An interconnected set of these nodes comprises one NFVI-Node and constitutes one NFVI-PoP. An NFV provider might maintain a number of NFVI-PoPs at distributed locations, providing service to a variety of users, each of whom could implement VNF software on compute domain nodes at various NFVI-PoP locations.

Hypervisor Domain

The hypervisor domain is a software environment that abstracts hardware and implements services, such as starting a VM, terminating a VM, acting on policies, scaling, live migration, and high availability. The principal elements in the hypervisor domain are as follows:

- **Compute/storage resource sharing/management:** This element manages these resources and provides virtualized resource access for VMs.
- **Network resource sharing/management:** This element manages these resources and provides virtualized resource access for VMs.
- **Virtual machine management and API:** This element provides the execution environment of a single VNFC instance ([Vn-Nf]/VM, interface 7 in Figure 8.8).
- **Control and admin agent:** This element connects to the virtualized infrastructure manager (VIM); refer to Figure 8.6.
- **vswitch:** The vswitch function, described below, is implemented in the hypervisor domain. However, functionally it forms an integral part of the infrastructure network domain.

A vswitch is an Ethernet switch implemented by the hypervisor that interconnects virtual network interface cards (NICs) of VMs with each other and with the NIC of the compute node. If two VNFs are on the same physical server, they are connected through the same vswitch. If two VNFs are on different servers, the connection passes through the first vswitch to the NIC and then to an external switch. This switch forwards the connection to the NIC of the desired server. Finally, this NIC forwards it to its internal vswitch and then to the destination VNF.

Infrastructure Network Domain

The infrastructure network domain (IND) performs a number of roles. It provides:

- The communication channel between the VNFCs of a distributed VNF
- The communications channel between different VNFs
- The communication channel between VNFs and their orchestration and management
- The communication channel between components of the NFVI and their orchestration and management
- The means of remote deployment of VNFCs
- The means of interconnection with the existing carrier network

Figure 8.10 illustrates key reference points defined for the IND. As mentioned earlier, Ha/CSr-Ha/Nr defines the interface between the IND and the servers/storage of the compute domain. Ex-Nf is the reference point between any existing and/or non-virtualized network (interface 13 in Figure 8.8). Reference point [VI-HA]/Nr is the interface between the hardware network resources of the IND and the virtualization layer. The virtualization layer provides container interfaces for virtual network entities. The [Vn-Nf]/N reference point (interface 7 in Figure 8.8) is the virtual network (VN) container interface (e.g., a link or a LAN) for carrying communication between VNFC instances. Note that a single VN can support communication between more than a single pairing of VNFC instances (e.g., a LAN).

We need to make an important distinction between the virtualization function provided by the hypervisor domain and that provided by the infrastructure network domain. Virtualization in the hypervisor domain uses virtual machine technology to create an execution environment for individual VNFCs. Virtualization in IND creates virtual networks for interconnection VNFCs with each other and with network nodes outside the NFV ecosystem. These latter types of nodes are called physical network functions (PNFs).

Virtual Networks

In general terms, a virtual network is an abstraction of physical network resources, as seen by some upper software layer. Virtual network technology enables a network provider to support multiple virtual networks that are isolated from one another. Users of a single virtual network are not aware of the details of the underlying physical network or of the other virtual network traffic sharing the physical network resources. Two common approaches for creating virtual networks are

(1) protocol-based methods, which define virtual networks based on fields in protocol headers, and (2) virtual machine–based methods, with which networks are created among a set of virtual machines by the hypervisor. The NFVI network virtualization combines both these forms.

The ISG NFV specification, Network Functions Virtualisation (NFV); Infrastructure; Network Domain (GS NFV-INF 005), deals with both L2 and L3 virtual networks, described in the following section.

L2 Versus L3 Virtual Networks

Protocol-based virtual networks can be classified based on whether they are defined at protocol Layer 2 (L2), which is typically the LAN medium access control (MAC) layer, or Layer 3 (L3), which is typically the Internet Protocol (IP) layer. With an L2 VN, a virtual LAN is identified by a field in the MAC header, such as the MAC address or a virtual LAN ID field inserted into the header. So, for example, within a data center, all the servers and end systems connected to a single Ethernet switch could support virtual LANs among the connected devices. Now suppose that there are IP routers connecting segments of the data center (see Figure 8.11). Normally, an IP router will strip off the MAC header of an incoming Ethernet frame and insert a new MAC header when forwarding the packet to the next network. The L2 VN could be extended across this router only if the router had additional capability to support the L2 VN, such as being able to reinsert the virtual LAN ID field in the outgoing MAC frame. Similarly, if an enterprise had two data centers connected by a router and a dedicated line, that router would need the L2 VN capability to extend a VN.

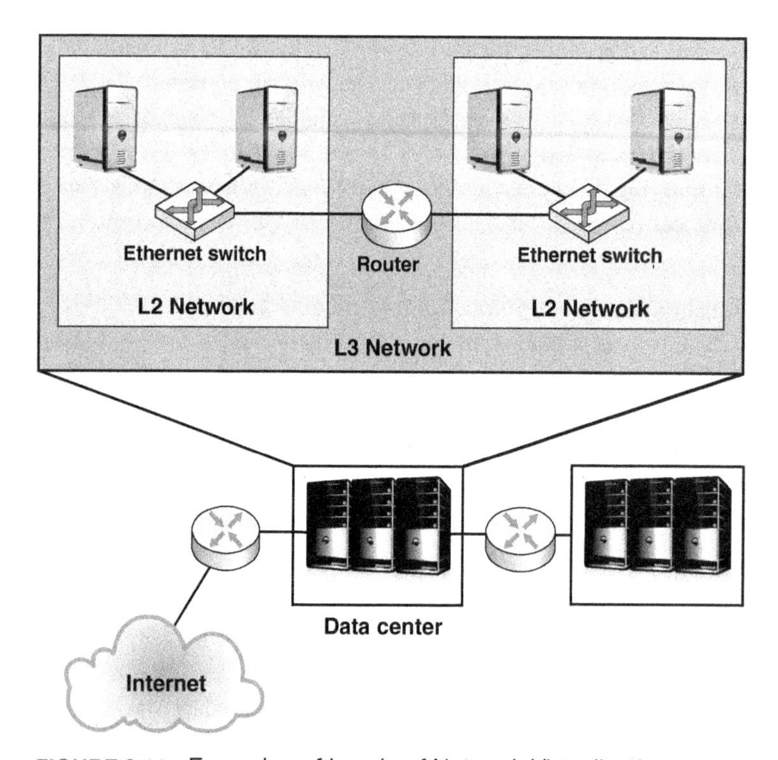

FIGURE 8.11 Examples of Levels of Network Virtualization

An L3 VN makes use of one or more fields in the IP header. A good example of this is a virtual private network (VPN) defined using IPsec. Packets traveling on a VPN are encapsulated in a new outer IP header, and the data is encrypted so that VPN traffic is isolated and protected as it transits a third-party network such as the Internet.

8.7 Virtualized Network Functions

A VNF is a virtualized implementation of a traditional network function. Table 8.3 provides examples of functions that could be virtualized.

TABLE 8.3 Potential Virtualized Network Functions

Network Element	Function
Switching elements	Broadband network gateways, carrier-grade Network Address Translation (NAT), routers
Customer premises equipment	Home routers, set-top boxes
Tunneling gateway elements	IPsec/SSL virtual private network gateways
Traffic analysis	Deep packet inspection (DPI), quality of experience (QoE) measurement
Assurance	Service assurance, service-level agreement (SLA) monitoring, testing and diagnostics
Signaling	Session border controllers, IP multimedia subsystem components
Control plane/access functions	AAA servers, policy control and charging platforms, Dynamic Host Configuration Protocol (DHCP) servers
Application optimization	Content delivery networks, cache servers, load balancers, accelerators
Security	Firewalls, virus scanners, intrusion detection systems, spam protection

VNF Interfaces

As discussed earlier in this chapter, a VNF is composed of one or more VNF components (VNFCs). The VNFCs of a single VNF are connected internally to the VNF. This internal structure is not visible to other VNFs or to the VNF user.

Figure 8.12 shows the interfaces relevant to a discussion of VNFs:

- **SWA-1:** The interface enables communication between a VNF and other VNFs, PNFs, and endpoints. Note that an interface is to the VNF as a whole and not to individual VNFCs. SWA-1 interfaces are logical interfaces that primarily make use of the network connectivity services available at the SWA-5 interface.

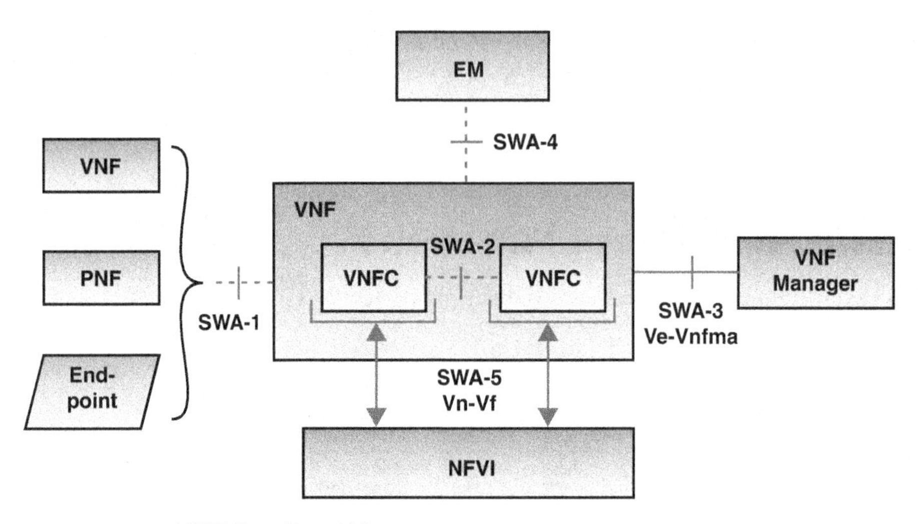

FIGURE 8.12 VNF Functional View

- **SWA-2:** This interface enables communications between VNFCs within a VNF. This interface is vendor specific and thus not a subject for standardization. This interface may also make use of the network connectivity services available at the SWA-5 interface. However, if two VNFCs within a VNF are deployed on the same host, other technologies may be used to minimize latency and enhance throughput, as described later in this chapter.
- **SWA-3:** This is the interface to the VNF manager within the NFV management and orchestration module. The VNF manager is responsible for life cycle management (creation, scaling, termination, etc.). The interface typically is implemented as a network connection using IP.
- **SWA-4:** This is the interface for runtime management of the VNF by the element manager.
- **SWA-5:** This interface describes the execution environment for a deployable instance of a VNF. Each VNFC maps to a virtualized container interface to a VM.

VNFC-to-VNFC Communication

You read earlier in this chapter that the internal structure of a VNF, in terms of multiple VNFCs, is not exposed externally. The VNF appears as a single functional system in the network it supports. However, internal connectivity between VNFCs within the same VNF or across collocated VNFs needs to be specified by the VNF provider, supported by the NFVI, and managed by the VNF manager. The VNF architectural framework specification GS NFV 002 describes a number of architecture design models that are intended to provide desired performance and QoS, such as access to storage or compute resources. One of the most important of these design models relates to communication between VNFCs.

Figure 8.13 illustrates six scenarios using different network technologies to support communication between VNFCs:

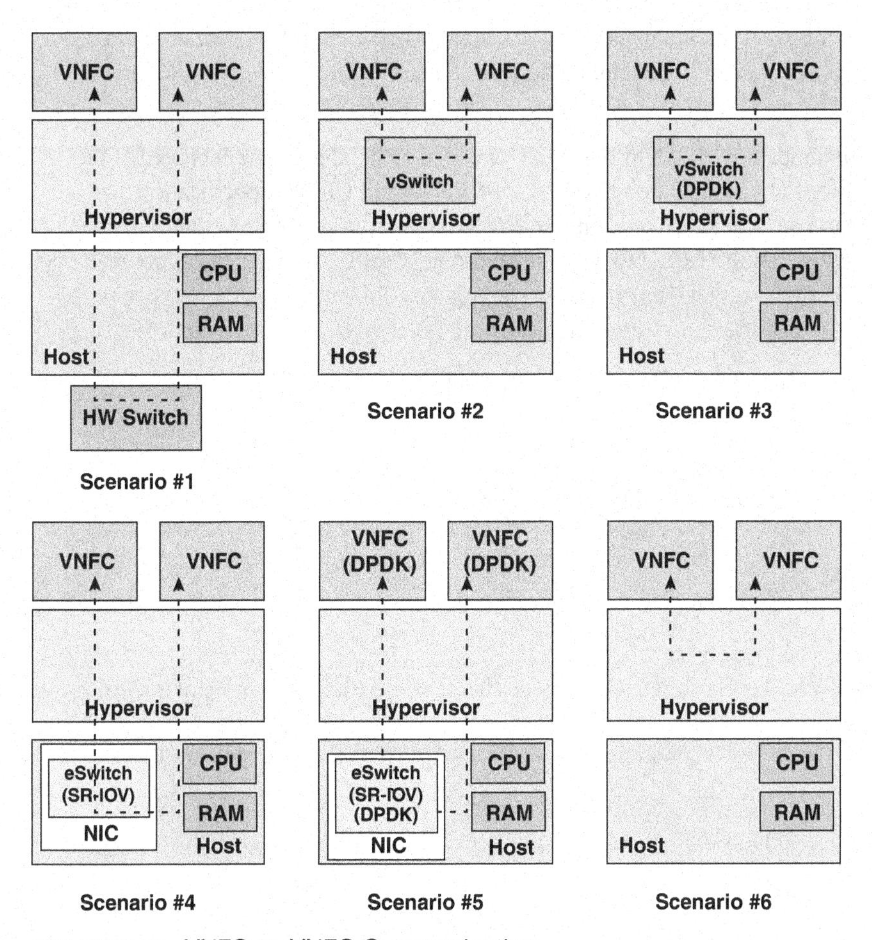

FIGURE 8.13 VNFC-to-VNFC Communication

- **Scenario 1:** Communication occurs through a hardware switch. In this case, the VMs supporting the VNFCs bypass the hypervisor to directly access the physical network interface controller (NIC). This provides enhanced performance for VNFCs on different physical hosts.
- **Scenario 2:** Communication occurs through the vswitch in the hypervisor. This is the basic method of communication between collocated VNFCs but does not provide the QoS or performance that may be required for some VNFs.

- **Scenario 3:** Greater performance can be achieved by using appropriate data processing acceleration libraries and drivers compatible with the CPU being used. The library is called from the vswitch. An example of a suitable commercial product is the Data Plane Development Kit (DPDK), which is a set of data plane libraries and network interface controller drivers for fast packet processing on Intel architecture platforms. Scenario 3 assumes a type 1 hypervisor (refer to Figure 8.2).
- **Scenario 4:** Communication occurs through an embedded switch (eswitch) deployed in the NIC with single root I/O virtualization (SR-IOV). SR-IOV is a PCI-SIG specification that defines a method to split a device into multiple PCI Express requester IDs (virtual functions) in a fashion that allows an I/O MMU to distinguish different traffic streams and apply memory and interrupt translations so that these traffic streams can be delivered directly to the appropriate VM and in a way that prevents nonprivileged traffic flows from impacting other VMs.
- **Scenario 5:** An embedded switch is deployed in the NIC hardware with SR-IOV and with data plane acceleration software deployed in the VNFC.
- **Scenario 6:** A serial bus directly connects two VNFCs that have extreme workloads or very low latency requirements. This is essentially an I/O channel means of communication rather than a NIC means.

VNF Scaling

An important property of VNFs is elasticity, which means being able to do one or more of the following:

- **Scale up:** Expand capability by adding resources to a single physical machine or virtual machine.
- **Scale down:** Reduce capability by removing resources from a single physical machine or virtual machine.
- **Scale out:** Expand capability by adding additional physical or virtual machines.
- **Scale in:** Reduce capability by removing physical or virtual machines.

Every VNF has associated with it an elasticity parameter of no elasticity, scale up/down only, scale out/in only, or both scale up/down and scale out/in.

A VNF is scaled by scaling one or more of its constituent VNFCs. Scale out/in is implemented by adding/removing VNFC instances that belong to the VNF being scaled. Scale up/down is implemented by adding/removing resources from existing VNFC instances that belong to the VNF being scaled.

8.8 SDN and NFV

Separate standards bodies are pursuing SDN and NFV, and a growing number of providers have announced or are working on products in the two fields. While these technologies can be implemented and deployed separately, there is clearly potential for added value through the coordinated use of the two technologies. Increasingly, especially within 5G core networks, SDN and NFV are deployed as a tightly interoperating capability. The combination provides a broad, unified software-based networking approach to abstract and programmatically control network equipment and network-based resources.

The relationship between SDN and NFV is perhaps best viewed as SDN functioning as an enabler of NFV. A major challenge with NFV is to enable the user to configure a network so that VNFs running on servers are connected to the network at the appropriate place, with the appropriate connectivity to other VNFs, and with desired QoS. With SDN, users and orchestration software can dynamically configure the network and the distribution and connectivity of VNFs. Without SDN, NFV requires much more manual intervention, especially when resources outside the scope of NFVI are part of the environment.

Consider load balancing, in which load balancer services are implemented as VNF entities. If demand for load balancing capacity increases, a network orchestration layer can rapidly spin up new load balancing instances and also adjust the network switching infrastructure to accommodate the changed traffic patterns. In turn, the load balancing VNF entity can interact with the SDN controller to assess network performance and capacity and use the additional information to balance traffic better or even to request provisioning of additional VNF resources.

ETSI believes that NFV and SDN complement each in a number of ways, including the following:

- The SDN controller fits well into the broader concept of a network controller in an NFVI network domain.
- SDN can play a significant role in the orchestration of the NFVI resources, both physical and virtual, enabling functionality such as provisioning, configuration of network connectivity, bandwidth allocation, automation of operations, monitoring, security, and policy control.
- SDN can provide the network virtualization required to support multitenant NFVIs.
- Forwarding graphs can be implemented using the SDN controller to provide automated provisioning of service chains and ensure strong and consistent implementation of security and other policies.
- The SDN controller can be run as a VNF, possibly as part of a service chain including other VNFs. For example, applications and services originally developed to run on the SDN controller could also be implemented as separate VNFs.

Figure 8.14, from [ETSI19], indicates the potential relationship between SDN and NFV. The arrows can be described as follows:

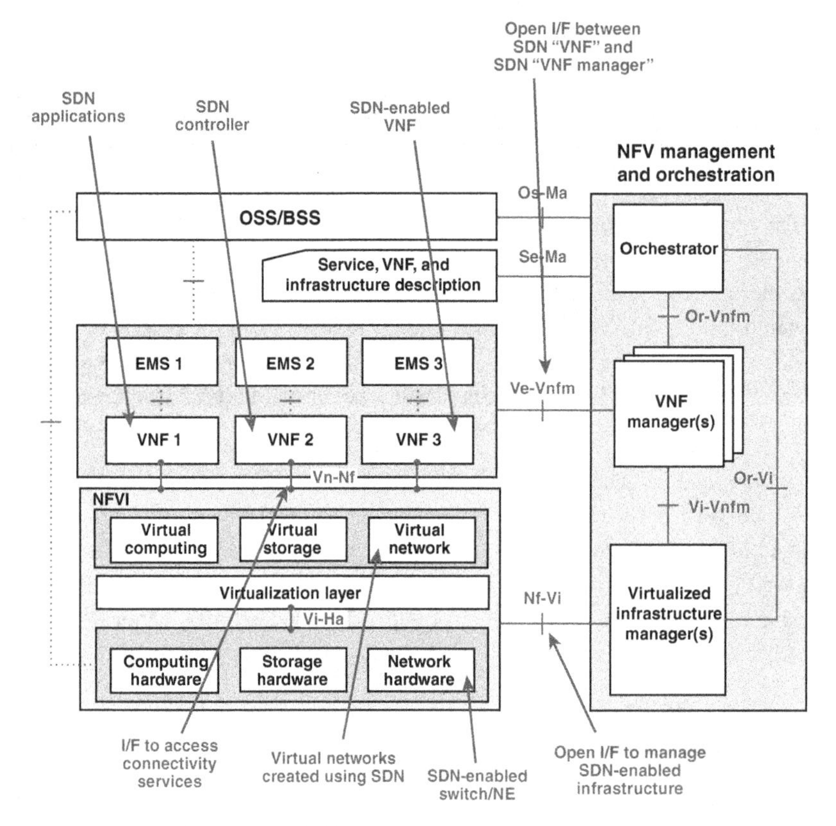

FIGURE 8.14 Mapping of SDN Components with the NFV Architecture

- SDN-enabled switches/NEs include physical switches, hypervisor virtual switches, and embedded switches on the NICs.
- Virtual networks created using an infrastructure network SDN controller provide connectivity between VNFC instances.
- The SDN controller can be virtualized, running as a VNF with its element management (EM) and VNF manager. Note that there may be SDN controllers for the physical infrastructure, the virtual infrastructure, and the virtual and physical network functions. Therefore, some of these SDN controllers may reside in the NFVI or MANO functional blocks (not shown in the figure).
- SDN-enabled VNF includes any VNF that may be under the control of an SDN controller (e.g., virtual router, virtual firewall).
- SDN applications, such as service-chaining applications, can be VNFs themselves.
- The Nf-Vi interface allows management of the SDN-enabled infrastructure.

- The Ve-Vnfm interface is used between the SDN VNFs (i.e., SDN controller VNF, SDN network functions VNF, SDN applications VNF) and their respective VNF manager for life cycle management.
- The Vn-Nf interface allows SDN VNFs to access connectivity services between VNFC interfaces.

8.9 Key Terms and Review Questions

Key Terms

business support system (BSS)	L2 virtual network
capital expenditure (CapEx)	L3 virtual network
commercial off-the-shelf (COTS)	layered virtual network
compute domain	network functions virtualization (NFV)
compute node	network interface card
consolidation ratio	network node
container	NFV management and orchestration (MANO)
container interface	NFV infrastructure (NFVI)
content delivery network (CDN)	NFV orchestrator
deep packet inspection	NFVI domain
element management	operational expenditure (OpEx)
element management system (EMS)	operations support system
eswitch	point of presence (PoP)
forwarding graph (FG)	reference points
functional block interface	scale down
gateway node	scale in
hardware virtualization	scale out
hypervisor	scale up
hypervisor domain	service chaining
infrastructure-based virtual network	storage node
infrastructure network domain (IND)	

type 1 hypervisor	virtualized infrastructure manager
type 2 hypervisor	virtualization
virtual machine (VM)	virtualization container
virtual machine monitor (VMM)	virtualized network function (VNF)
virtual network	VNF manager
virtual overlay	vswitch
virtual partition	

Review Questions

1. Briefly describe type 1 and type 2 virtualization.
2. Briefly describe container virtualization.
3. List the key reasons organizations use virtualization.
4. List three key NFV principles involved in creating practical network services.
5. What are the NFV domains of operation?
6. What types of relationships are supported between VNFs?
7. What is the difference between CapEx and OpEx?
8. List and briefly describe the major parts of the NFV reference architecture.
9. What is meant by the term *reference point* in the context of NFV?
10. List and briefly describe the main reference points in the NFV architecture.
11. What is the NFV infrastructure?
12. List and briefly describe the three NFVI domains.
13. What is the difference between an NFVI container and a virtualization container?
14. What are the principal elements in a typical computer domain?
15. Explain the concept of eswitch.
16. What is an NFVI node?
17. List and briefly describe the types of compute domain NFVI nodes.

18. What are the principal elements of the hypervisor domain?

19. Explain the concept of vswitches.

20. Explain the difference between L2 and L3 virtual networks.

21. List and briefly describe the VNF interfaces.

22. Define scale up, scale down, scale in, and scale out.

8.10 References and Documents

References

ETSI19 European Telecommunications Standards Institute. "ETSI NFV Announces New Features to Its Architecture to Support 5G." *ETSI Press Release*, July 1, 2019.

TOZZ20 Tozzi, C. "How Containers at the Edge Can Accelerate the 5G Rollout." *Platform 9 Blog*, May 21, 2020. https://platform9.com/blog/how-containers-at-the-edge-can-accelerate-the-5g-rollout/

Documents

ISG NFV *Network Functions Virtualization: An Introduction, Benefits, Enablers, Challenges & Call for Action*. ISG NFV white paper. October 2012.

ISG NFV *Network Functions Virtualisation (NFV), Architectural Framework*. ETSI GS NFV 002, December 2014.

ISG NFV *Network Functions Virtualisation (NFV), Infrastructure, Network Domain*. ETSI GS NFV-INF 005, December 2014.

ISG NFV *Network Functions Virtualisation (NFV), Infrastructure Overview*. ETSI GS NFV-INF 001, January 2015.

Chapter 9

Core Network Functionality, QoS, and Network Slicing

Learning Objectives

After studying this chapter, you should be able to:

- List and explain the 5G core network requirements defined by 3GPP
- Explain the relationship among priority, QoS, and policy control
- Present an overview of the concept of tunneling
- Present an overview of the PDU session establishment procedure
- Define 5QI and explain how it is used
- Explain the difference between QoS parameters and QoS characteristics
- Summarize the requirements for network slicing
- Give a functional description of network slice implementation

Chapters 7 and 8 covered the two essential enablers of 5G services provided by core networks: software defined networking (SDN) and network functions virtualization (NFV). With this foundation, this chapter presents an overview of the 5G core network functions and services.

Section 9.1 discusses the requirements for core networks, looking at requirements outlined by ITU-T and by 3GPP. Then, Section 9.2 examines the functional architecture of the core network. First, the section explains the concept of tunneling in 5G networks. Then it discusses the key operational aspect of core networks, which is PDU session establishment. Finally, Section 9.2 examines the central role of the policy control function.

Section 9.3 provides a detailed analysis of quality of service (QoS) and its role in 5G. Section 9.4 discusses network slicing, which is a critical capability for 5G.

9.1 Core Network Requirements

This section examines two sets of core network functional requirements: those defined by ITU-T and those defined by 3GPP.

Network Operational Requirements

As mentioned in Chapter 2, "5G Standards and Specifications," ITU-T Y.3101 (*Requirements of the IMT-2020 Network*, April 2018) defines network operational requirements. The requirements from that document are as follows:

- **Network flexibility and programmability:** The network should support a wide range of devices, users, and applications, with evolving requirements for each. Significant concepts in this regard are network functions virtualization (discussed in Chapter 8, "Network Functions Virtualization"), separation of user and control planes, and network slicing. The latter two concepts are discussed later in this chapter.
- **Fixed mobile convergence:** The focus of this requirement is to enable subscriber access through multi-access networks in seamless, integrated fashion.
- **Enhanced mobility management:** The network should support a wide variety of mobility options.
- **Network capability exposure:** The IMT-2020 network should provide suitable ways (e.g., via application program interfaces [APIs]) to expose network capabilities and relevant information (e.g., information for connectivity, QoS, and mobility) to third parties. This enables third parties to dynamically customize the network capabilities for diverse use cases within the limits set by the IMT-2020 network operator.
- **Identification and authentication:** There should be a unified approach to user and device identification and authentication mechanisms.
- **Security and personal data protection:** The IMT-2020 network must provide effective mechanisms to preserve security and personal data protection for different types of devices, users, and services, including rapid adaptation to dynamic network changes.
- **Efficient signaling:** There are two aspects to this requirement. The signaling mechanisms should be designed to mitigate risks of control and data traffic bottlenecks. Also, the network should provide lightweight signaling protocols and mechanisms to accommodate limited-resource devices.
- **Quality of service control:** The network should support different QoS levels for different services and applications.
- **Network management:** The network should provide a unified network management framework to support interworking of different providers and management of legacy networks.
- **Charging:** The IMT-2020 network needs to support different charging policies and requirements of network operators and service providers, including third parties that may be

involved in a given IMT-2020 network deployment. The charging models to be supported include, but are not limited to, charging based on volume, time, session, and application.

- **Interworking with non-IMT-2020 networks:** IMT-2020 networks should support user-transparent interworking with legacy networks.
- **IMT-2020 network deployment and migration:** The network design should accommodate incremental deployment with migration capabilities for services and related users.

For each of the 12 general requirements listed above, Y.3101 includes a number of specific, more detailed requirements. Figure 9.1, which repeats Figure 2.10, lists these requirements.

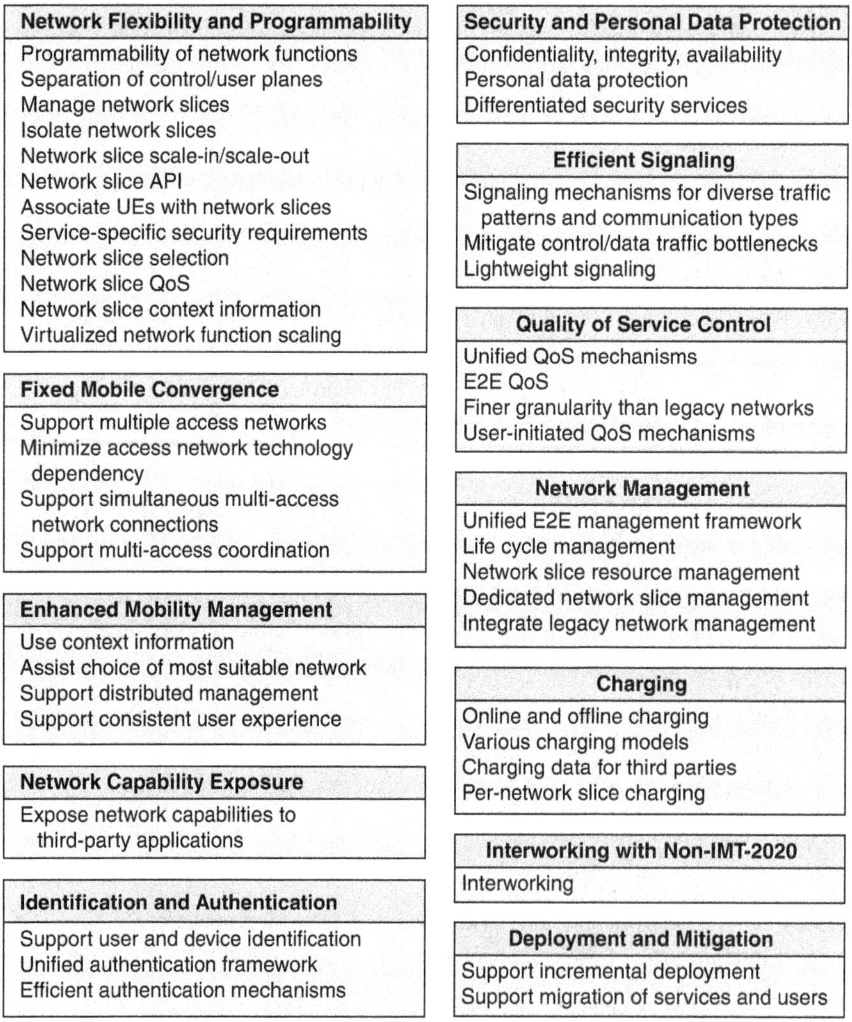

Network Flexibility and Programmability
Programmability of network functions
Separation of control/user planes
Manage network slices
Isolate network slices
Network slice scale-in/scale-out
Network slice API
Associate UEs with network slices
Service-specific security requirements
Network slice selection
Network slice QoS
Network slice context information
Virtualized network function scaling

Fixed Mobile Convergence
Support multiple access networks
Minimize access network technology dependency
Support simultaneous multi-access network connections
Support multi-access coordination

Enhanced Mobility Management
Use context information
Assist choice of most suitable network
Support distributed management
Support consistent user experience

Network Capability Exposure
Expose network capabilities to third-party applications

Identification and Authentication
Support user and device identification
Unified authentication framework
Efficient authentication mechanisms

Security and Personal Data Protection
Confidentiality, integrity, availability
Personal data protection
Differentiated security services

Efficient Signaling
Signaling mechanisms for diverse traffic patterns and communication types
Mitigate control/data traffic bottlenecks
Lightweight signaling

Quality of Service Control
Unified QoS mechanisms
E2E QoS
Finer granularity than legacy networks
User-initiated QoS mechanisms

Network Management
Unified E2E management framework
Life cycle management
Network slice resource management
Dedicated network slice management
Integrate legacy network management

Charging
Online and offline charging
Various charging models
Charging data for third parties
Per-network slice charging

Interworking with Non-IMT-2020
Interworking

Deployment and Mitigation
Support incremental deployment
Support migration of services and users

FIGURE 9.1 Requirements from the Network Operation Point of View

Basic Network Requirements

3GPP Technical Specification TS 22.261 (*Technical Specification Group Services and System Aspects, Service requirements for the 5G system, Stage 1 (Release 17)*, December 2020) defines requirements for 34 basic capabilities to be provided by a 5G network. Figure 9.2 lists these requirements (and repeats Figure 2.17). For each capability, TS 22.261 provides a description and elaborates on the requirements for that capability. The remainder of this section provides details on the key capabilities that are new to 5G.

Network slicing	Subscription aspects	Ethernet transport services
Diverse mobility management	Energy efficiency	Non-public networks
Multiple access technologies	Markets requiring minimal service levels	5G LAN-type service
Resource efficiency	Extreme long-range coverage in low-density areas	Positioning services
Efficient user plane	Multi-network connectivity and service delivery across operators	Cyber-physical control applications in vertical domains
Efficient content delivery	3 GPP access network selection	Messaging aspects
Priority, QoS, and policy control	eV2X aspects	Steering of roaming
Dynamic policy control	NG-RAN sharing	Minimization of service interruption
Connectivity models	Unified access control	UAV aspects
Network capability exposure	QoS monitoring	Video, imaging, and audio for professional applications
Context aware network		Critical medical applications
Self backhaul		
Flexible broadcast/multicast service		

eV2X = Enhanced Vehicle-to-Everything
UAV = Unmanned Aerial Vehicle

FIGURE 9.2 3GPP Basic Capability Requirements

Network Slicing

Network slicing enables operators to customize their network for different applications and customers. Slices can differ in services (e.g., priority, policy control, and security), in performance (e.g., latency, availability, reliability, and data rates), in the types and detail of assurance data, and in the type of failure diagnosis offered. Alternatively, a network slice can serve only specific users (e.g., public safety users, corporate customers, or industrial users). A network slice can provide the functionality of a complete network, including radio access network (RAN) and core network functions. Section 9.4 examines network slicing in detail.

Efficiency

TS 22.261 includes the following four capabilities related to efficiency:

- **Resource efficiency:** 5G networks need to be optimized for supporting diverse user equipment (UE) and services. Some of the underlying principles include bulk provisioning, resource efficient access, optimization for UE-originated data transfer, and efficiencies based on reduced needs related to mobility management for stationary UE and UE with restricted range of movement.
- **Efficient user plane:** Cloud-based applications can involve substantial computation that occurs far from the end user device, with substantial or time-sensitive data transfers. Such cases require

low end-to-end latencies and high data rates. 5G optimizes the user plane efficiency for such scenarios by locating applications in a service-hosting environment close to the end user. Video-based services (e.g., live streaming, virtual reality) and personal data storage applications have generated massive growth in mobile broadband traffic. In-network content caching—provided by the operator, third party, or both—can improve the user experience, reduce backhaul resource usage, and make more efficient use of radio resources for such applications.

- **Efficient content delivery:** Video-based services, such as live streaming and virtual reality, can place a considerable burden on the cellular network. To support such services, 5G networks emphasize caching content as much as possible near the end user, such as by using multi-access edge computing. In addition, 5G must support applications that involve a relatively small amount of data but have stringent latency requirements. Efficient delivery of such small packets requires the use of signaling protocols that do not require lengthy procedures and do not involve large amounts of control data.
- **Energy efficiency:** For mobile devices, energy efficiency translates directly into battery usage. Constrained IoT devices are especially of concern; such a device has a small battery, and this not only puts constrains on general power usage but also implies limitations on both the maximum peak power and continuous current drain. Thus, the 5G design must put minimal control signaling burden on such devices.

Diverse Mobility Management

Mobility management refers to a relationship between the mobile station and the RAN that is used to set up, maintain, and release the various physical channels. 5G supports different mobility management methods that minimize signaling overhead and optimizes access for user equipment with different mobility management needs. Devices may be:

- Stationary during their entire usable life (e.g., sensors embedded in infrastructure)
- Stationary during active periods but nomadic between activations (e.g., fixed access)
- Mobile within a constrained and well-defined space (e.g., in a factory)
- Fully mobile

Different applications have varying requirements for the network to mitigate the effects of mobility. Applications such as voice telephony rely on the network to ensure seamless mobility. Applications such as video streaming, on the other hand, have application layer functionality (e.g., buffering) to handle service delivery interruptions during mobility. These applications still require the network to minimize the interruption time.

Although mobility management is primarily a RAN responsibility, because of the much more distributed nature of 5G networks, mobility also has an impact in the core network. With IP traffic offload or service hosting close to the network edge, mobility of a device also implies that the anchor node in the network may need to be updated. Internet peering and service hosting have to follow the device when it is traveling across the network coverage area.

Priority, QoS, and Policy Control

Policy control is a generic term, and in a network, there are many different policies that could be implemented, such as policies related to security, mobility, and use of access technologies. When discussing policies, it is thus important to understand the context. In the context of the discussion of 5G capabilities and the policy control function, the following definitions are useful:

- **Policy:** A set of rules specifying the user plane services and functions available to a particular user, supplied by the network. In particular, a policy specifies the priority to be applied to a given user's traffic and the quality of service (QoS) to be provided to the user.
- **Policy control:** The process by which network resources are controlled to implement a given policy for a given user.
- **Priority:** A value assigned to specific packets transmitted to/from a user that determines the relative importance of transmitting those packets during the upcoming opportunity to use the medium.
- **Quality of service:** The measurable end-to-end performance properties of a network service, which can be guaranteed in advance by a service-level agreement between a user and a service provider, to satisfy specific customer application requirements. These properties may include throughput (bandwidth), transit delay (latency), error rates, security, packet loss, packet jitter, and so on.

Priority is typically included under the category of QoS, but it useful, when discussing policy control, to separate priority from other QoS parameters. The 5G network supports many commercial services and regulatory services (e.g., public safety communication) that need priority treatment. Some of these services share common QoS characteristics, such as latency and packet loss rate, but they may have different priority requirements. Mobile telephony and voice-based services for public safety share common QoS characteristics but may have different priority requirements.

Further, there are situations in which it is desirable to change the priority of a user connection but hold other QoS parameters constant and vice versa. As an example of the former, consider a healthcare patient monitored by wearable and/or implanted sensors, which provide periodic reports sent by a low-priority service. If the sensors detect a health emergency, such as the patient falling down, high-priority messages need to be sent to provide rapid response. Another example is a robot in an automated factory environment: During certain phases of the robot's operation, it may require lower latency or a higher data rate.

Connectivity Models

5G networks support both direct and indirect connectivity models for user equipment (UE). Figure 9.3 illustrates examples of three connection models:

- **Direct 3GPP connection:** An example is a sensor that communicates with an application server or with another device through a 5G network. Figure 9.3 shows a surveillance camera with 5G air interface capability.

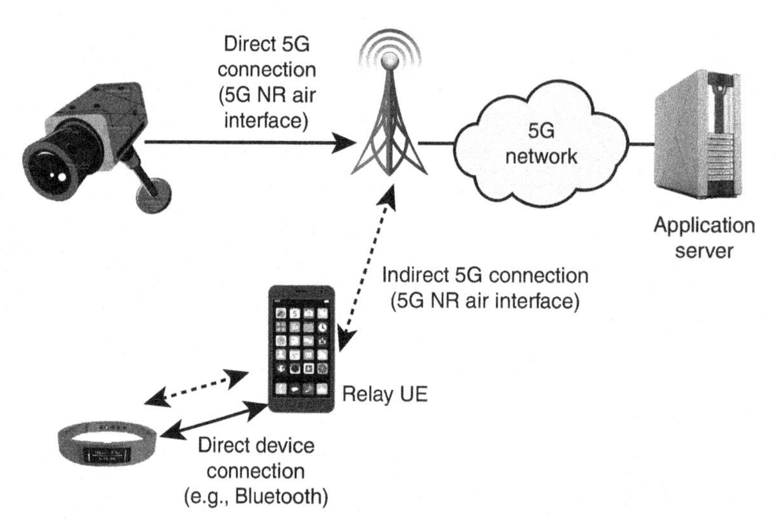

FIGURE 9.3 Connectivity Modes for Devices

- **Indirect 3GPP connection:** An example is a smart wearable that communicates through a smartphone to the 5G network. Figure 9.3 shows a fitness tracker that communicates with the user's smartphone via Bluetooth and through the phone to the 5G network. Another example is an IoT device that communicates using some wireless protocol with an IoT gateway or relay to a 5G network.
- **Direct device connection:** An example is a biometric device that communicates directly with other biometric devices or with a smartphone associated with the same patient. This is illustrated in Figure 9.3 by the connection between the fitness tracker and the smartphone.

Figure 9.3 shows an example of wearable health monitor device. In a remote setting, such as the patient's home, the device communicates indirectly to the application server. This is accomplished by a direct wireless connection to a local device that is capable of connecting to the 5G network. In an environment that provides direct support for multiple IoT devices, such as a hospital, the health monitoring devices can connect directly with the 5G network. In a large hospital setting, there may be thousands of wearable and implanted patient devices, as well as numerous other hospital-related IoT devices. This would require a massive machine type communications (mMTC) capability.

Network Capability Exposure and Context Awareness

In order to allow third parties to access information regarding capabilities provided by the 5G network (e.g., information for connectivity, QoS, and mobility) and to dynamically customize the network capabilities for diverse use cases, the 5G network should provide suitable ways (e.g., via APIs) to expose network capabilities and relevant information to third parties. Network capability exposure enables a third party to customize a dedicated network slice or allows the third party to manage an application in a service-hosting environment.

Applications may also provide the network with context awareness information. For example, radio resource management can be optimized if the network is informed about application characteristics (e.g., expected traffic over time). Other characteristics of a device, such as mobility, speed, and battery status, can be used to optimize allocation of functionality and content in the network.

Flexible Broadcast/Multicast Service

A flow, as implemented in SDN, is a distinguishable stream of related packets that results from a single user activity and requires the same QoS. The term **multicast** refers to a flow that has more than one recipient. If the group of recipients consists of all the potential recipients in some context, such as all the UE on a local network, or all the UE attached to a given virtual network, the term **broadcast** applies.

A high-capacity multicast/broadcast capability is an important requirement for 5G to meet the increasing demand for video services, ad hoc multicast/broadcast streams, software delivery over wireless, group communications, and broadcast/multicast IoT applications. A broadcast/multicast service should allow flexible and dynamic allocation of resources between unicast and multicast services within a network, and it should also allow the deployment of standalone broadcast networks. It should be possible to stream multicast/broadcast content efficiently over wide geographic areas as well as target the distribution of content to very specific geographic areas spanning only a limited number of base stations.

A white paper from 5G PPP (*View on 5G Architecture*, Version 3.0, June 2019) lists the following requirements for core network support of a flexible broadcast/multicast service:

- Enabling multicast and broadcast capabilities should require a small footprint on top of the existing unicast architecture.
- Wherever possible, treat multicast and broadcast as an internal optimization tool inside the network operator's domain.
- Consider terrestrial broadcast as a service offered also to UE without uplink capabilities that can be delivered as a self-containing service by a subset of functions of multicast and broadcast architecture.
- Simplify the system setup procedure to keep the system cost marginal. The goal is to develop an efficient system in terms of architecture/protocol simplicity and resource efficiency. Despite simplified procedures, the architecture should also allow flexible session management.
- Focus on the protocols that allow efficient IP multicast.
- Enable caching capabilities inside the network.

9.2 Core Network Functional Architecture

Chapter 3, "Overview of 5G Use Cases and Architecture," introduces the core network functional architecture. Figure 9.4, which repeats Figure 3.6, shows a service-based representation of the functional architecture defined in TS 23.501 (*Technical Specification Group Services and System Aspects, System Architecture for the 5G System [5GS], Stage 2 [Release 16]*, December 2020).

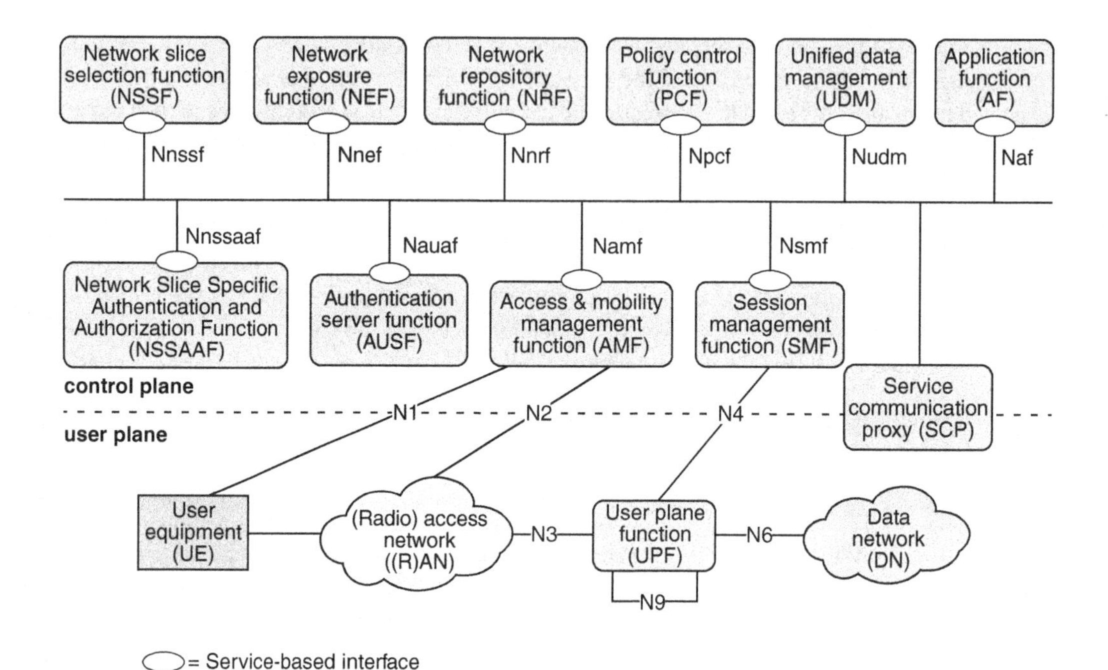

FIGURE 9.4 Non-Roaming 5G System Architecture

Table 9.1 summarizes the functionality of each network function (NF).

TABLE 9.1 Core Network Functions

Function	Description
Application function (AF)	Provides session-related information to the PCF so the SMF can use this information for session management.
Access and mobility management function (AMF)	Includes registration, reachability, and mobility management tasks.
Authentication server function (AUSF)	Performs authentication between UE and the network. The AMF initiates the UE authentication by invoking the AUSF. The AUSF selects an authentication method and performs UE authentication procedures.
Network exposure function (NEF)	Exposes capabilities of network functions and network slices as a service to third parties. In order to expose the capabilities, NEF stores the capability information and provides it upon capability discovery request.
Network repository function (NRF)	Assists the discovery and selection of required network functions (NFs). Each NF instance registers itself when instantiated and updates its status (i.e., activation/deactivation) so that the NRF can maintain information about the available network function instances. In general, each network slice instance has its own NRF, at least logically. In certain cases, such as when the network slice instances are in the same administrative domain, a single NFR instance can be shared by multiple network slice instances.

Function	Description
Network slice selection function (NSSF)	Selects appropriate network slice instances for UE. When UE requests registration with the network, the AMF sends a network slice selection request to the NSSF with preferred network slice selection information. The NSSF responds with a message including the list of appropriate network slice instances for the UE.
Policy control function (PCF)	Controls and manages policy rules, including rules for QoS enforcement, charging, and traffic routing. The PCF enables end-to-end QoS enforcement with QoS parameters (e.g., maximum bit rate, guaranteed bit rate, priority level) at the appropriate granularity (e.g., per UE, per flow, and per PDU session).
Session management function (SMF)	Provides connectivity (i.e., PDU session) for UE as well as control of the user plane for that connectivity (e.g., selection/re-selection of user plane network functions and user path, enforcement of policies including QoS policy and charging policy).
User plane function (UPF)	Performs traffic routing and forwarding, PDU session tunnel management, and QoS enforcement. The PDU session tunnels are used between access network and UPFs, as well as between different UPFs as user plane data transport for PDU sessions.
Unified data management (UDM)	Responsible for access authorization and subscription management. UDM works with the AMF and AUSF as follows: The AMF provides UE authentication, authorization, and mobility management services. The AUSF stores data for authentication of UE, and the UDM stores UE subscription data.

The interworking of these various NFs to implement the various procedures performed by the core network is extraordinarily complex. TS 23.502 (*Technical Specification Group Services and System Aspects, Procedures for the 5G System [5GS], Stage 2 [Release 16]*, December 2020) lists dozens of these procedures. The current version of the document is 603 pages long, suggesting the scale of the implementation task. This section is intended to provide insight into the functional operation of a 5G core network. The first two subsections examine two of the procedures defined in TS 23.502. The final subsection focuses specifically on the key role of the PCF.

Tunneling

Before discussing the session establishment process, we need to cover the concept of an IP tunnel. Referring back to Figure 3.8, the 5G base stations, designated *gNB*, are connected to the core network and specifically provide user plane and control plane protocol terminations toward the UE. That is, a protocol connection exists between the gNB and two elements of the core network: the AMF and the UPF.

Consider UE that wishes to send an IP packet to an endpoint attached to the Internet. The UE communicates with the Internet in three stages: (1) to/from the radio access network (RAN) that provides a wireless link between the UE and the gNB; (2) using a link, typically a wireline, between the gNB and the core network; and (3) using a link from the core network and the endpoint on the Internet.

Figure 9.5a illustrates the process of transmitting an IP packet from the UE. The UE creates an IP packet. The packet header includes the source IP address of the UE and the destination IP address of an endpoint on the Internet. This packet is sent directly over the RAN to the gNB. However, the gNB does not have a direct connection to the Internet. Instead, it has a connection to the core network—typically a fiber connection. (Chapter 15, "5G Radio Access Network," discusses other possibilities.) The gNB needs to send this packet to the UPF that is managing this session for the UE. To do this, the gNB encapsulates the entire IP packet from the UE by appending a new IP header with a source IP address of the gNB and a destination IP address of the UPF entity assigned to this session. In this operation, the original IP packet, including its header, from the UE is treated as a data block for the new, outer IP packet. This process is known as **tunneling,** and the path from the gNB to the UPF is referred to as a **tunnel**. In this case, the tunnel is known as a CN tunnel.

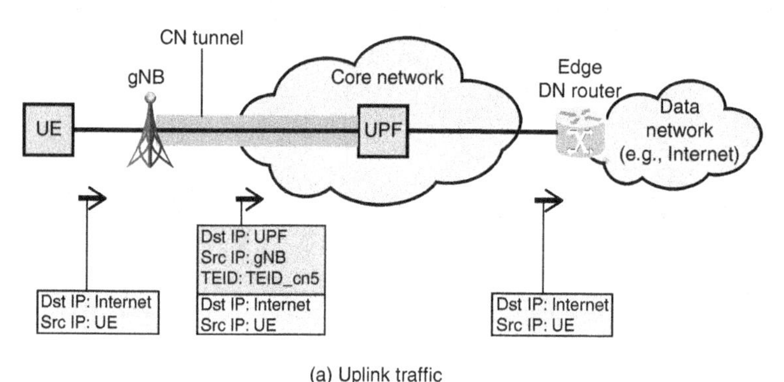

(a) Uplink traffic

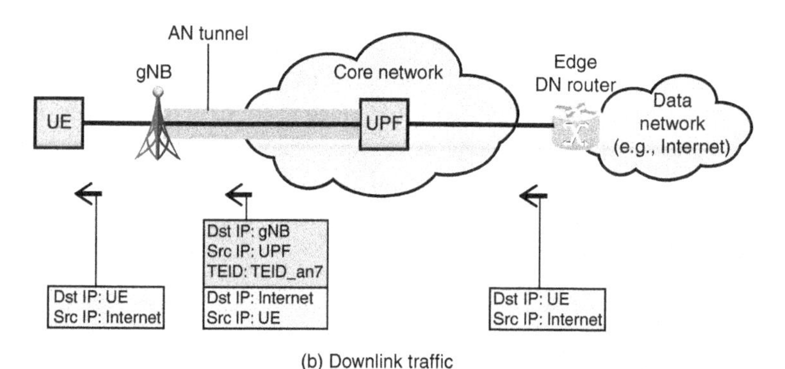

(b) Downlink traffic

FIGURE 9.5 CN and AN Tunnels

The header for the tunnel packet also includes a tunnel endpoint ID (TEID), in this case labeled TEID_cn5. There will be multiple UEs connected to the gNB, and each will generate one CN tunnel (possibly more) to the same UPF. The core network has to be able to distinguish which tunnel belongs to which UE, and that is the purpose of the TEID.

Once the tunnel packet reaches the UPF, it strips off the outer header and sends the original IP packet to a router on the edge of the DN. The DN then routes that packet to a destination device on the DN. The core network is not concerned with this final step. The core network establishes a session that runs from the UE through the RAN and the CN and terminates at the edge router of the DN.

Figure 9.5a and the preceding discussion somewhat simplify the potential configuration. It may be that the session travels through a UPF located near the gNB to a UPF located near the DN edge router. In that case, an additional tunnel is needed between the two UPFs.

The CN tunnel is unidirectional, providing an uplink for the UE. For bidirectional data exchange, a similar unidirectional tunnel, called an AN tunnel, is needed in the downlink direction, as shown in Figure 9.5b. In this case, it is the UPF that adds the encapsulating IP header with an AN tunnel TEID. The header is stripped off by the gNB before the packet is delivered to the UE.

PDU Session Establishment

This subsection looks at the simple case of PDU session establishment initiated by UE. A PDU session—which may simply be called a session—is an association between the UE and a data network that provides a PDU connectivity service. A PDU connectivity service is a service that provides for the exchange of PDUs between UE and a data network (e.g., the Internet). The objective of the UE's PDU session establishment is to establish a default QoS flow between the UE and the data network (DN). The UE can then use the default QoS flow inside the established PDU session to exchange traffic with the DN. In 5G, QoS flow is the lowest granularity of a traffic flow where QoS and charging can be applied.

Here we return to the session establishment process for a new PDU session, which is briefly examined in Chapter 3, "Overview of 5G Use Cases and Architecture," and illustrated in simplified form in Figure 3.10. Figure 9.6, from TS 23.502, shows the process in greater detail. The procedure assumes that the UE has already registered on the AMF and that the AMF has already retrieved the user subscription data from the UDM. The following steps are involved:

Step 1. From UE to AMF: In order to establish a new PDU session, UE generates a new PDU session ID and sends a message containing a PDU session establishment request to the AMF. The message contains a number of parameters, including a PDU session ID, request type, session management capabilities, protocol configuration option, data network name (DNN), and PDU data network (DN) request container (authorization information). The access network (AN) encapsulates the message sent by the UE in an RP-an message (where the RP-an is a reference point between the AN and AMF) and sends it, together with user location information and access type information, to the AMF.

Step 2. The AMF determines that the message corresponds to a request for a new PDU session based on the fact that the request type indicates "initial request" and that the PDU session ID is not used for any existing PDU session(s) of the UE. The AMF selects an SMF, considering the target data network, network slice instance (NSI), subscription information retrieved from the UDM, and access type information.

Step 3. From AMF to SMF: The AMF sends a Nsmf_PDUSession_CreateSMContext request to the SMF.

Step 4. Based on the data provided by UE, the SMF communicates with the UDM and PCF to get relevant information for PDU session creation and to determine whether the request is valid.

Step 5. From SMF to AMF: If the request is valid, the SMF returns a Nsmf_PDUSession_CreateSMContext response, which includes SM context information.

Step 6. Optional secondary authentication/authorization: If the session requires authentication and authorization, this is performed, as described in a separate part of TS 23.502.

Step 7. The purpose of step 7 is to receive policy and charging control (PCC) rules before selecting the UPF instance. Policy control is the process whereby the PCF indicates to the SMF how to control the QoS flow. Policy control includes QoS control and/or gating control. Gating control is the process of blocking or allowing packets that belong to a service data flow or detected application's traffic to pass through to the UPF. Charging control is the process of applying online charging and/or offline charging, as appropriate. Step 7 has two substeps:

Step 7a. The SMF selects a PCF instance for this session. The following factors may be considered at PCF discovery and selection for a PDU session:

- Local operator policies
- Selected data network name (DNN)
- The network slice instance of the PDU session

Step 7b. The SMF may perform an SM policy association establishment procedure to establish a session management (SM) policy association with the PCF and get the default PCC rules for the PDU session.

Step 8. If the request type in the PDU session establishment request is "initial request," the SMF selects one or more UPFs, as needed. For IP type PDU sessions, the SMF allocates an IP address (prefix) for the PDU session. If the request type is "existing PDU session," the SMF maintains the same IP address (prefix) that has already been allocated to the UE. The selection and reselection of the UPF are performed by the SMF by considering UPF deployment scenarios such as a centrally located UPF and a distributed UPF located close to or at the access network site. The selection of the UPF also enables deployment of the UPF with different capabilities (e.g., UPFs supporting no or a subset of optional functionalities).

Step 9. The SMF may perform an SMF-initiated SM policy association modification procedure in the event that a policy control request trigger condition is met. The policy control request triggers relevant for SMF define the conditions when the SMF shall interact again with the PCF after PDU session establishment. Examples of triggers include the UE moving to another operator's domain, a QoS change, and a routing information change. The PCF may provide updated policies to the SMF.

Step 10. The SMF initiates a session establishment procedure with the selected UPF. This involves the following two substeps:

Step 10a. The SMF sends a session establishment request to the UPF and provides packet detection, enforcement, and reporting rules to be installed on the UPF for this PDU session. If the SMF is configured to request IP address allocation from the UPF, then the SMF indicates to the UPF to perform the IP address/prefix allocation and includes the information required for the UPF to perform the allocation. If CN tunnel information is allocated by the SMF, the CN tunnel information is provided to UPF in this step. The SMF also determines a number of other services related to this session.

Step 10b. The UPF acknowledges by sending a session establishment response.

Step 11. SMF to AMF: The SMF requests the AMF to transfer SM information for the requested PDU session to the UE and AN. The SM message transfer signaling message to the AMF contains the PDU session ID, which allows the AMF to know which AN toward the UE to use. The message contains SM information to be forwarded to the AN by the AMF that includes CN tunnel information, QoS-related information, and other session-related information. The message also contains SM information to be forwarded to the UE by the AMF via the AN, which includes CN tunnel information, QoS-related information, and other session-related information.

Step 12. AMF to AN: The AMF sends session-related information to the AN that is related to the information received from the SMF.

Step 13. AN to UE: The AN allocates an AN tunnel (between the AN gNB and the core network) for the PDU session. The AN sends session-related information to the UE that is related to the information received from the SMF.

Step 14. AN to AMF: The AN sends a response message to the AMF with AN tunnel information and other SM-related information.

Step 15. AMF to SMF: The AMF forwards the SM information received from the AN to the SMF through a context request message.

Step 16. SMF-UPF exchange: The SMF initiates a session modification procedure with the UPF by sending a session modification request message. The SMF provides AN tunnel information to the UPF as well as corresponding forwarding rules. The UPF responds to the SMF with an RP-su session modification response message. If multiple UPFs are used in the PDU session, the UPF terminating the CN tunnel performs this procedure. After this step, the UPF delivers any downlink PDUs to the UE. If this is an authenticated UE, the SMF registers the PDU session with the UDM by providing SM information.

Step 16a. The SMF initiates a session modification procedure with the UPF. The SMF provides AN tunnel information to the UPF as well as the corresponding forwarding rules.

Step 17. SMF to AMF: Nsmf_PDUSession_UpdateSMContext response (cause): The SMF responds to the context request received from the AMF in step 15. After this step, the AMF forwards relevant events that the SMF subscribes to (e.g., location reporting, UE moving into or out of the area of interest).

Step 18. (Conditional) SMF to AMF: If any time after step 5 the PDU session establishment is not successful, the SMF informs the AMF by invoking Nsmf_PDUSession_SMContextStatusNotify (release). The SMF also releases any session(s) created, any PDU session address, if allocated (e.g., IP address), and the association with the PCF, if any. In this case, step 19 is skipped.

Step 19. SMF to UE: In the case of PDU session type IPv6 or IPv4v6, the SMF generates an IPv6 router advertisement and sends it to the UE.

Step 20. If the UE has indicated support for transferring port management information containers, the SMF informs the PCF that a manageable Ethernet port has been detected. The SMF also includes the port number, MAC address, and port management information container.

Step 21. If the PDU session establishment failed after step 4, the SMF unsubscribes the modification of SM subscription data.

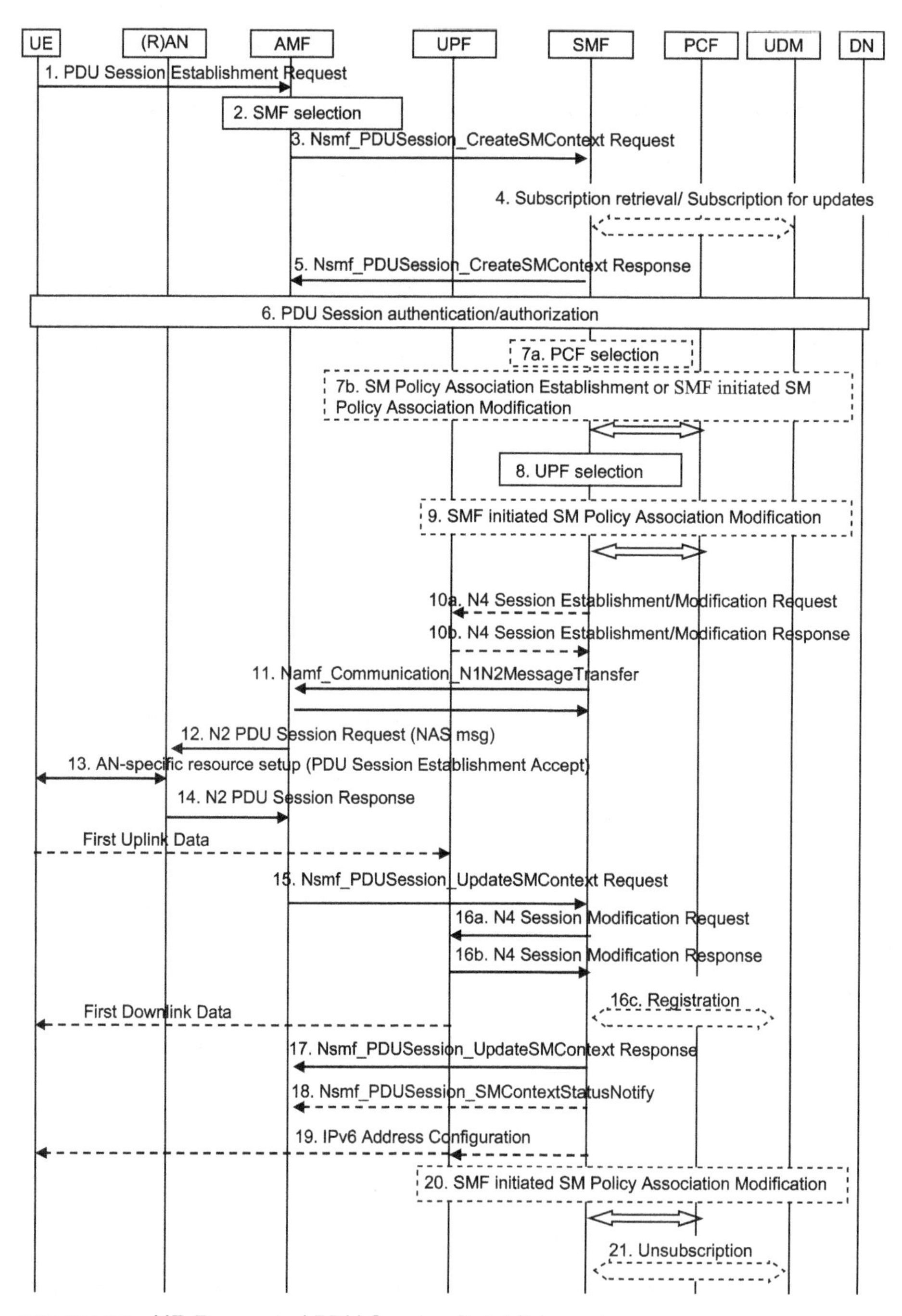

FIGURE 9.6 UE-Requested PDU Session Establishment

Figure 9.7 illustrates the message flows involved in PDU session establishment and gives some idea of the complexity of the operation. However, the preceding 21-step enumeration of tasks is only a summary overview. The specification in TS 23.502 occupies 21 pages, and it includes numerous references to other sections in the same TS as well as other TS documents. Thus, the full specification runs to well over 100 pages—and this is just one of dozens of procedures that must be implemented in the core network.

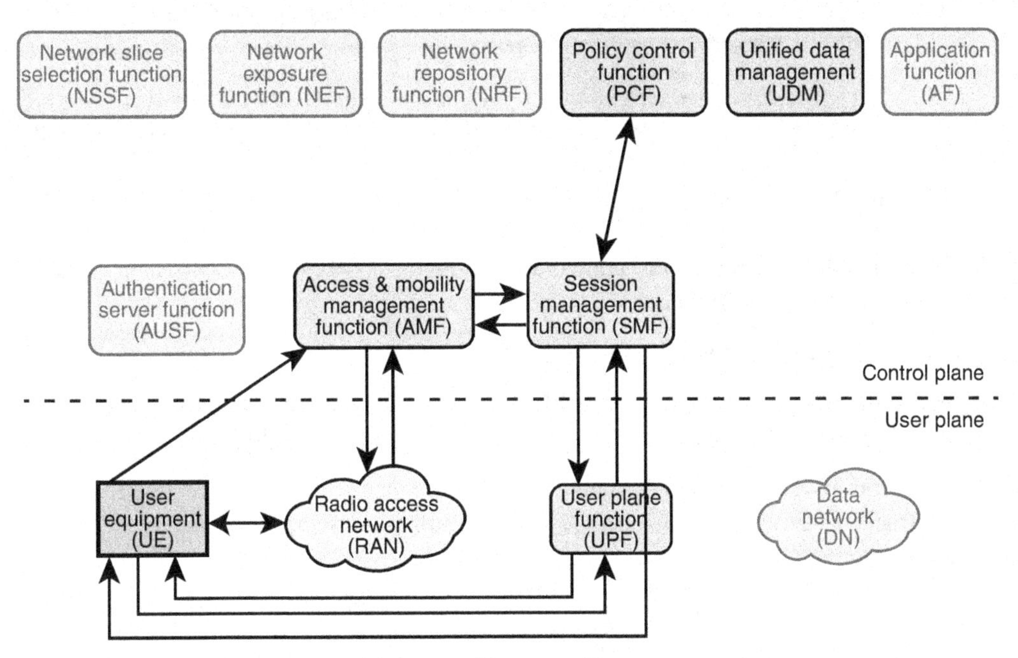

FIGURE 9.7 PDU Session Establishment Message Flow

Policy Control Function

The core network supports a common policy framework, together with network policies that allow UEs to choose the most suitable access network and access-agnostic QoS mechanisms. Thus a key element of the core network functional architecture is the policy control function (PCF).

Before discussing PCF functionality, consider the following related terms:

- **PDU session:** This is a logical connection that carries all the communication between UE and a data network (DN).
- **QoS flow:** This is the lowest level of granularity within the 5G system for defining policy and charging rules. A PDU session may contain multiple QoS flows.
- **Service data flow (SDF):** An SDF provides an end-to-end packet flow between UE and a specific application at the DN. One or more SDFs can be transported in the same QoS flow if they share the same policy and charging rules.

PCF Requirements

The overall requirement on the PCF function is to enable the core network to apply policy and charging control to UE accesses. As mentioned in the preceding subsection, policy control is the process whereby the PCF indicates to the SMF how to control the QoS flow. Policy control includes QoS control and/or gating control. Gating control is the process of blocking or allowing packets that belong to a service data flow or detected application's traffic to pass through to the UPF. Charging control is the process of applying online charging and/or offline charging, as appropriate.

3GPP TS 23.503 (*Technical Specification Group Services and System Aspects*; *Policy and charging control framework for the 5G System [5GS]*; *Stage 2 [Release 16]*, April 2020) breaks down specific PCF requirements into non-session management–related requirements and session management–related requirements. The non-session management–related requirements are as follows:

- **Access- and mobility-related policy control requirements:** These requirements support the AMF, as described later in this chapter.
- **UE policy control requirements:** These requirements provide policy information to the UE.
- **Network status analytics information requirements:** These requirements relate to collecting slice-specific network status analytic information and using that data in policy decisions.
- **Management of packet flow descriptions:** These descriptions provide the capability to create, update, or remove PFDs in the NEF (PFDF) and the distribution from the NEF (PFDF) to the SMF and finally to the UPF.
- **SMF selection management–related policy control requirements:** These requirements provide SMF selection management–related policies to the AMF.
- **Support for non-session management–related network capability exposure:** This support enables an AF to request non-session management–related policy control functionality from the NEF.

The session management–related requirements are as follows:

- **Charging-related requirements:** These requirements provide information to allow for charging control on each SDF.
- **Policy control requirements:** These requirements cover gating control and QoS control requirements:
  - **Gating control:** You apply gating control on a per-SDF basis. For example, session termination triggers gating control for each affected SDF.
  - **QoS control:** The PCF must support QoS control and the SDF, QoS flow, and PDU session levels.

- **Usage monitoring control requirements:** The PCF may use monitoring of both volume and time use to make dynamic policy decisions. It sends the applicable thresholds (of time or volume) to the SMF for monitoring and notification to the PCF. The monitoring is possible for an individual SDF or a group of SDFs or all traffic on a PDU session.
- **Application detection and control requirements:** The application detection and control feature comprises the request to detect the specified application traffic, the report from the SMF to the PCF on the start or stop of application traffic, and the application of the specified enforcement and charging actions.
- **Support for session management–related network capability exposure:** This support enables an AF to request session management–related policy control functionality for capability exposure.
- **Traffic steering control:** This is the capability to activate/deactivate traffic steering policies from the PCF in the SMF.

Interfaces with Other Network Functions

Figure 9.8 shows a reference point representation indicating how the PCF interfaces with other network functions.

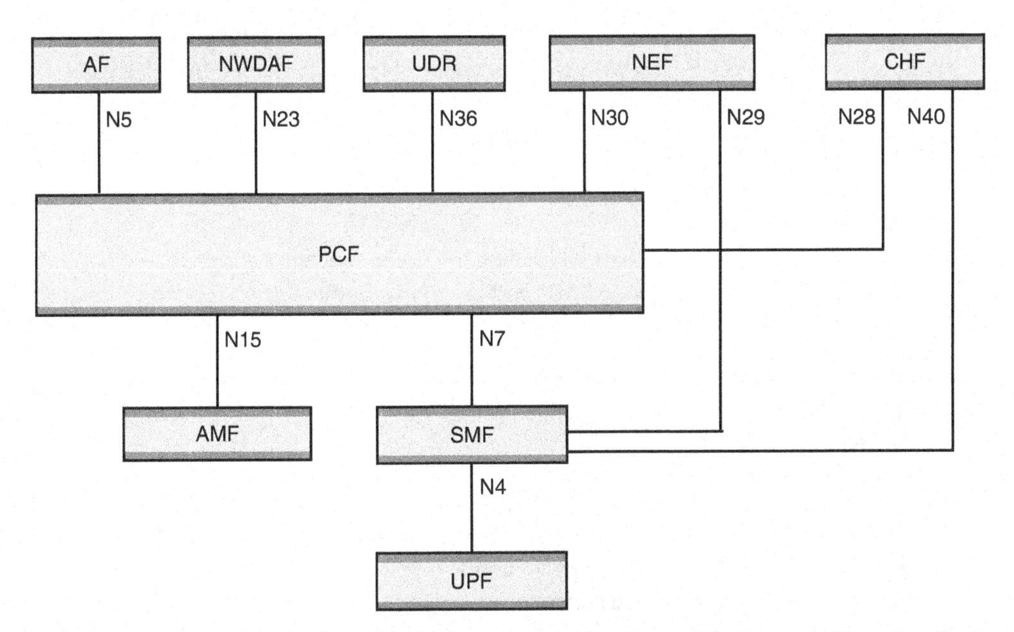

FIGURE 9.8 Overall Non-Roaming Reference Architecture of Policy and Charging Control of Framework for the 5G System (Reference Point Representation)

This figure introduces the following NFs that are not shown in Figure 9.4:

- **Network data analytics function (NWDAF):** Used for data collection and analytics for centralized as well as edge computing resources. It provides network slice–specific data analytics to the PCF and NSSF, which in turn use this data for policy decisions (PCF) and slice selections (NSSF).
- **Unified data repository (UDR):** Serves as a single repository of subscription data, application data, and policy data by integrating with NF consumers (including NEF, AMF, and PCF). It also notifies for the subscription data changes.
- **Charging function (CHF):** Provides an account balance management function, a rating function, and a charging gateway function.

The PCF interacts with other interfaces as follows:

- **PCF–AF interface:** This interface allows for the transport of application-level session information and Ethernet port management information from the AF to the PCF. This includes bandwidth requirements for QoS, identification of application service providers and applications, traffic routing based on applications access, and identification of application traffic for policy and charging control.
- **PCF–SMF interface:** This interface enables the PCF to have dynamic control over the policy and charging behavior at an SMF. The SMF receives control plane information from NFs and user plane information from the UPF. An SMF triggers the PCF to enforce policy decisions when the policy trigger related to session management is met.
- **PCF–SMF–UPF interface:** The PCF and UPF don't communicate directly with each other. They exchange policy actions/enforcements via the SMF. The SMF provisions the policy and threshold rules on the UPF for usage control based on the static/dynamic policy rules configured in the PCF, predefined rules in the SMF, and/or credit control triggers received from the CHF.
- **PCF–AMF interface:** The AMF acts as a single entry point for the UE connection. The PCF provides access and mobility management–related policies for the AMF in order to trigger policy rules on the UE or user sessions.
- **PCF–UDR interface:** The PCF retrieves the policy-/subscription-/application-specific data from the UDR. Policy control–related subscription and application-specific data gets provisioned into the UDR. The UDR can also generate notifications based on the changes in the subscription information, according to the operator's pricing model.
- **PCF–CHF interface:** This interface enables the PCF to access policy control status information related to subscriber spending. The CHF stores the policy counter information against the subscriber pricing plan and notifies the PCF whenever the subscriber breaches the policy thresholds, based on usage consumption. On receiving policy trigger information, the PCF

applies the policy decision by interacting with the SMF (which in turn informs the UPF for the policy enforcement).

- **PCF–NEF interface:** The NEF exposes network function services and resources to the external world. In terms of interaction with the PCF, it exposes the capabilities of network functions for supporting policy and charging.
- **PCF–NWDAF interface:** The PCF collects slice-specific network status analytic information from the NWDAF. The NWDAF provides network data analytics (i.e., load-level information) to the PCF on a network slice level. The PCF is able to use that data in its policy decisions.

9.3 Quality of Service

A wide variety of applications and devices use 5G networks, including cloud computing, big data, the pervasive use of mobile devices on enterprise networks, and the increasing use of video streaming. These factors together contribute to the increasing difficulty in maintaining satisfactory network performance. The key tool in characterizing and measuring the network performance that an enterprise desires to achieve is quality of service (QoS). QoS is the measurable end-to-end performance properties of a network service, which can be guaranteed in advance by a service-level agreement (SLA) between a user and a service provider in order to satisfy specific customer application requirements. QoS enables a network manager to determine whether the network is meeting user needs and to diagnose problem areas that require adjustment to network management and network traffic control.

This section begins by summarizing the QoS capabilities required in a 5G network, as defined by ITU-T. Then, it introduces a QoS architectural framework that provides insight into the scope and complexity of a QoS system. The remainder of the section covers QoS details defined by 3GPP.

QoS Capabilities

QoS capabilities and accompanying SLAs serve two purposes:

- Enable networks to offer different levels of QoS to customers on the basis of customer requirements.
- Allocate network resources efficiently, maximizing effective capacity.

ITU-T Y.3106 (*Quality of Service Functional Requirements for the IMT-2020 Network*, April 2019) defines a QoS life cycle management process that encompasses the entire range of capabilities involved in providing QoS. Figure 9.9, based on a figure in Y.3106, shows the four stages in the QoS management life cycle. As shown, a QoS capability encompasses the interface between the UE and the AN, where the QoS service is visible to the user; the functionality in the AN and in the CN to provide the QoS; and the ability to exchange QoS information and requirements with other networks.

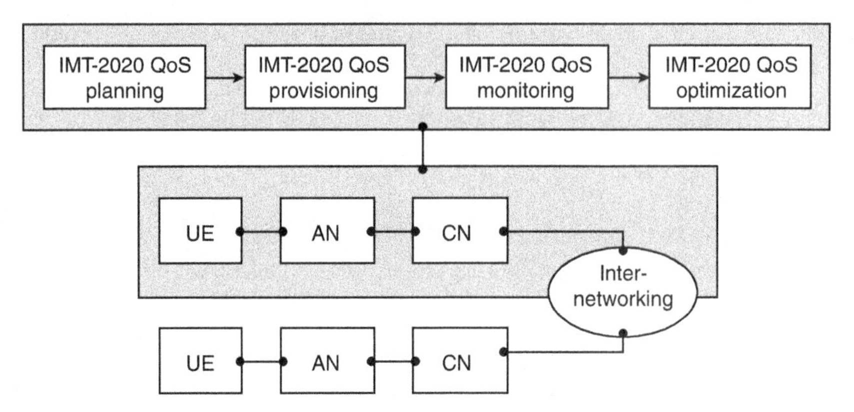

FIGURE 9.9 High-Level QoS Capabilities of the IMT-2020 Network

Table 9.2 defines the four QoS management categories.

TABLE 9.2 QoS Management Categories

Category	Definition
QoS planning	The process of determining the mechanisms and services to be implemented on the network.
QoS provisioning	The process of configuring and maintaining selected network elements based on customer SLAs and observed quality performance.
QoS monitoring	The process of collecting QoS statistics, faults, and warnings. This data is then used for generating analysis reports and making changes and upgrades to the network.
QoS optimization	The process of accessing monitored information, processing the data to determine service and network quality metrics, and initiating corrective actions when any of the quality levels is considered unsatisfactory.

Important requirements for QoS planning include the following:

- Support service-driven QoS planning for the IMT-2020 network.
- Support dynamical modeling of diversified IMT-2020 usage scenarios (e.g., eMBB, MTC, and URLLC).
- Convert service models to traffic models accurately.
- Support an accurate estimate of network coverage, capacity, resources, and network slice requirements.
- Estimate and allocate network resources in a way that efficiently maximizes utilization.
- Support QoS-aware routing to satisfy different service requirements for delay, bandwidth, throughput, load balance, cost, etc.

QoS provisioning requirements are as follows:

- Support E2E QoS for diversified IMT-2020 usage scenarios (eMBB, MMTC, and URLLC).
- Support translation of service-centric SLA to resource-facing network slice descriptions.
- Support efficient E2E QoS provisioning with the capabilities of a global network view, on-demand softwarized network functions, autonomous network slicing management, and orchestration.
- Support unified and access-agnostic (fixed or mobile access) QoS control from a core network (CN) perspective.
- Support proper QoS interworking and mapping among UE, AN, CN, and other data networks (DNs).
- Support a finer level of QoS granularity based on flows to meet different service requirements.
- Support QoS enforcement, which includes flow classification, marking, congestion avoidance, queue shaping, and queue scheduling based on QoS rules.

QoS monitoring requirements are as follows:

- Provide a mechanism for supporting real-time E2E QoS monitoring.
- Provide an interface to applications for QoS monitoring (e.g., to initiate QoS monitoring, request QoS parameters, events, or logging information).
- Respond to an authorized user request to provide real-time QoS monitoring information within a specified time after receiving the request.
- Provide real-time QoS parameters and events information to an authorized application or network entity.
- Support an update or refresh rate for real-time QoS monitoring within a specified time.
- Log the history of QoS events, including, for example, parts of the SLA that are not met and timestamps of the events and event positions (e.g., UE and radio access points associated with the events).
- Support different levels of granularity for QoS monitoring (e.g., per flow or per set of flows).

QoS optimization requirements are as follows:

- Support intelligent QoS anomaly detection based on the analysis of QoS data.
- Support traffic prediction based on the analysis of QoS data.
- Support QoS anomaly prediction based on the analysis of QoS data.
- Support QoS optimization to provide and ensure a desired service performance level during the life cycle of the service.

QoS Architectural Framework

Before we look at the Internet standards that deal with provision of QoS on the Internet and in private internetworks, in this section we consider an overall architectural framework that relates the various elements that go into QoS provision. Such a framework has been developed by ITU-T. Recommendation Y.1291 (*An Architectural Framework for Support of Quality of Service in Packet Networks*, May 2004) provides an overview of the mechanisms and services that comprise a QoS facility.

The Y.1291 framework consists of a set of generic network mechanisms for controlling the network service response to a service request, which can be specific to a network element, or for signaling between network elements, or for controlling and administering traffic across a network. Figure 9.10 shows the relationships among these elements, which are organized into three planes: data, control, and management. This architectural framework is an excellent overview of QoS functions and their relationships and provides a useful basis for summarizing QoS.

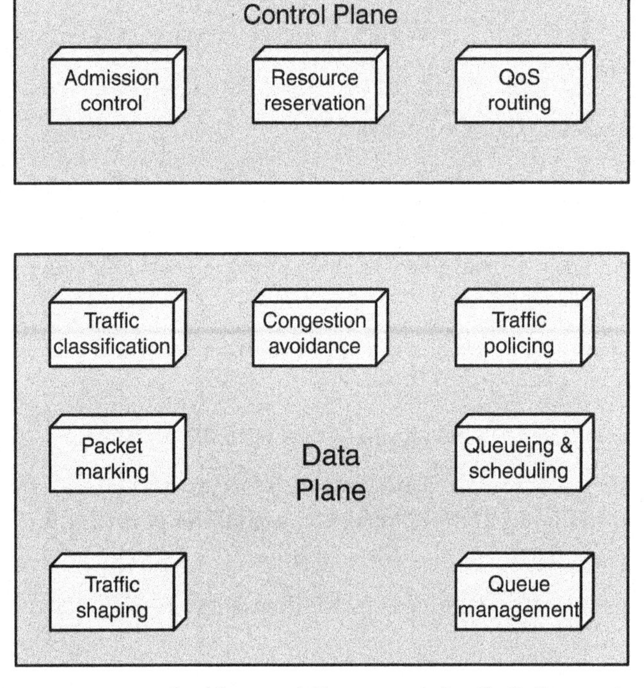

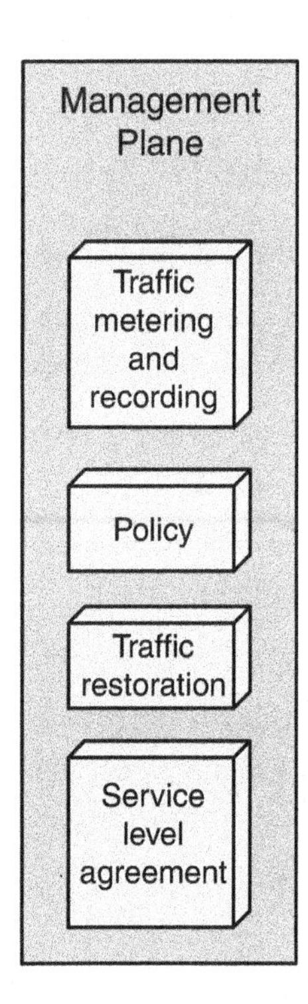

FIGURE 9.10 Architectural Framework for QoS Support

Data Plane

The data plane includes mechanisms that operate directly on flows of data. The following discussion briefly describes each mechanism in turn.

Traffic classification refers to the assignment of packets to a traffic class by the ingress router at the ingress edge of the network. Typically, the classification entity looks at multiple fields of a packet, such as source and destination address, application payload, and QoS markings, and determines the aggregate to which the packet belongs. This classification gives network elements a method to weigh the relative importance of one packet over another in a different class. All traffic assigned to a particular flow or other aggregate can be treated similarly. The flow label in the IPv6 header can be used for traffic classification. Other routers en route perform a classification function as well, but the classification does not change as the packets traverse the network.

Packet marking encompasses two distinct functions. First, packets may be marked by ingress edge nodes of a network to indicate some form of QoS that the packet should receive. Examples include the Differentiated Services (DS) field in IPv4 and IPv6 packets and the Traffic Class field in Multiprotocol Label Switching (MPLS) labels. An ingress edge node can set the values in these fields to indicate a desired level of QoS. Such markings may be used by intermediate nodes to provide differential treatment to incoming packets. Second, packet marking can be used to mark packets as nonconformant, either by the ingress node or intermediate nodes, so that they can be dropped later, if congestion is experienced.

Traffic shaping controls the rate and volume of traffic entering and transiting the network on a per-flow basis. The entity responsible for traffic shaping buffers nonconformant packets until it brings the respective aggregate into compliance with the traffic. The resulting traffic thus is not as bursty as the original and is more predictable.

Congestion avoidance deals with means for keeping the load of the network under its capacity such that it can operate at an acceptable performance level. The specific objectives are to avoid significant queuing delays and, especially, to avoid congestion collapse. In a typical congestion avoidance scheme, senders reduce the amount of traffic entering the network upon an indication that network congestion is occurring (or is about to occur). Unless there is an explicit indication, packet loss or timer expiration is normally regarded as an implicit indication of network congestion.

Traffic policing determines whether the traffic being presented is, on a hop-by-hop basis, compliant with prenegotiated policies or contracts. Nonconformant packets may be dropped, delayed, or labeled as nonconformant.

Queuing and scheduling algorithms, also referred to as queuing discipline algorithms, determine which packet to send next and are used primarily to manage the allocation of transmission capacity among flows.

Queue management algorithms manage the length of packet queues by dropping packets when necessary or appropriate. Active management of queues is concerned primarily with congestion avoidance.

Control Plane

The **control plane** is concerned with creating and managing the pathways through which user data flows. It includes admission control, QoS routing, and resource reservation.

Admission control determines what user traffic may enter the network. This may be in part determined by the QoS requirements of a data flow compared to the current resource commitment in the network. But beyond balancing QoS requests with available capacity to determine whether to accept a request, there are other considerations in admission control. Network managers and service providers must be able to monitor, control, and enforce use of network resources and services based on policies derived from criteria such as the identity of users and applications, traffic/bandwidth requirements, security considerations, and time of day/week.

QoS routing determines a network path that is likely to accommodate the requested QoS of a flow. This contrasts with the philosophy of the traditional routing protocols, which generally look for a least-cost path through the network.

Resource reservation is a mechanism that reserves network resources on demand for delivering desired network performance to a requesting flow.

Management Plane

The management plane is concerned with mechanisms that affect both control plane and data plane mechanisms. The control plane deals with the operation, administration, and management aspects of the network. It includes SLAs, traffic restoration, traffic metering and recording, and policy.

A **service-level agreement (SLA)** typically represents an agreement between a customer and a provider of a service that specifies the level of availability, serviceability, performance, operation, or other attributes of the service.

Traffic metering and recording concerns monitoring the dynamic properties of a traffic stream using performance metrics such as data rate and packet loss rate. It involves observing traffic characteristics at a given network point and collecting and storing the traffic information for analysis and further action. Depending on the conformance level, a meter can invoke necessary treatment (e.g., dropping or shaping) for the packet stream. Section 9.7 discusses the types of metrics that are used in this function.

Traffic restoration refers to the network response to failures. This encompasses a number of protocol layers and techniques.

Policy is a category that refers to a set of rules for administering, managing, and controlling access to network resources. These rules can be specific to the needs of the service provider or reflect an agreement between the customer and service provider, which may include reliability and availability requirements over a period of time and other QoS requirements.

QoS Classification, Marking, and Differentiation

The 3GPP document TS 23.501 uses the following terms:

- **Traffic classification:** Grouping traffic into classes based on user-defined QoS values
- **User plane marking:** Marking packets to indicate to which QoS classification they belong
- **QoS differentiation:** Using a different QoS set of values for different categories of traffic

Recall from Section 9.2 that a PDU session between UE and a DN may contain multiple QoS flows, and each QoS flow may contain one or more service data flows (SDFs). An SDF is associated with a particular application. A QoS flow is where QoS differentiation takes place and where packets are marked to indicate their traffic classification.

Figure 9.11, from TS 23.501, illustrates 5G principles for classification and marking of user plane traffic and mapping of QoS flows. The mapping happens two times. In the core, the UPF maps a QoS flow to a single tunnel. There is a one-to-many relationship between the tunnel on the AN core interface and the data radio bearers on the air interface. A RAN node (gNB) may map multiple QoS flows to one data radio bearer. Incoming application data packets are classified based on the QoS and service requirements of the service data flows of the application. The session management function (SMF) assigns the QoS flow ID (QFI) and derives its QoS profile from the information provided by the PCF.

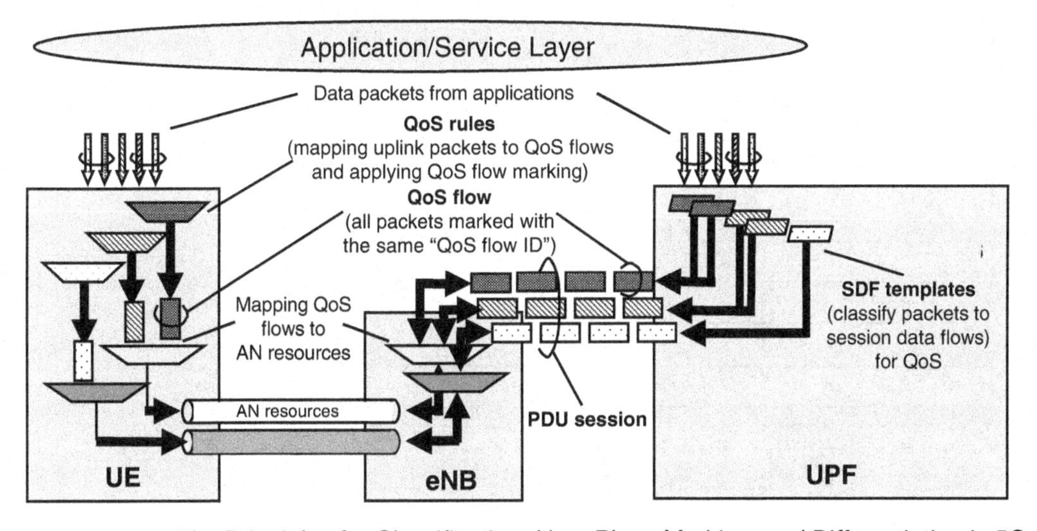

FIGURE 9.11 The Principles for Classification, User Plane Marking, and Differentiation in 5G

The SMF provides:

- The QFI together with the QoS profile to the AN
- QoS flow marking (i.e., the QFI) and the necessary information to enable classification
- Bandwidth enforcement and marking of user plane traffic to the UPF
- QoS rules enabling classification and marking of user plane traffic to the UE

The QoS capability includes reflective QoS, which is a method to reduce the signaling to the UE for uplink (UL) data classification. Reflective QoS applies the same QoS profile to both UL and downlink (DL), allowing a simple principle for applying the same QoS differentiation to application data in the DL and UL.

3GPP QoS Architecture

Figure 9.12, from TS 38.300 (*Technical Specification Group Radio Access Network*; *NR*; *NR and NG-RAN Overall Description*; *Stage 2 [Release 16]*; September 2020), shows a high-level view of the 3GPP QoS architecture, encompassing both the radio access network (RAN) and the core network (5GC). NG-RAN and 5GC ensure QoS by mapping packets to appropriate QoS flows.

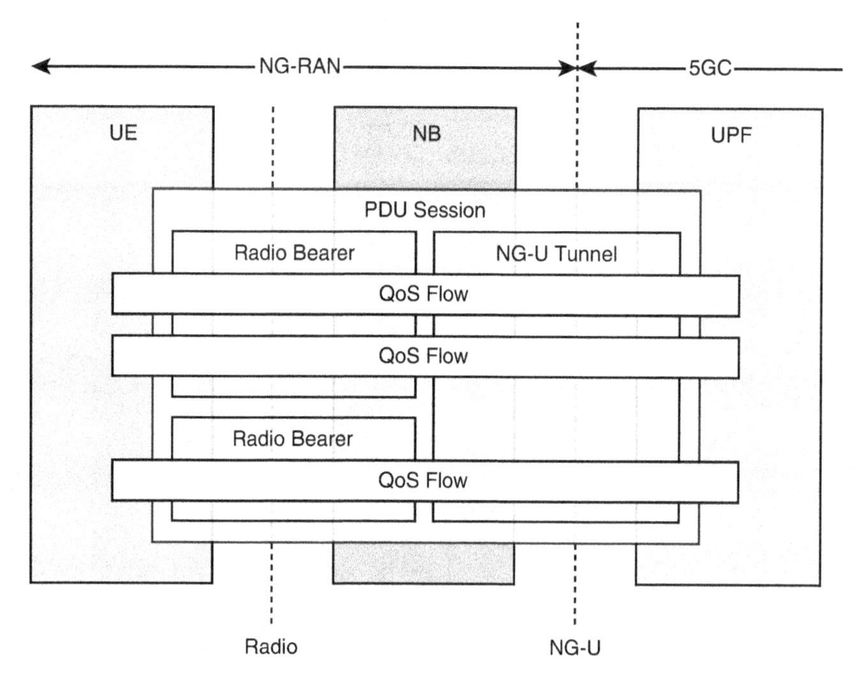

FIGURE 9.12 3GPP QoS Architecture

At the RAN, radio bearer paths may transmit one or more QoS flows, as long as the performance parameters of the radio bearer are sufficient for the flows. Recall that a radio bearer is an information transmission path of defined capacity, delay, bit error rate, and other parameters.

On the core network side, the flows in a PDU session are exchanged through the bidirectional tunnels to the UPF. The core network then performs the functions required to support the QoS for each flow.

QoS Parameters

The QoS model developed in TS 23.501 makes use of a key set of QoS parameters and QoS characteristics, as shown in Figure 9.13. Together, the parameters and characteristics define the requirements associated with a QoS flow.

QoS Parameters
5QI (5G QoS identifier)
ARP (allocation and retention priority)
RQA (reflective QoS attribute)
Notification control
Flow bit rates
Aggregate bit rates
Default values
Maximum packet loss rate
Wireline access network-specific 5G QoS parameters

QoS Characteristics
Resource type
Priority level
Packet delay budget
Packet error rate
Averaging window
Maximum data burst volume

FIGURE 9.13 Elements of 3GPP QoS Model

The following parameters are associated with a QoS flow:

- **5QI (5G QoS identifier):** This is an integer value used as a reference to a set of values assigned to QoS characteristics. Thus, a standardized combination of QoS characteristics can be preconfigured so that the AN and CN are informed of the QoS characteristics for a flow by means of the 5QI.
- **ARP (allocation and retention priority):** The ARP consists of three attributes:
  - **ARP priority level:** Defines the relative importance of a QoS flow. The range of the ARP priority level is 1 to 15, with 1 as the highest level of priority. In cases of congestion, when all QoS requirements cannot be fulfilled for one or more QoS flows, the priority level determines for which QoS flows the QoS requirements are prioritized. In cases where there is no congestion, the priority level determines the resource distribution between QoS flows.
  - **ARP pre-emption capability:** Defines whether a QoS flow may get resources that were already assigned to another QoS flow with a lower ARP priority level. It is set as either enabled or disabled.

- **ARP pre-emption vulnerability:** Defines whether a QoS flow may lose the resources assigned to it in order to admit a QoS flow with a higher ARP priority level. It is set as either enabled or disabled.

- **RQA (reflective QoS attribute):** Reflective QoS means that the UE uses the same QoS parameters on the uplink as obtained from the downlink QoS flow. RQA, when included, indicates that some (not necessarily all) traffic carried on this QoS flow is subject to reflective QoS.

- **Flow bit rates:** There are two categories of flow bit rates. A guaranteed bit rate (GBR) guarantees at least a minimum bit rate capacity for the flow. A non-GBR QoS flow does not guarantee the bit rate. GBR QoS flows include the following parameters:

  - **Guaranteed flow bit rate (GFBR) UL and DL:** Denotes the bit rate that is guaranteed to be provided by the network to the QoS flow over the averaging time window.

  - **Maximum flow bit rate (MFBR) UL and DL:** Limits the bit rate to the highest bit rate that is expected by the QoS flow (e.g., excess traffic may get discarded or delayed by a rate-shaping or policing function at the UE, RAN, or UPF). Bit rates above the GFBR value and up to the MFBR value may be provided with relative priority determined by the priority level of the QoS flows.

- **Notification control:** This parameter indicates whether notifications are requested from the NG-RAN when the GFBR can no longer (or can again) be guaranteed for a QoS flow during the lifetime of the QoS flow. Notification control may be used for a GBR QoS flow if the application traffic is able to adapt to the change in the QoS (e.g., if the AF is capable of triggering rate adaptation).

- **Aggregate bit rates:** Two parameters related to bit rates are associated with each UE:

  - **Per session aggregate maximum bit rate (Session-AMBR):** For each PDU session of a UE, this parameter limits the aggregate bit rate that can be expected to be provided across all non-GBR QoS flows for a specific PDU session. Session-AMBR is measured over an AMBR averaging window, which is a standardized value. Session-AMBR is not applicable to GBR QoS flows.

  - **Per UE aggregate maximum bit rate (UE-AMBR):** This parameter limits the aggregate bit rate that can be expected to be provided across all non-GBR QoS flows of a UE. Each AN sets its UE-AMBR to the sum of the Session-AMBR of all PDU sessions with an active user plane to this AN up to the value of the received UE-AMBR from the AMF. The UE-AMBR is a parameter provided to the AN by the AMF, based on the value of the subscribed UE-AMBR retrieved from UDM or the dynamic serving network UE-AMBR retrieved from the PCF (e.g., for a roaming subscriber). The AMF provides the UE-AMBR provided by the PCF to the AN, if available. The UE-AMBR is measured over an AMBR averaging window, which is a standardized value. The UE-AMBR is not applicable to GBR QoS flows.

- **Default values:** For each PDU session setup, these default values apply to one or more non-GBR flows. The SMF retrieves the subscribed Session-AMBR values as well as the subscribed default values for the 5QI and the ARP and, optionally, the 5QI priority level, from the UDM.
- **Maximum packet loss rate:** This parameter indicates the maximum rate for lost packets of the QoS flow that can be tolerated in the uplink and downlink directions.
- **Wireline access network-specific 5G QoS parameters:** There are additional parameters applicable only to wireline access networks.

Note that the guaranteed flow bit rate (GFBR) and maximum flow bit rate (MFBR) only apply to GBR flows. The per session aggregate maximum bit rate and the per UE aggregate maximum bit rate only apply to non-GBR flows.

QoS Characteristics

Each 5QI has associated with it a set of values of characteristics that describe the packet forwarding treatment that a QoS flow receives edge-to-edge between the UE and the UPF. The following characteristics are associated with a QoS flow:

- **Resource type:** There are three resource types: GBR, delay-critical GBR, and non-GBR. Both GBR types are typically authorized on demand, which requires dynamic policy and charging control. The definitions of PDB (packet delay budget) and PER (packet error rate) are different for GBR and delay-critical GBR resource types, and MDBV (maximum data burst volume) applies only to the delay-critical GBR resource type. A non-GBR QoS flow may be pre-authorized through static policy and charging control.
- **Priority level:** This characteristics indicates a priority in scheduling resources among QoS flows, and it has following characteristics:
  - The lowest numeric value corresponds to the highest priority.
  - The priority level is used to differentiate between QoS flows of the same UE, and it is also used to differentiate between QoS flows from different UEs.
  - When there is congestion such that all QoS requirements cannot be fulfilled for one or more QoS flows, the priority level is used to select which QoS flows the QoS requirements are prioritized such that a QoS flow with priority level value *N* is prioritized over QoS flows with higher priority level values. For example, the priority level serves as a tie-breaker when two packets compete for a given network resource at the same time.
  - In the absence of congestion, the priority level is used to define the resource distribution between QoS flows. In addition, the scheduler may prioritize QoS flows based on other parameters (e.g., resource type, radio condition) in order to optimize application performance and network capacity.

    - Every standardized 5QI is associated with a default value for the priority level, as specified in TS 23.501.
    - Priority level may also be signaled together with a standardized 5QI to the AN, and if it is received, it is used instead of the default value.
    - Priority level may also be signaled together with a preconfigured 5QI to the AN, and if it is received, it is used instead of the preconfigured value.
- **Packet delay budget (PDB):** This characteristic, which defines an upper bound for the time that a packet may be delayed between the UE and the UPF, has the following characteristics:
    - For some 5QI, the value of the PDB is the same in UL and DL.
    - In the case of 3GPP access, the PDB is used to support the configuration of scheduling and link layer functions (e.g., the setting of scheduling priority weights).
    - For GBR QoS flows using the delay-critical resource type, a packet delayed more than PDB is counted as lost if the data burst is not exceeding the MDBV within the period of PDB and the QoS flow is not exceeding the GFBR.
    - For GBR QoS flows with GBR resource type not exceeding GFBR, 98% of the packets do not experience delay exceeding the 5QI's PDB.
    - Services using a GBR QoS flow and sending at a rate smaller than or equal to the GFBR can in general assume that congestion-related packet drops will not occur.
    - Services using non-GBR QoS flows should be prepared to experience congestion-related packet drops and delays. In uncongested scenarios, 98% of the packets should not experience delay exceeding the 5QI's PDB.
    - PDB for non-GBR and GBR resource types denotes a "soft upper bound" in the sense that an "expired" packet (e.g., a link layer SDU that has exceeded the PDB) does not need to be discarded and is not added to the PER.
    - For a delay-critical GBR resource type, packets delayed more than the PDB are added to the PER and can be discarded or delivered depending on local decision.
- **Packet error rate (PER):** This characteristics defines an upper bound for the rate of PDUs (e.g., IP packets) that have been processed by the sender of a link layer protocol but that are not successfully delivered by the corresponding receiver to the upper layer. Equivalently, the PER defines an upper bound for a rate of non-congestion-related packet losses. For GBR QoS flows with delay-critical GBR resource type, a packet that is delayed more than PDB is counted as lost and is included in the PER unless the data burst exceeds the MDBV within the period of PDB or the QoS flow exceeds the GFBR.
- **Averaging window:** This characteristics represents the duration over which the GFBR and MFBR are calculated (e.g., in the RAN, UPF, and UE). Thus, the bit rate is calculated as B/W, where B is the number of bits transmitted during a window of size W seconds. Each GBR QoS

flow is associated with an averaging window. Every standardized 5QI (of GBR and delay-critical GBR resource types) is associated with a default value for the averaging window. The averaging window may also be signaled together with a standardized 5QI to the RAN and UPF, and if it is received, it is used instead of the default value. The averaging window may also be signaled together with a preconfigured 5QI to the RAN, and if it is received, it is used instead of the preconfigured value.

- **Maximum data burst volume (MDBV):** This characteristics denotes the largest amount of data that the 5G-AN is required to serve within a period of 5G-AN PDB (i.e., the 5G-AN part of the PDB). Each GBR QoS flow with a delay-critical resource type is associated with an MDBV. Every standardized 5QI of delay-critical GBR resource type is associated with a default value for the MDBV. The MDBV may also be signaled together with a standardized 5QI to the (R)AN, and if it is received, it is used instead of the default value. The MDBV may also be signaled together with a preconfigured 5QI to the (R)AN, and if it is received, it is used instead of the preconfigured value.

Standardized 5QI-to-QoS Characteristic Mapping

TS 23.501 provides a collection of predefined QoS profiles with associated 5QI values. These standardized 5G QoS identifier (5QI) values correspond to services that are likely to be frequently used in 5G networks and that would thus benefit from optimized signaling through the use of standardized QoS characteristics. From the user perspective, the standardized QoS profiles relieve the user from the necessity of designing a set of QoS characteristic values for services that are common. Nonstandardized 5QIs can be used in an operator network or by agreement between two or more operators.

For each of the 5QI entries, TS 23.501 lists examples of applications. These include a variety of traffic classes, which are as follows:

- **Conversational:** This interactive service provides for bidirectional communication by means of real-time (no store-and-forward) end-to-end information transfer from user to user. Examples of such flows include telephony speech and also VoIP and video conferencing. Sensitivity to delay is high because of the real-time nature of the flows. The time relationship between the stream entities has to be preserved (to maintain the same experience for all flows and all parties involved in the conversation).
- **Streaming:** The streaming class refers to flows in which the user is watching real-time video or listening to real-time audio (or both). The real-time data flow is always aiming at a live (human) destination. Streaming is both a real-time flow and a one-way transport. The delay sensitivity is lower than that of conversational flows because it is expected that the receiving end includes a time-alignment function (e.g., buffering). Because the flow is unidirectional, variations in delay do not adversely affect the user experience as long as the variation is within the alignment function boundaries.

- **Mission critical:** The failure or disruption of this type of service would result in serious damage to the users of the service. Mission-critical communications are secure, reliable, and readily available, and typically time is a vital factor.

The 5QI values fall into three groupings for the three resource types (i.e., GBR, delay-critical GBR, and non-GBR). Table 9.3 shows the values of characteristics for various QoS profiles of the GBR resource type and suggests examples of applications for each 5QI entry.

TABLE 9.3 Standardized 5QI to QoS Characteristics Mapping for GBR Resource Type*

5QI	Default Priority	PDB (ms)	PER	Default MDBV	Default Averaging Window	Examples of Services
1	20	100	10^{-2}	N/A	2 s	Conversational voice
2	40	150	10^{-3}	N/A	2 s	Conversational video
3	30	50	10^{-3}	N/A	2 s	Real-time gaming V2X messages Electricity distribution (medium voltage) Process automation monitoring
4	50	300	10^{-6}	N/A	2 s	Non-conversational video (buffered streaming)
65	7	75	10^{-2}	N/A	2 s	Mission-critical user plane push-to-talk voice
66	20	100	10^{-2}	N/A	2 s	Non-mission-critical user plane push-to-talk voice
67	15	100	10^{-3}	N/A	2 s	Mission-critical user plane video
71	56	150	10^{-6}	N/A	2 s	Live uplink streaming
72	56	300	10^{-4}	N/A	2 s	
73	56	300	10^{-8}	N/A	2 s	
74	56	500	10^{-6}	N/A	2 s	
76	56	500	10^{-4}	N/A	2 s	

**PDB = Packet Delay Budget; PER = Packet Error Rate; MDBV = Maximum Data Burst Volume*

The profiles show support for, among others, the following:

- Smart grid
- Process automation monitoring
- Autonomous vehicles (V2X messaging)
- Mission-critical public safety applications
- Low-latency enhanced mobile broadband (eMBB)

- Augmented reality
- Discrete automation
- Intelligent transport systems

Several observations should be made. The two mission-critical applications in Table 9.3 have significantly higher priority (i.e., lower priority number) than the other applications. Push-to-talk (PTT), also known as press-to-transmit, is a method of having conversations or talking on half-duplex communication lines, including two-way radio, using a momentary button to switch from voice reception mode to transmit mode. For example, the PTT feature on Zoom allows a user to remain muted throughout a Zoom meeting and hold down the spacebar to unmute and talk. Both mission-critical and non-mission-critical PTT have relatively high priority, though the mission-critical priority is higher.

Live uplink streaming is a service in which a user with a radio connection (e.g., reporter, drone) streams a live video feed into the network or to a second party. TS 23.501 defines five different characteristic value combinations for this service (5QI values 71, 72, 73, 74, and 76). TR 26.939 (*Technical Specification Group Services and System Aspects, Guidelines on the Framework for Live Uplink Streaming [FLUS] [Release 16]*, September 2019) defines eight different uses cases for live uplink streaming and provides guidance on the choice of QoS parameters and characteristics for each use case. The five QoS profiles in Table 9.3 support the likely choices to be made for the eight use cases.

Table 9.4 shows the values of characteristics for various QoS profiles of the delay-critical GBR resource type. All of these application examples are assigned relatively high priority.

TABLE 9.4 Standardized 5QI-to-QoS Characteristics Mapping for the Delay-Critical GBR Resource Type*

5QI	Default Priority	PDB (ms)	PER	Default MDBV	Default Averaging Window	Examples of Services
82	19	10	10^{-4}	255 bytes	2 s	Discrete automation
83	22	10	10^{-4}	1354 bytes	2 s	Discrete automation V2X messages (cooperative lane change)
84	24	30	10^{-5}	1354 bytes	2 s	Intelligent transport systems
85	21	5	10^{-5}	255 bytes	2 s	Electricity distribution (high voltage) V2X messages (remote driving)
86	18	5	10^{-4}	1354 bytes	2 s	V2X messages (advanced driving, collision avoidance)

**PDB = Packet Delay Budget; PER = Packet Error Rate; MDBV = Maximum Data Burst Volume*

Table 9.5 shows the values of characteristics for various QoS profiles of the non-GBR resource type. 5QI 70 is non-GBR, intended for mission-critical data, with a priority of 55, a PDB of 200 ms, and a

PER tolerance of at most 10^{-6}. The traffic types intended for 5QI 70 are the same as for 5QIs 6, 8, and 9: buffered streaming video and TCP-based traffic, such as www, email, chat, FTP, P2P, and other file sharing applications. However, 5QI 70 is specifically intended for applications that are mission critical. For this reason, 5QI 70 priority is higher than 6, 8, or 9 priorities (55 versus 60, 80, and 90, respectively).

TABLE 9.5 Standardized 5QI-to-QoS Characteristics Mapping for the Non-GBR Resource Type*

5QI	Default Priority	PDB (ms)	PER	Default MDBV	Default Averaging Window	Examples of Services
5	10	100	10^{-6}	N/A	N/A	IMS (IP multimedia subsystem) signaling
6	60	300	10^{-6}	N/A	N/A	Video (buffered streaming) TCP based (e.g., www, email, chat, FTP, P2P file sharing)
7	70	100	10^{-3}	N/A	N/A	Voice Video (live streaming) Interactive gaming
8	80	300	10^{-6}	N/A	N/A	Video (buffered streaming) TCP based (e.g., www, email, chat, FTP, P2P file sharing)
9	90	300	10^{-6}	N/A	N/A	Video (buffered streaming) TCP based (e.g., www, email, chat, FTP, P2P file sharing)
69	5	60	10^{-6}	N/A	N/A	Mission-critical delay-sensitive signaling
70	55	200	10^{-6}	N/A	N/A	Mission-critical data
79	65	50	10^{-2}	N/A	N/A	V2X messages
80	68	10	10^{-6}	N/A	N/A	Low-latency eMBB applications AR

**PDB = Packet Delay Budget; PER = Packet Error Rate; MDBV = Maximum Data Burst Volume*

Tables 9.3 through 9.5 provide some insight into the distinction between QoS parameters and QoS characteristics. Although there is some overlap, very broadly it can be said that the QoS parameters are used at configuration time to determine the network resources needed for creating a network slice for supporting this set of QoS parameter values. The QoS characteristics are more relevant to dynamic decisions made during the operation of the QoS flow, such as using the priority level as a tie-breaker when two flows compete for a resource.

Another perspective is based on the distinction between an application and a specific instance of an application. 3GPP has determined that the variables defined as characteristics are typically the same for a wide variety of instances of an application, and so it is efficient and useful to the user to provide standardized sets of values. Depending on the context, individual instances of an application may require different sets of values for some of the parameters, especially the flow bit rates (GFBR and MFBR) and the aggregate bit rates (session-AMBR and UE-AMBR).

9.4 Network Slicing

One of the most important features of 5G is network slicing. Network slicing involves virtualization technologies such as SDN and NFV, which enable a 5G network operator to provide customized networks by creating multiple virtual end-to-end networks, referred to as network slices. Each network slice can be defined according to different requirements on functionality, QoS, and specific users. Table 9.6 defines the network slicing terms used in 3GPP documents.

TABLE 9.6 3GPP Network Slicing Terminology

Term	Definition
Network slice	A logical network that provides specific network capabilities and network characteristics.
Network slice instance	A set of network function instances and the required resources (e.g., compute, storage, and networking resources) that form a deployed network slice.
Network slice instance identifier (NSI ID)	An identifier for identifying the core network part of a network slice instance when multiple network slice instances of the same network slice are deployed and there is a need to differentiate between them in the 5GC.
NF instance	An identifiable instance of the NF.
NF set	A group of interchangeable NF instances of the same type, supporting the same services and the same network slice(s). The NF instances in the same NF set may be geographically distributed but have access to the same context data.
Slice/service type (SST)	The expected network slice behavior in terms of features and services.
Slice differentiator (SD)	Optional information that complements the SST(s) to differentiate among multiple network slices of the same SST.
Single network slice selection assistance information (S-NSSAI)	Defines a unique network slice, composed of an SST and an SD.
Network slice selection assistance information (NSSAI)	A vector of up to eight S-NSSAI values that are used to identify and select slice instances associated with a UE.
Network slice selection function (NSSF)	A network function that selects the set of network slice instances to serve the UE. It also determines the allowed and configured NSSAI and, if necessary, maps to the subscribed S-NSSAIs and determines the AMF set to be used (or a candidate list) to serve the UE.
Network slice selection policy (NSSP)	A set of rules (at least one rule), where each rule attributes an application with a certain S-NSSAI. This is used by the UE to associate the matching application with the S-NSSAI.
Network slice-specific authentication and authorization (NSSAA)	A network function that allows a third party to add and remove users to and from a network slice instance.

[LI17] lists the following advantages of slice-based networking compared with traditional networking:

- Network slicing can provide logical networks with better performance than one-size-fits-all networks.
- A network slice can scale up or down as service requirements and the number of users change.
- Network slices can isolate the network resources of one service from the others; the configurations among various slices don't affect each other. Therefore, the reliability and security of each slice can be enhanced.
- A network slice is customized according to QoS requirements, which can optimize the allocation and use of physical network resources.

Network slicing is made possible by the softwarization techniques of network functions virtualization (NFV) and software-defined networking (SDN). NFV implements the network functions (NFs) in a network slice, enabling the isolation of each network slice from all other network slices. Isolation can be achieved by one or more of the following: (1) using a different physical resource, (2) separation by virtualization, which may allow sharing of physical resources, or (3) through sharing a resource with the guidance of a respective policy that defines the access rights for each tenant. Isolation assures QoS and security requirements for that slice, independent of other slices operating on the network from the same or different users.

Once a network slice is defined, SDN operates to monitor and enforce QoS requirements by controlling the behavior of the QoS flow for each slice.

Network Slicing Concepts

Network slicing permits a physical network to be separated into multiple virtual networks (logical segments) that can support different radio access networks or several types of services for certain customer segments, greatly reducing network construction costs by using communication channels more efficiently. In essence, network slicing allows the creation of multiple virtual networks atop a shared physical infrastructure. This virtualized network scenario devotes capacity to certain purposes dynamically, according to need. As needs change, so can the devoted resources. Using common resources such as storage and processors, network slicing permits the creation of slices devoted to logical, self-contained, and partitioned network functions. Network slicing supports the creation of virtual networks to provide a given level of QoS, such as guaranteed delay, throughput, reliability, and/or priority.

A network slice creates a partition of the core network consisting of virtualized network functions and resources running on some of the core network hardware resources. Figure 9.14, based on concepts in the Next Generation Mobile Networks (NGMN) document *5G End-to-End Architecture Framework* (August 2019), illustrates network slicing concepts. It shows a simple core network configuration composed of three types of devices.

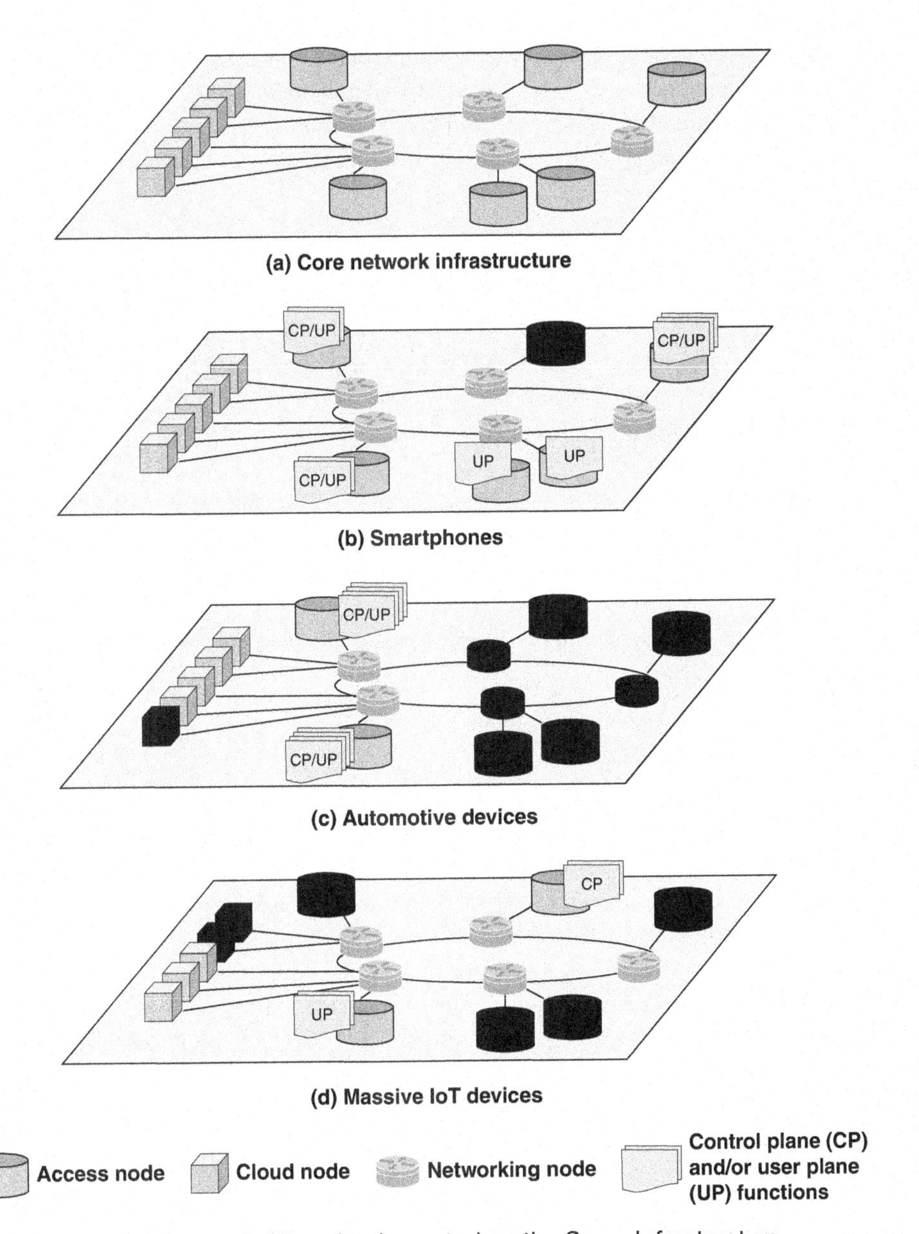

FIGURE 9.14 5G Network Slices Implemented on the Same Infrastructure

The devices are as follows:

- **Cloud nodes:** These nodes provide cloud services, software, and storage resources. There are likely to be one or more central cloud nodes that provide traditional cloud computing service. In addition, cloud-edge nodes provide low-latency and higher-security access to client devices at the edge of the network. All of these nodes include virtualization system software to support virtual machines and containers. NFV enables effective deployment of cloud resources to the appropriate edge node for a given application and given fixed or mobile user. The combination of SDN and NFV enables the movement of edge resources and services to dynamically accommodate mobile users.

- **Networking nodes:** These nodes are IP routers and other types of switches for implementing a physical path through the network for a 5G connection. SDN provides for flexible and dynamic creation and management of these paths.
- **Access nodes:** These nodes provide an interface to radio access networks (RANs), which in turn provide access to mobile UE. SDN creates paths that use an access node for one or both ends of a connection involving a wireless device.

The remainder of Figure 9.14 illustrates three use cases. The blacked-out core network resources represent resources not used to create the network slice. Cloud nodes that are part of the slice may include the following:

- Control plane functions associated with one or more user plane functions (e.g., a reusable or common framework of control)
- Service- or service category–specific control plane and user plane function pairs (e.g., a user-specific multimedia application session)

The first network slice depicted in Figure 9.14 is for a typical smartphone use case. Such a slice might have fully fledged functions distributed across the network. The second network slice in Figure 9.14 indicates the type of support that may be allocated for automobiles in motion. This use case emphasizes the need for security, reliability, and low latency. A configuration to achieve this would limit core network resources to nearby cloud edge nodes,[1] plus the recruitment of sufficient access nodes to support the use case. The final use case illustrated in Figure 9.14 is for a massive IoT deployment, such as a huge number of sensors. The slice can contain just some specific CP and UP functions with, for example, no mobility functions. The CP and UP functions might include filtering and preliminary data analysis at the edge and big data types of analysis at a more central node. This slice would only need to engage access nodes nearest to the IoT device deployment.

Requirements for Network Slicing

TS 22.261 lists requirements for network slicing in two categories: general requirements and management requirements.

1. Cloud-edge computing, or cloud-edge networking, refers to the deployment of cloud capabilities at the network edge. Chapter 10 explores this topic.

General Requirements

The general requirements for network slicing are as follows:

- Support is needed to provide connectivity to home and roaming users in the same network slice.
- In a shared 5G network configuration, operators can apply all the requirements to their allocated network resources.
- IMS needs to be supported as part of a network slice.
- IMS needs to be supported independent of network slices.

IP Multimedia Subsystem (IMS) is a standards-based architectural framework for delivering multimedia communications services such as voice, video, and text messaging over IP networks [KOUK06]. The IMS specifications were originally developed by 3GPP in the early 2000s to standardize access to multimedia services using cellular networks. The specifications define a complete framework and architecture that enables the convergence of video, voice, data, and mobile network technologies.

Management Requirements

The management requirements for network slicing are as follows:

- The operator should be able to create, modify, and delete a network slice.
- The operator should be able to define and update the set of services and capabilities supported in a network slice.
- The operator should be able to configure the information that associates UE to a network slice.
- The operator should be able to configure the information that associates a service to a network slice.
- The operator should be able to assign UE to a network slice, to move UE from one network slice to another, and to remove UE from a network slice based on subscription, UE capabilities, the access technology being used by the UE, the operator's policies, and services provided by the network slice.
- A mechanism is needed for the VPLMN (visited public land mobile network), as authorized by the HPLMN (home public land mobile network), to assign UE to a network slice with the needed services or to a default network slice.
- A UE should be able to be simultaneously assigned to and access services from more than one network slice of one operator.
- Traffic and services in one network slice should have no impact on traffic and services in other network slices in the same network.
- Creation, modification, and deletion of a network slice should have no or minimal impact on traffic and services in other network slices in the same network.

- A network slice should be able to scale (i.e., adapt its capacity).
- The network operator should be able to define a minimum available capacity for a network slice. Scaling of other network slices on the same network should have no impact on the availability of the minimum capacity for that network slice.
- The network operator should be able to define a maximum capacity for a network slice.
- The network operator should be able to define a priority order between different network slices in the event that multiple network slices compete for resources on the same network.
- The operator should be able to differentiate policy control, functionality, and performance provided in different network slices.

Identifying and Selecting a Network Slice

Single network slice selection assistance information (S-NSSAI) defines a single network slice. An S-NSSAI consists of two elements:

- **Slice/service type (SST):** An identifier that refers to the expected slice behavior in terms of features and services. Standardized SST values provide a way to establish global interoperability for slicing so that 5G networks can support the roaming use case more efficiently for the most commonly used SSTs. Table 9.7 lists the standardized SSTs.
- **Slice differentiator (SD):** Optional information that complements the SST to differentiate among multiple network slices of the same SST.

TABLE 9.7 Standardized Slice/Service Type Values

Slice/Service Type	SST Value	Characteristics
eMBB	1	Slice suitable for the handling of 5G enhanced mobile broadband
URLLC	2	Slice suitable for the handling of ultra-reliable and low-latency communications
MIoT	3	Slice suitable for the handling of massive IoT
V2X	4	Slice suitable for the handling of V2X services

UE may be served by up to eight network slices at a time, each identified by an S-NSSAI. The set of S-NSSAIs associated with a UE form a network slice selection assistance information (NSSAI) data object.

Functional Aspects of Network Slicing

Figure 9.15 illustrates how core network functions (NFs) are used to implement network slices. Some NF instances support multiple network slices serving UE, and others are specific to a given slice.

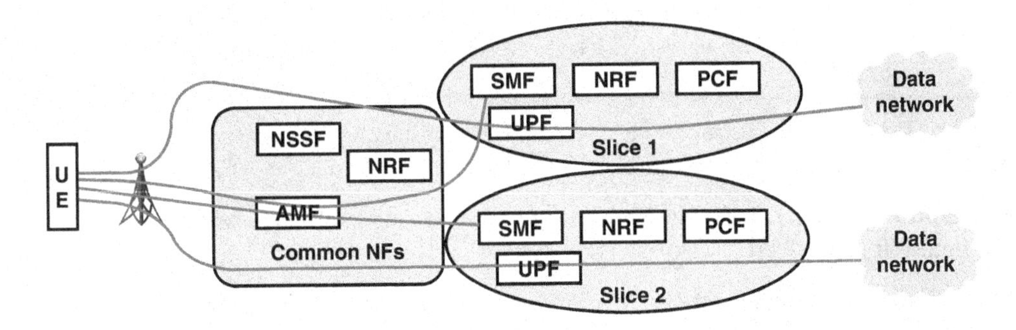

AMF = Access and Mobility Management Function
NF = Network Function
NRF = Network Repository Function
NSSF = Network Slice Selection Function
PCF = Policy Control Function
SMF = Session Management Function
UPF = User Plane Function

FIGURE 9.15 Network Functions That Support Network Slicing

The common NFs are:

- **Access and mobility management function (AMF):** Network slice instance selection is usually triggered as part of the registration procedure by the first AMF that receives the registration request from the UE. When UE accesses the network, the AMF provides functionalities to register and de-register the UE with the network, and it establishes the user context in the network. In the registration procedure, AMF performs (but is not limited to) network slice instance selection, UE authentication, authorization of network access and network services, and network access policy control. In addition, when a session establishment request message is received from UE, the AMF performs discovery and selection of the SMF that is the most appropriate to manage the session.
- **Network slice selection function (NSSF):** The AMF retrieves the slices that are allowed by the user subscription and interacts with the NSSF to select the appropriate network slice instance (e.g., based on allowed S-NSSAIs, 5G network ID, and other parameters). The NSSF responds with a message that includes the list of appropriate network slice instances for the UE. As a result, the registration process may switch to another AMF if needed.
- **Network repository function (NRF):** During the AMF-NSSF interaction, the NSSF may return the identity of one or more NRFs to be used to select NFs and services within the selected network slice instance(s).

The slice-specific NFs are:

- **Session management function (SMF):** The UE sends a message to the AMF, requesting that a PDU session be associated to one S-NSSAI and one data network (DN). The AMF selects the appropriate SMF, which manages the PDU session. The SMF sets up the PDU session for the UE and controls the user plane operation. The SMF selects the UPF and invokes enforcement of QoS and charging policies.
- **User plane function (UPF):** Once a PDU session is established, QoS flows for this PDU session over this network slice pass through the UPF.
- **Policy control function (PCF):** The SMF gets policy information related to session establishment from the PCF.
- **Network repository function (NRF):** The SMF uses the NRF to discover the required NFs for the individual network slice.

Generic Slice Template

3GPP TS 28.531 (*Technical Specification Group Services and System Aspects, Management and orchestration, Provisioning [Release 16]*, December 2020) includes a description of the Generic Network Slice Template (GST) concept, which is specified by the GSM Association (GSMA). The GST provides a standardized list of attributes that can be used to characterize different types of network slices [GSMA20]. A network slice type (NEST) is a GST filled with (ranges of) values. There may be two kinds of NESTs:

- **Standardized NEST (S-NEST):** Attributes are assigned (ranges of) values by standards-developing organization (SDOs), working groups, forums, and so forth, such as 3GPP, GSMA, 5G Automotive Association (5GAA), and 5G Alliance for Connected Industry and Automation (5G-ACIA).
- **Private NEST (P-NEST):** Attributes are assigned (ranges of) values by the network slice providers that are different from those assigned in S-NESTs.

Network slice providers can build their network slice product offerings based on S-NESTs and/or P-NESTs.

GSMA has developed the GST to be a list of attributes sufficient for describing a wide range of NESTs that can be fully constructed by allocating values (or ranges of values) to each relevant attribute in the GST. A network operator can use a NEST to identify the network resources and functions needed to instantiate network slices. The process to fill in the GST and to create a NEST involves three steps:

Step 1. Study use cases and derive service requirements based on discussions with the slice customers, such as vertical industries or specific enterprises.

Step 2. Convert the service requirements identified in step 1 into technical requirements.

Step 3. Document the technical requirements produced in step 2 using the NEST by filling in the values of each of the attributes of the GST.

The current version of the GST lists 35 attributes, shown in Figure 9.16.

Availability Area of service Delay tolerance Deterministic communication Downlink throughput per network slice Downlink maximum throughput per UE Energy efficiency Group communication support Isolation level Maximum supported packet size Mission-critical support MMTel support NB-IoT support	Network functions owned by network slice customer Maximum number of PDU sessions Maximum number of UEs Performance monitoring Performance prediction Positioning support Radio spectrum Root cause investigation Session and service continuity support Simultaneous use of the network slice Slice quality of service parameters	Support for non-IP traffic Supported device velocity Synchronicity UE density Uplink throughput per network slice Uplink maximum throughput per UE User management openness User data access V2X communication mode Latency from (last) UFP to application server Network Slice Specific Authentication and Authorization (NSSAA) required

FIGURE 9.16 Generic Network Slice Template Attributes

9.5 SDN and NFV Support for 5G

The 5G Infrastructure Public Private Partnership refers to SDN and NFV as the fundamental pillars that support the wide range of key performance indicators (KPIs) for the new 5G use cases in a cost-efficient way [5GPP20]. With the SDN and NFV framework of 5G, mobile network operators are able to offer new services to consumers, enterprises, verticals, and third-party tenants by addressing their respective requirements.

The European Telecommunications Standards Institute document ETSI GS NFV-EVE 005 (*Network Functions Virtualisation [NFV], Ecosystem, Report on SDN Usage in NFV Architectural Framework*, December 2015) examines the manner in which SDN can be incorporated in the NFVI to provide connectivity services. Figure 9.17, based on a figure in GS NFV-EVE 005, illustrates the placement of SDN controllers to achieve appropriate cooperation between SDN and NFV. The framework incorporates two controllers: one logically placed at the tenant level and another at the NFVI level. Each controller centralizes the control plane functionalities and provides an abstract view of all the connectivity-related components it manages.

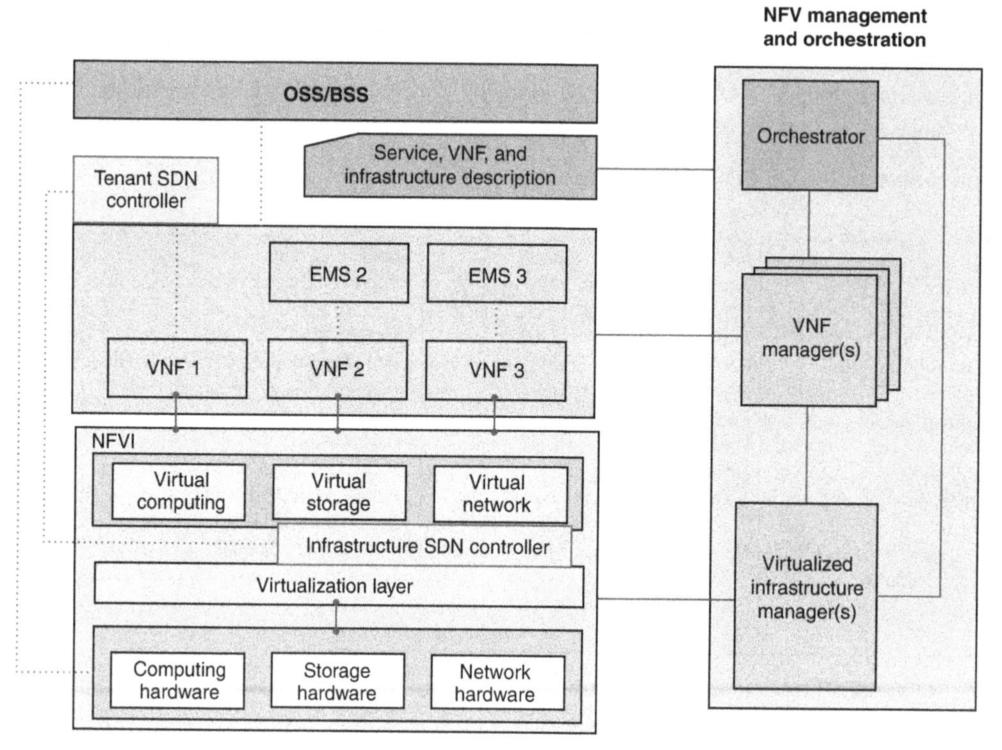

FIGURE 9.17 Integrating SDN Controllers into the Reference NFV Architectural Framework

The controllers are as follows:

- **Infrastructure SDN controller (IC):** This controller enables communication among VNFs and among their components, including the cases when those VNFs are instantiated in separated PoPs, reachable through a WAN connection. Managed by the VIM, this controller may change infrastructure behavior on demand according to VIM specifications, adapted from tenant requests.

- **Tenant SDN controller (TC):** Instantiated in the tenant domain as one of the VNFs or as part of the NMS, this second controller dynamically manages the pertinent VNFs used to realize the tenant's network service(s). These VNFs are the underlying forwarding plane resources of the TC. The operation and management tasks that the TC carries out are triggered by the applications running on top of it (e.g., the OSS).

[ORDO17] describes the manner in which the architecture of Figure 9.17 supports network slicing. The two controllers manage and control their underlying resources via programmable southbound interfaces, implementing protocols such as OpenFlow. The two controllers provide different levels of abstraction. The IC provides an underlay to support the deployment and connectivity of VNFs, and the TC provides an overlay comprising tenant VNFs that, properly composed, define the network service(s) that such a tenant independently manages on its slice(s). The IC is not aware of the number of slices that utilize the VNFs it connects or the tenant or tenants that operate such slices. For the TC, the network is abstracted in terms of VNFs, without notions of how those VNFs are physically deployed. Despite their different abstraction levels, both controllers have to coordinate and synchronize their actions.

Figure 9.18, from [LI17], provides a more concrete depiction of how SDN and NFV cooperate to support 5G network slices. The framework is constructed as three layers:

- **Application and service layer:** This layer contains a heterogenous collection of service instances. The example of Figure 9.17 shows three instances: connected vehicles, virtual reality, and mobile broadband (MBB). A service instance may serve multiple tenants or users. The service instance informs the slicing MANO module, described later in this section, of its service requirements, which are then mapped to a network slice.

- **Virtual resource layer:** This layer provides all the virtual resources required for network slices, such as radio, computing, storage, and network bandwidth. These resources are provided as virtual network functions residing on virtual machines (VMs). In this example, the three VMs on the left support the connected vehicles application. The next two VMs plus a shared VM support the virtual reality application. The shared VM and the last two VMs support the MBB application.

- **Software-defined infrastructure layer:** This layer is composed of a software-defined infrastructure with software-based control and management encompassing multiple SDN domains and cloud-edge networks. Each SDN domain has a local controller. There is a global SDN controller to coordinate the local controllers.

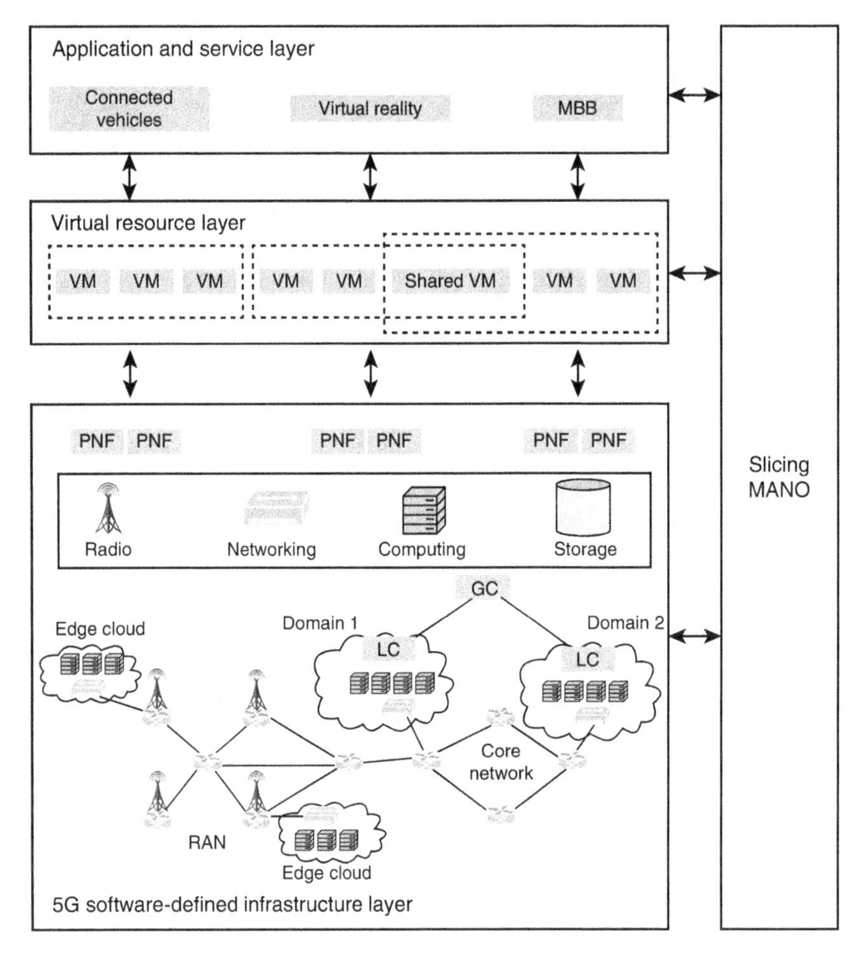

FIGURE 9.18 5G Network Slicing Framework

The final component of this architecture is the slicing MANO, which is essentially an enhanced NFV MANO that manages NFV, SDN, and slicing resources. Its tasks include:

- Creating and managing VM instances by using the infrastructure resources
- Mapping network functions to virtual resources and connecting network functions to create service chains
- Managing the life cycle of network slices by interacting with the application and service layer (e.g., automated creation of service-oriented slices, dynamic maintenance by monitoring service requirements and virtual resources)

9.6 Key Terms and Review Questions

Key Terms

admission control	QoS parameters
broadcast	QoS routing
congestion avoidance	queuing
conversational	quality of service (QoS)
core network	resource reservation
mission critical	scheduling
multicast	service-level agreement (SLA)
network slice	service data flow (SDF)
network slicing	streaming
packet marking	traffic classification
PDU session	traffic policing
policy	traffic restoration
policy control	traffic shaping
priority	tunnel
QoS characteristics	tunneling
QoS differentiation	user plane marking
QoS flow	

Review Questions

1. List and define the IMT-2020 network operational requirements.
2. List and define the basic network requirements for 5G specified by 3GPP.
3. What are the differences between priority, QoS, and policy control?
4. What is 5G tunneling?
5. Define the two types of tunnels and explain the purpose of each of them.
6. What are the differences among PDU session, QoS flow, and service data flow?

7. What are the main purposes of QoS capabilities?
8. What are the four stages in the QoS life cycle?
9. Define the common functions used to provide QoS in the data plane.
10. Define the common functions used to provide QoS in the management plane.
11. What are the differences between QoS classification, marking, and differentiation?
12. List and briefly define the QoS parameters in the TS 23.501 model.
13. List and briefly define the QoS characteristics in the TS 23.501 model.
14. What is the difference between QoS parameters and characteristics?
15. List and define the general requirements for network slicing specified by 3GPP.
16. List and define the management requirements for network slicing specified by 3GPP.
17. What NFs support multiple slices for UE?
18. What NFs support a single slice instance?

9.7 References and Documents

References

5GPP20 5G Infrastructure Public Private Partnership. *View on 5G Architecture*. 5G PPP Architecture Working Group white paper, February 2020.

GSMA20 GSM Association. *Generic Network Slice Template Version 3.0.* May 2020.

KOUK06 Koukal, M., and Bestak, R. "Architecture of IP Multimedia Subsystem." *Proceedings ELMAR Symposium*, June 2006.

LI17 Li, X., et al. "Network Slicing for 5G: Challenges and Opportunities." *IEEE Internet Computing*, September/October 2017.

ORDO17 Ordonez-Lucena, J., et al. "Network Slicing for 5G with SDN/NFV: Concepts, Architectures and Challenges." *IEEE Communications Magazine*, May 2017.

Documents

3GPP TR 22.891 *Technical Specification Group Services and System Aspects, Feasibility Study on New Services and Markets Technology Enablers, Stage 1 (Release 14)*. September 2016.

3GPP TR 26.939 *Technical Specification Group Services and System Aspects, Guidelines on the Framework for Live Uplink Streaming (FLUS) (Release 16)*. September 2019.

3GPP TS 22.261 *Technical Specification Group Services and System Aspects, Service requirements for the 5G system, Stage 1 (Release 17).* December 2020.

3GPP TS 23.501 *Technical Specification Group Services and System Aspects, System architecture for the 5G System (5GS), Stage 2 (Release 16).* December 2020.

3GPP TS 23.502 *Technical Specification Group Services and System Aspects, Procedures for the 5G System (5GS), Stage 2 (Release 16).* December 2020.

3GPP TS 23.503 *Technical Specification Group Services and System Aspects, Policy and charging control framework for the 5G System (5GS), Stage 2 (Release 16).* April 2020.

3GPP TS 28.531 *Technical Specification Group Services and System Aspects, Management and orchestration, Provisioning (Release 16).* December 2020.

3GPP TS 38.300 *Technical Specification Group Radio Access Network, NR, NR and NG-RAN Overall Description, Stage 2 (Release 16).* September 2020.

5G PPP *View on 5G Architecture.* Version 3.0. June 2019.

ETSI GS NFV-EVE 005 *Network Functions Virtualisation (NFV), Ecosystem, Report on SDN Usage in NFV Architectural Framework.* December 2015.

ITU-T Y.1291 *An Architectural Framework for Support of Quality of Service in Packet Networks.* May 2004.

ITU-T Y.3101 *Requirements of the IMT-2020 Network.* April 2018.

ITU-T Y.3106 *Quality of Service Functional Requirements for the IMT-2020 Network.* April 2019.

NGMN *5G End-to-End Architecture Framework.* August 2019.

Chapter 10

Multi-Access Edge Computing

Learning Objectives

After studying this chapter, you should be able to:

- Define MEC
- Explain why MEC is essential to all three usage scenarios supported by 5G
- Present an overview of the ETSI MEC architecture
- Explain the difference between a MEC host and a MEC platform
- Understand the relationship between MEC and NFV
- Describe how MEC supports network slicing
- List and define the three categories of MEC use cases described in ETSI documents
- Discuss the role MEC can play in the factory of the future
- Explain how MEC can target a specific subscriber or group of subscribers
- Explain the importance of video analytics in a number of 6G use cases and the role of MEC

This chapter addresses an essential element of 5G: multi-access edge computing. It is worthwhile to start with the following definitions:

- **Cloud computing:** A loosely defined term for any system providing access via the Internet (or other networks) to processing power, storage, network, software, or other computing services, often via a web browser. Often, these services are rented from an external company that hosts and manages them.
- **Edge computing:** A strategy to deploy processing capability at the network edge, where end terminals are connected, and to perform the bulk of processing of data that is derived from and fed to the end terminals.

- **Cloud-edge computing:** A form of edge computing that offers application developers and service providers cloud computing capabilities, as well as an IT service environment, at the edge of a network. The aim is to deliver compute, storage, and bandwidth much closer to data inputs and/or end users.
- **Multi-access edge computing (MEC):** Cloud-edge computing that provides an IT service environment and cloud computing capabilities at the edge of an access network that contains one or more types of access technology and in close proximity to its users. It is characterized by either ultra-low latency or high data rate capacity or both. For wireless access (e.g., in a radio access network), MEC provides real-time access to radio network information that can be leveraged by applications.
- **Mobile edge computing:** A term that was formerly used, with the acronym MEC, to denote what is now referred to as multi-access edge computing but limited to wireless access to a cellular network.

The chapter begins with a discussion of MEC support for 5G usage scenarios. The following two sections describe MEC architectural concepts and the ETSI MEC architecture. Then, Section 10.4 describes the incorporation of MEC into an NFV framework. Section 10.5 explains MEC support for network slicing. Finally, Section 10.6 discusses numerous MEC use cases.

10.1 MEC and 5G

Multi-access edge computing is a vital element in 5G deployment. MEC enables operator and third-party services to be hosted close to the user equipment's (UE's) access point of attachment. MEC enables application developers and content providers to use cloud computing capabilities and IT services at the network edge. Using virtualization technology, MEC systems provide compute, storage, and network resources, as well as the functions needed for applications. As a result, MEC provides essential support in all three 5G usage scenarios (refer to Figure 2.2 in Chapter 2, "5G Standards and Specifications"):

- **Enhanced mobile broadband (eMBB):** eMBB applications require high data rates—both peak data rates and overall capacity rates. Moving much of the UE communications to a near edge relieves the core network of a significant burden. In particular, QoS flows supported by network slices generally follow a much shorter path, making satisfaction of QoS more achievable.
- **Massive machine type communications (MMTC):** 5G must support massive Internet of Things (IoT) deployments with high connection density. An edge computing paradigm enables the collection of huge amounts of data at local edge processors. The edge processors can do some processing and consolidation of the data before transmitting the results across the 5G network to a central repository.
- **Ultra-reliable and low-latency communications (URLLC):** URLLC applications by definition demand very low latencies. Low latency levels can be achieved only if the interaction

between the UE and the URLLC application is local. Transmission across the breadth of a core network would prohibit providing the quality of service (QoS) required of such applications.

MEC also offers additional privacy and security and provides potential significant cost-efficiency.

10.2 MEC Architectural Concepts

Figure 10.1 illustrates the distinction between cloud computing and edge computing in the context of wireless network access to the Internet. In traditional cloud computing, cloud services—including applications, storage, and virtual platforms—are maintained centrally by the cloud service provider. Edge computing is a distributed, decentralized architecture that brings cloud-based services closer to the user, eliminating the need for bidirectional network traffic across the wireless core network.

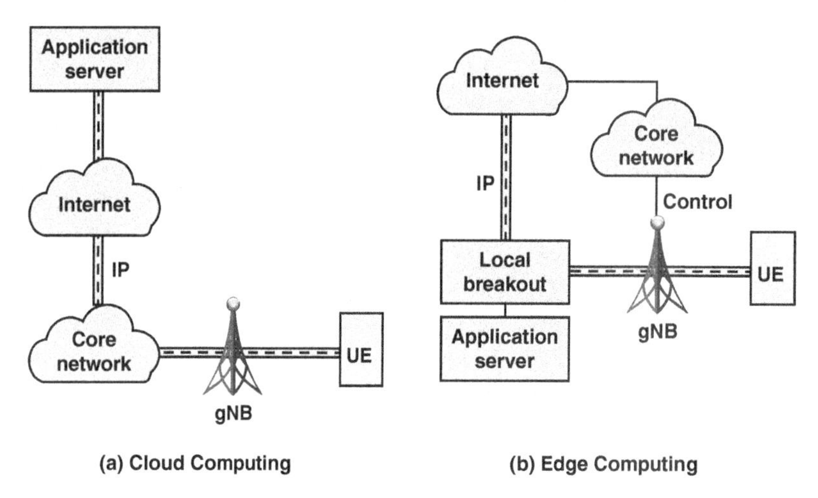

FIGURE 10.1 Cloud Versus Edge Computing

The essential architectural element that supports edge computing is a **local breakout** device. Local breakout is a concept in which the data plane traffic is routed locally to cloud services (compute and storage) without having to cross the breadth of the core network. This offloading solution saves core network load and reduces end-to-end latency. Creating a local breakout of some mobile data traffic enables content and applications to be processed as close as possible to the edge of the core network and closer to the mobile user.

Figure 10.2 (which repeats Figure 3.4 from Chapter 3, "Overview of 5G Use Cases and Architecture") suggests how edge devices fit into the 5G network architecture. The core network includes a number of routers and other network switches for moving data within and through the network. There are also access points, which provide functionality for access from the radio access network (RAN) and wired networks. Finally, there are cloud computing platforms, many of which are located at the edge of the core network. The edge nodes may be considered part of the core network or separate devices attached at the edge of the core network.

There are several networking aspects of the local breakout to consider:

- In addition to providing access to local cloud services, the local breakout generally provides a connection to the Internet, bypassing the core network of the cellular provider.
- The local breakout may be to local networking as well as local compute and storage services. For example, in vehicle-to-infrastructure applications, there is often a roadside infrastructure with a network and different application servers. With local breakout, the mobile device can locally connect to that data network. As another example, a 5G-enabled factory could be all 5G mobile devices talking to a factory IT application server implemented on a MEC platform. But it is more likely for the 5G mobile devices to talk to a factory IT network that includes multiple servers and devices connected using other network technology (e.g., Wi-Fi).

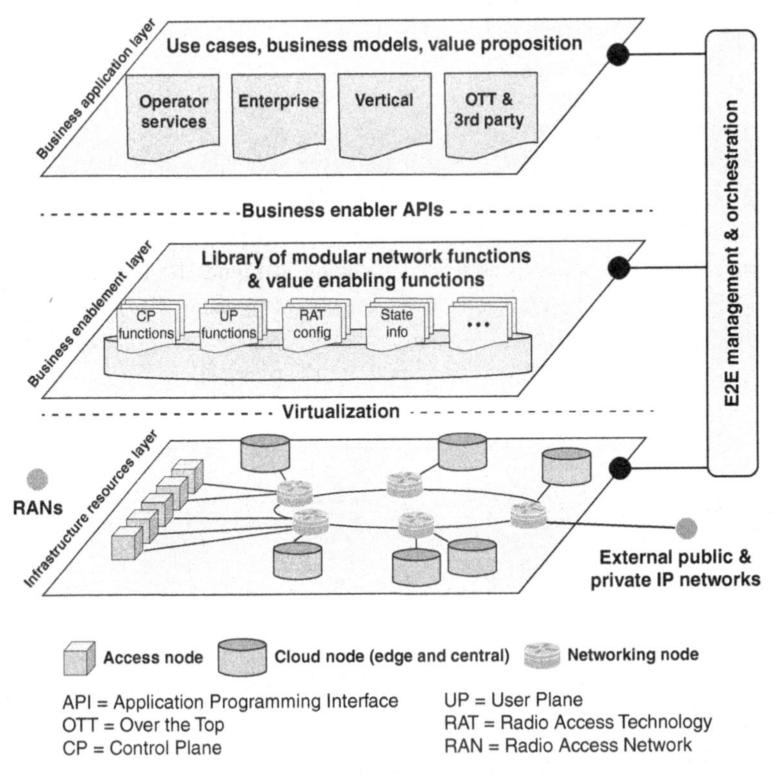

FIGURE 10.2 5G Architecture (Next Generation Mobile Networks)

Figure 10.3 provides another view of a MEC-based architecture and gives an indication of the functional relationships among the elements. Some of the literature uses the terms *device edge* and *infrastructure edge*. The device edge is the outermost component; it includes end user devices and sensors located in the "last mile" of the network. The infrastructure edge is connected to the device edge by an access network. The infrastructure edge typically sits in close proximity to the device edge but has characteristics of a traditionally hosted data center. The infrastructure edge may, in turn, be interconnected across many sites, using an aggregation edge layer of networking, or through a core network to a public cloud provider or regional data center.

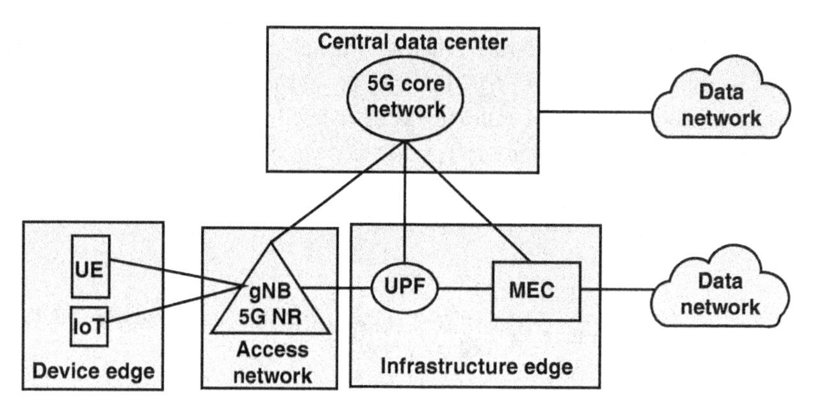

FIGURE 10.3 MEC Integrated into a 5G Network

Whether MEC edge devices are thought of as part of the core network or as edge systems attached to the core network, network functions virtualization (NFV) and the use of virtualized network functions (VNFs) integrate the MEC devices with the 5G core network, as shown in Figure 10.4. This figure shows key network functions (NFs) that are implemented on virtual platforms using NFV. The user plane function (UPF) is distributed into multiple MEC systems. A MEC system hosts one or more applications and other cloud services and supports the implementation of network slices.

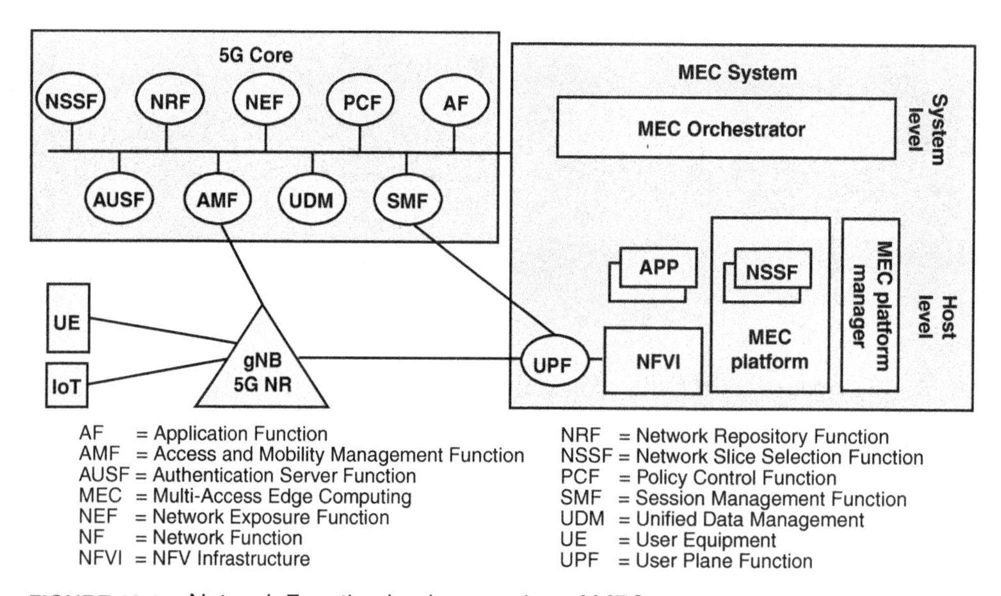

FIGURE 10.4 Network Function Implementation of MEC

10.3 ETSI MEC Architecture

The European Telecommunications Standards Institute (ETSI), which has developed a family of standards for NFV, is also the body developing standards for MEC. The most important ETSI documents on MEC, from the perspective of this book, are as follows:

- **ETSI GS MEC 001:** *Multi-Access Edge Computing (MEC), Terminology*
- **ETSI GS MEC 002:** *Multi-Access Edge Computing (MEC), Phase 2: Use Cases and Requirements*
- **ETSI GS MEC 003:** *Multi-Access Edge Computing (MEC), Framework and Reference Architecture*
- **ETSI GR MEC 024:** *Multi-Access Edge Computing (MEC), Support for Network Slicing*

Table 10.1 provides definitions for a number of terms that are used in the ETSI MEC documents.

TABLE 10.1 MEC Terminology

Term	Definition
Multi-access edge computing (MEC)	A system that provides an IT service environment and cloud computing capabilities at the edge of an access network that contains one or more types of access technology and in close proximity to its users.
MEC application	An application that can be instantiated on a MEC host within the MEC system and can potentially provide or consume MEC services.
MEC host	An entity that contains a MEC platform and a virtualization infrastructure that provides compute, storage, and network resources to MEC applications.
MEC host-level management	Components that handle the management of the MEC-specific functionality of a particular MEC platform, a MEC host, and the MEC applications running on it.
MEC management	MEC system-level and host-level management.
MEC platform	A collection of functionality that is required to run MEC applications on a specific MEC host virtualization infrastructure and to enable them to provide and consume MEC services and that can provide a number of MEC services.
MEC service	A service provided via a MEC platform either by the MEC platform itself or by a MEC application.
MEC system	A collection of MEC hosts and MEC management necessary to run MEC applications.
MEC system-level management	Management components that have an overview of the complete MEC system.

Design Principles

ETSI GS MEC 002 lists the following general principles that guide the design and development of MEC edge systems:

- **NFV alignment:** A MEC edge host system dynamically supports a number of applications. Thus, the resources required vary over time, and a virtualized environment provides the needed flexibility. The ETSI architecture employs NFV. If an edge system is part of the core network, this fits naturally with the NFV implementation of the core network. If the edge system is considered to be outside the core, the use of NFV enables hosting and management of virtual network functions (VNFs) using the same management infrastructure.
- **Mobility support:** A mobile user may require that support of edge-based applications move from one MEC system to another dynamically. Thus, the application needs to be implemented on a virtual machine such that the application can be moved from one virtual environment to another.
- **Deployment independence:** MEC capability should support deployment in variety of ways, including:
  - Deployment at the radio node.
  - Deployment at an aggregation point. An **aggregation point** is a location in a physical network deployment that is intermediate between the core network and a number of homogeneous or heterogeneous network termination points (base station, cable modems, LAN access points, etc.) and that can act as a location for a MEC host.
  - Deployment at the edge of the core network.
- **Simple and controllable APIs:** APIs should be easy to access and provide an effective means for controlling underlying resources to enable rapid development of applications.
- **Smart application location:** A MEC application needs to run at the appropriate physical location at any point in time, taking into account compute, storage, network resource, and latency requirements.
- **Application mobility to/from an external system:** The MEC architecture should support the movement of applications between a MEC host and an external cloud environment.
- **Representation of features:** GS MEC 002 introduces the concept of features. A feature is a group of related requirements that are assigned a unique name. The MEC architectural framework needs to support mechanisms to identify whether a specific feature is supported. Table 10.2 lists the MEC features.

TABLE 10.2 MEC Features

Feature	Description
UserApps	Provides support for creating a MEC application on one or more MEC hosts and establishment of connectivity between UE and a specific instance of the application.
SmartRelocation	Provides support for UserApps, for moving a MEC application from one MEC host to another, and for relocation of an application to/from a MEC host and a cloud environment outside the MEC system.
RadioNetworkInformation	Provides current radio network information related to the user plane. Information includes QoS and actual throughput for specific connections.
LocationService	Provides information about the locations of specific UEs currently served by the radio node(s) associated with the MEC host.
BandwidthManager	Enables MEC applications to register (statically and/or dynamically) their bandwidth and/or priority requirements.
UEIdentity	Enables a MEC to register a tag representing UE. This makes it possible to set packet filters for routing traffic based on a tag representing the UE.
WLANInformation	Exposes current wireless LAN information based on information received from external sources and/or generated locally.
V2XService	Supports the capability to provide feedback information from the network to a vehicle in support of V2X functions, which helps with predicting whether a communication channel is currently reliable (e.g., in terms of fulfilling latency requirements and 100% packet arrival).
5GCoreConnect	Provides information to the MEC system from the 5G network exposure function or other 5G core network function. Based on this information, the MEC system should support selection of a MEC host or MEC hosts and the instantiation of an application on the selected MEC host or hosts.

MEC System Reference Architecture

ETSI has developed a multi-access system reference architecture, as shown in Figure 10.5 (which is from ETSI GS MEC 003). There are four main elements of this architecture: the virtualization infrastructure manager (VIM), the MEC host, the MEC platform manager, and the MEC orchestrator.

Virtualization Infrastructure Manager

The VIM corresponds to the VIM in the NFV architecture, as shown in Figure 8.6 in Chapter 8, "Network Functions Virtualization." It controls and manages the interaction of an app with the virtualized compute, storage, and network resources under its authority. It is responsible for allocating, maintaining, and releasing virtual resources of the virtualization infrastructure. The VIM also maintains software images for fast app instantiation.

The VIM facilitates fault and performance monitoring by reporting information on virtualized resource usage to the MEC orchestrator, described later in this chapter.

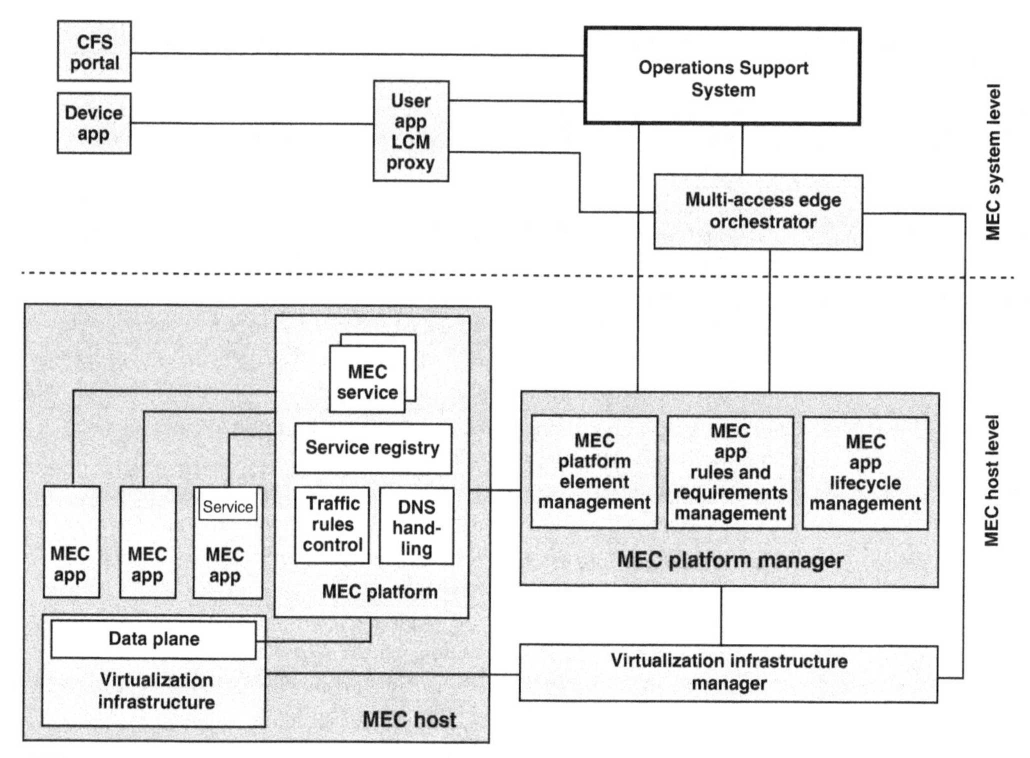

FIGURE 10.5 Multi-Access Edge System Reference Architecture

MEC Host

The mobile edge host is a logical construct that facilitates mobile edge applications (apps), offering a virtualization infrastructure that provides computation, storage, and network resources, as well as a set of fundamental functionalities (mobile edge services) required to execute apps, known as the mobile edge platform. A MEC host includes the virtualization infrastructure, MEC applications, and the MEC platform, which are discussed in the following paragraphs.

The **virtualization infrastructure** corresponds to the NFV infrastructure (NFVI) described in Section 8.6 in Chapter 8. The virtualization infrastructure includes a data plane that executes the traffic rules received by the MEC platform and routes the traffic among applications, services, the Domain Name System (DNS) server/proxy, the 5G network, other access networks, local networks, and external networks.

The main elements of the virtualization infrastructure are:

- **Hardware resources:** These include compute, storage, and network resources.
- **Virtualization layer:** This layer mediates the hardware resources to the virtual machines of the software appliances, providing an abstraction of the hardware.
- **Virtual resources:** Compute, storage, and network resources are instantiated as virtual machines.

MEC applications are virtual network functions that run on top of the virtual machines. These include all user and network applications that run on the edge host. MEC applications can have a certain number of rules and requirements associated with them, such as required resources, maximum latency, and required or useful services. These requirements are validated by the MEC system-level management and can be assigned to default values, if values are missing.

MEC apps interact with the MEC platform to obtain services, to indicate availability, and to perform app relocation when required. Apps can also provide services to other peer apps.

The **MEC platform** is the collection of essential functionality required to run MEC applications on a particular virtualization infrastructure and enable them to discover, advertise, and consume edge services. The MEC platform includes the following elements:

- **MEC service:** Network-related APIs are exposed by the MEC service to MEC applications. MEC applications are network-aware, and the MEC service exposes information through these APIs. GS MEC 003 defines the following MEC services:
  - **Radio network information:** This information includes radio network conditions, measurement and statistics information related to the user plane, and information related to the UE served by the radio nodes associated with the MEC host.
  - **Location:** This is the location of UEs currently served by the radio nodes associated with the MEC host.
  - **Bandwidth manager:** The bandwidth manager allows allocation of bandwidth to certain traffic.
- **Service registry:** This is the repository of information that can be exposed by the MEC service. It includes:
  - Radio network conditions
  - Location information, such as the location of UE
- **Bandwidth manager:** The bandwidth manager allows allocation of bandwidth to certain traffic routed to and from MEC applications, and it handles prioritization.
- **Traffic rules control:** The MEC platform can assign priorities to competing application traffic and can provide the virtualization infrastructure with a set of forwarding rules for the data plane. The forwarding rules are based on the policies received by the MEC platform manager and mobile applications.

- **DNS handling:** MEC applications support routing of all DNS traffic received from UE to a local DNS server or proxy. This enables DNS redirection to be handled by a local DNS server instead of being sent across the core network and the Internet.

MEC Platform Manager

The MEC platform manager corresponds to the VNF manager in the NFV architecture (refer to Figure 8.6 in Chapter 8). The MEC platform manager oversees the management of MEC applications and services. It includes the following elements:

- **MEC platform element management:** Handles FCAPS (fault, configuration, accounting, performance, and security) management for the MEC platform.
- **MEC application rules and requirements management:** Manages the application rules, traffic rules, and DNS configuration.
- **MEC application life cycle management:** Handles instantiation, maintenance, and deletion of MEC applications on VMs.

Multi-Access Edge Orchestrator

The multi-access edge orchestrator corresponds to the Orchestrator in the NFV architecture (refer to Figure 8.6 in Chapter 8). It performs the following functions:

- Life cycle management of MEC applications, achieved by talking to the application through the MEC platform manager
- On-boarding of application packages, including checking of the integrity and authenticity of the packages
- Selecting the appropriate MEC host(s) for application instantiation based on constraints, such as latency, available resources, and available services

Related Elements

Figure 10.5 shows several additional elements that are not part of the system architecture but play related roles. The following sections describe these elements.

Operations Support System (OSS)

The OSS consists of applications that support back-office activities in a network and that provision and maintain customer services. Network planners, service designers, operations, architects, support, and engineering teams of a service provider typically use operations support systems.

There are two categories of OSS:

- **Network OSS:** This is a traditional OSS that is dedicated to providers of telecommunication services. The processes supported by network OSS include service management and maintenance of the network inventory, configuration of particular network components, and fault management.
- **Cloud OSS:** OSS of cloud infrastructure is the system dedicated to providers of cloud computing services. Cloud OSS supports processes for the maintenance, monitoring, and configuration of cloud resources.

In a 5G/MEC environment, the OSS encompasses the functionality of both these categories.

CFS Portal

The customer-facing service (CFS) portal enables customers of a mobile operator to order new MEC applications and monitor service-level agreements (SLAs).

User APP LCM Proxy

The user application life cycle management proxy is an optional capability for a MEC system and requires the system to support the UserApps feature. This capability provides support for creating MEC applications on one or more MEC hosts and establishing connectivity between a UE and a specific instance of the application. In essence, a user can trigger specific applications in a MEC system from his or her device.

10.4 MEC in NFV

MEC and NFV are complementary concepts. ETSI has designed the MEC architecture to make it possible to instantiate MEC applications and NFV virtualized network functions (VNFs) in the same virtualization infrastructure. Thus, MEC can be readily implemented as part of a 5G core network.

Figure 10.6, from [KHAN20], illustrates the mapping of MEC functionality onto an NFV infrastructure. Many of the MEC architecture components are VNFs and thus can be moved to NFV as VNFs.

MEC Components Implemented as VNFs

All of the elements of the MEC architecture at the MEC host level are VNFs or management elements that can be supported in the NFV architecture. Figure 10.6 indicates how the components are mapped. As you can see in this mapping, the following elements are deployed as VNFs:

- The MEC platform
- All MEC applications

- The data plane component of the virtualization infrastructure
- Two components of the MEC platform manager: The MEC platform element manager and the MEC application rules and requirements manager

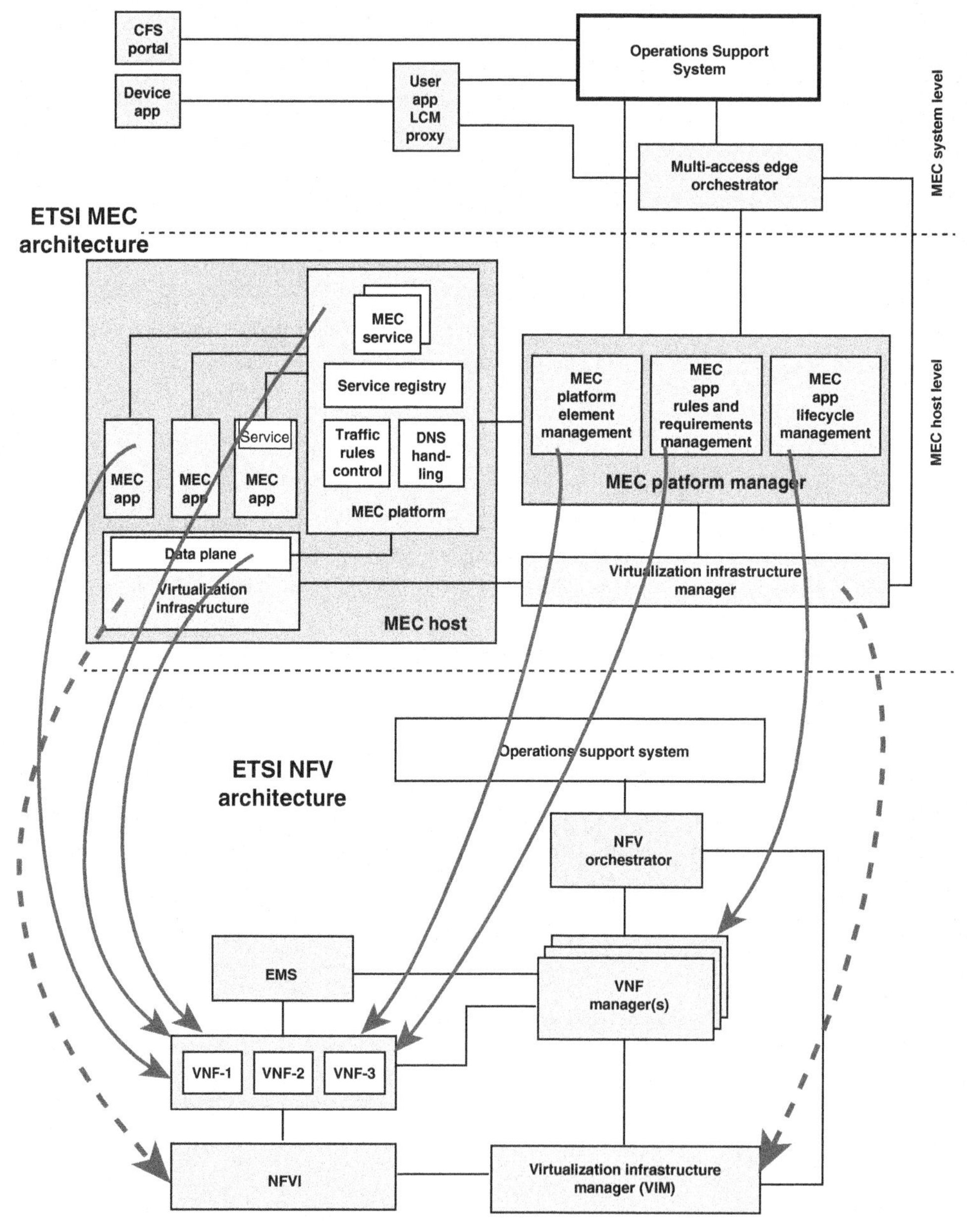

FIGURE 10.6 MEC Mapped to NFV

MEC Components Replaced by NFV Components

The MEC architecture includes two virtualization components—the virtualization infrastructure and the virtualization infrastructure manager (VIM)—that duplicate functions found in the NFV architectures. Figure 10.6 uses dashed arrows to indicate the mapping. In essence, the virtualization infrastructure is deployed as an NFVI and is managed by a VIM, as defined in the ETSI NFV documentation.

MEC System-Level Components

The elements at the MEC system level are retained as they are in NFV as new components. The MEC orchestrator (MEO) is replaced by a MEC application orchestrator (MEAO) that relies on the NFV orchestrator (NFVO) for resource orchestration and for orchestration of the set of MEC application VNFs as one or more NFV network services (NSs).

10.5 MEC Support for Network Slicing

Network slicing is an essential enabling technology for providing differentiated quality of service (QoS) to various applications and end users. With network slicing, the radio access network and the core network can allocate resources and implement traffic policies on a per-slice basis, such that different slices consume different amounts of resources and provide different QoS levels.

In order to meet demanding QoS requirements, such as ultra-low latency, high bandwidth, or support for massive numbers of UEs, network slices defined with such requirements must use MEC. Thus, it is imperative to develop technical solutions for providing MEC support for network slicing. ETSI GR MEC 024 (*Multi-Access Edge Computing (MEC), Support for Network Slicing*, November 2019) addresses this need.

Figure 10.7, from GR MEC 024, illustrates the allocation of MEC components on a per-slice basis in an NFV environment. Each network slice is implemented as a separate MEC platform and MEC platform manager supporting one or more apps plus a data plane VNF that defines the user traffic rules and priorities. In the figure, the lighter and darker shaded boxes indicate dedicated MEC instances for the two slices. The MEC components in black boxes are shared across the two network slices. The white boxes are MEC components not directly involved in supporting different slices, although these components are slice aware.

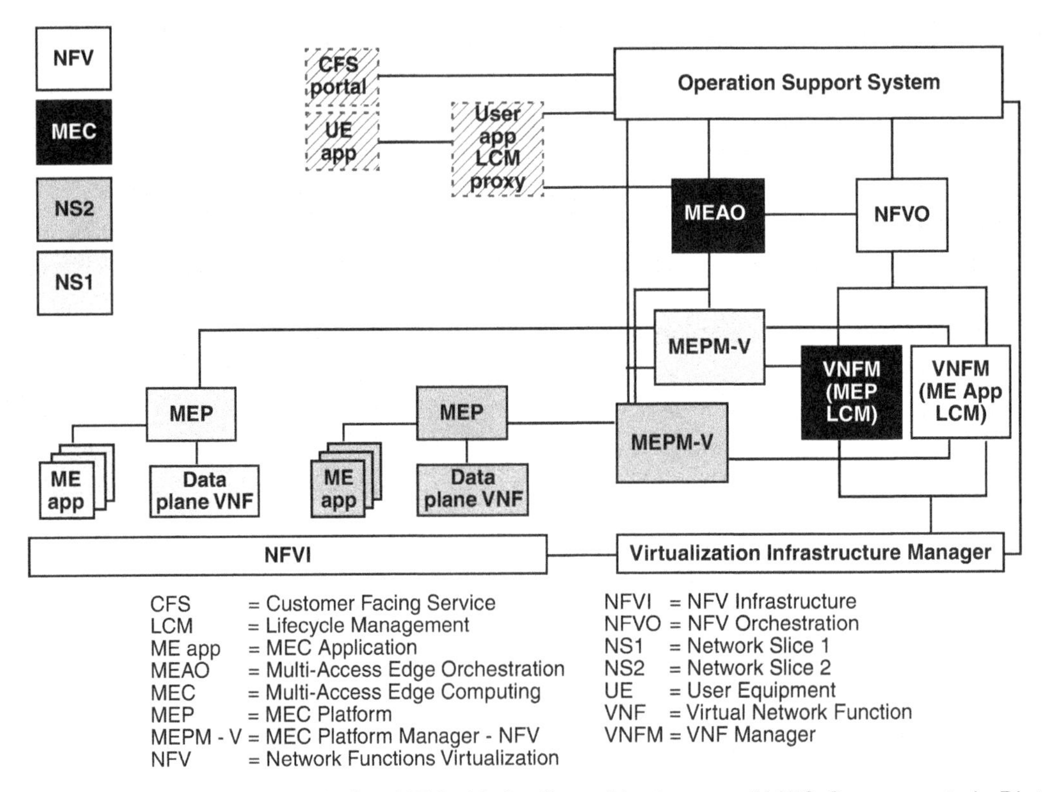

FIGURE 10.7 Example of MEC in NFV with Dedicated Instances of MEC Components in Distinct Network Slice Instances (NSIs)

10.6 MEC Use Cases

ETSI GS MEC 002 describes 35 MEC use cases for MEC hosts at the operator network edge. These use cases are useful for indicating the range of ways in which MEC can be employed to provide technical benefits and to serve as a guide for deriving MEC requirements.

ETSI groups the use cases into three categories, such that requirements on the architecture are generally quite similar for use cases within a category and quite different between the categories, as shown in Figure 10.8.

The categories are as follows:

- **Consumer-oriented services:** Services that directly benefit the end user (i.e., the user using the UE).
- **Operator and third-party services:** Services that take advantage of computing and storage facilities close to the edge of the operator's network. These services are usually not directly benefiting the end user, but can be operated in conjunction with third-party service companies.

- **Network performance and quality of experience (QoE) improvements:** Services aimed at improving performance of the network, either via application-specific or generic improvements. The user experience is generally improved, but these are not new services provided to the end user.

Customer-oriented services

- Augmented reality, assisted reality, virtual reality, cognitive assistance
- Gaming and low-latency cloud applications
- MEC edge video orchestration
- Location-based service recommendation
- Radio network information generation in aggregation point
- Application computation offloading
- Camera as a service
- Video production and delivery in a stadium environment
- Multi-user, multi-network

Operator and third-party services

- Security, safety, data analytics
- Active device location tracking
- Application portability
- Vehicle-to-infrastructure communication
- Flexible development with containers
- Third-party cloud provider
- IPTV over WTTx
- MEC platform consuming information from operator-trusted MEC application

Network performance and QoE improvements

- Mobile video delivery optimization using throughput guidance for TCP
- Local content caching at the mobile edge
- SLA management
- Mobile backhaul optimization
- Direct interaction with MEC application
- Traffic deduplication
- Bandwidth allocation manager for applications
- Radio access bearer monitoring
- MEC host deployment in dense-network environment
- Unified enterprise communications
- Optimizing QoE and resource utilization in multi-access network
- Media delivery optimizations at the edge
- Multi-RAT application computation offloading
- MEC system deployment in 5G

FIGURE 10.8 MEC Use Cases

Consumer-Oriented Services

ETSI GS MEC 002 lists the following use cases in the consumer-oriented services category:

- **Augmented reality, assisted reality, virtual reality, cognitive assistance:** The four variants of this use case are:
    - Augmented reality (AR): AR allows the user to see the real world with virtual objects composited with the real world. Therefore, AR supplements reality rather than completely replacing it.

- Assisted reality: Assisted reality is similar to AR, but its purpose is to actively inform the user of matters of particular interest to that user (danger warnings, ongoing conversations, etc.). This might be helpful, for example, in supporting people with disabilities to improve their interactions with their surroundings.
- Virtual reality (VR): VR services give users the ability to access the sights and sounds of remotely located complex systems in real time.
- Cognitive assistance: Cognitive assistance takes the concept of augmented reality one step further by providing personalized feedback to the user on activities the user might be performing (e.g., cooking, recreational activities, furniture assembly).

These applications require massive computational capability, high communication bandwidth, and ultra-low latency [SUKH19]. A MEC host near the user can satisfy these requirements by offloading some of the computational burden from the user device and caching information from remote databases.

- **Gaming and low-latency cloud applications:** Gaming applications require low-latency connection to the game server. This can be accomplished by placing the gaming software on a nearby MEC host. More generally, any user activity that requires low-latency access to a cloud-based application can be supported by a MEC app on a nearby edge host.
- **MEC edge video orchestration:** This use case refers to visual content that is produced and consumed at the same location close to consumers in a densely populated and clearly limited area. Such a case could be a sports event or a concert where a remarkable number of consumers are using their handheld devices to access user-selected tailored content. The overall video experience is combined from multiple sources, including locally produced video and additional information as well as master video from a central production server. The user is given an opportunity to select tailored views from sets of local video sources. A MEC architecture is ideally suited to this use case.
- **Location-based service recommendation:** This use case consists of one or more apps that provide useful information to a mobile user in real time, based on the user's current location, such as a shopping mall, a museum, or a tourist site. A 5G network can determine a user's location and connect the user to a local MEC host that can provide location-based information to the user.
- **Radio network information generation in aggregation point:** To provide the required QoS to certain applications that are hosted at a MEC host, the MEC platform includes a service that provides radio information, as discussed in Section 10.3. When the MEC host is located at a base station, it may easily gather the required information. When the MEC host is at an aggregation point or the edge of the core network, the radio network information service must implement a program that is able to determine the needed information. The program may use network traffic information and self-learning methods to determine or estimate the required radio information.

- **Application computation offloading:** This use case consists of hosting an end user application on a nearby MEC host rather than on the user device. This use case is effective when the application is computation intensive, allowing the application to be used regardless of the capability of the end user device. Examples include graphical rendering (e.g., high-speed browser, artificial reality, 3D games), intermediate data processing (e.g., sensor data cleansing, video analyzing), and value-added services (e.g., translation, log analytics).

- **Camera as a service:** This use case, also referred to as video surveillance as a service (VSaaS), refers to hosted cloud-based video surveillance. The service typically includes video recording, storage, remote viewing, management alerts, and cybersecurity. The video processing and management are performed offsite, using the cloud. Incorporating MEC in such a service has several advantages. It enables local storage of much of the video content, reducing the burden on the network. For some applications, the system detects and tracks objects and may take some action, such as triggering an alarm. The processing for these applications may present a burden if assigned to the surveillance camera system but may need rapid response, which is best achieved by a local MEC host.

- **Video production and delivery in a stadium environment:** This use case is discussed in Chapter 6, "Ultra-Reliable and Low-Latency Communications," and illustrated in Figure 6.5. A MEC system deployed inside the venue enables visual content to be produced, composed, processed, and consumed locally. Examples include events such as sports, concerts, public meetings, and conferences. Consumers can select tailored content using their handheld devices. This may include a specific viewing angle, multi-viewing angles, slow motion replay, analytics and statistics, side-by-side comparison between players, and other features. Users may request a viewing angle or a shot from a location that is not available from the user's physical seat or section. Running the video applications at the edge allows easy control of service quality and improvement in performance of video delivery and consumption.

- **Multi-user, multi-network applications:** Applications involving multiple users across multiple networks are becoming very popular. An example is online gaming. Users play games inside their homes as well as when they are outside (e.g., when traveling, waiting in a shopping mall, waiting for train or plane). They play on multiple devices and demand rich applications with very low latency. To provide such rich gaming applications with very low latency, game service providers are using edge computing services. Such edge computing services may be provided by network operators or third-party service providers.

- **In-vehicle MEC hosts supporting automotive workloads:** Section 6.2 in Chapter 6 discusses the intelligent transportation URLLC use case, which includes assisted driving, autonomous driving, and tele-operated driving. 5G-connected vehicles will be able to exchange messages with each other, with the roadside infrastructure, with back-end servers, and with the Internet. They will do so with reduced latency, increased reliability, and large throughput and with high mobility and user density. As the number of supported vehicles increases, researchers and developers are demonstrating that the use of MEC hosts based in the vehicles provides superior performance and reliability compared to the use of MEC hosts on the edge of the core network or the use of cloud-based apps [GIAN20, GRAM19].

- **Factories of the future:** This use case is discussed in the following subsection.

Factories of the Future

Section 6.4 in Chapter 6 covers Industry 4.0, also referred to as smart factories or factories of the future, in some detail. This is one of the most demanding usage scenarios for 5G support. As Figure 6.12 and Table 6.4 indicate, full support of a smart factory requires eMBB, mMTC, and URLLC capabilities. To meet this wide range of requirements, the networking and computing infrastructure needs not only the full capabilities of a 5G network but also extensive use of a MEC strategy.

Figure 10.9, which repeats Figure 6.9 from Chapter 6, suggests how the various elements of an automated factory relate to one another in a hierarchical networked system. The lower portion of the figure depicts the operational technology (OT) domain, which consists of the hardware and software that detect or cause a change through direct monitoring and/or control of physical devices, processes, and events in the enterprise. The upper section is the information technology domain. IT refers to the entire spectrum of technologies for information processing, including software, hardware, communications technologies, and related services.

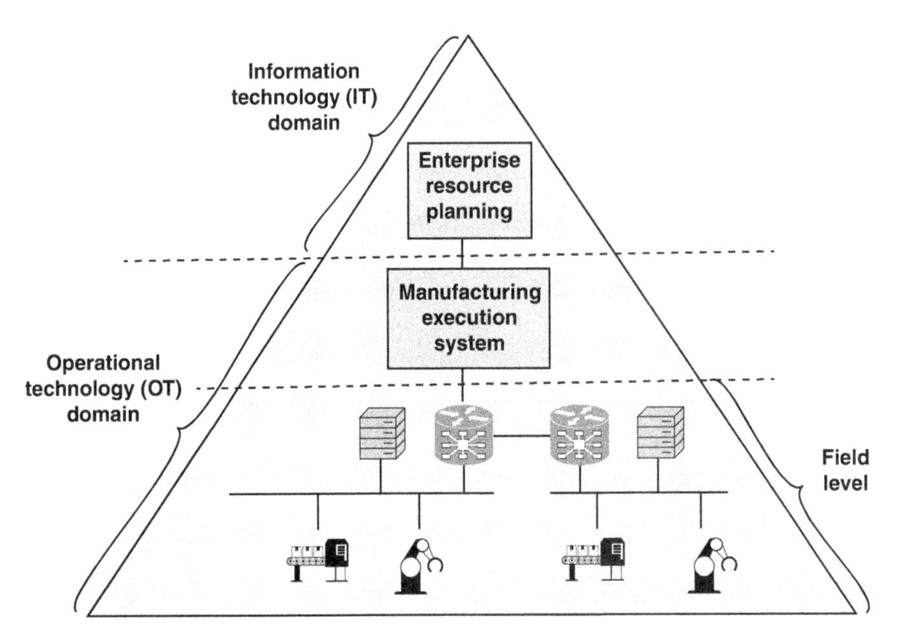

FIGURE 10.9 Hierarchical Network Design Based on the Industrial Automation Pyramid

Figure 10.10, based on a figure in GS MEC 002, illustrates a distributed MEC approach to satisfying the requirements for factory automation. In this example, some sensors are connected by fixed connections to a wireless LAN (WLAN) access point. A collection of sensors and actuators form an IoT that is connected to a field-level LAN by an IoT gateway. A MEC host at the field level provides processing and storage support for the sensor and actuator deployment. At a higher level, a MEC host could be deployed within the enterprise to support enterprise-wide applications and apps that support the operational domain (refer to Figure 10.9). This MEC host could also provide access to enterprise-wide databases and applications that are used to control actuators and consolidate and interpret sensor data.

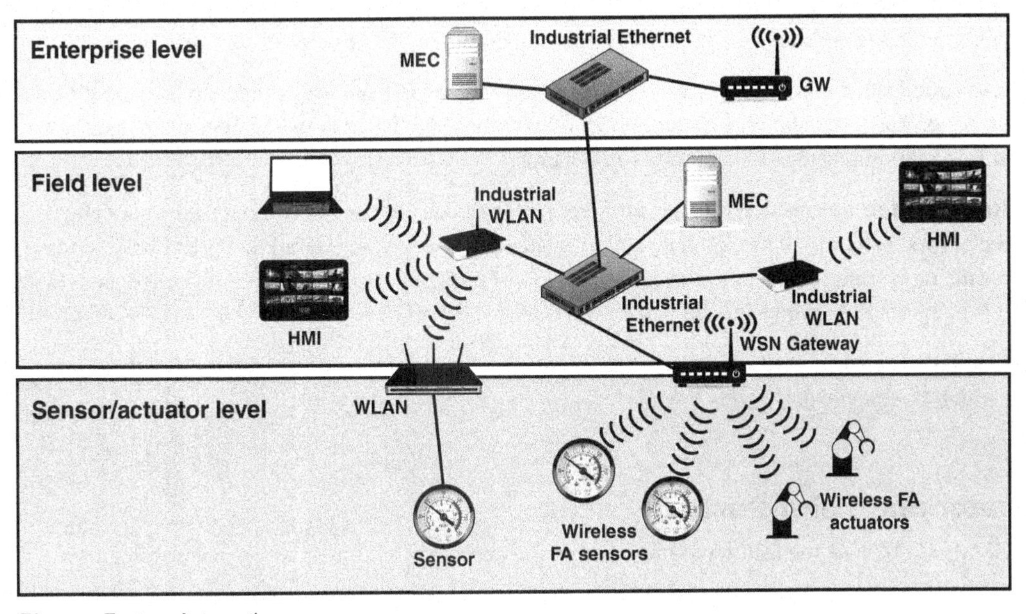

FA = Factory Automation
GW = Gateway
HMI = Human-Machine Interface
WLAN = Wireless Local Area Network
WSN = Wireless Sensor Network

FIGURE 10.10 MEC Deployment Scenario for Factory Automation

The use of this distributed MEC architecture yields a number of advantages, including the following [NICH20, ISMA18]:

- **Lower latency:** This is perhaps the most important benefit. Without low latency, smart manufacturing is unable to experience the full benefits of the IoT. If a connected machine on an assembly line recognizes a malfunction, any delay in transmitting that signal could be costly: It could lead to damaged parts or even injury.

- **Increased cybersecurity:** The centralized nature of traditional cloud computing makes it vulnerable to hacking, including denial of service attacks. If a factory spreads processing and storage functions throughout the edge, however, no single attack could bring down the whole network. When computing happens closer to the data source, less data is at risk at any given moment.
- **More manageable data analytics:** Big data is a foundational technology for Industry 4.0. Massive deployments of IoT devices generate huge amounts of data, and analysis of this data can improve operations and make it possible to derive the most value from sensor data. However, analyzing all this information requires a considerable amount of storage, bandwidth, and computing power. Edge computing distributes the burden in a structured fashion.
- **Expanded interoperability:** MEC hosts can function as IoT gateways that can provide the necessary protocol translation for communications to be established between devices that are not able to communicate with each other in a factory.
- **Reduced storage costs:** With edge storage, factories can choose to send only relevant data to their cloud solutions. The edge can act as a gateway by analyzing data locally and only sending results or summarized data to the cloud. On top of reducing the strain on cloud-based analytics, this helps save on storage costs.
- **Quality of service:** MEC hosts can maximize the efficient use of bandwidth while minimizing endpoint bottlenecks.

Operator and Third-Party Services

ETSI GS MEC 002 lists the following use cases in the operator and third-party services category:

- **Security, safety, data analytics:** This use case refers to an application that collects a very large amount of information from an mMTC IoT configuration where all devices are serviced by a single local radio access network (RAN). The application runs on a MEC host deployed close to the RAN. The application processes the information and extracts the significant summary data, which it sends to a central server. A subset of the data might be stored locally for a certain period for later cross-check verification.
- **Active device location tracking:** A number of end user applications require that the network keep track of the current locations of mobile devices. Applications include mobile advertising, student location monitoring on campus, personnel management, and various smart city applications. To effectively keep up with the movement of a large number of UEs in a given area, a location tracking app on a nearby MEC host is an efficient solution.
- **Application portability:** This use case simply means that MEC hosts should supply standardized APIs to ensure that apps are fully portable on MEC hosts from different vendors.

- **Vehicle-to-infrastructure communication:** This use case refers to communication between vehicles and roadside sensors, and it falls into the general category of assisted driving, which is discussed in Section 6.2 in Chapter 6. The roadside application incorporates algorithms that use data received from vehicles and roadside sensors to recognize high-risk situations in advance, and it sends alerts and warnings to the vehicles in the area. The drivers of the vehicles can immediately react (e.g., by avoiding the lane hazard, slowing down, or changing route). MEC hosts placed near clusters of sensors are needed to satisfy the requirement for ultra-low latency.
- **Flexible development with containers:** MEC hosts should support both containers and virtual machines as virtualization technologies available to third parties. This use case simply calls out the need to support containerized applications provided by third parties. In cases where a smaller footprint is one of the key design factors, containers are the preferred solution.
- **Third-party cloud provider:** This use case is related to an emerging business model in which computational resources for edge cloud service are provided by alternative facility providers that are nontraditional network operators. These providers are referred to as third-party edge owners (TEOs). A TEO may be a property management company or any holder of real estate that provides cloud resources to network operators and traditional cloud service providers. MEC hosts play a role in this business model by providing to TEOs localized service such as radio network and traffic information.
- **IPTV over WTT*x*:** WTT*x* (wireless to the *x*) is a 4G- and 5G-based broadband access solution that uses wireless technology to provide fiber-like broadband access for household. WTT*x* boasts superior network performance, low cost, fast deployment, easy maintenance, and rich services [HUAW16]. IPTV (which involves delivery of television content over Internet Protocol networks) over WTT*x* provides operators with fast access to home entertainment markets over existing cellular networks. MEC deployment allows a substantial amount of offloading of traffic from the core network. This is especially relevant for video on demand, where video content can be prepositioned at edge MEC hosts for transmission on demand to local customers.
- **MEC platform consuming information from operator-trusted MEC application:** This use case is discussed in the following subsection.

MEC Platform Consuming Information from Operator-Trusted MEC Application

This use case allows an application to target a specific subscriber or a group of subscribers. For example:

- Allowing an anonymous group of flat-rate billing subscribers access to content locally from the MEC host
- Sending targeted advertising for a certain group of users within the mobile network

- Providing content to a specific group of users that might be, for example, in the same club, association, or public service group
- Providing enterprise services to company employees

The MEC host supports this use case by routing traffic to a MEC application based on the UE IP address rather than the destination IP address. For this purpose, the MEC host needs a mapping of the UE IP address to a specific subscriber or subscriber group. This mapping information is provided by an external source, such as the core network. A MEC application that is trusted by the operator receives this mapping information and provides it to the MEC platform.

An operator-trusted application is more than just an application running on the MEC host. In essence, it is an extension of the MEC platform functionality. Such an application has advanced privileges to provide information to the MEC platform securely. Trust is achieved by requiring that such applications and the platform are mutually authenticated and authorized.

Figure 10.11, from GS MEC 002, illustrates an example use case for this concept. Subscriber traffic that has been registered with a specific tag that maps to the UE IP address is routed to the local MEC application rather than across the 5G network.

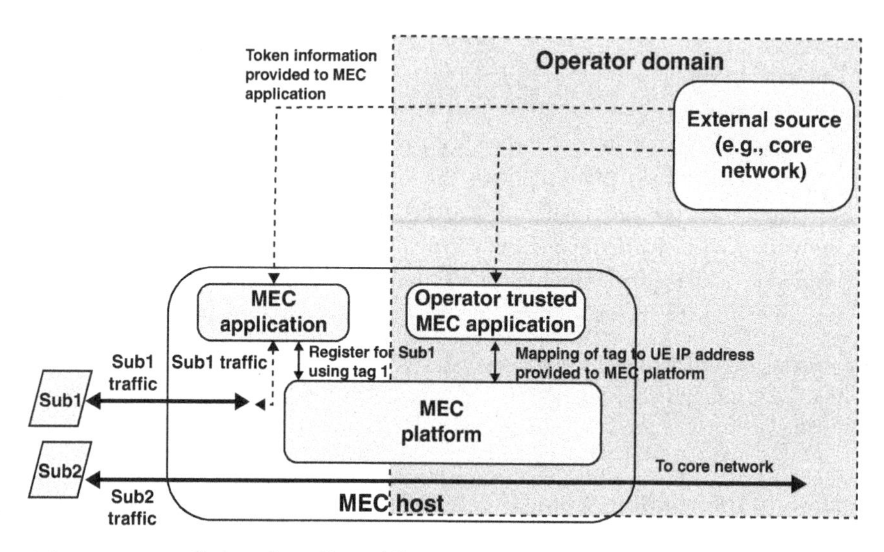

FIGURE 10.11 Subscriber-Based Routing

Network Performance and QoS Improvements

ETSI GS MEC 002 lists the following use cases in the network performance and QoS improvements category:

- **Mobile video delivery optimization using throughput guidance for TCP:** Mobile throughput guidance (MTG) is a potential means to improve customer experience during mobile Internet sessions; it makes explicit the range of bandwidth that the mobile access link is likely to sustain in the near future [GMSA17]. This information can improve the performance of Transmission Control Protocol (TCP), which uses a complex congestion control algorithm to regulate the flow of competing TCP sessions over shared networks or links. In general, TCP attempts to determine the current state of network capacity based on the round-trip time required to acknowledge packets. TCP responds to perceived congestion by drastically slowing the flow rate and gradually recovering. However, in mobile networks, the capacity available over the radio access network is highly volatile. Factors include variation in signal quality due to the mobility of the UE and environmental factors, as well as congestion management procedures at the air interface. For this specific use case, a radio analytics MEC application provides a back-end video server with a near-real-time indication on the throughput estimated to be available at the radio downlink interface in the next instant. The video server can use this information to assist with TCP congestion control decisions. With this additional information, TCP does not need to overload the network when probing for available resources, nor does it need to rely on heuristics to reduce its sending rate after a congestion episode.

- **Local content caching at the mobile edge:** This use case is one of the most obvious and widely deployed benefits of MEC. Local content caching is essential for realizing enhanced mobile broadband (eMBB) services on 5G networks. Storing and processing video, high-resolution graphic, and other content on MEC hosts greatly reduces the traffic burden on the core network.

- **SLA management:** MEC host system providers typically offer service-level agreements (SLAs) to third-party application developers. An SLA specifies data plane traffic performance requirements and virtualized resource requirements. The MEC host enables the application provider to collect data to monitor SLA compliance. SLA management refers to the capability of exposing specific parameters for monitoring.

- **Mobile backhaul optimization:** The term *backhaul* refers to the network path between a base station in the radio access network (RAN) and the core network. Typically, there is little or no coordination between the RAN and the backhaul network. When there is capacity degradation in the backhaul, the RAN is not informed about it; in addition, when the RAN needs less capacity, the backhaul is not aware of it. This use case combines information from the RAN with information from the backhaul network to optimize the resources in the backhaul. In cases where the backhaul network connects to a core network via a MEC host, a traffic analytics application on the MEC host can compute traffic information based on radio network information it obtains from a MEC service available via the MEC platform and the backhaul information it obtains from the monitoring application. The traffic analytics can use the traffic monitoring service to get the user plane traffic and identify the applications that the user uses.

- **Direct interaction with a MEC application:** In essence, this use case defines the capability needed to reroute UE-application traffic from the application hosted in a remote cloud server to that application now hosted on the MEC platform.

- **Traffic deduplication:** The majority of the traffic on the Internet is video, and much of this is redundant video-on-demand content. This use case employs a traffic deduplication technique that relies on compression and decompression of redundant traffic ([LE12], [SPRI00]). In essence, a server near the source of a redundant stream of traffic sends one copy of a block of traffic plus an index to a MEC host near the recipients. Subsequently, the source server just sends the index, reducing the load on the core network.
- **Bandwidth allocation manager for applications:** This use case is for the common situation in which multiple different applications and/or multiple instances of the same application are running concurrently on a single MEC host. A bandwidth allocation manager on the MEC platform collects bandwidth resource requirements and available bandwidth resources and allocates bandwidth to each session/application according to static/dynamic requirements.
- **Radio access bearer monitoring:** A UE can have multiple dedicated bearers carrying traffic with different QoS requirements and different QoS class identifier (QCI) values. This use case refers to monitoring traffic on different bearers between the UE and different applications hosted on the MEC host.
- **MEC host deployment in a dense network environment:** In order to identify wireless network congestion, a MEC service available via the MEC platform provides radio network information to a dedicated application. When network congestion is identified, the MEC application can communicate with counterpart applications running on devices to request that they activate direct device-to-device communication network capabilities through application-specific means.
- **Unified enterprise communications:** Communication within an enterprise increasingly relies on mobile devices both for IT communication and telephony. To provide effective support, many enterprises, particularly larger enterprises, deploy multiple base stations on the enterprise premises to create a coverage pattern of small cells. These cells offer contiguous small cell coverage within the enterprise environment. As connected mode users move within the enterprise environment, their sessions are handed over between neighboring enterprise small cells. As with any other base stations, those within the enterprise need to connect via a radio access network to a core network for external communications. The role of a MEC host in such a configuration is to provide a path to the core network and on to the Internet on the one hand and a breakout to the enterprise LAN on the other.
- **Optimizing QoE and resource utilization in a multi-access network:** As discussed at the beginning of this chapter, multi-access edge computing is a form of cloud-edge computing that expands the capability of the early mobile edge computing scheme. With multi-access edge computing, UE has access to the MEC platform through multiple types of network connections, which may include wireless techniques such as cellular and Wi-Fi, as well as wired techniques such as Ethernet and DSL. The quality of experience (QoE) of the end user depends, in part, on how effectively these access network resources are used. In this use case, the MEC host uses network information to dynamically select network paths based on knowledge of current

conditions in the relevant access networks. For this purpose, the MEC host includes software based on Multi-Access Management Services (MAMS), defined in RFC 8743 (*Multi-Access Management Services (MAMS)*, March 2020). MAMS consists of the following functions:

- **Client Connection Manager (CCM):** Negotiates the network path usage with NCM, based on clients' needs and capabilities.
- **Network Connection Manager (NCM):** Uses information obtained from the access network and, based on policy, current conditions, and the information exchanged with clients, configures the user plane paths for the multiconnectivity device.
- **Client Multiple Access Data Proxy (C-MADP):** Handles the user plane selection procedures at the client.
- **Network Multiple Access Data Proxy (D-MADP):** Handles the user plane selection procedures at the network.

- **Media delivery optimizations at the edge:** This use case optimizes the delivery of multimedia content via mobile and fixed networks to end users via the MEC host. For this purpose, the MEC server hosts an application conforming to the specification for SAND (Server and Network-Assisted Dynamic Adaptive Streaming over Hypertext Transfer Protocol [HTTP] [DASH]). SAND is defined in ETSI TR 126 957 (*Study on Server and Network-Assisted Dynamic Adaptive Streaming over HTTP [DASH] [SAND] for 3GPP Multimedia Services*, July 2018). SAND offers standardized interfaces for service providers and operators to enhance the streaming experience. In order to enhance the delivery of DASH content, SAND introduces messages between DASH clients and network elements or between various network elements in order to improve efficiency of streaming sessions by providing information about real-time operational characteristics of networks, servers, proxies, caches, content delivery networks (CDNs), and DASH client performance and status. SAND realizes the following:

  - Streaming enhancements via intelligent caching, processing, and delivery optimizations on the server and/or network side, based on feedback from clients on anticipated media segments, accepted alternative media content, client buffer level, and requested bandwidth.
  - Improved adaptation on the client side, based on network/server-side information such as cached segments, alternative segment availability, recommended media rate, and network throughput/QoS.

  A SAND edge server can leverage network-/link status–related information from a MEC server for determining the assistance messages to be sent to a streaming client.

- **Multi-RAT application computation offloading:** In an environment in which mobile devices can employ more than one radio access technology (e.g., Wi-Fi, 5G NR), the opportunity exists to reduce energy consumption by a mobile device application, either by switching to a different RAT when possible or offloading the application to a different mobile device under the user's control. MEC systems can also help an application select the most power-efficient RAT for the UE to improve the user experience in the network with multi-RAT coverage, in addition to considering other performance indicators (e.g., offloading latency).

- **MEC system deployment in a 5G environment:** Incorporating support for applications running on MEC systems is essential for meeting 5G performance requirements. This use case addresses the functional and architectural issues involved in deploying MEC systems in a 5G environment. Section 10.4 addresses this use case.
- **Video caching, compression, and analytics service chaining:** This use case is discussed in the following subsection.

Video Caching, Compression, and Analytics Service Chaining

Video analytics, also referred to as video content analysis (VCA) or intelligent video, involves the extraction of meaningful and relevant information from digital video [GAGV08]. Whereas video compression attempts to exploit the redundancy in digital video for the purpose of reducing size, analytics is concerned with understanding the content of video. Video analytics builds on research in computer vision, pattern analysis, and machine intelligence. Video analytics is a key technology for a number of 5G use cases, including the following:

- **Surveillance and public safety:** Processing live video streams almost instantaneously at the edge can lead to better surveillance and help in enforcing law and order. Two examples of this use case are face detection and incident identification and triggering, which allow law enforcement officers to take immediate actions involving an incident.
- **Autonomous driving:** Real-time video of the scene as seen by a self-driving car needs to be analyzed in a very short time to determine the actions to be taken by the car. A self-driving vehicle could already contain resources to process the scene instantaneously. Edge video analytics can help with processing (or preprocessing) further scenes or postprocessing video scenes for continual training and feedback.
- **Smart cities and IoT:** Video analytics at the edge is an important element in enabling smart cities. For example, traffic video analysis can be used to route traffic in the most efficient way. Fire or smoke detection in an area can be identified instantaneously to ensure that no traffic continues toward the danger zone; feedback can be sent to both the city infrastructure and connected cars in an area.
- **Enhanced infotainment services:** Video analytics at the edge can be used to enhance the real-life experience of event audiences such as those at sporting events, concerts, and other shows. Videos from different camera angles at an event can be analyzed and applied with AR/VR functions and presented to a live audience through large screens, smartphones, and VR devices.

Figure 10.12, from GS MEC 002, illustrates a scheme in which video is first stored on a local MEC host in a video content cache. The video content is then processed by a video compression algorithm followed by a video analytics application. Typically the video compression scheme is the one standardized by MPEG-4. This provides an ideal data representation for supporting indexing and retrieval schemes. It also simplifies the task of video structure parsing and keyframe extraction because many of the necessary content features (e.g., object motion) are readily available [DIMI02].

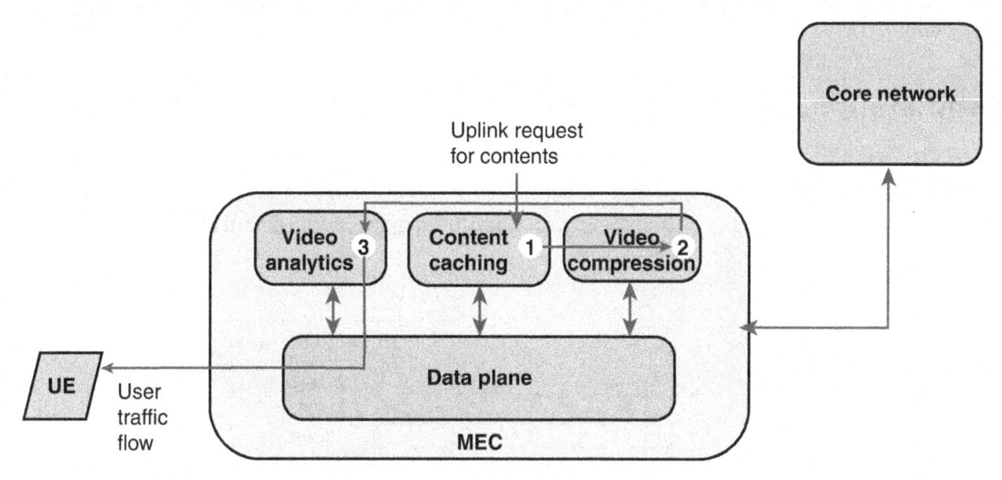

FIGURE 10.12 Video Content Delivery

10.7 3GPP Architecture for Enabling Edge Applications

In 2020, 3GPP released a draft version of an architecture, referred to as EDGEAPP, for enabling edge applications (TS 23.558 *Technical Specification Group Services and System Aspects, Architecture for Enabling Edge Applications [Release 17]*, November 2020). The architecture and procedures in TS 23.558 very much focus on how the UE and edge work together to discover edge applications and/or provision new edge applications. The document includes architectural requirements for enabling edge applications, an application layer architecture fulfilling the architecture requirements, and procedures to enable the deployment of edge applications.

Key requirements are:

- **UE application portability:** Changes in application clients compared to the existing cloud environment are avoided.
- **Service differentiation:** The mobile operator is able to provide service differentiation (e.g., by enabling/disabling the edge computing functionalities).

- **Flexible deployment:** There can be multiple edge computing providers within a single PLMN (Public Land Mobile Network) operator network. The edge data network can be a subarea of a PLMN.
- **Integration with 5G network:** The core network provides capability exposure, such as location service, QoS, and application function traffic influence, to the edge applications.
- **Service continuity:** There is support for continuation of application context across edge deployments.

EDGEAPP Functional Architecture

Figure 10.13 shows the functional architecture for enabling edge applications. The architecture defines a generic environment in which UE connects to an edge data network through the core network. The edge data network is a local (to the UE) data network, and thus application data traffic only traverses the local edge of the core network.

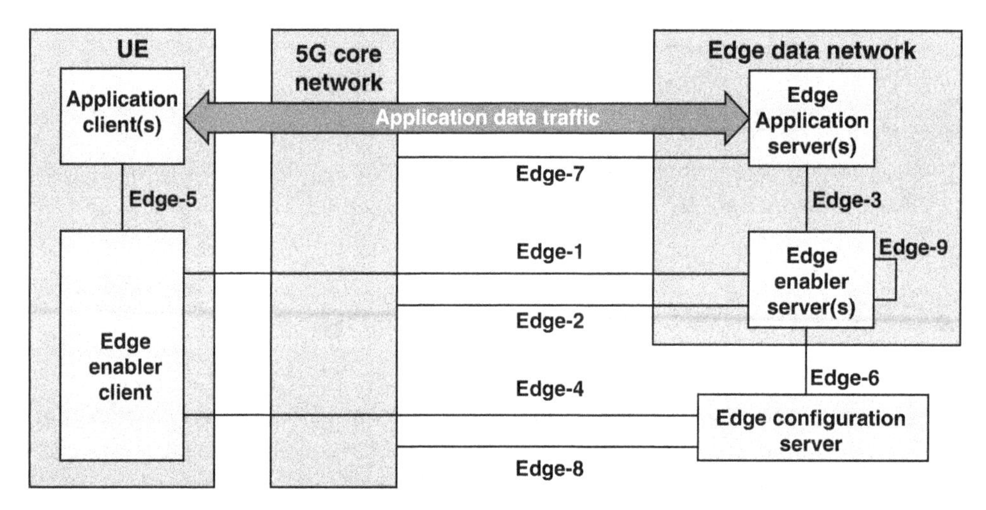

FIGURE 10.13 Architecture for Enabling Edge Applications

Functional Entities

There are five defined functional entities within the EDGEAPP functional architecture:

- **Edge enabler server (EES):** A server that provides information to the EEC related to the edge applications, such as availability/enablement and related configuration. The EES exposes capabilities of the 5G network to edge applications.
- **Edge enabler client (EEC):** A client that provides supporting functions needed for the application client(s). The EEC enables discovery of edge applications and provisioning of configuration data.

- **Edge configuration server (ECS):** A server that provides supporting functions needed for the EEC to connect with an EES. The ECS provides edge data network configuration information to the EEC. It also interacts with the 5G core network for accessing the capabilities of network functions.
- **Application client:** The application resident in the UE that is performing the client function.
- **Edge application server (EAS):** The application server resident in the edge data network that is performing the server functions. The application client connects to the EAS in order to access services of the application with the benefits of edge computing.

Reference Points

Chapter 3 introduces the concept of the reference point, which is defined by ITU-T and 3GPP as a conceptual point at the conjunction of two non-overlapping functional groups. In architecture diagrams, a reference point is depicted as a labeled line between two functional entities between a functional entity and an external functional module or network. A reference point enables interactions between the two connected entities.

TS 23.588 defines nine reference points:

- **Edge 1:** Between the EES and the EEC. It supports:
  - Registration and de-registration of the EEC to the EES
  - Retrieval and provisioning of EAS configuration information
  - Discovery of EASs available in the EDN
- **Edge 2:** Between the EES and the 5G core network. It supports access to 5G core network functions and APIs.
- **Edge 3:** Between the EES and EASs. It supports:
  - Registration of EASs with availability information (e.g., time constraints, location constraints)
  - De-registration of EASs from the EES
  - Discovery of target EAS information to support application context transfer
  - Provision of access to network capability information (e.g., location information, QoS-related information)
  - Requests for setup of a data session between the application client and EAS with a specific QoS
- **Edge 4:** Between the ECS and the EEC. It supports provisioning of edge configuration information to the EEC.

- **Edge 5:** Between application clients and the EEC. The reference point is beyond the scope of the specification.
- **Edge 6:** Between the ECS and EES. It supports registration of EES information to the ECS.
- **Edge 7:** Between the EAS and the 5G core network. It supports access to 5G core network functions and APIs.
- **Edge 8:** Between the ECS and the 5G core network. It supports access to 5G core network functions and APIs.
- **Edge 9:** Between two EEEs. Edge 9 may be provided between EESs within different EDNs and within the same EDN. It supports discovery of target EAS information to support application context relocation.

The bulk of TS 23.558 specifies procedures and information flows for each reference point.

Synergized Mobile Edge Cloud Architecture

Figure 10.14 appears in both TS 23.558 and ETSI white paper 36 (*Harmonizing Standards for Edge Computing: A Synergized Architecture Leveraging ETSI ISG MEC and 3GPP Specifications*, July 2020), and the white paper refers to this as a synergized mobile edge cloud architecture. Figure 10.14 indicates how the EDGEAPP architecture and ETSI MEC architecture can complement each other. In the figure, the shaded areas correspond to the 3GPP elements that are illustrated in Figure 10.13.

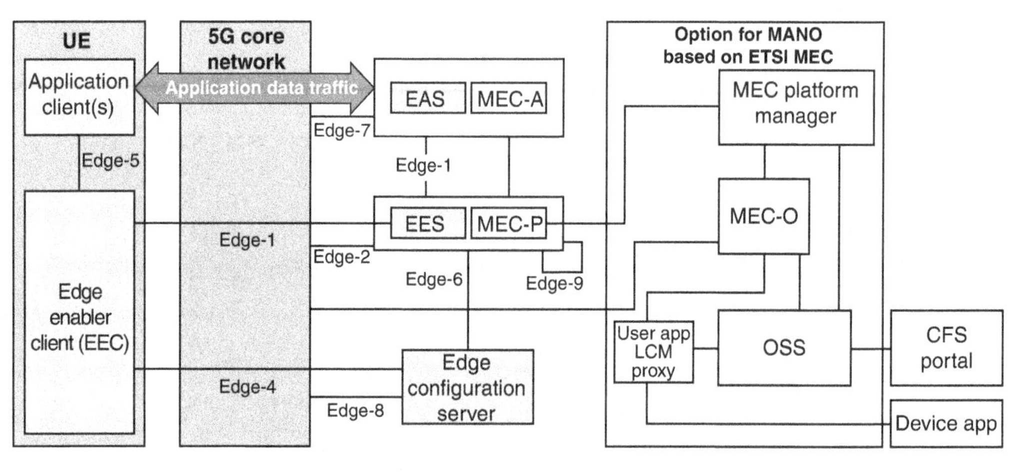

FIGURE 10.14 Relationship Between 3GPP Architecture for Enabling Edge Applications and ETSI MEC Architecture

The specific ways in which the synergized architecture and its underlying 3GPP and ETSI specification are realized are application and implementation dependent, and this is an area of further study within both 3GPP and ETSI.

10.8 Key Terms and Review Questions

Key Terms

aggregation point	mobile edge computing
cloud computing	multi-access edge computing (MEC)
cloud-edge computing	network function (NF)
device edge	network functions virtualization (NFV)
edge computing	network slicing
infrastructure edge	user plane function (UPF)
IPTV	video content analysis
local breakout	video analytics
MEC application	virtual network function (VNF)
MEC platform	virtualization infrastructure
MEC service	wireless to the *x* (WTT*x*)

Review Questions

1. Explain the difference between cloud computing, edge computing, cloud-edge computing, and MEC.
2. How does MEC support each of the three 5G usage scenarios?
3. What is a local breakout?
4. List and briefly explain the general design and development principles defined by ETSI.
5. List and give a brief summary of each of the main elements of the MEC system reference architecture.
6. What are the three main components of a MEC host?
7. What functions are performed by the MEC platform manager?
8. What elements of a MEC architecture can be deployed as VNFs?

9. How can MEC support network slicing?
10. List and briefly define the three categories of use cases specified by ETSI.
11. Describe the role for MEC in the factory of the future.
12. Give some examples of the ways in which an application may target a group of subscribers.
13. Describe the role of video analytics in 5G use cases.
14. How can MEC be used to support video analytics?

10.9 References and Documents

References

DIMI02 Dimitrova, N., et al. "Applications of Video-Content Analysis and Retrieval." *IEEE Multimedia*, July–September 2002.

GAGV08 Gagvani, N. "Introduction to Video Analytics." *EE Times*, August 22, 2008.

GIAN20 Giannone, F., et al. "Orchestrating Heterogeneous MEC-Based Applications for Connected Vehicles." *Computer Networks*, October 2020.

GRAM19 Grammatikos, P., and Cottis, P. "A Mobile Edge Computing Approach for Vehicle to Everything Communications." *Communications and Network*, vol. 11, 2019.

GMSA17 Global Mobile Suppliers Association. *Mobile Throughput Guidance.* White paper WWG.17. March 2017. https://www.gsma.com/newsroom/resources/ig-17-mobile-throughput-guidance/

HUAW16 Huawei Technologies. *WTTX Capacity White Paper*. March 2016. https://carrier.huawei.com/~/media/CNBG/Downloads/Technical%20Topics/Wireless-Network/WTTx%20Capacity%20White%20Paper.pdf

ISMA18 Ismail, N. "Edge Computing Is the Gateway to Smart Manufacturing." *Information Age*, January 31, 2018.

KHAN20 Khan, F. "MEC in NFV: How Does MEC Fit in NFV Architecture?" *TelcoCloud Bridge Blog*, June 3, 2020. https://www.telcocloudbridge.com/blog/mec-in-nfv-how-does-mec-fit-in-nfv-architecture/

LE12 Le, F., Srivatsa, M., and Iyengar, A. "Byte Caching in Wireless Networks." *International Conference on Distributed Computing Systems*, May 2012. https://researcher.watson.ibm.com/researcher/files/us-aruni/ByteCachingicdcs2012.pdf

NICH20 Nichols, M. "Edge Computing Is Essential for Smart Manufacturing Success." i-Scoop. https://www.i-scoop.eu/edge-computing-explained/edge-manufacturing-industry/

SPRI00 Spring, N., and Wetherall, D. "A Protocol-Independent Technique for Eliminating Redundant Network Traffic." *Computer Communication Review*, August 2000.

SUKH19 Sukhamani, S., et al. "Edge Caching and Computing in 5G for Mobile AR/VR and Tactile Internet." *IEEE Multimedia*, January–March 2019.

Documents

3GPP TS 23.558 *Technical Specification Group Services and System Aspects, Architecture for Enabling Edge Applications (Release 17)*. November 2020.

ETSI GS MEC 001 *Multi-Access Edge Computing (MEC), Terminology*. January 2019.

ETSI GS MEC 002 *Multi-Access Edge Computing (MEC), Phase 2: Use Cases and Requirements*. October 2018.

ETSI GS MEC 003 *Multi-Access Edge Computing (MEC), Framework and Reference Architecture*. January 2019.

ETSI GR MEC 024 *Multi-Access Edge Computing (MEC), Support for Network Slicing*. November 2019.

ETSI TR 126 957 *Study on Server and Network-Assisted Dynamic Adaptive Streaming over HTTP (DASH) (SAND) for 3GPP Multimedia Services*. July 2018.

ETSI White Paper #36 *Harmonizing Standards for Edge Computing: A Synergized Architecture Leveraging ETSI ISG MEC and 3GPP Specifications*. July 2020.

RFC 8743 *Multi-Access Management Services (MAMS)*. March 2020.

Chapter 11

Wireless Transmission

Learning Objectives

After studying this chapter, you should be able to:

- Explain the difference between the theoretical channel capacity values derived from the Nyquist and Shannon formulas
- Understand the mechanism of refraction
- Understand the difference between optical and radio line of sight
- Present an overview of line-of-sight transmission
- Understand the difference between S/N and E_b/N_0
- Present an overview of the concept of fading
- Describe the different forms of propagation impairments for millimeter wave transmission

This chapter and Chapters 12, "Antennas," 13, "Air Interface Physical Layer," and 14, "Air Interface Channel Coding," deal with aspects of the air interface between user equipment (UE) and the radio access network (RAN). Figure 1.6a in Chapter 1, "Cellular Networks: Concepts and Evolution," shows the air interface in the context of a cellular system.

The first four sections of this chapter discuss concepts related to the data rate that can be achieved over a wireless transmission link. These concepts are critical to understanding the specifications for the air interface for 5G networks. Section 11.1 examines the theoretical data rate that can be achieved over a wireless link of a given bandwidth. The next three sections consider the type of transmission impairments, as a function of frequency band, that limit the data rate to less than the theoretical maximum. Section 11.5 focuses on transmission in the millimeter wavelength region of the spectrum, which is crucial to the success of 5G.

11.1 Channel Capacity

A variety of impairments can distort or corrupt a signal. A common impairment is noise, which is any unwanted signal that combines with and hence distorts the signal intended for transmission and reception. Noise and other impairments are discussed later in this chapter. For the purposes of this section, it suffices to say that noise is something that degrades signal quality. For digital data, the issue that then arises is to what extent these impairments limit the data rate that can be achieved. The maximum rate at which data can be transmitted over a given communication path, or channel, under given conditions is referred to as the **channel capacity**.

There are four concepts here that relate to one another:

- **Data rate:** This is the rate, in bits per second (bps), at which data can be communicated.
- **Bandwidth:** Bandwidth is defined as the difference, in Hertz, between the limiting (i.e., upper and lower) frequencies of a spectrum. This is the bandwidth of the transmitted signal as constrained by the transmitter and the nature of the transmission medium, expressed in cycles per second, or Hertz.
- **Noise:** For this discussion, the concern is with the average level of noise over the communications path.
- **Error rate:** This is the rate at which errors occur, where an error is the reception of a 1 when a 0 was transmitted or the reception of a 0 when a 1 was transmitted.

The problem to be addressed is this: Communications facilities are expensive and, in general, the greater the bandwidth of a facility, the greater the cost. Furthermore, all transmission channels of any practical interest are of limited bandwidth. The limitations arise from the physical properties of the transmission medium or from deliberate limitations at the transmitter on the bandwidth to prevent interference from other sources. Accordingly, the objective is to make the most efficient use possible of a given bandwidth. For digital data, this means obtaining the highest data rate possible at a particular limit of error rate for a given bandwidth. The main constraint on achieving this efficiency is noise.

Nyquist Bandwidth

Consider the case of a channel that is noise free. In this environment, the limitation on data rate is simply the bandwidth of the signal. The Nyquist formulation of this limitation states that if the rate of signal transmission is $2B$, then a baseband[1] signal with frequencies no greater than B is sufficient to carry the signal rate. The converse is also true: Given a bandwidth of B, the highest signal rate that can

1. The term *baseband* refers to the spectral band occupied by an unmodulated signal. Baseband transmission is usually characterized by being much lower in frequency than the signal that results if the baseband signal is used to modulate a carrier frequency.

be carried is $2B$. This limitation is due to the effect of intersymbol interference, such as is produced by delay distortion.[2]

Note that the preceding paragraph refers to **signal rate**. If the signals to be transmitted are binary (i.e., take on only two values), then the data rate that can be supported by B Hz is $2B$ bps. As an example, consider a voice channel being used, via modem, to transmit digital data with one frequency level representing binary 1 and another frequency level representing binary 0. Assume a bandwidth of 3100 Hz. Then the capacity, C, of the channel is $2B = 6200$ bps. However, as discussed in Chapter 12, signals with more than two levels can be used; that is, each signal element can represent more than one bit. For example, if four possible frequencies are used as signals, then each signal element can represent two bits. With multilevel signaling, the Nyquist formulation becomes:

$$C = 2B \log_2 M$$

where M is the number of discrete signal elements or voltage levels. Thus, for $M = 8$, a value used with some modems, a bandwidth of $B = 3100$ Hz yields a capacity $C = 18{,}600$ bps.

So, for a given bandwidth, the data rate can be increased by increasing the number of different signal elements. However, this places an increased burden on the receiver: Instead of distinguishing one of two possible signal elements during each signal time, the receiver must distinguish one of M possible signals. Noise and other impairments on the transmission line will limit the practical value of M.

Shannon Capacity Formula

Nyquist's formula indicates that, all other things being equal, doubling the bandwidth doubles the data rate. Now consider the relationship among data rate, noise, and error rate. The presence of noise can corrupt one or more bits. If the data rate is increased, then the bits become "shorter" in time, so that more bits are affected by a given pattern of noise. Thus, at a given noise level, the higher the data rate, the higher the error rate.

Figure 11.1 provides an example of the effect of noise on a digital signal. Here the noise consists of a relatively modest level of background noise plus occasional larger spikes of noise. The digital data can be recovered from the signal by sampling the received waveform once per bit time. As can be seen, the noise is occasionally sufficient to change a 1 to a 0 or a 0 to a 1.

All of these concepts can be tied together neatly in a formula developed by the mathematician Claude Shannon. As just illustrated, the higher the data rate, the more damage that unwanted noise can do. For a given level of noise, we would expect that a greater signal strength would improve the ability to receive data correctly in the presence of noise. The key parameter involved in this reasoning is the **signal-to-noise ratio (SNR, or S/N)**,[3] which is the ratio of the power in a signal to the power contained

2. Delay distortion of a signal occurs when the propagation delay for the transmission medium is not constant over the frequency range of the signal.

3. Some of the literature uses SNR; other literature uses S/N. Also, in some cases, the dimensionless quantity is referred to as SNR or S/N, and the quantity in decibels is referred to as SNR_{db} or $(S/N)_{db}$. Other sources use just SNR or S/N to mean the dB quantity. This text uses SNR and SNR_{db}.

in the noise that is present at a particular point in the transmission. Typically, this ratio is measured at a receiver because it is at this point that an attempt is made to process the signal and eliminate the unwanted noise. For convenience, this ratio is often reported in decibels[4]:

$$\text{SNR}_{\text{dB}} = 10\log_{10}\frac{\text{signal power}}{\text{noise power}}$$

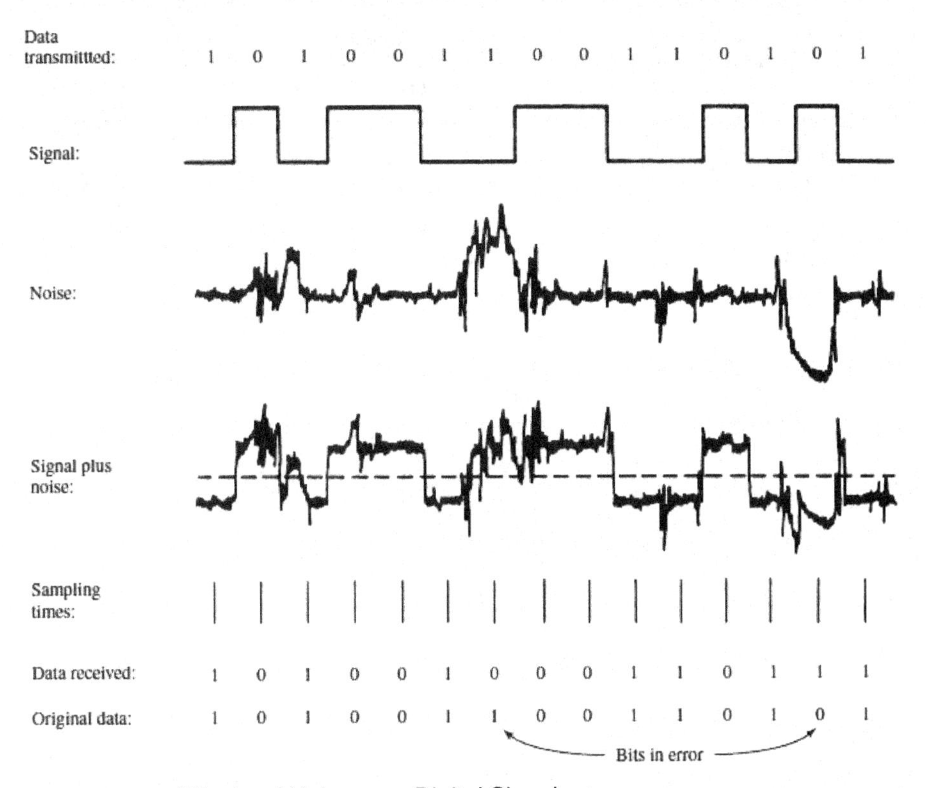

FIGURE 11.1 Effects of Noise on a Digital Signal

This expresses the amount, in decibels, that the intended signal exceeds the noise level. A high SNR means a high-quality signal.

The signal-to-noise ratio is important in the transmission of digital data because it sets the upper bound on the achievable data rate. Shannon's result is that the maximum channel capacity, in bits per second, obeys the equation:

$$C = B\log_2(1 + \text{SNR})$$

where C is the capacity of the channel in bits per second, and B is the bandwidth of the channel in Hertz. The Shannon formula represents the theoretical maximum that can be achieved for a channel. In

4. Annex 11A explains the concept of decibels.

practice, however, only much lower rates are achieved. One reason for this is that the formula assumes white noise (i.e., thermal noise). Impulse noise is not accounted for, nor are attenuation distortion or delay distortion. Various types of noise and distortion are discussed later in this chapter.

The capacity indicated in the preceding equation is referred to as the **error-free capacity**. Shannon proved that if the actual information rate on a channel is less than the error-free capacity, then it is theoretically possible to use a suitable signal code to achieve error-free transmission through the channel. Shannon's theorem unfortunately does not suggest a means for finding such codes, but it does provide a yardstick by which the performance of practical communication schemes may be measured.

Chapter 2, "5G Standards and Specifications," defines the **spectral efficiency**, also called **bandwidth efficiency**, of a digital transmission as the number of bits per second of data that can be supported by each Hertz of bandwidth. The theoretical peak spectral efficiency can be expressed using the preceding equation and moving the bandwidth B to the left-hand side, resulting in $C/B = \log_2 (1 + \text{SNR})$. C/B has the dimensions bps/Hz. Figure 11.2 shows the results on a log/log scale. At SNR = 1, we have $C/B = 1$. For SNR < 1 (signal power is less than noise power), the plot is linear; above SNR = 1, the plot flattens but continues to increase with increasing SNR.

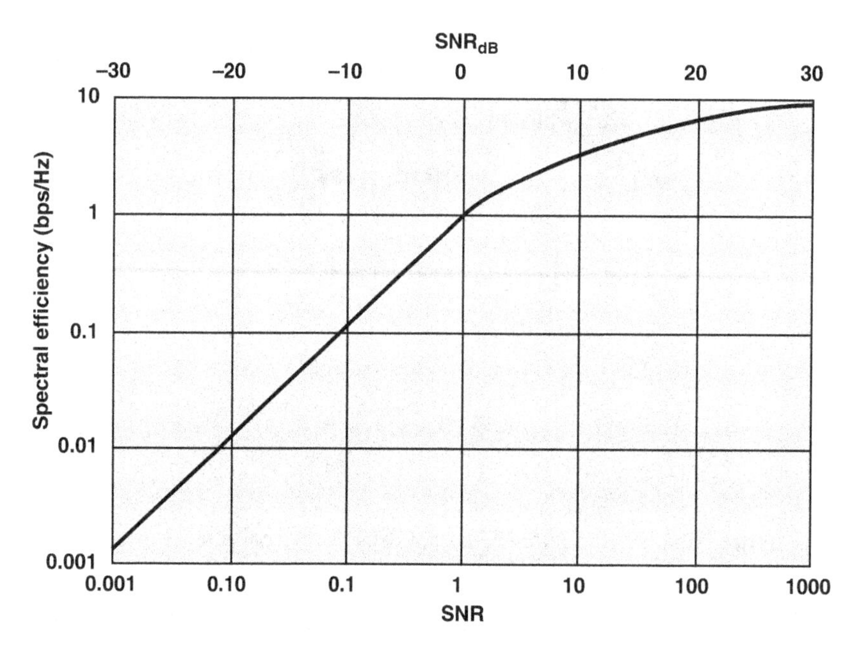

FIGURE 11.2 Spectral Efficiency Versus SNR

Note that below 0 dB SNR, noise is the dominant factor in the capacity of a channel. Shannon's theorem shows that communications are possible in this region but at a relatively low data rate—a rate that is reduced in proportion to the SNR (on a log/log scale). In the region of at least 6 dB above 0 dB SNR, noise is no longer the limiting factor in communications speed. In this region, there is little ambiguity in a signal's relative amplitude and phase, and achieving a high channel capacity depends on the design of the signal, including factors such as modulation type and coding. These topics are explored in Chapters 12 and 13.

Several other observations concerning the preceding equation may be instructive. For a given level of noise, it would appear that the data rate could be increased by increasing either signal strength or bandwidth. However, as the signal strength increases, so do the effects of nonlinearities in the system, leading to an increase in intermodulation noise. Note also that, because noise is assumed to be white, the wider the bandwidth, the more noise is admitted to the system. Thus, as *B* increases, SNR decreases.

Example

Let us consider an example that relates the Nyquist and Shannon formulations. Suppose that the spectrum of a channel is between 3 MHz and 4 MHz and $SNR_{dB} = 24$ dB. Then:

$$B = 4\text{ MHz} - 3\text{ MHz} = 1\text{ MHz}$$

$$SNR_{dB} = 24\text{ dB} = 10\log_{10}(SNR)$$

$$SNR = 251$$

Using Shannon's formula:

$$C = 10^6 \times \log_2(1 + 251) \approx 10^6 \times 8 = 8\text{ Mbps}$$

This is a theoretical limit and, as we have said, is unlikely to be reached. But assume that we can achieve the limit. Based on Nyquist's formula, how many signaling levels are required? We have:

$$C = 2B\log_2 M$$

$$8 \times 10^6 = 2 \times (10^6) \times \log_2 M$$

$$4 = \log_2 M$$

$$M = 16$$

11.2 Line-of-Sight Transmission

A signal radiated from an antenna travels along one of three routes:

- **Ground wave:** This mode follows the curvature of the earth and can propagate considerable distances, well over the visual horizon. This effect is found in frequencies up to about 2 MHz. A well-known example of ground wave communication is AM (amplitude modulation) radio.
- **Sky wave:** With this mode, a signal from an earth-based antenna can be viewed as being reflected from the ionized layer of the upper atmosphere (ionosphere) back down to earth. Although it appears that the wave is reflected from the ionosphere, as if the ionosphere were a hard reflecting surface, the effect is in fact caused by a phenomenon known as refraction.

This mode is effective in the range of about 3 to 30 MHz. Sky wave propagation is used for amateur radio, CB (citizens band) radio, and international broadcasts such as BBC and Voice of America.

- **Line of sight (LOS):** This mode is the only mode that is usable above 30 MHz. For ground-based LOS communication, the transmitting and receiving antennas must be within an effective line of sight of each other. The term *effective* is used because microwaves are bent or refracted by the atmosphere. The amount and even the direction of the bend depends on conditions, but generally microwaves are bent with the curvature of the earth and will therefore propagate farther than the optical line of sight.

Refraction

Refraction occurs because the velocity of an electromagnetic wave is a function of the density of the medium through which it travels. In a vacuum, an electromagnetic wave (such as light or a radio wave) travels at approximately 3×10^8 m/s. This is the constant, c, commonly referred to as the speed of light, which actually refers to the speed of light in a vacuum. In air, water, glass, and other transparent or partially transparent media, electromagnetic waves travel at speeds less than c.

When an electromagnetic wave moves from a medium of one density to a medium of another density, its speed changes. The effect is to cause a one-time bending of the direction of the wave at the boundary between the two media. This is illustrated in Figure 11.3. If moving from a less dense medium to a more dense medium, the wave will bend toward the more dense medium. This phenomenon is easily observed by partially immersing a stick in water. The result will look much like Figure 11.3, with the stick appearing shorter and bent.

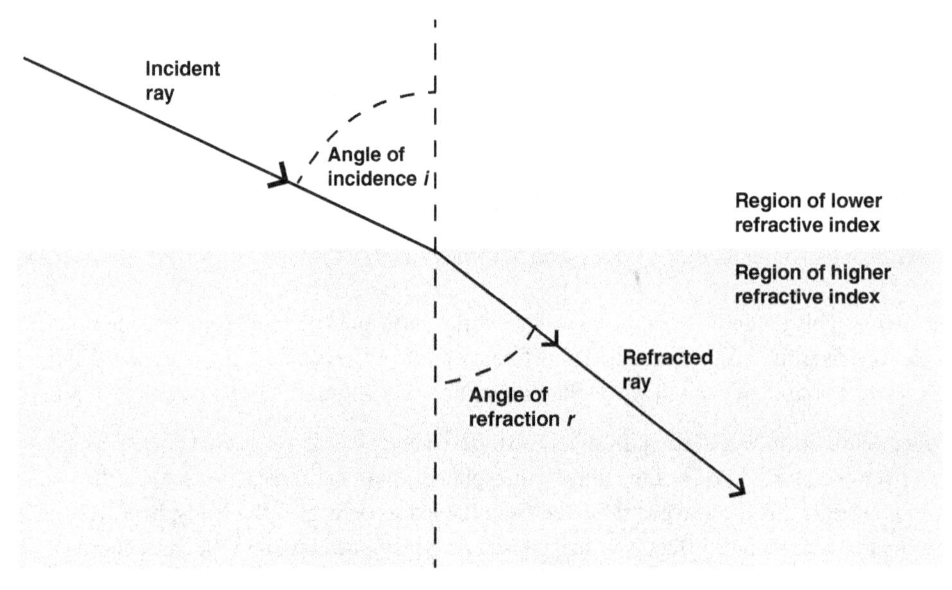

FIGURE 11.3 Refraction of an Electromagnetic Wave

The **index of refraction** of one medium relative to another is the sine of the angle of incidence, θ_i, divided by the sine of the angle of refraction, θ_r. The index of refraction is also equal to the ratio of the respective velocities in the two media. The absolute index of refraction of a medium is calculated in comparison with that of a vacuum. Refractive index varies with wavelength so that refractive effects differ for signals with different wavelengths.

Although Figure 11.3 shows an abrupt, one-time change in direction as a signal moves from one medium to another, a continuous, gradual bending of a signal will occur if it is moving through a medium in which the index of refraction gradually changes. Under normal propagation conditions, the refractive index of the atmosphere decreases with height so that radio waves travel more slowly near the ground than at higher altitudes. The result is a slight bending of the radio waves toward the earth. With sky waves, the density of the ionosphere and its gradual change in density cause the waves to be refracted back toward the earth.

Optical and Radio Line of Sight

With no intervening obstacles, the optical line of sight is influenced by the curvature of the earth and can be expressed as:

$$d = 3.57\sqrt{h}$$

where d is the distance between an antenna and the horizon in kilometers, and h is the antenna height in meters. The effective, or radio, line of sight to the horizon is expressed as:

$$d = 3.57\sqrt{\mathrm{K}h}$$

where K is an adjustment factor to account for the refraction. A good rule of thumb is K = 4/3. Thus, the maximum distance between two antennas for LOS propagation is $3.57\left(\sqrt{\mathrm{K}h_1} + \sqrt{\mathrm{K}h_2}\right)$, where h_1 and h_2 are the heights of the two antennas. Figure 11.4 illustrates the difference between optical and radio horizons.

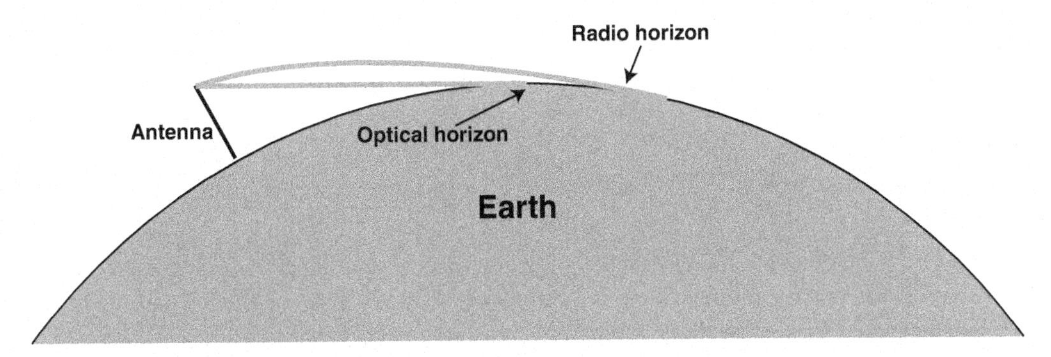

FIGURE 11.4 Optical and Radio Horizons

Example

The maximum distance between two antennas for LOS transmission if one antenna is 100 m high and the other is at ground level is:

$$d = 3.57\sqrt{Kh} = 3.57\sqrt{133} = 41\,\text{km}$$

Now suppose that the receiving antenna is 10 m high. To achieve the same distance, how high must the transmitting antenna be? The result is:

$$41 = 3.57\left(\sqrt{Kh_1} + \sqrt{13.3}\right)$$

$$\sqrt{Kh_1} = \frac{41}{3.57} - \sqrt{13.3} = 7.84$$

$$h_1 = 7.84^2/1.33 = 46.2\,\text{m}$$

This is a savings of over 50 m in the height of the transmitting antenna. This example illustrates the benefit of raising receiving antennas above ground level to reduce the height required for the transmitter.

11.3 Line-of-Sight Transmission Impairments

With any communications system, the signal that is received will differ from the signal that is transmitted due to various transmission impairments. This section examines the various impairments and comments on their effects on the information-carrying capacity of a wireless communications link. For LOS wireless transmission, the most significant impairments are:

- Attenuation and attenuation distortion
- Free space loss
- Noise
- Atmospheric absorption
- Multipath
- Refraction

Attenuation

The strength of a signal degrades with distance over any transmission medium. For wireless transmission, reduction in strength, or attenuation, is a complex function of distance and the makeup of the atmosphere. Attenuation introduces three factors for a transmission engineer:

- A received signal must have sufficient strength so that the electronic circuitry in the receiver can detect and interpret the signal.
- The signal must maintain a level sufficiently higher than noise to be received without error.

- Attenuation is greater at higher frequencies, causing distortion because signals are typically composed of many frequency components.

The first and second factors are dealt with by attention to signal strength and the use of amplifiers or repeaters. For a point-to-point transmission (i.e., one transmitter and one receiver), the signal strength of the transmitter must be strong enough to be received intelligibly but not so strong as to overload the circuitry of the transmitter or receiver, which would cause distortion. Regulatory requirements also limit transmission power. Beyond a certain distance, the attenuation becomes unacceptably great, and repeaters or amplifiers are used to boost the signal at regular intervals. These problems are more complex when there are multiple receivers, where the distance from transmitter to receiver is variable.

The third factor is known as **attenuation distortion**. Because the attenuation varies as a function of frequency, the received signal is distorted, reducing intelligibility. Specifically, the frequency components of the received signal have different relative strengths than the frequency components of the transmitted signal. To overcome this problem, techniques are available for equalizing attenuation across a band of frequencies. One approach is to use amplifiers that amplify high frequencies more than lower frequencies.

Free Space Loss

For any type of wireless communication, the signal disperses with distance. Energy dispersal can be viewed as radiating in a sphere, with a receiver on the surface extracting energy on part of the surface area. A larger and larger sphere occurs as distance from the transmitter increases, so there is less energy per each unit of surface area. Therefore, an antenna with a fixed area will receive less signal power the farther it is from the transmitting antenna. For satellite communication, this is the primary mode of signal loss. Even if no other sources of attenuation or impairment are assumed, a transmitted signal attenuates over distance because the signal is being spread over a larger and larger area. This form of attenuation is known as **free space loss**, and it can be expressed in terms of the ratio of the radiated power P_t to the power P_r received by the antenna or, in decibels, by taking 10 times the log of that ratio. For the ideal isotropic antenna (one that radiates power in all directions equally), free space loss is:

$$\frac{P_t}{P_r} = \frac{(4\pi d)^2}{\lambda^2} = \frac{(4\pi fd)^2}{c^2}$$

where:

P_t = signal power at the transmitting antenna

P_r = signal power at the receiving antenna

λ = carrier wavelength

f = carrier frequency

d = propagation distance between antennas

c = speed of light (3×10^8 m/s)

where d and λ are in the same units (e.g., meters).

This equation can be recast in decibels as:

$$L_{dB} = 10\log\frac{P_t}{P_r} = 20\log\left(\frac{4\pi d}{\lambda}\right) = -20\log(\lambda) + 20\log(d) + 21.98\text{ dB}$$
$$= 20\log\left(\frac{4\pi fd}{c}\right) = 20\log(f) + 20\log(d) - 147.56\text{ dB} \qquad \textbf{(Equation 11.1)}$$

Figure 11.5 illustrates the free space loss equation.[5]

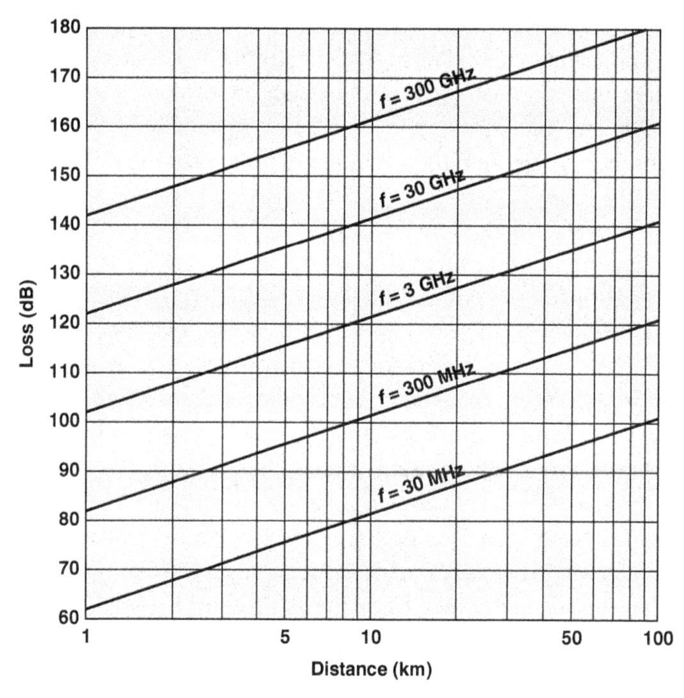

FIGURE 11.5 Free Space Loss

For other antennas, we must take into account the gain of the antenna, which yields the following free space loss equation:

$$\frac{P_t}{P_r} = \frac{(4\pi)^2(d)^2}{G_r G_t \lambda^2} = \frac{(\lambda d)^2}{A_r A_t} = \frac{(cd)^2}{f^2 A_r A_t}$$

5. There is some inconsistency in the literature over the use of the terms *gain* and *loss*. For example, when there is a reduction of power by half, it is common to say that this is a gain of –3 dB or a loss of 3 dB. However, some of the literature would say that this is a loss of –3 dB. This text follows the former convention. See Annex 11A.

where:

G_t = gain of the transmitting antenna

G_r = gain of the receiving antenna

A_t = effective area of the transmitting antenna

A_r = effective area of the receiving antenna

The third fraction is derived from the second fraction using the relationship between antenna gain and the effective area of the antenna. The concept of effective area is discussed in Chapter 12. We can recast this equation as:

$$\begin{aligned} L_{dB} &= 20 \log(\lambda) + 20 \log(d) - 10 \log(A_t A_r) \\ &= -20 \log(f) + 20 \log(d) - 10 \log(A_t A_r) + 169.54 \text{ dB} \end{aligned}$$ **(Equation 11.2)**

Thus, for the same antenna dimensions and separation, the longer the carrier wavelength (i.e., the lower the carrier frequency f), the higher the free space path loss. It is interesting to compare Equations (11.1) and (11.2). Equation (11.1) indicates that as the frequency increases, the free space loss also increases, which would suggest that at higher frequencies, losses become more burdensome. However, Equation (11.2) shows that we can easily compensate for this increased loss with antenna gains. As discussed in Chapter 12, an antenna produces increased gain at higher frequencies. Thus, there is a net gain at higher frequencies, other factors remaining constant. Equation (11.2) shows that at a fixed distance, an increase in frequency results in an increased loss measured by $20 \log(f)$. However, if we take into account antenna gain and fix antenna area, then the change in loss is measured by $-20 \log(f)$; that is, there is actually a decrease in loss at higher frequencies.

Example

Determine the isotropic free space loss at 4 GHz for the shortest path to a synchronous satellite from earth (35,863 km). At 4 GHz, the wavelength is $(3 \times 10^8)/(4 \times 10^9) = 0.075$ m. Then:

$$L_{dB} = -20 \log(0.075) + 20 \log(35.853 \times 10^6) + 21.98 = 195.6 \text{ dB}$$

Now consider the antenna gain of both the satellite- and ground-based antennas. Typical values are 44 dB and 48 dB, respectively. The free space loss is:

$$L_{dB} = 195.6 - 44 - 48 = 103.6 \text{ dB}$$

Now assume a transmit power of 250 W at the earth station. What is the power received at the satellite antenna? A power of 250 W translates into 24 dBW, so the power at the receiving antenna is $24 - 103.6 = -79.6$ dBW, where dBW is the decibel-watt. This signal is approximately 10^{-8} W, but it is still usable by receiver circuitry.

Path Loss Exponent in Practical Cellular Systems

Cellular systems involve many types of obstructions that cause reflections, scattering, etc. Both theoretical and measurement-based models have shown that beyond a certain distance, the average received signal power decreases logarithmically with distance according to a $10n\log(d)$ relationship, where n is known as the **path loss exponent** [RAPP02]. Such models have been used extensively. Both Equations (11.1) and (11.2) showed a $20\log(d)$ term, which came from a d^2 distance relationship, hence a path loss exponent of $n = 2$. The term $20\log(d)$ should be replaced with the more general $10n\log(d)$ term, as follows:

$$\frac{P_t}{P_r} = \frac{(4\pi d)^n}{\lambda^2} = \frac{(4\pi fd)^n}{c^2}$$

$$\begin{aligned} L_{dB} &= 10\log\frac{P_t}{P_r} = 20\log\left(\frac{4\pi d}{\lambda}\right) = -20\log(\lambda) + 10n\log(d) + 21.98\text{ dB} \\ &= 20\log\left(\frac{4\pi fd}{c}\right) = 20\log(f) + 10n\log(d) - 147.56\text{ dB} \end{aligned}$$

(Equation 11.3)

Using effective areas and the general path loss exponent, n:

$$\begin{aligned} L_{dB} &= 20\log(\lambda) + 10n\log(d) - 10\log(A_t A_r) \\ &= -20\log(f) + 10n\log(d) - 10\log(A_t A_r) + 169.54\text{ dB} \end{aligned}$$

(Equation 11.4)

Table 11.1 shows typical path loss exponents obtained for various environments. Note that in-building LOS is better than $n = 2$ because reflections help keep the signal stronger than if it decayed with distance, as is the case in free space.

TABLE 11.1 Path Loss Exponents for Different Environments

Environment	Path Loss Exponent, n
Free space	2
Urban area cellular radio	2.7 to 3.5
Shadowed cellular radio	3 to 5
In-building LOS	1.6 to 1.8
Obstructed in building	4 to 6
Obstructed in factories	2 to 3

Example

Compare the path loss in dB for two possible cellular environments where there is (1) free space between mobiles and base stations and (2) urban area cellular radio with $n = 3.1$. Use 1.9 GHz at a distance of 1.5 km and assume isotropic antennas.

For free space using $n = 2.0$:

$$L_{dB} = 20 \log (1.9 \times 10^9) + 10 \times 2.0 \log(1.5 \times 10^3) - 147.56 = 101.53 \text{ dB}$$

For urban cellular radio using $n = 3.1$:

$$L_{dB} = 20 \log (1.9 \times 10^9) + 10 \times 3.1 \log(1.5 \times 10^3) - 147.56 = 136.47 \text{ dB}$$

Example

Compare the range of coverage for two possible cellular environments where there is (1) free space between mobiles and base stations and (2) urban area cellular radio with $n = 3.1$. Use 1.9 GHz and assume isotropic antennas. Assume that the transmit power is 2 W, and the received power must be above −110 dBW:

$$P_t \text{ in dB} = 10 \log (2) = 3.0$$

The requirement is, therefore, $L_{dB} < 113$ dB.

For free space using $n = 2.0$:

$$L_{dB} = 20 \log (1.9 \times 10^9) + 10 \times 2.0 \log(d) - 147.56 < 113 \text{ dB}$$

$$10 \times 2.0 \log(d) < 74.99 \text{ dB}$$

$$d < 5.61 \text{ km}$$

For an urban area using $n = 3.1$:

$$L_{dB} = 20 \log (1.9 \times 10^9) + 10 \times 3.1 \log(d) - 147.56 < 113 \text{ dB}$$

$$10 \times 3.1 \log(d) < 74.99 \text{ dB}$$

$$d < 262 \text{ m}$$

Noise

For any data transmission event, the received signal consists of the transmitted signal modified by the various distortions imposed by the transmission system plus additional unwanted signals that are inserted somewhere between transmission and reception. These unwanted signals are referred to as noise. Noise is the major limiting factor in communications system performance.

Noise may be divided into four categories:

- Thermal noise
- Intermodulation noise
- Crosstalk
- Impulse noise

Thermal Noise

Thermal noise is due to thermal agitation of electrons. It is present in all electronic devices and transmission media and is a function of temperature. Thermal noise is uniformly distributed across the frequency spectrum and hence is often referred to as **white noise**. Thermal noise cannot be eliminated and therefore places an upper bound on communications system performance. Because of the weakness of the signal received by satellite earth stations, thermal noise is particularly significant for satellite communication.

The amount of thermal noise to be found in a bandwidth of 1 Hz in any device or conductor is:

$$N_0 = \mathrm{k}T \text{ (W/Hz)}$$

where[6]:

N_0 = noise power density in watts per 1 Hz of bandwidth

k = Boltzmann's constant = $1.38\ [\tau\sigma]\ 10^{-23}$ J/K

T = temperature, in kelvins (absolute temperature)

Example

Room temperature is usually specified as T = 17°C, or 290 K. At this temperature, the thermal noise power density is:

$$N_0 = (1.3803\ [\tau\sigma]\ 10^{-23})\ [\tau\sigma]\ 290 = 4\ [\tau\sigma]\ 10^{-21} \text{ W/Hz} = -204 \text{ dBW/Hz}$$

The noise is assumed to be independent of frequency. Thus the thermal noise in watts present in a bandwidth of B Hertz can be expressed as:

$$N = k\text{TB}$$

6. A Joule (J) is the International System (SI) unit of electrical, mechanical, and thermal energy. A watt is the SI unit of power, equal to 1 Joule per second. The kelvin (K) is the SI unit of thermodynamic temperature. For a temperature in degrees kelvin of T, the corresponding temperature in degrees Celsius is equal to $T - 273.15$.

or, in decibel-watts:

$$N = 10 \log k + 10 \log T + 10 \log B$$
$$= -228.6 \text{ dBW} + 10 \log T + 10 \log B$$

Example

Given a receiver with an effective noise temperature of 294 K and a 10-MHz bandwidth, the thermal noise level at the receiver's output is

$$N = -228.6 \text{ dBW} + 10 \log (294) + 10 \log 10^7$$
$$= -228.6 + 24.7 + 70$$
$$= -133.9 \text{ dBW}$$

Intermodulation Noise

When signals at different frequencies share the same transmission medium, the result may be **intermodulation noise**. Intermodulation noise produces signals at a frequency that is the sum or difference of the two original frequencies or multiples of those frequencies. For example, the mixing of signals at frequencies f_1 and f_2 might produce energy at the frequency $f_1 + f_2$. This derived signal could interfere with an intended signal at the frequency $f_1 + f_2$.

Intermodulation noise is produced when there is some nonlinearity in the transmitter, receiver, or intervening transmission system. Normally, these components behave as linear systems; that is, the output is equal to the input times a constant. In a nonlinear system, the output is a more complex function of the input. Such nonlinearity can be caused by component malfunction, the use of excessive signal strength, or just the nature of the amplifiers used. It is under these circumstances that the sum and difference frequency terms occur.

Crosstalk

Crosstalk has been experienced by anyone who, while using the telephone, has been able to hear another conversation; it is an unwanted coupling between signal paths. It can occur due to electrical coupling between nearby twisted pairs or, rarely, coax cable lines carrying multiple signals. Crosstalk can also occur when unwanted signals are picked up by microwave antennas; although highly directional antennas are used, microwave energy does spread during propagation. Typically, crosstalk is of the same order of magnitude as, or less than, thermal noise. However, in the unlicensed ISM (industrial, scientific, and medical) bands, crosstalk often dominates.

Impulse Noise

All of the types of noise discussed so far have reasonably predictable and relatively constant magnitudes. Thus it is possible to engineer a transmission system to cope with them. Impulse noise, however, is noncontinuous, consisting of irregular pulses or noise spikes of short duration and of relatively high amplitude. It is generated by a variety of causes, including external electromagnetic disturbances, such as lightning, and faults and flaws in the communications system.

Impulse noise is generally only a minor annoyance for analog data. For example, voice transmission may be corrupted by short clicks and crackles with no loss of intelligibility. However, impulse noise is the primary source of error in digital data transmission. For example, a sharp spike of energy of 0.01 s duration would barely be noticed for voice conversation but would wash out about 10,000 bits of data being transmitted at 1 Mbps.

The Expression E_b/N_0

Section 11.1 introduced the signal-to-noise ratio (SNR). There is a parameter related to SNR that is more convenient for determining digital data rates and error rates and that is the standard quality measure for digital communications system performance. The parameter is the ratio of signal *energy* per bit to noise power density per Hertz, E_b/N_0. Consider a signal, digital or analog, that contains binary digital data transmitted at a certain bitrate R. Recalling that 1 watt = 1 J/s, the energy per bit in a signal is given by $E_b = ST_b$, where S is the signal power and T_b is the time required to send one bit. The data rate R is just $R = 1/T_b$. Thus:

$$\frac{E_b}{N_0} = \frac{S/R}{N_0} = \frac{S}{\mathrm{k}TR} \qquad \textbf{(Equation 11.5)}$$

or, in decibel notation:

$$\left(\frac{E_b}{N_0}\right)_{\mathrm{dB}} = S_{\mathrm{dBW}} - 10\log R - 10\log \mathrm{k} - 10\log T$$
$$= S_{\mathrm{dBW}} - 10\log R + 228.6\ \mathrm{dBW} - 10\log T$$

The ratio E_b/N_0 is important because the bit error rate (BER) for digital data is a (decreasing) function of this ratio. Figure 11.6 illustrates the typical shape of a plot of BER versus E_b/N_0. Such plots are commonly found in the literature, and several examples appear in this text. For any particular curve, as the signal strength relative to the noise increases (i.e., increasing E_b/N_0), the BER at the receiver decreases. This makes intuitive sense. However, there is not a single unique curve that expresses the dependence of BER on E_b/N_0. Instead, the performance of a transmission/reception system, in terms of BER versus E_b/N_0, also depends on the way in which the data is encoded onto the signal. As an illustration of this, Figure 11.6 shows two curves, one of which gives better performance than the other. A curve below and to the left of another curve defines superior performance. At the same BER for two signals, the curve to the left shows how much less E_b/N_0 is needed (really E_b since N_0 is constant) to achieve that BER. For two signals using the same E_b/N_0, the curve below and to the left shows the better BER.

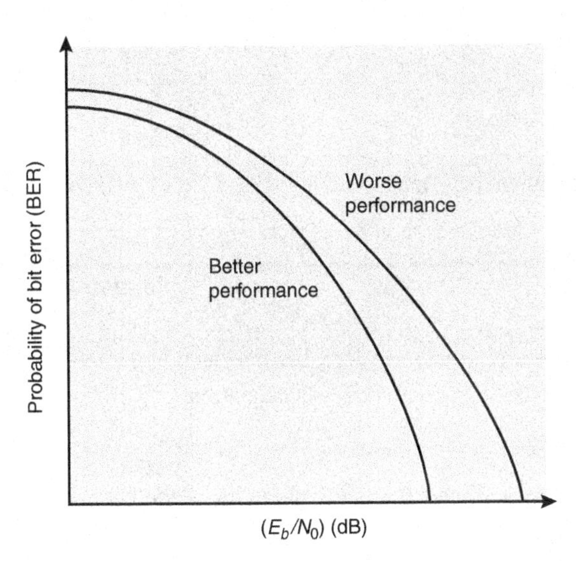

FIGURE 11.6 Typical Shapes of BER and E_b/N_0 Curves

Given a value of E_b/N_0 needed to achieve a desired error rate, the parameters in Equation (11.5) may be selected. Note that as the bitrate R increases, the transmitted signal power S, relative to noise, must increase to maintain the required E_b/N_0.

To grasp this result intuitively, consider again Figure 11.1. The signal here is digital, but the reasoning would be the same for an analog signal. In several instances, the noise is sufficient to alter the value of a bit. If the data rate were doubled, the bits would be more tightly packed together, and the same passage of noise might destroy two bits. Thus, for constant signal and noise strength, an increase in data rate increases the error rate. Also, some decoding schemes average out the noise effects over the bit period; if the bit period is shorter, there is less time for this averaging process, and it is less effective.

The advantage of E_b/N_0 compared to SNR is that the latter quantity depends on the bandwidth.

Example

Suppose a signal encoding technique requires that E_b/N_0 = 8.4 dB for a bit error rate of 10^{-4} (i.e., 1 bit error out of every 10,000 bits). If the effective noise temperature is 290 K (i.e., room temperature) and the data rate is 100 kbps, what received signal level is required to overcome thermal noise?

We have:

$$8.4 = S_{dBW} - 10 \log 100000 + 228.6 \text{ dBW} - 10 \log 290$$
$$= S_{dBW} - (10)(5) + 228.6 - (10)(2.46)$$
$$S = -145.6 \text{ dBW}$$

We can relate E_b/N_0 to SNR as follows:

$$\frac{E_b}{N_0} = \frac{S}{N_0 R}$$

The parameter N_0 is the noise power density, in watts/hertz. Hence, the noise in a signal with bandwidth B is $N = N_0 B$. Substituting, we have:

$$\frac{E_b}{N_0} = \frac{S}{N}\frac{B}{R} \qquad \textbf{(Equation 11.6)}$$

Another formulation of interest relates to E_b/N_0 spectral efficiency. Recall from Section 11.1 Shannon's result that the maximum channel capacity, in bits per second, obeys the equation:

$$C = B \log_2(1 + S/N)$$

where C is the capacity of the channel in bits per second, and B is the bandwidth of the channel in Hertz. This can be rewritten as:

$$\frac{S}{N} = 2^{C/B} - 1$$

Using Equation (11.6), and R with C, we have:

$$\frac{E_b}{N_0} = \frac{B}{C}\left(2^{C/B} - 1\right)$$

This is a useful formula that relates the achievable spectral efficiency C/B to E_b/N_0.

Example

Suppose we want to find the minimum E_b/N_0 required to achieve a spectral efficiency of 6 bps/Hz. Then $E_b/N_0 = (1/6)(2^6 - 1) = 10.5 = 10.21$ dB.

Atmospheric Absorption

An additional loss between the transmitting and receiving antennas is atmospheric absorption. Water vapor and oxygen contribute most to attenuation. A peak attenuation occurs in the vicinity of 22 GHz due to water vapor. At frequencies below 15 GHz, the attenuation is less. The presence of oxygen results in an absorption peak in the vicinity of 60 GHz but contributes less at frequencies below 30 GHz. Rain and fog (suspended water droplets) cause scattering of radio waves that results in attenuation. This can be a major cause of signal loss. Thus, in areas of significant precipitation, either path lengths have to be kept short or lower-frequency bands should be used.

Multipath

For wireless facilities where there is a relatively free choice of where antennas are to be located, they can be placed so that if there are no nearby interfering obstacles, there is a direct line-of-sight path from transmitter to receiver. This is generally the case for many satellite facilities and for point-to-point microwave. In other cases, such as mobile telephony, there are obstacles in abundance. The signal can be reflected by such obstacles so that multiple copies of the signal with varying delays can be received. In fact, in extreme cases, the receiver may capture only reflected signals and not the direct signal. Depending on the differences in the path lengths of the direct and reflected waves, the composite signal can be either larger or smaller than the direct signal. Reinforcement and cancellation of the signal resulting from the signal following multiple paths can be controlled for communication between fixed, well-sited antennas and between satellites and fixed ground stations. One exception is when the path goes across water where the wind keeps the reflective surface of the water in motion. For mobile telephony and communication to antennas that are not well sited, multipath considerations can be paramount.

Figure 11.7 illustrates, in general terms, the types of multipath interference typical in terrestrial (i.e., fixed microwave) and in mobile communications. For fixed microwave, in addition to the direct line of sight, the signal may follow a curved path through the atmosphere due to refraction, and the signal may also reflect from the ground. For mobile communications, structures and topographic features provide reflection surfaces.

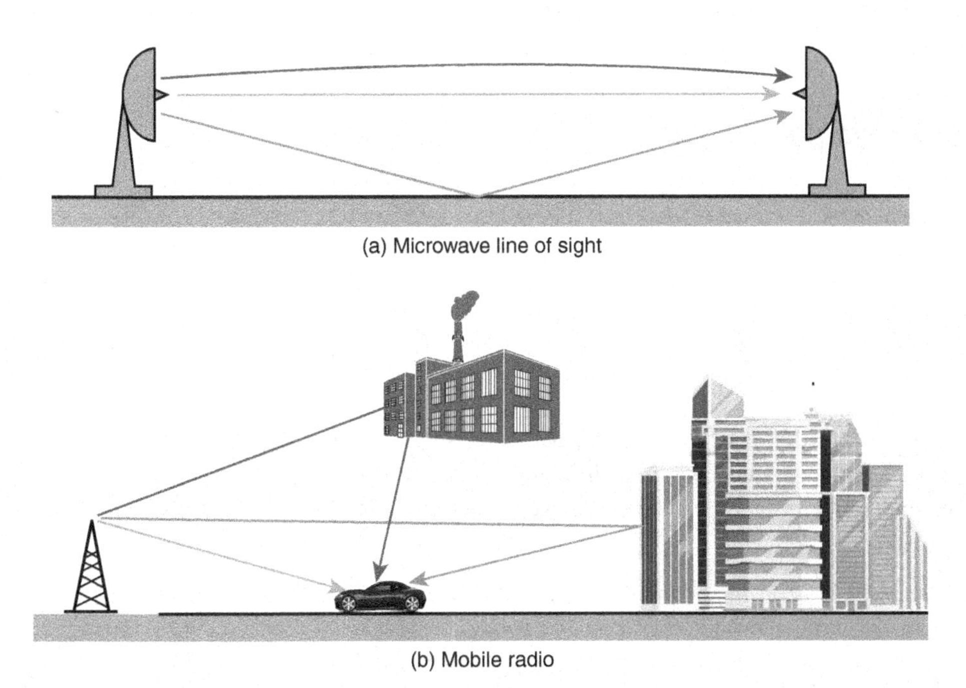

FIGURE 11.7 Examples of Multipath Interference

Refraction

Radio waves are refracted (or bent) when they propagate through the atmosphere. The refraction is caused by changes in the speed of the signal with altitude or by other spatial changes in the atmospheric conditions. Normally, the speed of the signal increases with altitude, causing radio waves to bend downward. However, on occasion, weather conditions may lead to variations in speed with height that differ significantly from the typical variations. This may result in a situation in which only a fraction or no part of the line-of-sight wave reaches the receiving antenna.

11.4 Fading in the Mobile Environment

Perhaps the most challenging technical problem facing communications systems engineers is fading in a mobile environment. **Fading** is defined as a time variation in phase, polarization, and/or signal strength of a received signal. In wireless communication, the term *fading* generally refers to the fluctuation in the strength of the signal received at the receiver caused by changes in the transmission medium or path(s). **Multipath** is the propagation phenomenon that results in a transmitted radio signal reaching the receiving antenna by two or more paths. **Multipath fading** refers to fading that occurs in any environment where there is multipath propagation. In a fixed environment, multipath fading is affected by changes in atmospheric conditions, such as rainfall. But in a mobile environment, where one of the two antennas is moving relative to the other, the relative locations of various obstacles change over time, creating complex transmission effects.

Multipath Propagation

Three propagation mechanisms, illustrated in Figure 11.8, play roles in fading. **Reflection** occurs when an electromagnetic signal encounters a surface that is large relative to the wavelength of the signal. For example, suppose a ground-reflected wave near the mobile unit is received. Because the ground-reflected wave has a 180° phase shift after reflection, the ground wave and the line-of-sight (LOS) wave may tend to cancel, resulting in high signal loss.[7] Further, because the mobile antenna is lower than most human-made structures in the area, multipath interference occurs. These reflected waves may interfere constructively or destructively at the receiver.

Diffraction occurs at the edge of an impenetrable body that is large compared to the wavelength of the radio wave. When a radio wave encounters such an edge, waves propagate in different directions, with the edge as the source. Thus, signals can be received even when there is no unobstructed LOS from the transmitter.

If the size of an obstacle is on the order of the wavelength of the signal or less, **scattering** occurs. An incoming signal is scattered into several weaker outgoing signals. At typical cellular microwave frequencies, there are numerous objects, such as lamp posts and traffic signs, that can cause scattering. Thus, scattering effects are difficult to predict.

---

7. On the other hand, the reflected signal has a longer path, which creates a phase shift due to delay relative to the unreflected signal. When this delay is equivalent to half a wavelength, the two signals are back in phase.

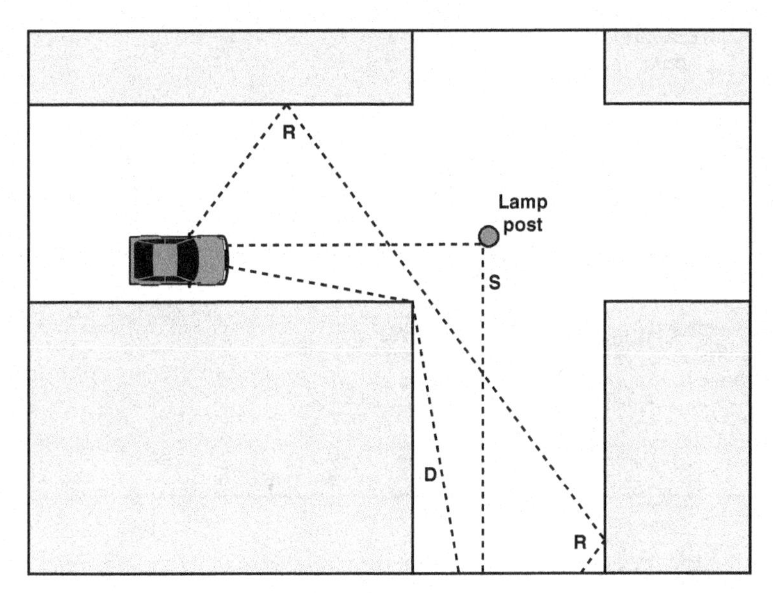

FIGURE 11.8 Sketch of Three Important Propagation Mechanisms: Reflection (R), Scattering (S), Diffraction (D)

These three propagation effects influence system performance in various ways, depending on local conditions and as the mobile unit moves within a cell. If a mobile unit has a clear LOS to the transmitter, then diffraction and scattering are generally minor effects, although reflection may have a significant impact. If there is no clear LOS, such as in an urban area at street level, then diffraction and scattering are the primary means of signal reception.

The Effects of Multipath Propagation

As just noted, one unwanted effect of multipath propagation is that multiple copies of a signal may arrive at different phases. If these phases add destructively, the signal level relative to noise declines, making signal detection at the receiver more difficult.

A second phenomenon, of particular importance for digital transmission, is **intersymbol interference (ISI)**. Consider that we are sending a narrow pulse at a given frequency across a link between a fixed antenna and a mobile unit. Figure 11.9 shows what the channel may deliver to the receiver if the impulse is sent at two different times. The upper line shows two pulses at the time of transmission. The lower line shows the resulting pulses at the receiver. In each case, the first received pulse is the desired LOS signal. The magnitude of that pulse may change because of changes in atmospheric attenuation. Further, as the mobile unit moves farther away from the fixed antenna, the amount of LOS attenuation increases. But in addition to this primary pulse, there may be multiple secondary pulses due to reflection, diffraction, and scattering. Now suppose that this pulse encodes one or more bits of data. In that case, one or more delayed copies of a pulse may arrive at the same time as the primary pulse for a subsequent bit. These delayed pulses act as a form of noise to the subsequent primary pulse, making recovery of the bit information more difficult.

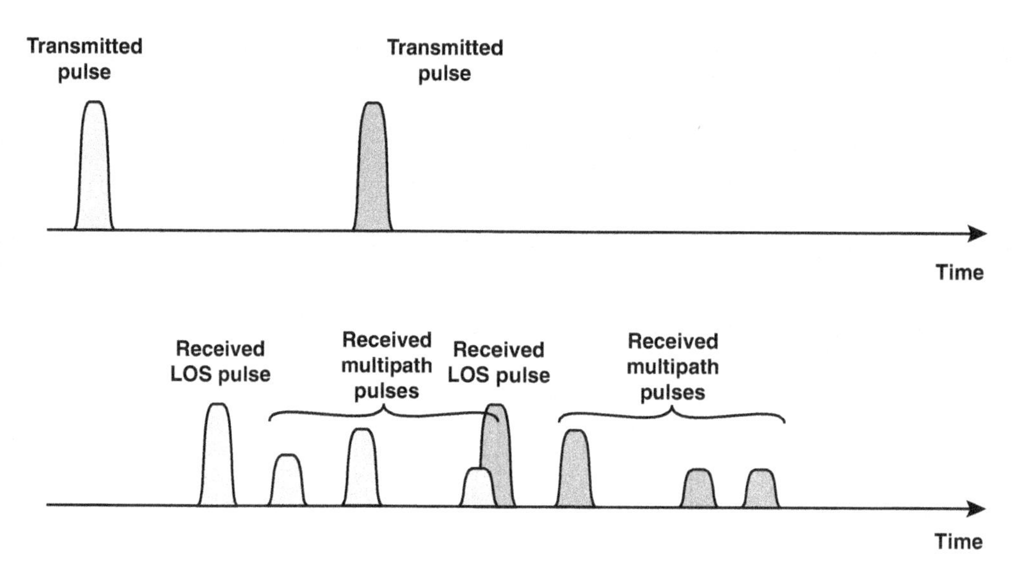

FIGURE 11.9 Two Pulses in Time-Variant Multipath

As the mobile antenna moves, the locations of various obstacles change; hence the number, magnitude, and timing of the secondary pulses change. This makes it difficult to design signal processing techniques that can filter out multipath effects so that the intended signal is recovered with fidelity.

Types of Fading

Fading effects in a mobile environment can be classified as either fast or slow. Refer to Figure 11.8; as the mobile unit moves down a street in an urban environment, rapid variations in signal strength occur over distances of about one-half a wavelength. At a frequency of 900 MHz, which is typical for mobile cellular applications, a wavelength is 0.33 m. The rapidly changing waveform in Figure 11.10 shows an example of the spatial variation of received signal amplitude at 900 MHz in an urban setting. Note that changes of amplitude can be as much as 20 or 30 dB over a short distance. This type of rapidly changing fading phenomenon, known as **fast fading**, affects not only mobile phones in automobiles but even mobile phone users walking down an urban street.

As a mobile user covers distances well in excess of a wavelength, the urban environment changes: The user passes buildings of different heights, vacant lots, intersections, and so forth. Over these longer distances, there is a change in the average received power level about which the rapid fluctuations occur. This is indicated by the slowly changing waveform in Figure 11.10 and is referred to as **slow fading**.

Fading effects can also be classified as flat or selective. **Flat fading**, or nonselective fading, is the type of fading in which all frequency components of the received signal fluctuate in the same proportions simultaneously. **Selective fading** affects unequally the different spectral components of a radio signal. The term *selective fading* is usually significant only relative to the bandwidth of the overall communications channel. If attenuation occurs over a portion of the bandwidth of the signal, the fading is considered to be selective; nonselective fading implies that the signal bandwidth of interest is narrower than, and completely covered by, the spectrum affected by the fading.

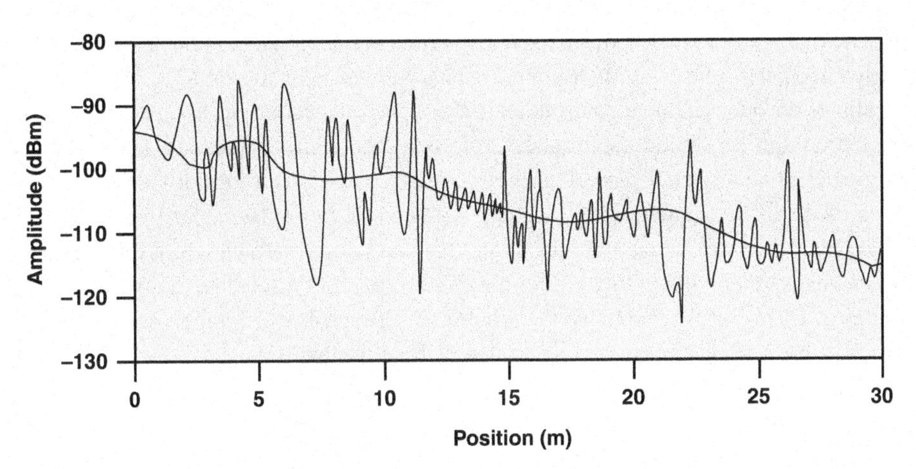

FIGURE 11.10 Typical Slow and Fast Fading in an Urban Mobile Environment

The Fading Channel

In designing a communications system, the communications engineer needs to estimate the effects of multipath fading and noise on the mobile channel. The simplest channel model, from the point of view of analysis, is the **additive white Gaussian noise (AWGN)** channel. In this channel, the desired signal is degraded by thermal noise associated with the physical channel itself as well as electronics at the transmitter and receiver (and any intermediate amplifiers or repeaters). This model is fairly accurate in some cases, such as for space communications and some wire transmissions, such as coaxial cable. For terrestrial wireless transmission, particularly in the mobile situation, AWGN is not a good guide for a designer.

Rayleigh fading occurs when there are multiple indirect paths between transmitter and receiver and there is no distinct dominant path, such as an LOS path. This represents a worst-case scenario. Fortunately, Rayleigh fading can be dealt with analytically, providing insights into performance characteristics that can be used in difficult environments, such as downtown urban settings.

Rician fading best characterizes a situation where there is a direct LOS path in addition to a number of indirect multipath signals. The Rician model is often applicable in an indoor environment, whereas the Rayleigh model characterizes outdoor settings. The Rician model also becomes more applicable in smaller cells or in more open outdoor environments. The channels can be characterized by a parameter K, defined as follows:

$$K = \frac{\text{power in the dominant path}}{\text{power in the scattered paths}}$$

When $K = 0$ the channel is Rayleigh (i.e., the numerator is zero) and when $K = \infty$, the channel is AWGN (i.e., the denominator is zero). Figure 11.11 shows system performance in the presence of noise. Here bit error rate is plotted as a function of the ratio E_b/N_0. Of course, as that ratio increases, the bit error rate drops. The figure shows that with a reasonably strong signal, relative to noise, an AWGN exhibit provides fairly good performance, as do Rician channels with larger values of K, roughly corresponding to microcells or an open country environment. The performance would be adequate for a digitized voice application, but for digital data transfer, efforts to compensate would be needed. The Rayleigh channel provides relatively poor performance; this is likely to be seen for flat fading and for slow fading; in these cases, error compensation mechanisms become more desirable. Finally, some environments produce fading effects worse than the so-called worst case of Rayleigh. Examples are fast fading in an urban environment and fading within the affected band of a selective fading channel. In these cases, no level of E_b/N_0 will help achieve the desired performance, and compensation mechanisms are mandatory. Chapters 12 and 13 discuss a number of these mechanisms.

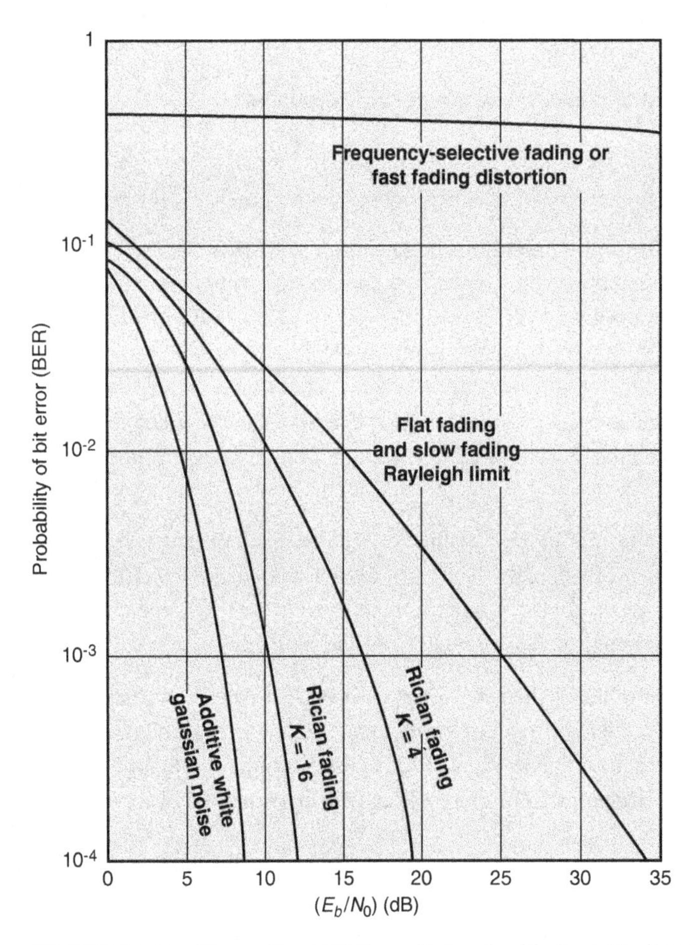

FIGURE 11.11 Theoretical Bit Error Rates for Various Fading Conditions

11.5 Millimeter Wave Transmission for 5G

The provision of high-capacity, high-data-rate services by 5G requires the use of high-bandwidth radio bearer channels. To meet this requirements, the 5G air interface makes use of channels in the millimeter wave (mmWave) range.

Traditionally, the term *millimeter wave* has been equated with the EHF (extremely high frequency) range of 30 to 300 GHz, with wavelengths between 10 mm and 1 mm, as shown in Figure 11.12. In a 5G context, millimeter waves refer to frequencies between 24 and 86 GHz, which encompasses the frequency bands contemplated for 5G use.

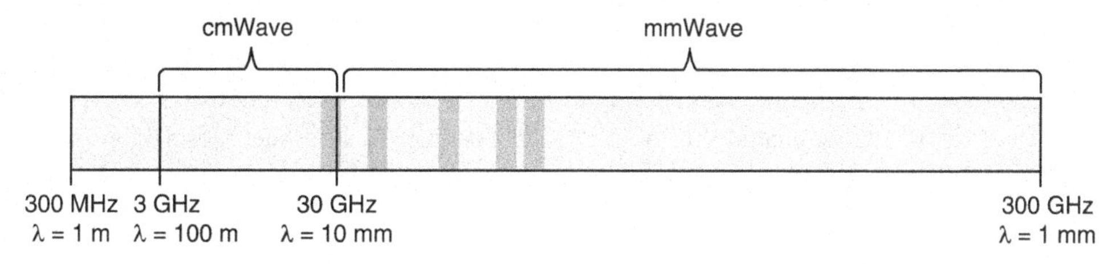

FIGURE 11.12 Millimeter Wave Spectrum

Propagation Impairments

Up through the fourth generation of cellular networks, the mmWave frequencies were not employed because of a number of propagation impairments that rendered these bands impractical for 4G and earlier cellular networks. These include [FCC97]:

- Free space loss
- Atmospheric gaseous losses
- Rain losses
- Foliage losses
- Blocking

Free Space Loss

The discussion of free space loss in Section 11.3 indicates that loss is an increasing function of frequency and distance and a decreasing function of the effective area of the antenna. In the mmWave range, the losses can be considerable.

With respect to antenna size, the following consideration is important. As discussed in Chapter 12, for the simplest antennas, the size of the antenna is proportional to the wavelength, so that for decreasing wavelength (increasing frequency), the antenna size decreases. There are several methods for holding the antenna size constant as the wavelength decreases. An important technique is to use an array of small antenna elements to form a directed beam.

Even with advanced antenna design, however, the free space loss in the mmWave range is such as to limit practical transmission distances in mobile communications to the tens or perhaps hundreds of meters—and not into the kilometer or more range.

Atmospheric Gaseous Losses

A separate component of loss that is significant in the mmWave range is atmospheric gaseous loss, which is the absorption of signal energy by molecules of oxygen, water vapor, and other gaseous atmospheric conditions. At frequencies below the mmWave range, these losses are insignificant. Figure 11.13 shows that above 20 GHz, the absorption loss becomes increasingly significant. Further, there are certain regions in which there is an additional peak of loss due to either H_2O or O_2. These regions are indicated by the darker shading in Figure 11.12. For 5G, the important absorption peaks occur at 24 and 60 GHz. The spectral regions between the peaks provide windows where propagation conditions are more favorable.

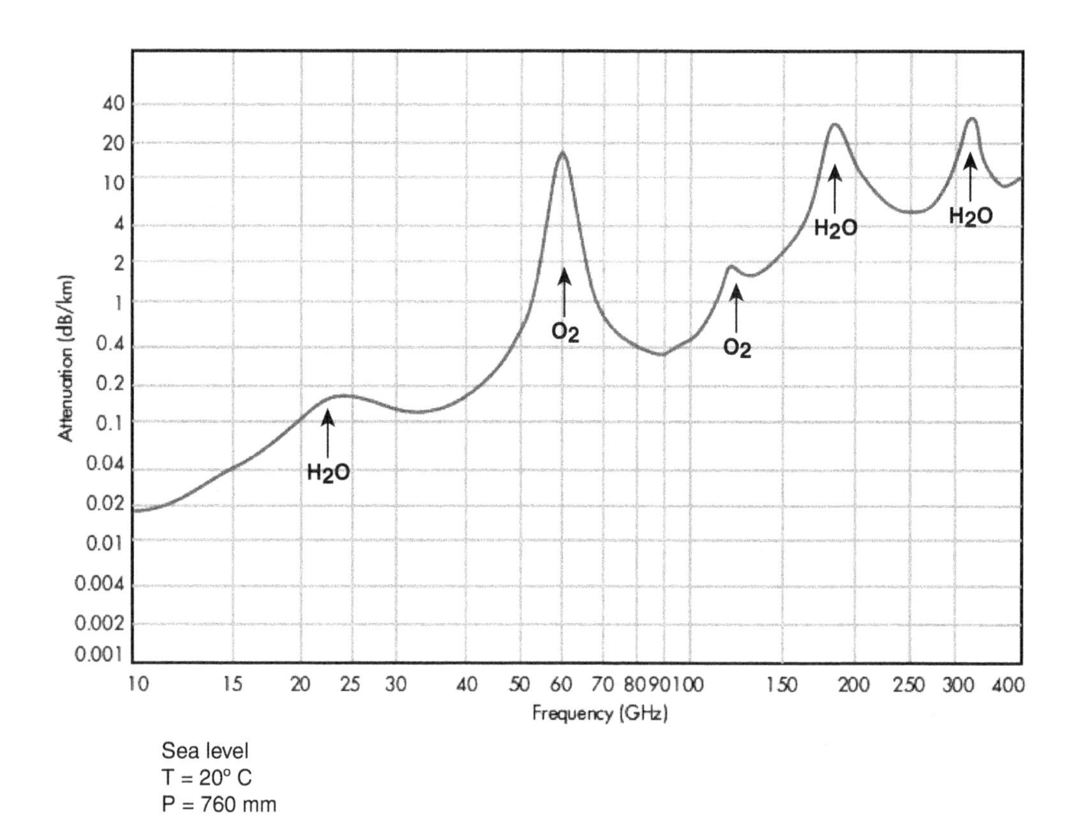

FIGURE 11.13 Average Atmospheric Absorption of Millimeter Waves

Rain Losses

Rain is another source of loss for mmWave propagation. Raindrops are roughly the same size as the millimeter radio wavelengths, and they therefore cause scattering of the radio signal. Figure 11.14 shows the attenuation per kilometer as a function of rain rate. As would be expected, the heavier the rain, the greater the loss of signal strength. This is an important consideration, especially in urban areas. Higher rainfall typically correlates with higher rates of vehicle accidents and other emergencies. Smart city applications and V2X applications that require high data rates operate in a mmWave band, and these applications may be needed most when there is heavy rainfall. This is an important consideration in system design.

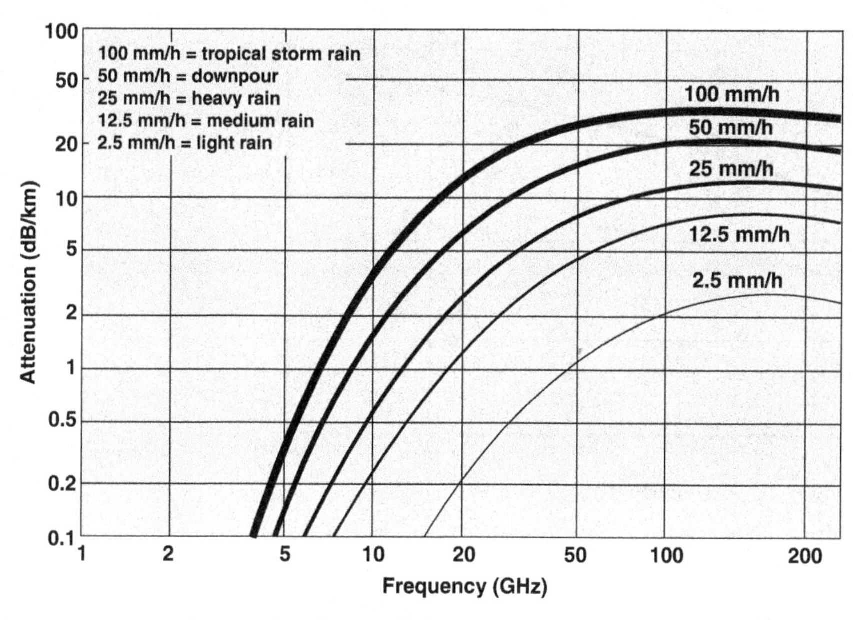

FIGURE 11.14 Attenuation Due to Rain

Foliage Losses

In the mmWave bands, foliage can introduce significant loss—or even a dominant amount of loss. The term *foliage depth* comes into play: If a cluster of vegetation is along the path between a radio transmitter and receiver, the foliage depth is the length of the portion of the path that passes through the foliage. The following empirical relationship estimates the loss L as a function of foliage depth R, applicable for a foliage depth of less than 400 m [HEMA18]:

$$L = 0.2 f^{0.3} R^{0.6} \text{ dB}$$

where f is frequency in MHz.

Figure 11.15 illustrates foliage loss. For example, the foliage loss at 40 GHz for a penetration of 10 meters (which is about equivalent to a large tree or two trees in tandem) is about 19 dB. This is a significant loss, especially when combined with the various other losses that are sustained in the mmWave region. Note that foliage loss exhibits seasonal variations.

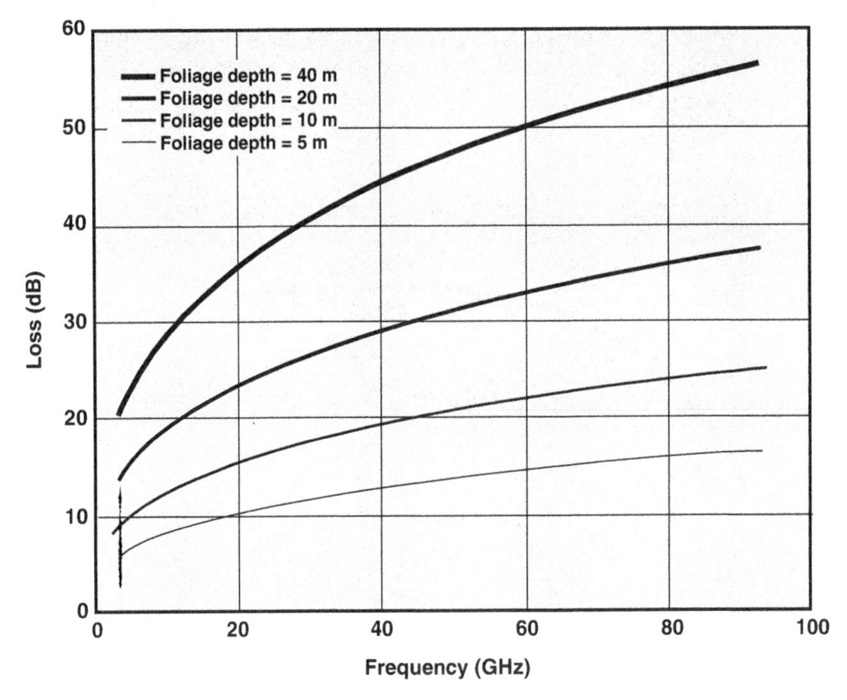

FIGURE 11.15 Foliage Penetration Loss

Blocking

Blocking, or **blockage**, is related to the amount of loss sustained by a wireless signal due to a solid object on the LOS path. If the loss is sufficiently high, the propagation transitions from primarily LOS to primarily non-line-of-sight (NLOS), and blocking occurs.

Millimeter waves are more vulnerable to being blocked by certain types of solid objects than lower-frequency radio waves. High levels of attenuation for certain building materials (e.g., brick and concrete) may keep millimeter waves transmitted from outdoor base stations confined to streets and other outdoor structures, although some signals might reach inside the buildings through glass windows and wood doors. The indoor coverage in this case can be provided by other means, such as indoor millimeter wave femtocell or Wi-Fi solutions. Interior walls and partitions within offices and homes generally do not impose significant losses.

This reflective property works both ways. LOS with a 5G antenna is not necessary to receive the signals. 5G networks use beamforming to direct waves off and around obstacles to your phone. This works in part because 5G equipment uses multiple antennas to send and receive signals, combining the data from multiple streams to strengthen the overall signal and increase the bandwidth. This works both outdoors, by reflecting signals off buildings, and indoors, by reflecting signals off walls. Carriers could definitely install beamforming transmitters inside stadiums or large malls.

Table 11.2, derived from various sources, indicates the effect of different materials on mmWave propagation, compared to that in a lower frequency range.

TABLE 11.2 Attenuation for Different Materials

Material	Thickness (cm)	< 3 GHz [ANDE04, ALLE94]	40 GHz [ALEJ08]	60 GHz [ANDE04]
Drywall	2.5	5.4	—	6.0
Office whiteboard	1.9	0.5	—	9.6
Clear glass	0.3/0.4	6.4	2.5	3.6
Mesh glass	0.3	7.7	—	10.2
Particle board	1.6	—	.6	—
Wood	0.7	5.4	3.5	—
Plasterboard	1.5	—	2.9	—
Mortar	10	—	160	—
Brick wall	10	—	178	—
Concrete	10	17.7	175	—

Implications for 5G

Millimeter wave communications systems are a natural fit for 5G networks. Specifically, systems that combine mmWave, advanced antenna design, and ultra-dense networks of very small cells marry new spectrum with greater spectral and spatial efficiency and promise a long-term solution for the spectrum crunch. mmWave antennas are also small enough to support electronically steerable antenna arrays not only at the base stations but also in the handsets; the larger bandwidths are well suited to support broadband device-to-device communications in meshed networks, and the shorter wavelengths lead naturally to small cell sizes. However, technology challenges are greater at mmWave frequencies, and much research and development remains to be done.

A presentation by Qualcomm [QUAL19] lists four areas of challenge for mmWave deployment:

- Limited coverage
- Difficulties with non-line-of-site (NLOS)
- Difficulties with mobile use
- Large form factor

Limited Coverage

As illustrated in Figures 11.13 through 11.15, mmWave propagation is subject to significant path loss, limiting distance and resulting in very small cell size. If the cell size is limited to a few tens of meters or at most a very few hundreds of meters, deployment requires many cells and therefore many base station antennas. This not only results in significant expenses but also may meet public resistance. The use of advanced antennas that can produce a narrow beam width to overcome loss to some extent allows larger cell sizes.

An additional means of cost reduction is co-siting. **Co-siting**, also called **collocation**, is the practice of sharing site resources between base station radios. On a practical level, this generally means multiple technologies, radios, and/or channel elements sharing space on a single tower. For this discussion, co-siting involves placing 5G mmWave antennas on non-mmWave cell towers. This reduces the number of cell towers that are needed. An additional advantage is the ability to share backhaul capacity to the RAN.

Although operators are using co-siting, there is increasing use of separate mmWave antennas to provide inexpensive coverage using numerous base stations. Examples are antennas in outdoor light poles [TOMA19] and in indoor ceiling lighting fixtures [REIC18].

Difficulties with Non-Line-of-Site (NLOS)

Blockage due to hands, bodies, walls, and other obstacles severely limits signal propagation, as indicated in Table 11.2. One means to mitigate this is to utilize steerable, directional antenna arrays. These arrays steer the beam in a direction that achieves a link through access to an alternate base station node that is LOS or access to the main, or alternate, base stations with a suitably low loss reflection.

Difficulties with Mobile Use

Commercial mmWave deployments for fixed wireless backhaul links and for satellite links have been around for some time. When the UE is mobile, however, path loss and blockage effects can change dramatically over a short period of time due to blockage from the user's hand and body for a handheld UE. A sophisticated infrastructure must employ adaptive beam steering and dynamic handoff techniques to maintain signal quality.

Another solution is dual connectivity. This can combine a connection at lower bandwidth with a mmWave connection. If the mmWave connection is temporarily lost (e.g., if a hand is waved), the connection only drops in throughput and is not completely lost. Chapter 15, "5G Radio Access Network," explores dual connectivity.

Large Form Factor

Transmission in the mmWave region consumes more power than at lower frequencies due to the use of a wider bandwidth. This poses a thermal challenge for small devices. Essentially, this problem is addressed through advances in antenna design and transmission technology to reduce power consumption in small devices with mmWave capability.

11.6 Key Terms and Review Questions

Key Terms

additive white Gaussian noise (AWGN)	free space loss
atmospheric absorption	ground wave
atmospheric gaseous loss	impulse noise
attenuation	index of refraction
attenuation distortion	intermodulation noise
bandwidth	intersymbol interference (ISI)
bandwidth efficiency	line of sight (LOS)
blockage	millimeter wave (mmWave)
blocking	multipath
channel capacity	noise
collocation	non-line-of-sight (NLOS)
co-siting	Nyquist bandwidth
crosstalk	optical LOS
data rate	path loss exponent
decibel (dB)	radio LOS
diffraction	rain loss
error-free capacity	Rayleigh fading
error rate	refraction
fading	reflection
fast fading	Rician fading
flat fading	scattering
foliage loss	selective fading
form factor	Shannon capacity

signal rate	spectral efficiency
signal-to-noise ratio	thermal noise
sky wave	white noise
slow fading	

Review Questions

1. Define channel capacity.
2. What is the difference between the theoretical channel capacity values derived from the Nyquist and Shannon formulas?
3. What key factors affect channel capacity?
4. What is the difference between optical and radio LOS?
5. What factors related to attenuation does a transmission engineer need to consider?
6. What variables are factors in free space loss?
7. Define the path loss exponent.
8. Name and briefly define four types of noise.
9. What is refraction?
10. What is multipath fading?
11. What is the difference between diffraction and scattering?
12. What is the difference between fast and slow fading?
13. What is the difference between flat and selective fading?
14. What is the frequency range for millimeter wave?
15. List and briefly define the main sources of loss for mmWave transmission.

11.7 References

ALEJ08 Alejos, A., Sanchez, M., and Cuinas, I. "Measurement and Analysis of Propagation Mechanisms at 40 GHz: Viability of Site Shielding Forced By Obstacles." *IEEE Transactions on Vehicular Technology*, November 2008.

ALLE94 Allen, L., et al., "Building Penetration Loss Measurements at 900 MHz, 11.4 GHz, and 28.8 GHz." *NTIA Report 94-306*, May 1994.

ANDE04 Anderson, C., and Rappaport, T. "In-Building Wideband Partition Loss Measurements at 2.5 and 60 GHz," *IEEE Transactions on Wireless Communications*, vol. 3, no. 3, May 2004.

FCC97 Federal Communications Commission. *Millimeter Wave Propagation: Spectrum Management Implications.* Office of Engineering and Technology Bulletin Number 70, July 1997.

HEMA18 Hemadeh, I., et al. "Millimeter-Wave Communications: Physical Channel Models, Design Considerations, Antenna Constructions, and Link-Budget." *IEEE Communications Surveys & Tutorials*. Second Quarter, 2018.

QUAL19 Qualcomm. Breaking the wireless barriers to mobilize 5G NR mmWave. Qualcomm Presentation, January 2019. https://www.qualcomm.com/invention/5g/5g-nr/mmwave

RAPP02 Rappaport, T. *Wireless Communications: Principles and Practice.* Upper Saddle River, NJ: Prentice Hall, 2002.

REIC18 Reichert, C. "Ericsson Adds to 5G Portfolio with Radio Dot." *ZDNet*, January 18, 2018.

TOMA19 Tomas, J. "Signify Launches New Smart Pole with IoT Apps, Connectivity." *Enterprise IoT Insights*, June 20, 2019.

ANNEX 11A: Decibels and Signal Strength

An important parameter in any transmission system is the signal strength. As a signal propagates along a transmission medium, there is loss, or *attenuation*, of signal strength. To compensate, amplifiers may be inserted at various points to impart a gain in signal strength.

It is customary to express gains, losses, and relative levels in decibels because:

- Signal strength often falls off exponentially, so loss is easily expressed in terms of the decibel, which is a logarithmic unit.
- The net gain or loss in a cascaded transmission path can be calculated with simple addition and subtraction.

The decibel is a measure of the ratio between two signal levels. The decibel gain is given by:

$$G_{dB} = 10\log_{10}\frac{P_{out}}{P_{in}}$$

where:

G_{dB} = gain, in decibels

P_{in} = input power level

P_{out} = output power level

$\log_{10}$ = logarithm to the base 10 (from now on, we will simply use log to mean $\log_{10}$)

Table 11.3 shows the relationship between decibel values and powers of 10.

TABLE 11.3 Decibel Values

Power Ratio	dB
2	3
10^1	10
10^2	20
10^3	30
10^4	40
10^5	50
10^6	60

Power Ratio	dB
0.5	–3
10^{-1}	–10
10^{-2}	–20
10^{-3}	–30
10^{-4}	–40
10^{-5}	–50
10^{-6}	–60

There is some inconsistency in the literature over the use of the terms *gain* and *loss*. If the value of G_{dB} is positive, this represents an actual gain in power. For example, a gain of 3 dB means that the power has doubled. If the value of G_{dB} is negative, this represents an actual loss in power. For example, a gain of –3 dB means that the power has halved, and this is a loss of power. Normally, this is expressed by saying there is a loss of 3 dB. However, some of the literature would say that this is a loss of –3 dB. It makes more sense to say that a negative gain corresponds to a positive loss. Therefore, we define a decibel loss as:

$$L_{dB} = -10\log_{10}\frac{P_{out}}{P_{in}} = 10\log_{10}\frac{P_{in}}{P_{out}} \quad \textbf{(Equation 11.7)}$$

Example

If a signal with a power level of 10 mW is inserted onto a transmission line, and the measured power some distance away is 5 mW, the loss can be expressed as:

L_{dB} = 10 log (10/5) = 10(0.3) = 3 dB

Note that the decibel is a measure of relative, not absolute, difference. A loss from 1000 mW to 500 mW is also a loss of 3 dB.

The decibel is also used to measure the difference in voltage, taking into account that power is proportional to the square of the voltage:

$$P = \frac{V^2}{R}$$

where:

P = power dissipated across resistance R

V = voltage across resistance R

Thus:

$$L_{\text{dB}} = 10 \log \frac{P_{\text{in}}}{P_{\text{out}}} = 10 \log \frac{V_{\text{in}}^2/R}{V_{\text{out}}^2/R} = 20 \log \frac{V_{\text{in}}}{V_{\text{out}}}$$

Example

Decibels are useful in determining the gain or loss over a series of transmission elements. Consider a series in which the input is at a power level of 4 mW, the first element is a transmission line with a 12-dB loss (i.e., –12-dB gain), the second element is an amplifier with a 35-dB gain, and the third element is a transmission line with a 10-dB loss. The net gain is (–12 + 35 – 10) = 13 dB. To calculate the output power P_{out}:

$$G_{\text{dB}} = 13 = 10 \log (P_{\text{out}}/4 \text{ mW})$$

$$P_{\text{out}} = 4 \times 10^{1.3} \text{ mW} = 79.8 \text{ mW}$$

Decibel values refer to relative magnitudes or changes in magnitude, not to absolute levels. It is convenient to be able to refer to an absolute level of power or voltage in decibels so that gains and losses with reference to an initial signal level may be calculated easily. The **dBW (decibel-Watt)** is used extensively in microwave applications. The value 1 W is selected as a reference and defined to be 0 dBW. The absolute decibel level of power in dBW is defined as:

$$\text{Power}_{\text{dBW}} = 10 \log \frac{\text{Power}_{\text{W}}}{1 \text{ W}}$$

Example

A power of 1000 W is 30 dBW, and a power of 1 mW is –30 dBW.

Another common unit is the **dBm (decibel-milliWatt)**, which uses 1 mW as the reference. Thus 0 dBm = 1 mW. The formula is:

$$\text{Power}_{\text{dBm}} = 10 \log \frac{\text{Power}_{\text{mW}}}{1\,\text{mW}}$$

Note the following relationships:

$$+30\ \text{dBm} = 0\ \text{dBW}$$

$$0\ \text{dBm} = -30\ \text{dBW}$$

Chapter 12

Antennas

Learning Objectives

After studying this chapter, you should be able to:

- Summarize the various approaches to compensating for errors and distortions introduced by multipath fading and other wireless channel impairments
- Understand the principle of operation of a parabolic antenna
- Explain antenna gain
- Discuss the principles of MIMO antennas
- Contrast single-user and multiple-user MIMO
- Make a presentation on the concept of beamforming
- Explain the key characteristics of active antennas
- Understand the distinguishing characteristics of massive MIMO

This chapter provides a survey of antenna technology. To begin, Section 12.1 provides an overview of the various approaches used to compensate for impairments. The remainder of the chapter discusses issues of antenna design that relate to improving the spectral efficiency of a wireless channel. Section 12.2 provides an overview of antenna technology. Sections 12.3 and 12.4 discuss multiple-input/multiple-output antennas, a technology that is critical to 5G.

12.1 Channel Correction Mechanisms

The efforts to compensate for the errors and distortions introduced by multipath fading and other wireless channel impairments fall into four general categories: adaptive equalization, diversity techniques, adaptive modulation and coding, and forward error correction. In the typical mobile wireless environment, techniques from all four categories are combined to combat the error rates encountered.

Adaptive Equalization

Adaptive equalization can be applied to transmissions that carry analog information (e.g., analog voice or video) or digital information (e.g., digital data, digitized voice or video) and is used to combat intersymbol interference. The process of equalization involves some method of gathering the dispersed symbol energy back together into its original time interval. Equalization is a broad topic; techniques include the use of so-called lumped analog circuits as well as sophisticated digital signal processing algorithms. Here we look at the digital signal processing approach.

Figure 12.1 illustrates a common approach using a linear equalizer circuit. In this specific example, for each output symbol, the input signal is sampled at five uniformly spaced intervals of time, separated by a delay τ. These samples are individually weighted by the coefficients C_i and then summed to produce the output. The circuit is referred to as adaptive because the coefficients are dynamically adjusted. Typically, the coefficients are set using a *training sequence*, which is a known sequence of bits. The training sequence is transmitted. The receiver compares the received training sequence with the expected training sequence and, on the basis of the comparison, calculates suitable values for the coefficients. Periodically, a new training sequence is sent to account for changes in the transmission environment.

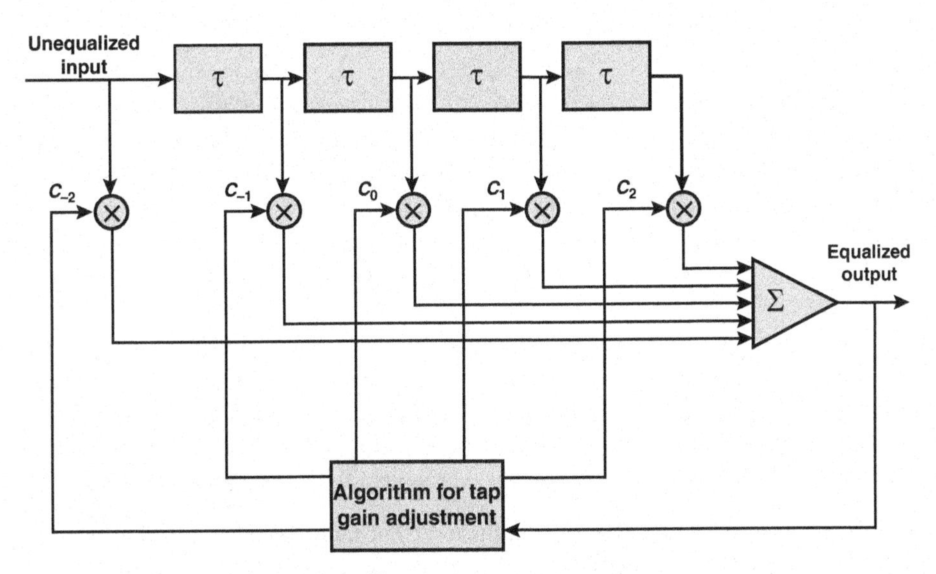

FIGURE 12.1 Linear Equalizer Circuit

For Rayleigh channels or worse, it may be necessary to include a new training sequence with every single block of data. Again, this represents considerable overhead, but it is justified by the error rates encountered in a mobile wireless environment.

Diversity Techniques

Diversity is based on the fact that individual channels experience independent fading events. For example, multiple antennas that are spaced far enough apart will have independent fading. It is therefore possible to compensate for error effects by providing multiple channels in some sense between transmitter and receiver, and sending either a duplicate of the signal over each channel or a part of the signal over each channel. This technique does not eliminate errors, but it reduces the error rate because it spreads out the transmission to avoid being subjected to the highest error rate that might occur. The other techniques—equalization, coding, and forward error correction—can then cope with the reduced error rate.

Space Diversity

Space diversity, or **spatial diversity**, means using different physical paths for the signal, at a single frequency. For example, multiple nearby antennas, if spaced far enough apart, may be used to receive the message with the signals combined in some fashion to reconstruct the most likely transmitted signal. Another example is the use of collocated multiple directional antennas, each oriented to a different reception angle, with the incoming signals again combined to reconstitute the transmitted signal. The important case of multiple-input/multiple-output antennas is covered later in this chapter.

Frequency Diversity

With **frequency diversity**, the signal is spread out over a larger frequency bandwidth or carried on multiple frequency carriers. The most important examples of this approach are orthogonal frequency-division multiplexing (OFDM) and spread spectrum. Traditional communications—both wireline and wireless—simply modulate a baseband signal up to a required transmission channel and frequency. No change to the original signal occurs. Two methods, however, have been used to overcome wireless channel impairments in which the signals are significantly modified for transmission.

- **Orthogonal frequency-division multiplexing (OFDM):** OFDM splits a signal into many lower-bit-rate streams that are transmitted over carefully spaced frequencies. This can overcome frequency selective fading by using significantly lower bandwidth per stream with longer bit times. Each of these frequencies can then be amplified separately. Various versions of OFDM are important in 4G and 5G. Chapter 13, "Air Interface Physical Layer," covers OFDM.

- **Spread spectrum:** This is a technique in which a signal is transmitted in a bandwidth considerably greater than the frequency content of the original information. Spread spectrum generally makes use of a sequential noise-like signal structure to spread the normally narrowband signal over a relatively wide band of frequencies. The receiver correlates the signals to retrieve the original information signal. This technique decreases potential interference to other receivers while achieving data confidentiality and increasing immunity of spread spectrum receivers to noise and interference. In addition, proper encoding enables multiple users to independently use the same higher bandwidth with very little interference. 2G and 3G systems employ spread spectrum techniques.

Time Diversity

Time diversity techniques spread out the data over time so that a noise burst affects fewer bits. Time diversity can be quite effective in a region of slow fading. If a mobile unit is moving slowly, it may remain in a region of a high level of fading for a relatively long interval. The result would be a long burst of errors even though the local mean signal level is much higher than the interference. Even powerful error correction codes may be unable to cope with an extended error burst. If digital data is transmitted in a synchronous time-division multiplexing (TDM) structure, in which multiple users share the same physical channel through the use of time slots, then block interleaving can be used to provide time diversity. Figure 12.2a illustrates the concept. Note that the same number of bits are still affected by the noise surge, but they are spread out over a number of logical channels. If each channel is protected by forward error correction, the error-correcting code may be able to cope with the smaller number of bits that are in error in a particular logical channel. If TDM is not used, time diversity can still be applied by viewing the stream of bits from the source as a sequence of blocks and then shuffling the blocks. In Figure 12.2b, blocks are shuffled in groups of four. Again, the same number of bits is in error, but the error-correcting code is applied to sets of bits that are spread out in time. Even greater diversity is achieved by combining TDM interleaving with block shuffling.

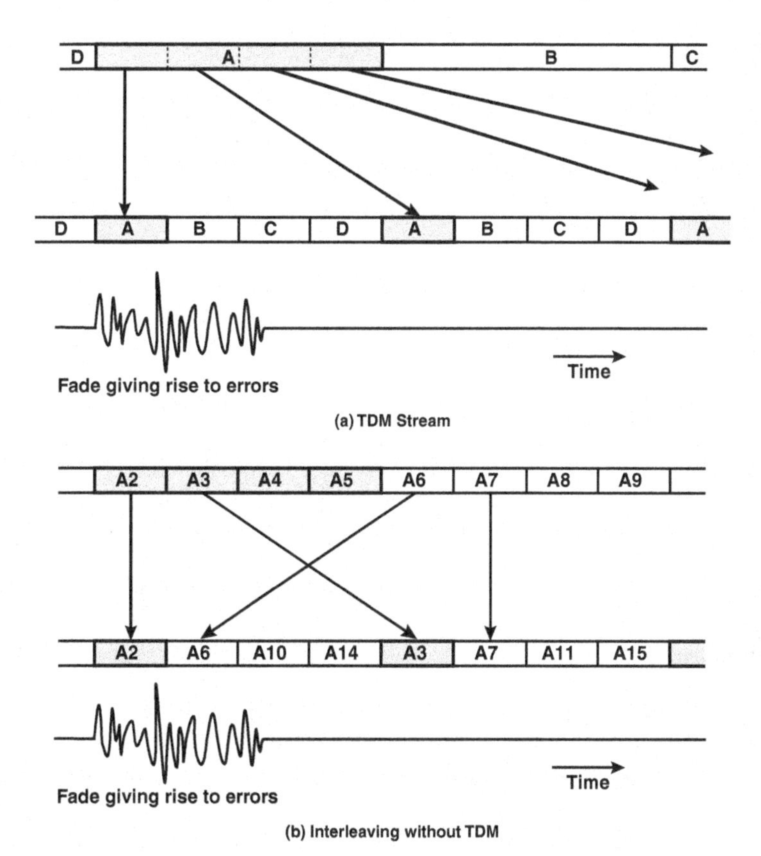

FIGURE 12.2 Interleaving Data Blocks to Spread the Effects of Error Bursts

The trade-off with time diversity is delay. The greater the degree of interleaving and shuffling used, the longer the delay in reconstructing the original bit sequence at the receiver.

Adaptive Modulation and Coding

The characteristics of a wireless channel can change hundreds of times per second due to fading, and modern systems use adaptive modulation and coding (AMC) to adjust their schemes just as quickly. This essentially involves creating signals that send as much information as possible for a given received signal strength and noise and then detecting and correcting the errors. To adapt hundreds of times per second, two features must be present in the protocols for a system:

- Mechanisms to measure the quality of the wireless channel. These might include monitoring packet loss rates or sending special pilot signals expressly for measurement purposes.
- Messaging mechanisms to communicate the signal quality indicators between transmitters and receivers and also to communicate the new modulation and coding formats.

Forward Error Correction

Forward error correction is applicable in digital transmission applications: those in which the transmitted signal carries digital data or digitized voice or video data. The term *forward* refers to procedures whereby a receiver, using only information contained in the incoming digital transmission, corrects bit errors in the data. This is in contrast to backward error correction, in which the receiver merely detects the presence of errors and then sends a request back to the transmitter to retransmit the data in error. Backward error correction is not practical in many wireless applications. For example, in satellite communications, the amount of delay involved makes retransmission undesirable. In mobile communications, the error rates are often so high that there is a high probability that the retransmitted block of bits will also contain errors. In these applications, forward error correction is required. In essence, forward error correction is achieved as follows:

Step 1. Using a coding algorithm, the transmitter adds a number of additional, redundant bits to each transmitted block of data. These bits form an error-correcting code and are calculated as a function of the data bits.

Step 2. For each incoming block of bits (data plus error-correcting code), the receiver calculates a new error-correcting code from the incoming data bits. If the calculated code matches the incoming code, then the receiver assumes that no error has occurred in this block of bits.

Step 3. If the incoming and calculated codes do not match, then one or more bits are in error. If the number of bit errors is below a threshold that depends on the length of the code and the nature of the algorithm, it is possible for the receiver to determine the bit positions in error and correct all errors.

Typically in mobile wireless applications, the ratio of total bits sent to data bits sent is between 2 and 3. This may seem an extravagant amount of overhead, in that the capacity of the system is cut to one-half or one-third of its potential, but the mobile wireless environment is such a challenging medium that such levels of redundancy are necessary.

Chapter 13 provides an introduction to forward error correction techniques.

12.2 Introduction to Antennas

An antenna can be defined as an electrical conductor or system of conductors used either for radiating electromagnetic energy or for collecting electromagnetic energy. For transmission of a signal, radio-frequency electrical energy from the transmitter is converted into electromagnetic energy by the antenna and radiated into the surrounding environment (atmosphere, space, water). For reception of a signal, electromagnetic energy impinging on the antenna is converted into radio-frequency electrical energy and fed into the receiver.

In two-way communication, the same antenna can be and often is used for both transmission and reception. This is possible because any antenna transfers energy from the surrounding environment to its input receiver terminals with the same efficiency used to transfer energy from the output transmitter terminals into the surrounding environment, assuming that the same frequency is used in both directions. Put another way, antenna characteristics are essentially the same whether an antenna is sending or receiving electromagnetic energy.

Radiation Patterns

An antenna radiates power in all directions but, typically, does not perform equally well in all directions. A common way to characterize the performance of an antenna is based on the radiation pattern. When the radiated power of a transmitting antenna is measured around the antenna, a shape called the radiation pattern emerges. The pattern is a graphical representation of the radiation properties of an antenna as a function of space coordinates.

An idealized antenna known as an isotropic antenna produces the simplest radiation pattern. An **isotropic antenna** is a point in space that radiates power in all directions equally. The actual radiation pattern for the isotropic antenna is a sphere with the antenna at the center. However, a radiation pattern is almost always depicted as a two-dimensional cross section of the three-dimensional pattern. The pattern for an isotropic antenna is shown in Figure 12.3a. The distance from the antenna to each point on the radiation pattern is proportional to the power radiated from the antenna in that direction. Figure 12.3b shows the radiation pattern of another idealized antenna. This is a directional antenna in which the preferred direction of radiation is along one axis.

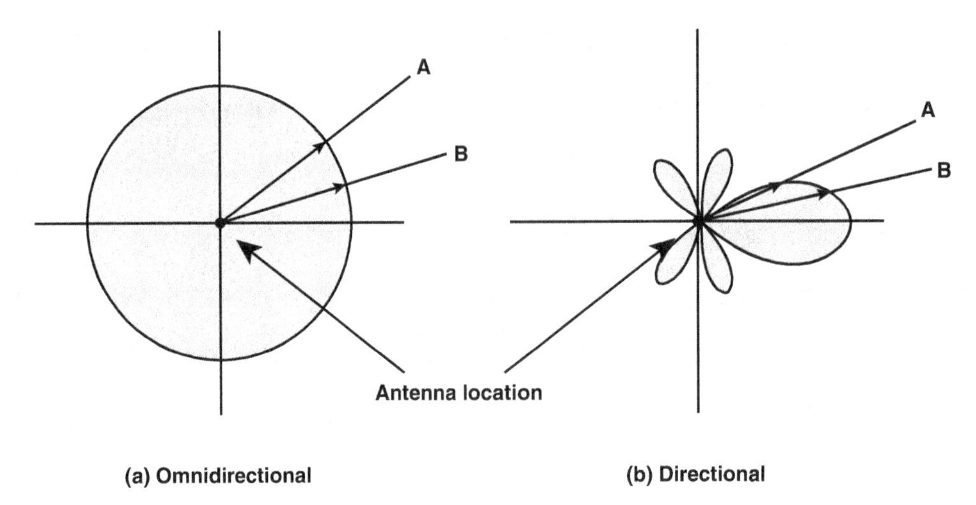

FIGURE 12.3 Idealized Radiation Patterns

The actual size of a radiation pattern is arbitrary. What is important is the *relative* distance from the antenna position in each direction. The relative distance determines the relative power. To determine the relative power in a given direction, a line is drawn from the antenna position at the appropriate angle, and the point of intercept with the radiation pattern is determined. Figure 12.3 shows a comparison of two transmission angles, A and B, drawn on the two radiation patterns. The isotropic antenna produces an omnidirectional radiation pattern of equal strength in all directions, so the A and B vectors are of equal length. For the antenna pattern in Figure 12.3b, the B vector is longer than the A vector, indicating that more power is radiated in the B direction than in the A direction, and the relative lengths of the two vectors are proportional to the amount of power radiated in the two directions. Please note that this type of diagram shows relative *antenna gain* in each direction, not relative distance of coverage, although they are, of course, related.

The radiation pattern provides a convenient means of determining the **beamwidth** of an antenna, which is a common measure of the directivity of an antenna. The beamwidth, also referred to as the half-power beamwidth, is the angular separation in which the magnitude of the radiation pattern decreases by 50% (–3 dB) from the peak of the main beam. Figure 12.4 illustrates this quantity. Another commonly quoted beamwidth is the **null-to-null beamwidth**. This is the angular separation from which the magnitude of the radiation pattern decreases to zero (negative infinity dB) away from the main beam.

When an antenna is used for reception, the radiation pattern becomes a **reception pattern**. The longest section of the pattern indicates the best direction for reception.

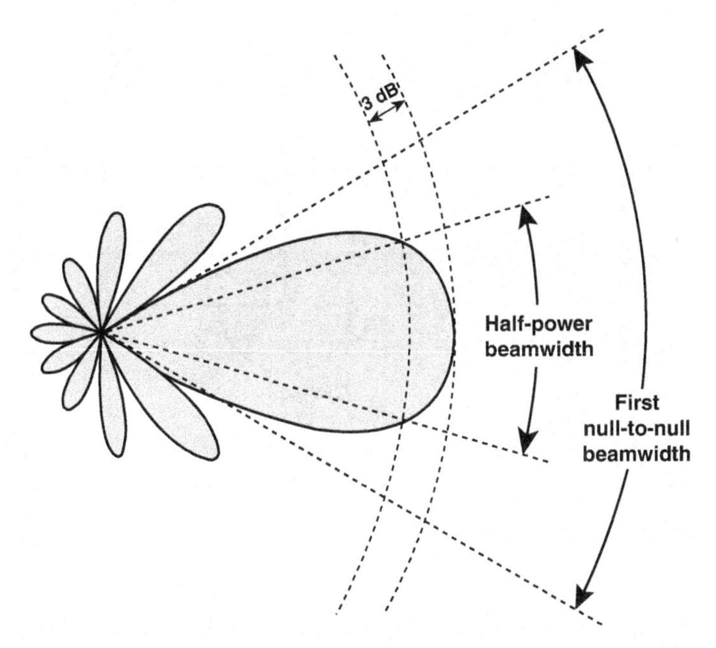

FIGURE 12.4 Half-Power Beamwidth and First Null-to-Null Beamwidth of an Antenna Radiation Pattern

Antenna Types

This subsection introduces the common basic types of antennas.

Dipoles and Monopoles

The two most fundamental antennas that can be implemented are the half-wave dipole, or Hertz, antenna (see Figure 12.5a) and the quarter-wave vertical, or Marconi, antenna (see Figure 12.5b). The half-wave dipole consists of two straight collinear conductors of equal length, separated by a small gap. The length of the antenna is one-half the wavelength of the signal that can be transmitted most efficiently. A half-wave dipole has a uniform or omnidirectional radiation pattern in one dimension and a figure-eight pattern in the other two dimensions (see Figure 12.5a). This means the energy is directed along the ground. Much less energy is expended vertically (and lost) than with an isotropic antenna.

The operating principle of the quarter-wave monopoles is based on the fact that the current distribution on an antenna structure that is only a quarter wavelength long is identical to that on a half-wave dipole if the antenna element missing from the dipole is replaced by a highly conducting surface. As a result of this reflection principle, vertical quarter-wave antennas on conducting ground have basically the same radiation pattern as half-wave dipole antennas (see Figure 12.5b). There is, of course, no radiation into the shadowed half of the space. A vertical quarter-wave antenna is the type commonly used for automobile radios and portable radios.

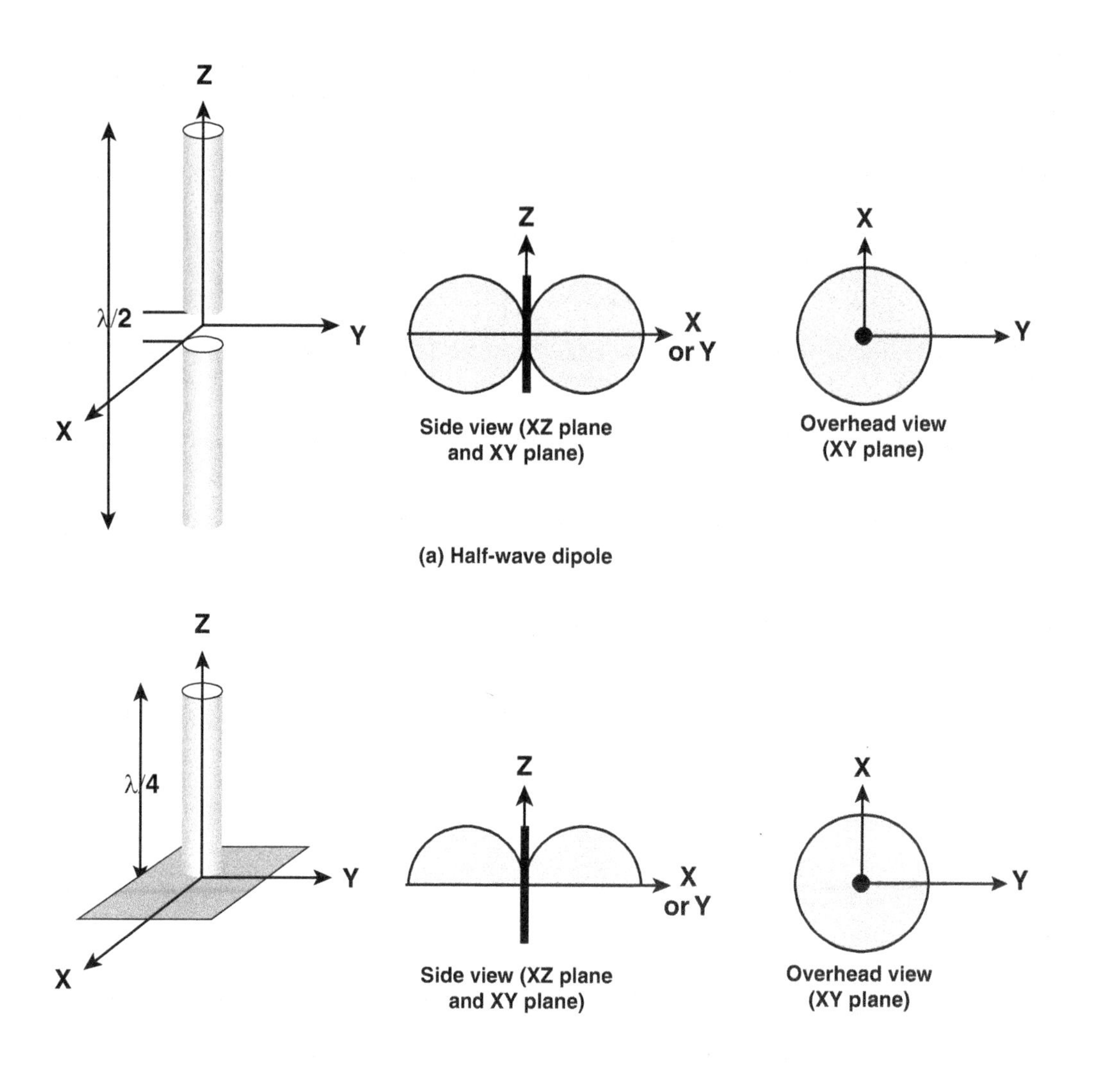

FIGURE 12.5 Simple Antennas

If multiple antennas are configured in an array of antennas, these multiple antennas can produce a directional beam. The radiation of the electric field from a single antenna is, using complex mathematics representation:

$$E = \frac{E_0}{d}\mathrm{Re}\left[\exp\left(j(\omega t - \frac{2\pi d}{\lambda}\right)\right] = \frac{E_0}{d}\cos\left(\omega t - \frac{2\pi d}{\lambda}\right) \quad \textbf{(Equation 12.1)}$$

With multiple antennas, the signals to each antenna can be adjusted with complex weights z_k to impose certain phase, amplitude, and time delay such that the sum of the antenna patterns sends or listens more strongly in a certain direction. This results in:

$$E = \mathrm{Re}\left[E_0 \exp(j\omega t)\sum_{k=1}^{N} z_k \frac{1}{d_k}\exp\left(-\frac{j2\pi d_k}{\lambda}\right)\right] \quad \textbf{(Equation 12.2)}$$

where d_k is the distance from each antenna element to the receiver. The weights are optimized according to different criteria. For example, for antennas placed in a **linear array**, a typical directional radiation pattern is shown in Figure 12.6. This pattern produces a **main lobe** that is 60° wide. This requires four antennas and is produced from a linear array where the antennas are spaced apart by a half of a wavelength. In this example, the main strength of the antenna is in the x direction. Notice that some energy is sent to the sides and back of the antenna in what are called the *sidelobes*. There are also, however, nulls in the patterns where very little signal energy is sent in those directions.

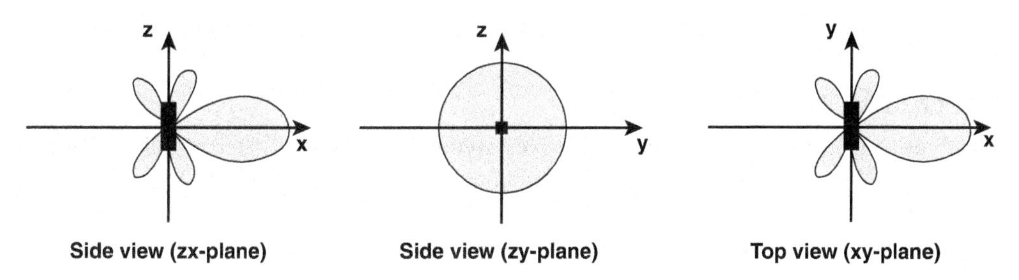

FIGURE 12.6 Radiation Patterns in Three Dimensions

Directional Antennas

Directional antennas are becoming increasingly practical and useful in modern systems, but they have actually been used for many years. For example, a typical cellular coverage area is split into three 120° sectors using three sets of directional antennas in a triangular configuration of antennas. For modern applications, directional antennas can be dynamically configured to follow individuals or groups of users to provide strong gain in intended directions and nulls toward interferers. These would be considered adaptive antenna arrays or switched antenna arrays.

Beginning with 4G, base stations with the capability of six sectors per cell have appeared, as shown in Figure 12.7. At the cost of additional complexity, the six-sector cell provides for an increased number of simultaneous users [HAQU11].

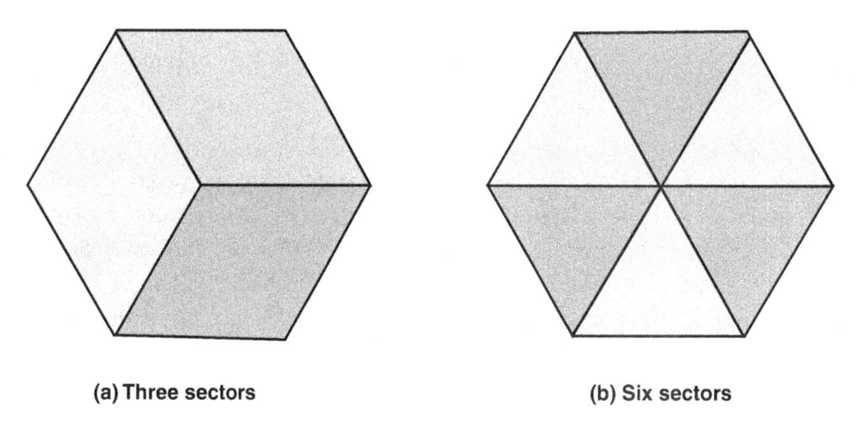

FIGURE 12.7 Different Cell Sectorization Schemes

Parabolic Reflective Antenna

An important type of antenna is the **parabolic reflective antenna**, which is used in terrestrial microwave and satellite applications. A parabola is the locus of all points equidistant from a fixed line and a fixed point not on the line. The fixed point is called the *focus*, and the fixed line is called the *directrix* (see Figure 12.8a). If a parabola is revolved about its axis, the surface generated is called a *paraboloid.* A cross section through the paraboloid parallel to its axis forms a parabola, and a cross section perpendicular to the axis forms a circle. Such surfaces are used in automobile headlights, optical and radio telescopes, and microwave antennas because of the following property: If a source of electromagnetic energy (or sound) is placed at the focus of the paraboloid, and if the paraboloid is a reflecting surface, the wave will bounce back in lines parallel to the axis of the paraboloid; Figure 12.8b shows this effect in cross section. In theory, this effect creates a parallel beam without dispersion. In practice, there is some dispersion because the source of energy must occupy more than one point. The converse is also true: If incoming waves are parallel to the axis of the reflecting paraboloid, the resulting signal is concentrated at the focus.

Figure 12.8c shows a typical radiation pattern for the parabolic reflective antenna, and Table 12.1 lists beamwidths for antennas of various sizes at a frequency of 12 GHz. Note that the larger the diameter of the antenna, the more tightly directional is the beam.

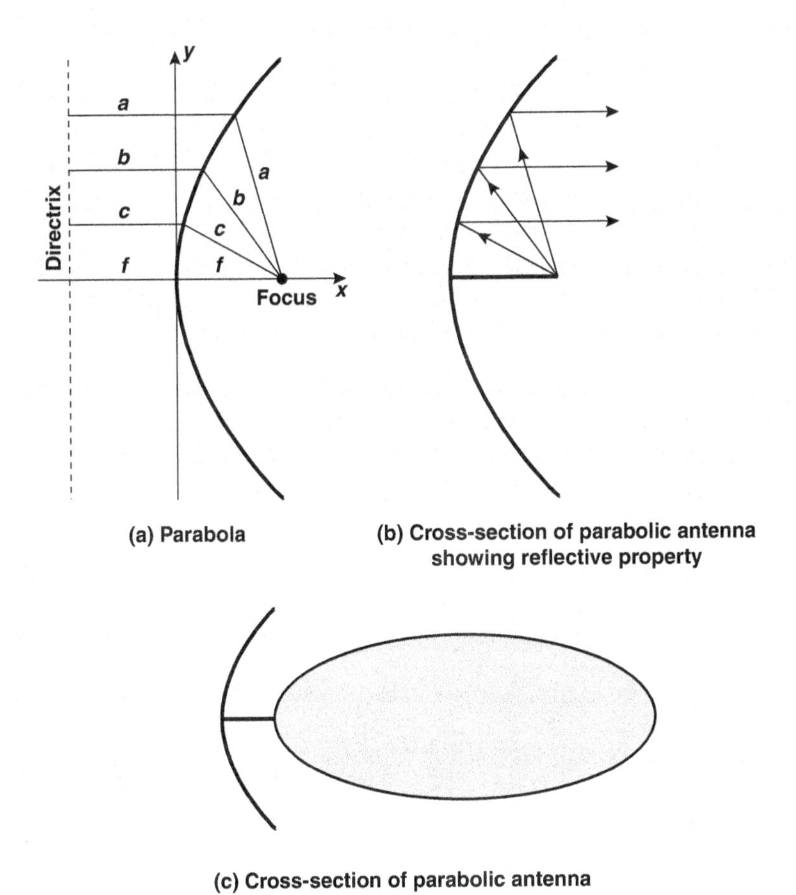

FIGURE 12.8 Parabolic Reflective Antenna

TABLE 12.1 Antenna Beamwidths for Various Diameter Parabolic Reflective Antennas at f = 12 GHz

Antenna Diameter (m)	Beamwidth (degrees)
0.5	3.5
0.75	2.33
1.0	1.75
1.5	1.166
2.0	0.875
2.5	0.7
5.0	0.35

Antenna Gain

Antenna gain is a measure of the directionality of an antenna. Antenna gain is defined as the power output, in a particular direction, compared to that produced in any direction by a perfect omnidirectional antenna (i.e., isotropic antenna). For example, if an antenna has a gain of 3 dB, that antenna improves upon the isotropic antenna in that direction by 3 dB, or a factor of 2. The increased power radiated in a given direction occurs at the expense of other directions. In effect, increased power is radiated in one direction by reducing the power radiated in other directions. It is important to note that antenna gain does not refer to obtaining more output power than input power but rather to directionality of output power.

A concept related to antenna gain is the **effective area** of an antenna. If we picture energy radiating outward in a spherical shape, the effective area is the surface area on that sphere where energy can be harvested. The effective area of an antenna is related to the physical size of the antenna and to its shape. The relationship between antenna gain and effective area is:

$$G = \frac{4\pi A_e}{\lambda^2} = \frac{4\pi f^2 A_e}{c^2}$$ **(Equation 12.3)**

where:

G = antenna gain

A_e = effective area

f = carrier frequency

c = speed of light ($\approx 3 \times 10^8$ m/s)

λ = carrier wavelength

Table 12.2 shows the antenna gains and effective areas for some typical antenna shapes.

TABLE 12.2 Antenna Gains and Effective Areas

Type of Antenna	Effective Area A_e (m^2)	Power Gain (relative to isotropic)
Isotropic	$\lambda^2/4\pi$	1
Infinitesimal dipole or loop	1.5 $\lambda^2/4\pi$	1.5
Half-wave dipole	1.64 $\lambda^2/4\pi$	1.64
Horn, mouth area *A*	0.81 *A*	10 A/λ^2
Parabolic, face area *A*	0.56 *A*	7 A/λ^2
Turnstile (two crossed, perpendicular dipoles)	1.15 $\lambda^2/4\pi$	1.15

Example

For a parabolic reflective antenna with a diameter of 2 m, operating at 12 GHz, what is the effective area and the antenna gain? According to Table 12.2, for a parabolic antenna, we use the face area, which is circular. We have an area of $A = \pi r^2 = \pi$ and an effective area of $A_e = 0.56\pi$. The wavelength is $\lambda = c/f = (3 \times 10^8)/(12 \times 10^9) = 0.025$ m. Then:

$$G = (7A)/\lambda^2 = (7 \times \pi)/(0.025)^2 = 35{,}186$$

$$G_{dB} = 45.46 \text{ dB}$$

12.3 Multiple-Input/Multiple-Output (MIMO) Antennas

If a transmitter and receiver implement a system with multiple antennas, this is called a **multiple-input/multiple-output (MIMO)** system. Such systems make it possible to implement several of the mechanisms discussed in Section 12.1. 5G systems make extensive use of MIMO antenna systems. Key features are base station antennas consisting of large arrays of antennas, the use of beamforming, and the use of beam management, all of which are described subsequently.

Section 12.2 describes the use of an array of antennas to provide a directional antenna pattern. As illustrated in Figure 12.9, three other important uses of antenna arrays are also possible.

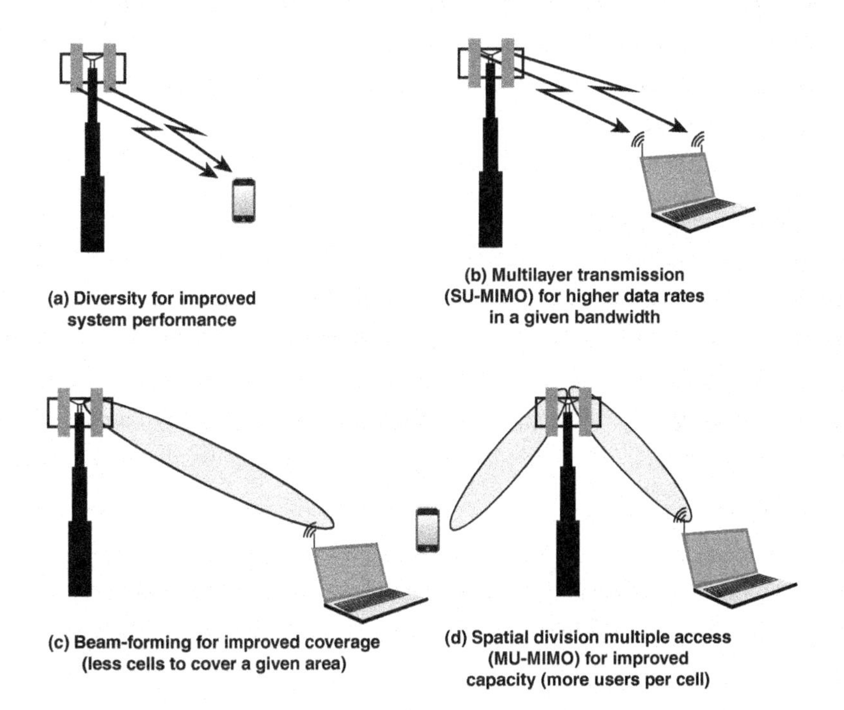

FIGURE 12.9 Four Uses of MIMO

The four uses of MIMO are as follows:

- **Diversity:** Space diversity can be accomplished in order to allow multiple received signals through multiple transmit and/or receive antennas. If spacing cannot be achieved for full signal independence, some benefits of space diversity can still be achieved.
- **Multiple streams:** Multiple parallel data streams can flow between pairs of transmit and receive antennas.
- **Beamforming:** Multiple antennas can be configured to create directional antenna patterns to focus and increase energy to intended recipients.
- **Multiple-user MIMO:** With enough MIMO antennas, directional antenna beams can be established to multiple users simultaneously.

Modern systems implement up to 4×4 (4 input, 4 output) and 8×8 MIMO configurations. Antenna systems have been approved in specifications for as many as 8 per antenna array, and two-dimensional arrays of 64 antennas or more are being envisioned for 5G.

The MIMO antenna architecture has become a key technology in evolving high-speed wireless networks, including IEEE 802.11 Wi-Fi LANs and 4G and 5G. MIMO exploits the space dimension to improve wireless systems in terms of capacity, range, and reliability. Together, MIMO and OFDM technologies form the cornerstone of emerging broadband wireless networks.

MIMO Principles

In a MIMO scheme, the transmitter and receiver employ multiple antennas. The source data stream is divided into *n* substreams, one for each of the *n* transmitting antennas. The individual substreams are the input to the transmitting antennas (multiple input). At the receiving end, *m* antennas receive the transmissions from the *n* source antennas via a combination of line-of-sight transmission and multipath, as shown in Figure 12.10. The output signals from all of the *m* receiving antennas (multiple output) are combined. With a lot of complex math, the result is a much better receive signal than can be achieved with either a single antenna or multiple frequency channels. Note that the terms *input* and *output* refer to the input to the transmission channel and the output from the transmission channel, respectively.

MIMO systems are characterized by the number of antennas at each end of the wireless channel. Thus, a 8×4 MIMO system has 8 antennas at one end of the channel and 4 at the other end. In configurations with a base station, the first number typically refers to the number of antennas at the base station. There are two types of MIMO transmission schemes:

- **Spatial diversity:** The same data is coded and transmitted through multiple antennas, which effectively increases the power in the channel proportionally to the number of transmitting antennas. This improves the signal-to-noise ratio (SNR) for cell edge performance. Further, diverse multipath fading offers multiple "views" of the transmitted data at the receiver, thus increasing robustness. In a multipath scenario where each receiving antenna would experience a different interference environment, there is a high probability that if one antenna is suffering a high level of fading, another antenna has sufficient signal level.

- **Spatial multiplexing:** A source data stream is divided among the transmitting antennas. The gain in channel capacity is proportional to the available number of antennas at the transmitter or receiver, whichever is less. Spatial multiplexing can be used when transmitting conditions are favorable and for relatively short distances compared to spatial diversity. The receiver must do considerable signal processing to sort out the incoming substreams, all of which are transmitting in the same frequency channel, and to recover the individual data streams.

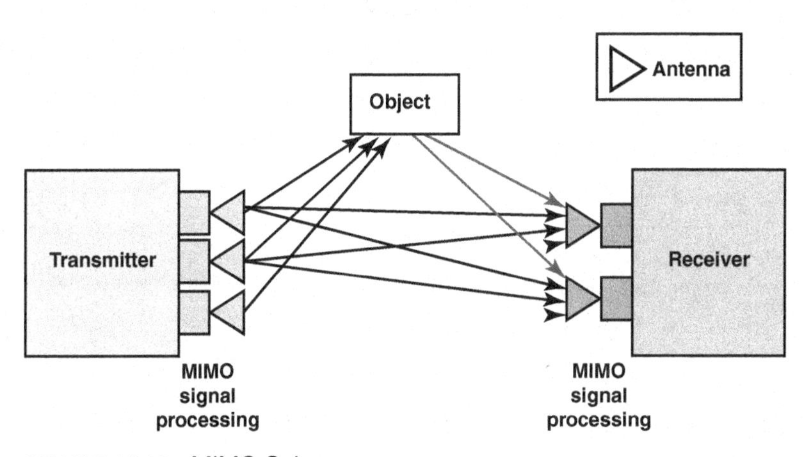

FIGURE 12.10 MIMO Scheme

For spatial multiplexing, there is a multilink channel that can be expressed as $\mathbf{y} = \mathbf{Hc} + \mathbf{n}$, where $\mathbf{y}$ is the vector of received signals, $\mathbf{c}$ is the vector of transmitted signals, $\mathbf{n}$ is an additive noise component, and $\mathbf{H}$ = [hij] is an $r \times t$ channel matrix, with r being the number of receiving antennas and t the number of transmitting antennas. The number of spatial data streams is min[r, t]. Figure 12.11 shows a channel with three transmitters and four receivers. The equation for this configuration is:

$$\begin{bmatrix} y_1 \\ y_2 \\ y_3 \\ y_4 \end{bmatrix} = \begin{bmatrix} h_{11} & h_{12} & h_{13} \\ h_{21} & h_{22} & h_{23} \\ h_{31} & h_{32} & h_{33} \\ h_{41} & h_{42} & h_{43} \end{bmatrix} \begin{bmatrix} c_1 \\ c_2 \\ c_3 \end{bmatrix} + \begin{bmatrix} n_1 \\ n_2 \\ n_3 \end{bmatrix}$$

h_{ij} are complex numbers x+jz that represent both the mean amplitude attenuation (x) over the channel and the path-dependent phase shift (z), and the n_i are additive noise components. The receiver measures the channel gains based on training fields containing known patterns in the packet preamble and can estimate the transmitted signal. The mathematics of the estimation is beyond the scope of this book.

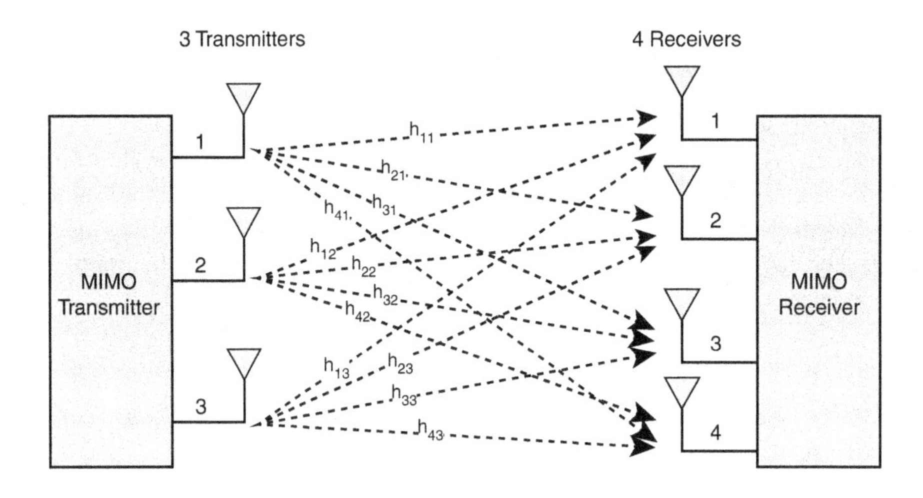

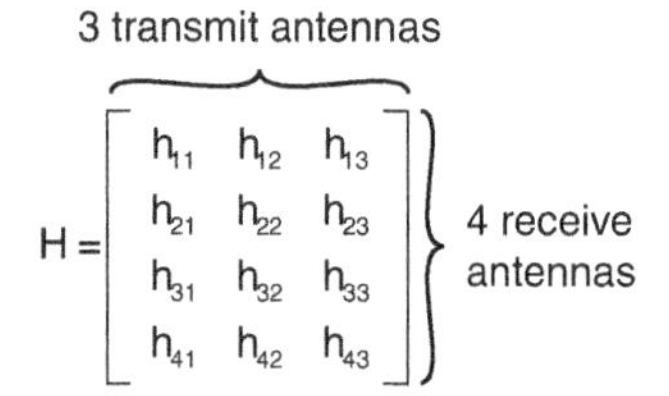

FIGURE 12.11 3×4 MIMO Scheme

Multiple-User MIMO

Multiple-user MIMO (MU-MIMO) extends the basic MIMO concept to multiple endpoints, each with multiple antennas. The advantage of MU-MIMO compared to single-user MIMO is that the available capacity can be shared to meet time-varying demands. MU-MIMO techniques are used in both Wi-Fi and 4G and 5G cellular networks.

There are two applications of MU-MIMO:

- **Uplink—multiple access channel (MAC):** With MIMO-MAC, multiple end users transmit simultaneously to a single base station.
- **Downlink—broadcast channel (BC):** With MIMO-BC, the base station transmits separate data streams to multiple independent users.

MIMO-MAC is used on the uplink channel to provide multiple access to subscriber stations. In general, MIMO-MAC systems outperform point-to-point MIMO, particularly if the number of receiver antennas is greater than the number of transmit antennas at each user. A variety of multiuser detection techniques are used to separate the signals transmitted by the users.

MIMO-BC is used on the downlink channel to enable the base station to transmit different data streams to multiple users over the same frequency band. MIMO-BC is more challenging to implement than MIMO-MAC. The techniques employed involve processing of the data symbols at the transmitter to minimize interuser interference.

12.4 Advanced Cellular Antennas

This section examines advanced cellular antennas used in 5G networks. The section begins with an overview of the evolution of cellular antenna technology. The remainder of the section examines key aspects of advanced antenna technology: beamforming, active antenna systems, and massive MIMO.

Evolution of Cellular Antennas

Figure 12.12, from [SAUN18], indicates that cellular antenna technology has in essence gone through four generations[1]:

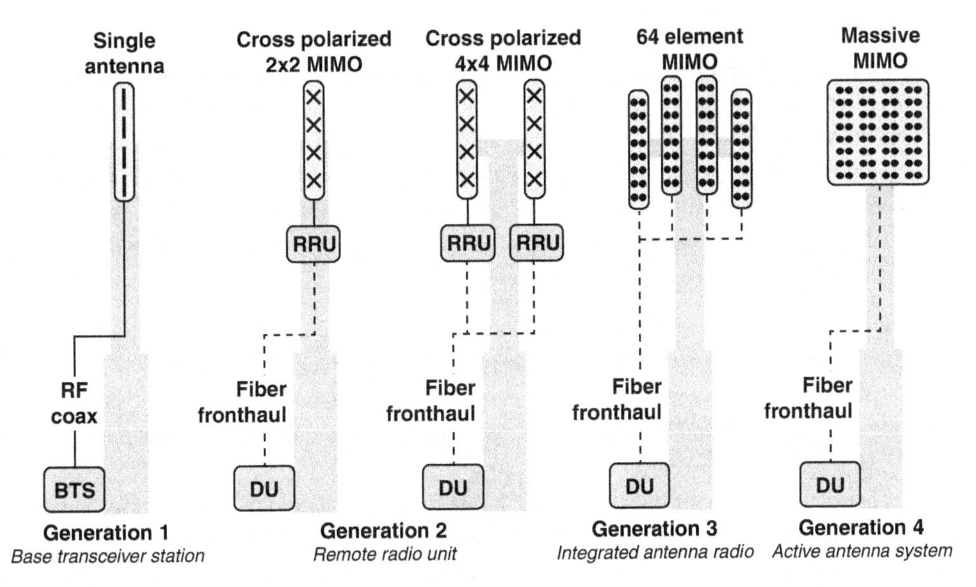

FIGURE 12.12 Progression of Antenna Technology

1. This designation of antenna technology generations does not have a one-to-one correspondence with the overall designation of cellular network generations.

The generations are as follows:

- **Generation 1:** Traditional cellular networks have mainly used sectorized antennas (three sectors) that offer frequency performance adequate to provide specific service for a specific carrier's frequency band. A typical configuration uses a base transceiver station that transmits radio-frequency signals through coaxial cable to a single antenna unit.
- **Generation 2:** The advent of 4G cellular networks brought the introduction of passive MIMO technologies and carrier aggregation, which allowed for devices to connect over multiple spatial multiplexed paths and over several frequency bands. In many countries, it is still the case that only 2×2 MIMO and limited carrier aggregation (due to user equipment [UE] limitations) are deployed [MILL20]. Generation 2 replaces the radio unit from the indoor enclosure at the base of a tower with a remote radio unit (RRU) at the tower top below the antenna. RRU replaces coaxial feeder cables with fiber-optic cable interconnects. The RRU enables the system to deploy one or more antennas at a distance from the base station. In the case of 4×4 MIMO and 8×8 MIMO, multiple remote units are connected to multiple MIMO antennas.
- **Generation 3:** The Generation 2 arrangement can, in principle, scale linearly with the increase in MIMO complexity. The disadvantages are cell tower clutter and cost and complexity issues. To overcome these disadvantages, the third generation of antenna technology provides more compact antenna systems. Integrated antenna radio (IAR) integrates the radio unit and antenna within the radome[2] where the radio interfaces with an antenna array. This arrangement does away with separate remote units, instead providing integrated transceivers, MIMO, and beamforming hardware in the same assembly as the antenna. The arrangement also eliminates the need for separate coax cables for each antenna element.
- **Generation 4**: To support mmWave, 5G requires the use of massive MIMO antennas of at least 64×64. Generation 4 integrates multiple radio transceivers inside the antenna, where each radio interfaces with a dedicated antenna element to form an array. Massive MIMO provides more sophisticated beamforming and beam management capability and narrower antenna patterns.

Each base station generation provides improvements in one or more critical areas: better radio performance, lower operating power, reduced size, or faster installation time [5GAM19]. For example, the transition from BTS to RRU saw a 50% cut in power consumption and 3 dB reduction in downlink loss. The transition from RRU to IAR saw a 40% reduction in size, 8% lower power, and 1 dB improvement in downlink loss. Generation 4 AAS achieves yet a higher level of performance.

Beamforming

Beamforming is one of the essential technologies in developing advanced cellular antenna systems. This subsection provides an overview.

---

2. A radome (radar dome) is a structural, weatherproof enclosure that protects a radar system or antenna and is constructed of material that minimally attenuates the electromagnetic signal transmitted or received by the antenna. Radomes protect antenna surfaces from weather and conceal antenna electronic equipment from public view.

Basic Principles

Beamforming is a technique by which an array of antennas can be steered to transmit radio signals in a specific direction. Rather than simply broadcasting energy/signals in all directions, the antenna arrays that use beamforming determine the direction of interest and send/receive a stronger beam of signals in that specific direction.

In this technique, each antenna element is fed separately with the signal to be transmitted. The phase and amplitude of each signal are then added constructively and destructively in such a way that they concentrate the energy into a narrow beam or lobe. The various transmitted signals merge in the air by normal coherence of the electromagnetic waves, thereby forming a virtual beam in a predetermined direction. To understand how this procedure works, consider a signal that is fed to different antenna elements shifted in phase different amounts for each element. Now picture the transmitted energy from each element at an angle of 45°. At any point along that 45° line, the distance traveled by electromagnetic waves from different antenna elements is not equal. If the phase shifting is such that at 45°, signals from all antenna elements arrive at the same phase, then the beam is strongest in that direction.

Figure 12.13 indicates the effect. There is one main lobe in the direction where the multiple signals reinforce each other to the maximum extent. In other directions, there is more interference and cancelation between signals, to a greater or lesser degree, forming weaker side lobes. In Figure 12.13, only three side lobes are depicted, although there are in fact numerous such lobes, each significantly weaker than the main lobe. In general, the main lobe becomes more dominant and narrower as more antenna elements are added to the array.

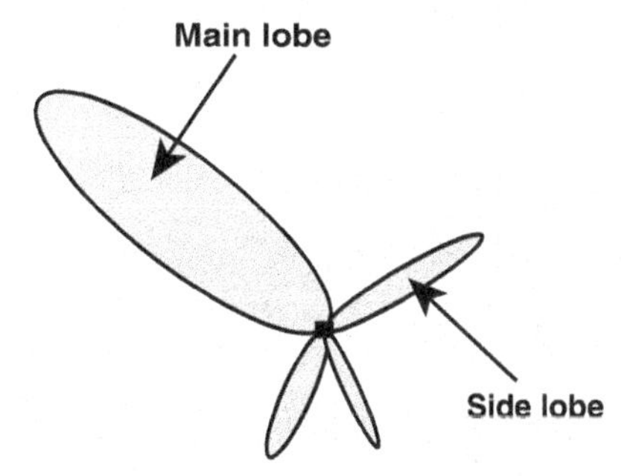

FIGURE 12.13 Illustration of Beamforming

Beamforming has the following advantages:

- **Higher SNR:** The highly directional transmission enhances the link budget, improving the range for both open-space and indoor penetration.
- **Interference prevention and rejection:** Beamforming prevails over cochannel interference (CCI), explained subsequently, by taking advantage of the antennas' spatial properties.

- **Higher network efficiency:** By substantially minimizing CCI, beamforming allows much denser deployments than are possible with single-antenna systems.

Cochannel interference can occur when the same frequency band is used in two cells that are relatively near each other. If conditions are right, the energy from a signal in one cell may be sufficient to be noticeable in the other cell and interfere with the local signal, generating errors. With highly directional beams, the possibility of interference is dramatically reduced.

Beam Management

Beam management refers to techniques and processes used to achieve the transmission and reception of data over relatively narrow beams. Beamforming and beam management are essential for using the mmWave region over the 5G air interface. Narrow beams are needed to compensate for high path loss and blockage. With the use of narrow beams, and especially if the UE is mobile, beam management provides the means for both the base station antenna and the onboard UE antenna to "lock on" to a beam that provides an optimal path from transmitter to receiver.

By adjusting the phase and amplitude parameters, a MIMO antenna can generate multiple beams, with each beam covering part of the cell area. For downlink transmission, the objective of beam management is to select a transmit beam to a UE so that the UE can receive the signal with the highest power and best SNR. For uplink transmission, the base station attempts to choose the receive beam for a UE with the best receive beamforming gain. Similarly, if the UE antenna system is capable of beamforming, the UE can utilize the beams to improve link quality.

The beam management procedure involves beamforming, beam sweeping, beam measurement, beam determination, and beam reporting, as shown in Figure 12.14, from [YUE17], which indicates the following elements in the context of downlink transmission:

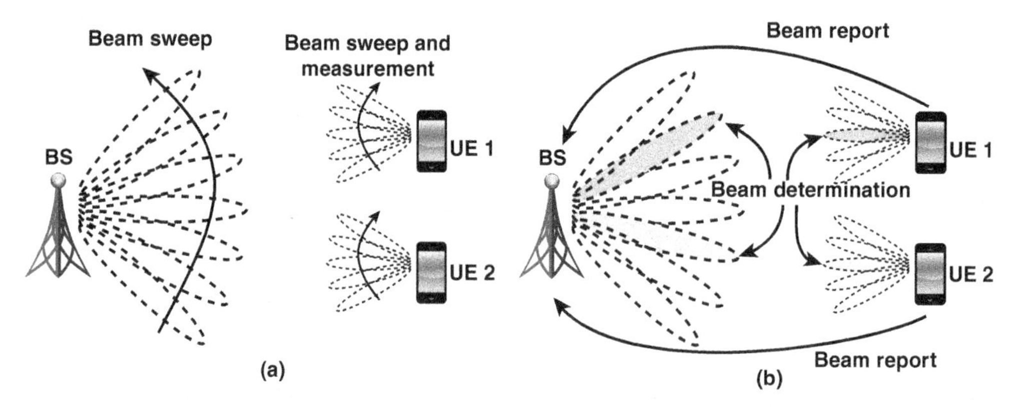

FIGURE 12.14 Beam Management Procedures with Downlink Transmissions

- **Beam sweeping:** The base station antenna (i.e., the 5G radio access network node gNB) transmits beams in a predetermined sequence for beam measurement at the UE side.
- **Beam measurement:** The UE measures the characteristics of received beamformed signals.
- **Beam determination:** The UE selects the optimal beam. In essence, the UE isolates the receive beam, which affords the best reception. The best results are obtained when the transmitting and receiving beam pair is optimal for the location of the UE at the time.
- **Beam reporting:** The UE reports back to the gNB the information based on beam measurement.

Beam management is an ongoing dynamic process that involves selecting an initial beam pair and then modifying the selection as transmit/receive conditions change.

FD-MIMO

The term full-dimension MIMO (FD-MIMO), or 3D-MIMO, refers to a MIMO antenna system that is capable of varying the direction of a beam in both horizontal (azimuth) and vertical (elevation) dimensions, as shown in Figure 12.15. Thus, FD-MIMO can project a beam in any direction in three-dimensional space.

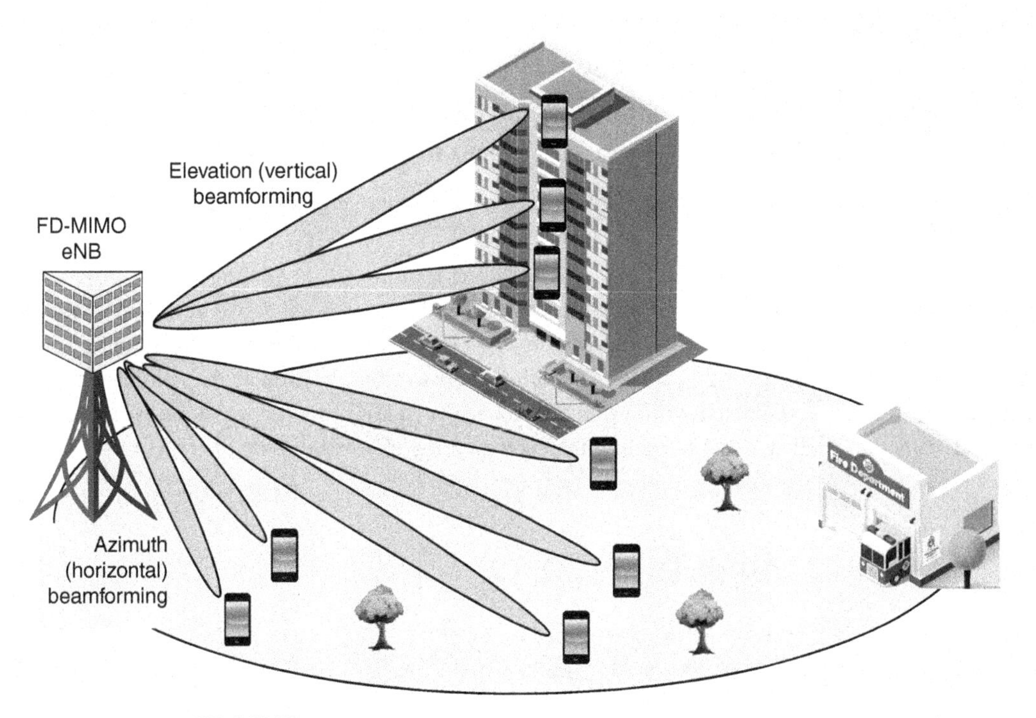

FIGURE 12.15 FD-MIMO

Earlier MIMO schemes are based on two-dimensional horizontal beamforming. The increase in the number of antenna elements in more recent MIMO systems makes it possible to exploit the vertical dimension for beamforming. This has advantages, especially in dense urban environments. The ability to adjust transmitted beams in the vertical dimension can improve the received signal power of terminals deep inside high-rise buildings and help overcome some of the building penetration loss. FD beamforming is also advantageous in indoor deployments in high-rise buildings, where a single base station may be able to optimize coverage over more than one floor. Such techniques directly increase spectral efficiency.

A distinguishing, and necessary, characteristic of an FD-MIMO antenna is that the antenna elements are arranged in a two-dimensional (2D) array, in contrast to the linear layout of earlier MIMO systems. An advantage of the 2D antenna array is that it reduces the form factor of the antenna compared to a linear antenna array used in earlier systems.

Active Antenna Systems

An active antenna system (AAS) is one in which radio-frequency (RF) components, such as power amplifiers and transceivers, are integrated with an array of antenna elements. There are a number of benefits to this, including the following:

- The site footprint is reduced.
- The distribution of radio functions to the individual antennas within the radome results in built-in redundancy and improved thermal performance.
- Distributed transceivers support advanced beamforming features and enable FD-MIMO.
- Integrating the active transceiver array and passive antenna array into one radome reduces cable losses.

Figure 12.16 provides a simplified view of the difference between passive and active antenna systems. In the passive system, only the phase shifter is incorporated with the radiating antenna element in the radome. The oscillator, or exciter, that provides the reference waveform as well as the auxiliary high-frequency clock signals is collocated with a transmitter/receiver (T/R) module. In an active antenna system, the T/R module is distributed to each of the antenna array elements inside the radome.

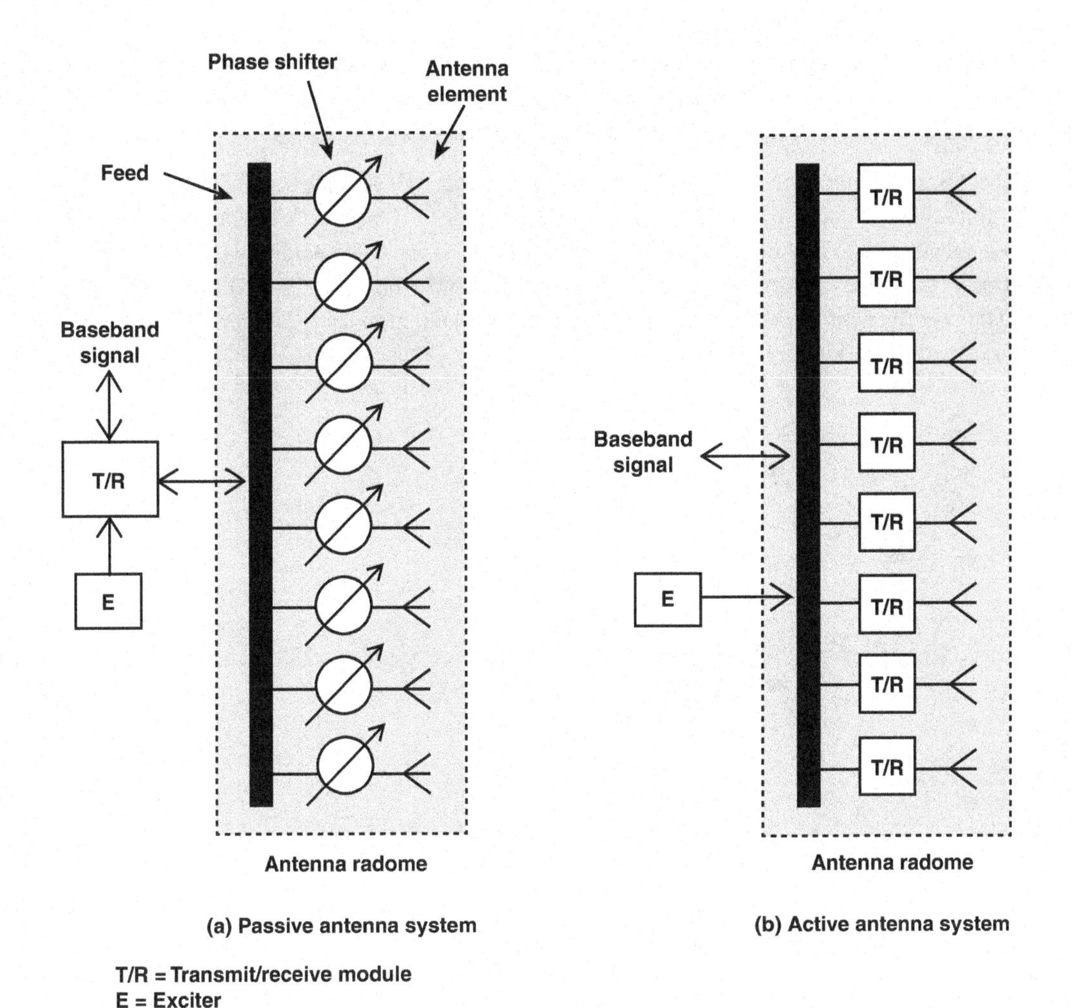

FIGURE 12.16 Simplified View of Passive and Active Antenna Systems

Massive MIMO

By using AAS technology, it is possible to deploy 2D arrays with large numbers of antennas placed on a plane; this is referred to as a massive MIMO (mMIMO) configuration. A greater number of transceivers (TRx) on an antenna means more degrees of freedom to modify the radiation pattern of the transmitted signal based on where the receiver is located.

Massive MIMO is a scaled-up version of the conventional small-scale MIMO systems. As shown in Figure 12.17, from [ALBR19], an mMIMO system incorporates a large number (practically some dozens or hundreds—and theoretically up to thousands) of base station antenna elements to serve simultaneously multiple users with a flexibility to determine which users to schedule for reception at any given time. The most common mMIMO concept assumes that the user terminals have just a single antenna and that the number of antennas at the base station is significantly larger than the number of served users.[3] However, mobile phones are increasingly equipped with multiple antennas, providing more flexibility.

Massive MIMO is implemented in two different duplexing approaches—frequency-division duplexing (FDD) and time-division duplexing (TDD)—both introduced in Section 1.6 in Chapter 1, "Cellular Networks: Concepts and Evolution." FDD uses different frequencies for downlink (DL) and uplink (UL). TDD uses the same frequency for both UL and DL, with a block of time allocated in each direction. TDD benefits from channel reciprocity: The channel estimation using UL can be utilized for DL beamforming, thus leading to less overhead.

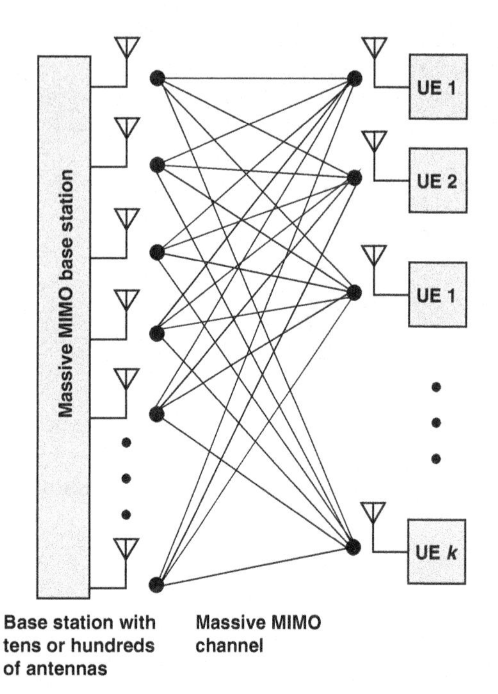

FIGURE 12.17 Massive MIMO Logical Architecture

As the number of antennas in a MIMO system increases, there is an increase in three types of gain [BARR16, BUSA18]:

- **Antenna gain:** Increasing the number of antennas increases the antenna gain, making it possible to radiate energy in a more directional manner toward the user, reducing system interferences and increasing capacity. For multiuser MIMO, array gain is proportional to the number of antennas, provided that the base station/UE channels exhibit orthogonality.

---

3. Thus, the term *massive MIMO* is often a misnomer, due to the single-antenna UE. Often the system is actually a multiuser single-input/multiple-output (SIMO) uplink or multiple-input/single-output (MISO) downlink. As is customary in the literature, this text refers to both single- and multiple-antenna terminal cases as massive MIMO.

- **Diversity gain:** Diversity gain refers to the ability to exploit spatial diversity. As discussed in Section 12.3, spatial diversity is achieved by sending multiple versions of the same signal through different antennas and combining these at the receiver, improving signal robustness. For a MIMO system with M antennas at the transmitter and N antennas at the receiver, the maximum diversity gain is $M \times N$. The maximum gain is achieved if the $M \times N$ fading coefficients are independent.
- **Multiplexing gain:** As discussed in Section 12.3, spatial multiplexing is achieved by transmitting different portions of a data stream between different transmitter/receiver pairs. Thus spatial multiplexing reuses the frequency band, creating multiple communications channels and increasing capacity. Depending on SNR characteristics, MIMO can produce up to min[M, N] independent data streams.

However, increasing the number of antennas introduces the following technical challenges [LU14, BUSA18]:

- **Computational complexity:** The computational complexity for channel estimation, signal processing, and other tasks at the base station increases with increasing numbers of antennas and users.
- **Channel estimation:** For multiuser MIMO, channel estimation refers to determining the transmission characteristics for the air link between each UE and the base station. MIMO systems typically use TDD, especially as the number of antennas increases. Figure 12.18 shows a simplified view of the time division used on a given link, forming a frame with three phases of operation. The pilot subframe is a sequence of predefined symbols used by the base station to estimate channel status information. Each UE sends a different sequence of pilot symbols so that the base station can discriminate among users.

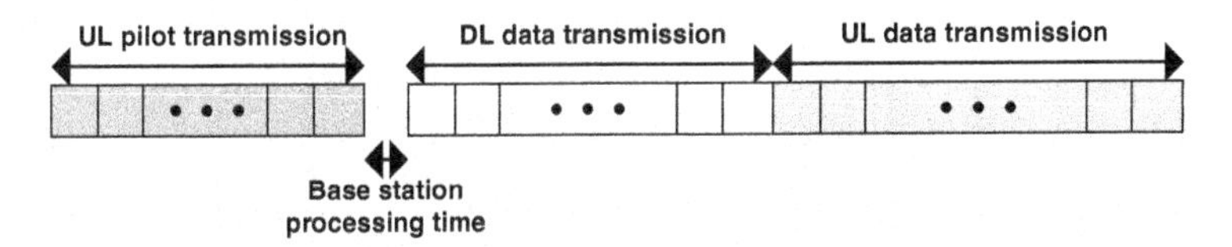

FIGURE 12.18 Simplified Structure for MIMO TDD Frame

- **Pilot contamination:** Within each sector or cell, the pilots from the various UEs are orthogonal to prevent interference. However, there is the potential for interference between pilots in adjacent cells, called pilot contamination. [LU14] discusses a number of possible mitigation methods.

Table 12.3, from [BUSA18], summarizes the basic benefits and challenges of the different antenna technologies.

TABLE 12.3 Benefits and Challenges for Antenna Technologies*

Antenna Technology	SISO	SU-MIMO	MU-MIMO	mMIMO
Diversity gain	×	✓	✓✓	✓✓✓✓
Multiplexing gain	×	✓✓	✓✓✓	✓✓✓✓
Array gain	×	✓✓	✓✓	✓✓✓✓
mmWave bandwidth	×	×	×	×
Computational complexity	×	××	×××	××××
Channel estimation challenge	×	××	×××	××××
Pilot contamination issue	×	××	×××	××××

* × = challenge
✓ = benefit

The number of symbols signifies the normalized quantity relative to SISO.

12.5 Key Terms and Review Questions

Key Terms

3D MIMO	isotropic antenna
active antenna system	linear array
adaptive equalization	main lobe
adaptive modulation and coding	massive MIMO (mMIMO)
antenna	millimeter wave (mmWave)
antenna gain	multiple-input/multiple-output (MIMO) antenna
beamwidth	multiple-user MIMO
beamforming	orthogonal frequency-division multiplexing (OFDM)
dipole	parabolic reflective antenna
directional antenna	radiation pattern
effective area	radome
error-correcting code	reception pattern
forward error correction	sidelobe
frequency diversity	
full-dimension MIMO (FD-MIMO)	

space diversity	spread spectrum
spatial diversity	time diversity
spatial multiplexing	time-division multiplexing (TDM)

Review Questions

1. Define adaptive equalization.
2. What is the difference between space, frequency, and time diversity?
3. What two functions are performed by an antenna?
4. What is an isotropic antenna?
5. What information is available from a radiation pattern?
6. What is the advantage of a parabolic reflective antenna?
7. What factors determine antenna gain?
8. What is meant by the term multiple-input/multiple-output antenna?
9. What distinguishes single-user MIMO from multiple-user MIMO?
10. Explain the two types of MIMO transmission schemes.
11. What are two applications of multiple-user MIMO?
12. What is beamforming?
13. Describe the advantages of beamforming.
14. What is beam management?
15. Describe the four elements of beam management for downlink transmission.
16. What is FD-MIMO?
17. Describe the benefits of an active antenna system compared to a passive antenna system.
18. What are the alternative duplexing approaches for implementing massive MIMO?

12.6 References

5GAM19 5G Americas. *Advanced Antenna Systems for 5G.* White paper, August 2019.

ALBR19 Albreem, M., Juntti, M., and Shahabuddin, S. "Massive MIMO Detection Techniques: A Survey." *IEEE Communications Surveys and Tutorials*, Fourth Quarter, 2019.

BARR16 Barreto, A., et al. "5G—Wireless Communications for 2020." *Journal of Communication and Information Systems*, no. 1, 2016.

BUSA18 Busan, S., et al. "Millimeter-Wave Massive MIMO Communication for Future Wireless Systems: A Survey." *IEEE Communications Surveys and Tutorials*, Second Quarter, 2018.

HAQU11 Haque, A., et al. "Performance Analysis of UMTS Cellular Network Using Sectorization Based on Capacity and Coverage." *International Journal of Advanced Computer Science and Applications*, vol. 2, no. 6, 2011.

LU14 Lu, L., et al. "An Overview of Massive MIMO: Benefits and Challenges." *IEEE Journal of Selected Topics in Signal Processing*, October 2014.

MILL20 Miller, M. "Antennas Evolve to Meet 5G Requirements." *High Frequency*, February 2020.

SAUN18 Saunders, J. *The Rise & Outlook of Antennas in 5G.* ABI Research white paper, June 2018. https://go.abiresearch.com/lp-rise-and-outlook-of-antennas-in-5g

YUE17 Yue, G., et al. "MIMO Technologies in 5G New Radio." *GetMobile*, March 2017.

Chapter 13

Air Interface Physical Layer

Learning Objectives

After studying this chapter, you should be able to:

- Describe the three major ways digital data can be encoded onto an analog signal
- Determine performance of modulation schemes from E_b/N_0 curves
- Understand the distinction between PSK and differential PSK
- Explain $\pi/2$-BPSK
- Present an overview of the concept of multicarrier modulation
- Understand the concepts of fast Fourier transform and inverse fast Fourier transform
- Discuss the nature and importance of peak-to-average power ratio
- Distinguish between OFDM, OFDMA, and SC-FDMA
- Explain the term *numerology* in the context of 5G

This chapter continues the treatment of the air interface, with a look at the two most important aspects of the physical layer specification: modulation and waveform definition. Section 13.1 provides an overview of modulation schemes, as well as a treatment of the specific techniques standardized for 5G by 3GPP. Section 13.2 introduces various waveform definitions based on orthogonal frequency-division multiplexing, and Section 13.3 discusses the specific OFDM techniques specified by 3GPP.

13.1 Modulation Schemes

3GPP TS 38.211 (*Technical Specification Group Radio Access Network, NR, Physical Channels and Modulation [Release 16]*, June 2020) specifies the modulation schemes supported for 5G, as shown in Figure 13.1. This section provides an overview of these schemes after first reviewing more elementary modulation techniques.

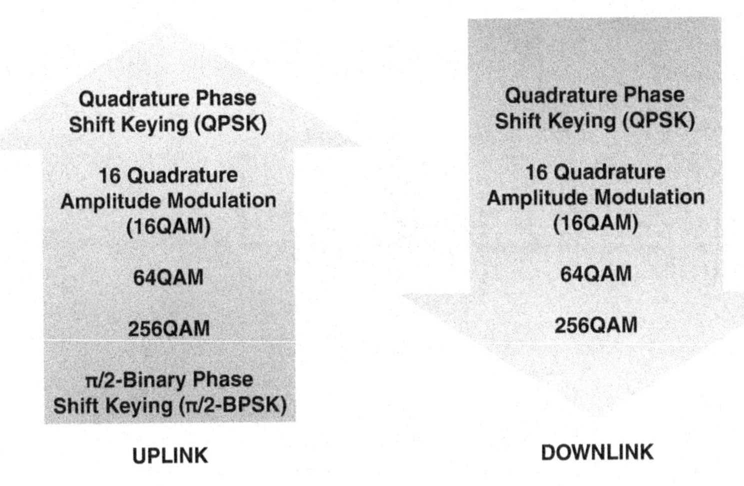

FIGURE 13.1 3GPP Standard Modulation Schemes for 5G

Modulation

The basis for transmission of data over a wireless channel is a continuous constant-frequency signal known as the **carrier signal**. The frequency of the carrier signal is chosen to be compatible with the transmission channel being used. Data are transmitted using a carrier signal and modulation. **Modulation** is the process of varying the amplitude, phase, and/or frequency of the carrier signal to convey analog or digital data. The modulating function, consisting of the analog or digital data, is called the **baseband signals**. There are three basic encoding or modulation techniques for transforming digital data into analog signals, as illustrated in Figure 13.2: amplitude-shift keying (ASK), frequency-shift keying (FSK), and phase-shift keying (PSK). In all these cases, the resulting signal occupies a bandwidth centered on the carrier frequency.

Amplitude-Shift Keying

In ASK, the two binary values are represented by two different amplitudes of the carrier frequency. Commonly, one of the amplitudes is zero; that is, one binary digit is represented by the presence, at constant amplitude, of the carrier, and the other is represented by the absence of the carrier (see Figure 13.2a). The resulting transmitted signal for one bit time is:

$$\textbf{ASK:}\quad s(t) = \begin{cases} A\cos(2\pi f_c t) & \text{binary 1} \\ 0 & \text{binary 0} \end{cases} \qquad \textbf{(Equation 13.1)}$$

where the carrier signal is $A\cos(2\pi f_c t)$. ASK is susceptible to sudden gain changes and is a rather inefficient modulation technique.

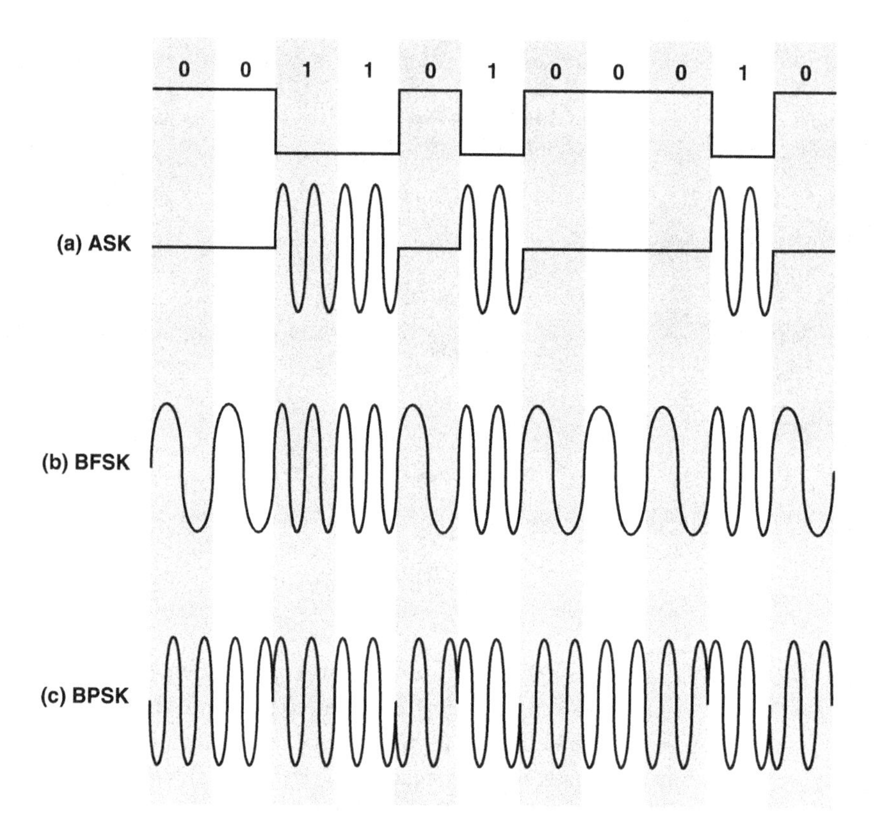

FIGURE 13.2 Modulation of Analog Signals for Digital Data

Frequency-Shift Keying

The most common form of FSK is binary FSK (BFSK), in which the two binary values are represented by two different frequencies near the carrier frequency (see Figure 13.2b). The resulting transmitted signal for one bit time is:

$$\textbf{BFSK:} \quad s(t) = \begin{cases} A\cos(2\pi f_1 t) & \text{binary 1} \\ A\cos(2\pi f_2 t) & \text{binary 0} \end{cases} \qquad \textbf{(Equation 13.2)}$$

where f_1 and f_2 are typically offset from the carrier frequency f_c by equal but opposite amounts.

Phase-Shift Keying

In PSK, the phase of the carrier signal is shifted to represent data.

Two-Level PSK

The simplest PSK scheme uses two phases to represent the two binary digits (see Figure 13.2c) and is known as binary phase-shift keying (BPSK). The resulting transmitted signal for one bit time is:

$$\textbf{BPSK:} \quad s(t) = \begin{cases} A\cos(2\pi f_c t) \\ A\cos(2\pi f_c t + \pi) \end{cases} = \begin{cases} A\cos(2\pi f_c t) & \text{binary 1} \\ -A\cos(2\pi f_c t) & \text{binary 0} \end{cases} \qquad \textbf{(Equation 13.3)}$$

Because a phase shift of 180° (π) is equivalent to flipping the sine wave or multiplying it by –1, the rightmost expressions in Equation (13.3) can be used. This leads to a convenient formulation. If we have a bit stream, and we define $d(t)$ as the discrete function that takes on the value of +1 for one bit time if the corresponding bit in the bit stream is 1 and the value of –1 for one bit time if the corresponding bit in the bit stream is 0, then we can define the transmitted signal as:

$$\textbf{BPSK:} \quad s_d(t) = A\, d(t) \cos(2\pi f_c t) \qquad \textbf{(Equation 13.4)}$$

An alternative form of two-level PSK is **differential PSK (DPSK)**. Figure 13.3 shows an example. In this scheme, a binary 0 is represented by sending a signal burst of the same phase as the previous signal burst sent. A binary 1 is represented by sending a signal burst of opposite phase to the preceding one. This term *differential* refers to the fact that the phase shift is with reference to the previous bit transmitted rather than to some constant reference signal. In differential encoding, the information to be transmitted is represented in terms of the changes between successive data symbols rather than the signal elements themselves. DPSK avoids the requirement for an accurate local oscillator phase at the receiver that is matched with the transmitter. As long as the preceding phase is received correctly, the phase reference is accurate.

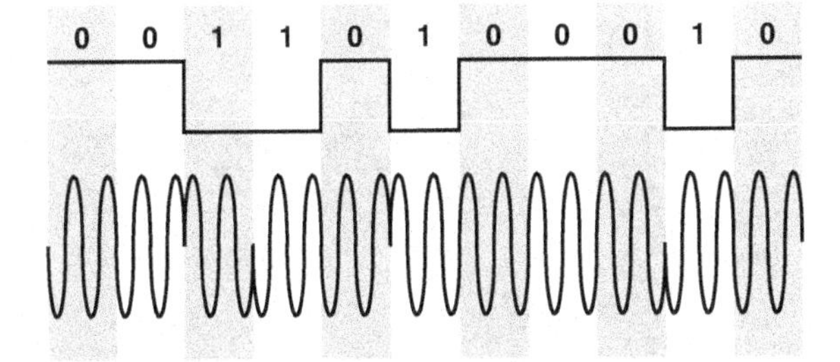

FIGURE 13.3 Differential Phase-Shift Keying (DPSK)

Four-Level PSK

More efficient use of bandwidth can be achieved if each signaling element represents more than one bit. For example, instead of a phase shift of 180°, as is allowed in BPSK, a common encoding technique known as quadrature phase-shift keying (QPSK) uses phase shifts separated by multiples of $\pi/2$ (90°).

$$\textbf{QPSK:}\quad s(t)=\begin{cases} A\cos\left(2\pi f_c t+\dfrac{\pi}{4}\right) & 11\\ A\cos\left(2\pi f_c t+\dfrac{3\pi}{4}\right) & 01\\ A\cos\left(2\pi f_c t-\dfrac{3\pi}{4}\right) & 00\\ A\cos\left(2\pi f_c t-\dfrac{\pi}{4}\right) & 10 \end{cases} \qquad \textbf{(Equation 13.5)}$$

Thus, each signal element represents two bits rather than one.

Figure 13.4 shows the QPSK modulation scheme in general terms, and a variant known as offset QPSK (OQPSK). For QPSK, the input is a stream of binary digits with a data rate of $R = 1/T_b$, where T_b is the width of each bit. This stream is converted into two separate bit streams of $R/2$ bps each, by taking alternate bits for the two streams. The two data streams are referred to as the I (in-phase) and Q (quadrature phase) streams. In Figure 13.4, the upper stream is modulated on a carrier of frequency f_c by multiplying the bit stream by the carrier. For convenience of modulator structure, we map binary 1 to $\sqrt{1/2}$ and binary 0 to $-\sqrt{1/2}$. Thus, a binary 1 is represented by a scaled version of the carrier wave, and a binary 0 is represented by a scaled version of the negative of the carrier wave, both at a constant amplitude. This same carrier wave is shifted by 90° and used for modulation of the lower binary stream. The two modulated signals are then added together and transmitted.

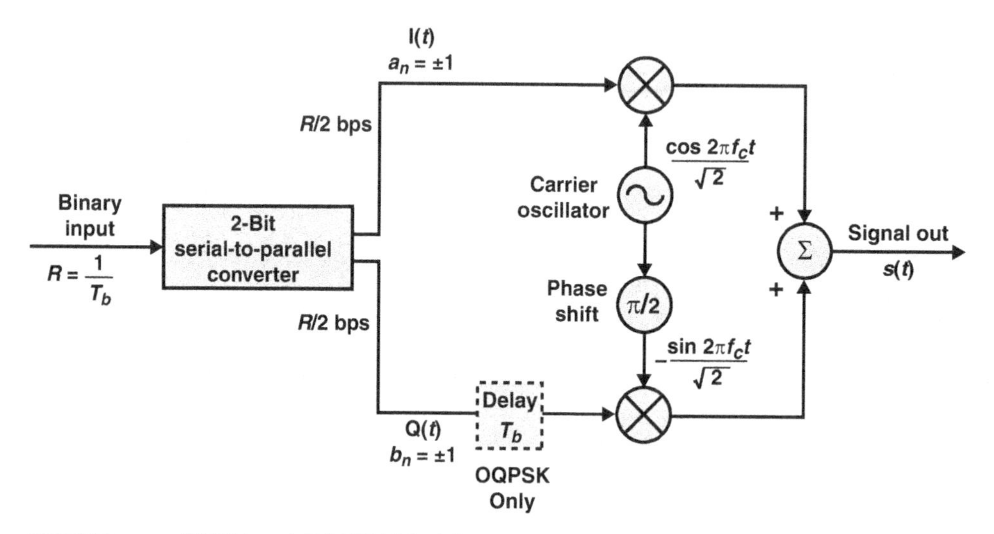

FIGURE 13.4 QPSK and OQPSK Modulators

The transmitted signal can be expressed as follows:

$$\textbf{QPSK:}\quad s(t)=\frac{1}{\sqrt{2}}I(t)\cos 2\pi f_c t-\frac{1}{\sqrt{2}}Q(t)\sin 2\pi f_c t$$

Figure 13.5 shows an example of QPSK coding. Each of the two modulated streams is a BPSK signal at half the data rate of the original bit stream. Thus, the combined signals have a symbol rate that is half the input bitrate. Note that from one symbol time to the next, a phase change of as much as 180° (π) is possible.

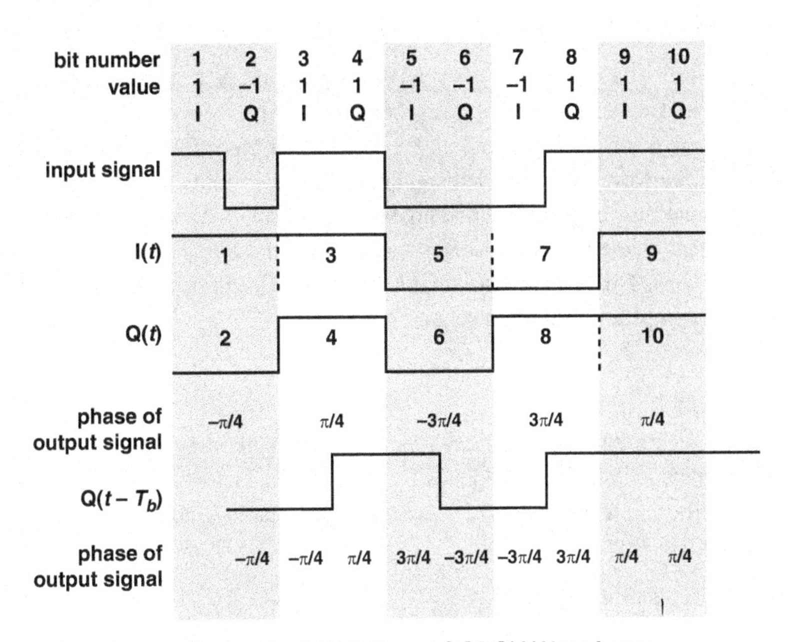

FIGURE 13.5 Example of QPSK and OQPSK Waveforms

Figures 13.4 and 13.5 also depict OQPSK. The difference between QPSK and OQPSK is that OQPSK introduces a delay of one bit time in the Q stream, resulting in the following signal:

$$\textbf{OQPSK:}\quad s(t) = \frac{1}{\sqrt{2}} I(t)\cos 2\pi f_c t - \frac{1}{\sqrt{2}} Q(t - T_b)\sin 2\pi f_c t$$

Because OQPSK differs from QPSK only by the delay in the Q stream, its spectral characteristics and bit error performance are the same as for QPSK. From Figure 13.5, we can observe that only one of two bits in the pair can change sign at any time, and thus the phase change in the combined signal never exceeds 90° (π/2). This can be an advantage because physical limitations on phase modulators make large phase shifts at high transition rates difficult to perform. OQPSK also provides superior performance when the transmission channel (including transmitter and receiver) has significant nonlinear components. The effect of nonlinearities is a spreading of the signal bandwidth, which may result in adjacent channel interference. It is easier to control this spreading if the phase changes are smaller—hence the advantage of OQPSK over QPSK.

Multilevel PSK

The use of multiple levels can be extended beyond taking bits two at a time. It is possible to transmit bits three at a time by using eight different phase angles. Further, each angle can have more than one amplitude. For example, a standard 9600-bps modem uses 12 phase angles, 4 of which have two amplitude values, for a total of 16 different signal elements.

This latter example points out very well the difference between the **data rate *R*** (in bps) and the **modulation rate *D*** (in baud) of a signal. Let us assume that this scheme is being employed with digital input in which each bit is represented by a constant voltage pulse: one level for binary 1 and one level for binary 0. The data rate is $R = 1/T_b$. However, the encoded signal contains $L = 4$ bits in each signal element, using $M = 16$ different combinations of amplitude and phase. The modulation rate can be seen to be $R/4$ because each change of signal element communicates 4 bits. Thus the line signaling speed is 2400 baud, but the data rate is 9600 bps. This is the reason that higher bitrates can be achieved over voice-grade lines by employing more complex modulation schemes.

Performance

In looking at the performance of various digital-to-analog modulation schemes, the first parameter of interest is the bandwidth of the modulated signal. This depends on a variety of factors, including the definition of bandwidth used and the filtering technique used to create the bandpass signal. We will use some straightforward results from [COUC13].

The transmission bandwidth B_T for ASK is of the form:

$$\textbf{ASK:} \quad B_T = (1 + r)R \qquad \textbf{(Equation 13.6)}$$

where R is the bitrate, and r is related to the technique by which the signal is filtered to establish a bandwidth for transmission—typically $0 < r < 1$. Thus the bandwidth is directly related to the bitrate. Equation (13.6) is also valid for PSK and, under certain assumptions, FSK.

With multilevel PSK (MPSK), significant improvements in bandwidth can be achieved. In general:

$$\textbf{MPSK:} \quad B_T = \left(\frac{1+r}{L}\right)R = \left(\frac{1+r}{\log_2 M}\right)R \qquad \textbf{(Equation 13.7)}$$

where L is the number of bits encoded per signal element, and M is the number of different signal elements.

Table 13.1 shows the ratio of data rate, R, to transmission bandwidth for various schemes. This ratio is also referred to as the **bandwidth efficiency**. As the name suggests, this parameter measures the efficiency with which bandwidth can be used to transmit data. The advantage of multilevel signaling methods now becomes clear.

TABLE 13.1 Bandwidth Efficiency (R/B_T) for Various Digital-to-Analog Encoding Schemes

	$r = 0$	$r = 0.5$	$r = 1$
ASK	1.0	0.67	0.5
PSK	1.0	0.67	0.5
Multilevel PSK			
$M = 4, L = 2$	2.00	1.33	1.00
$M = 8, L = 3$	3.00	2.00	1.50
$M = 16, L = 4$	4.00	2.67	2.00
$M = 32, L = 5$	5.00	3.33	2.50

Of course, the preceding discussion refers to the spectrum of the input signal to a communications line. Nothing has yet been said of performance in the presence of noise. Figure 13.6 summarizes some results based on reasonable assumptions concerning the transmission system [COUC13]. Here bit error rate is plotted as a function of the ratio E_b/N_0 defined in Chapter 11, "Wireless Transmission." Of course, as that ratio increases, the bit error rate drops. Further, DPSK and BPSK are about 3 dB superior to ASK and FSK.

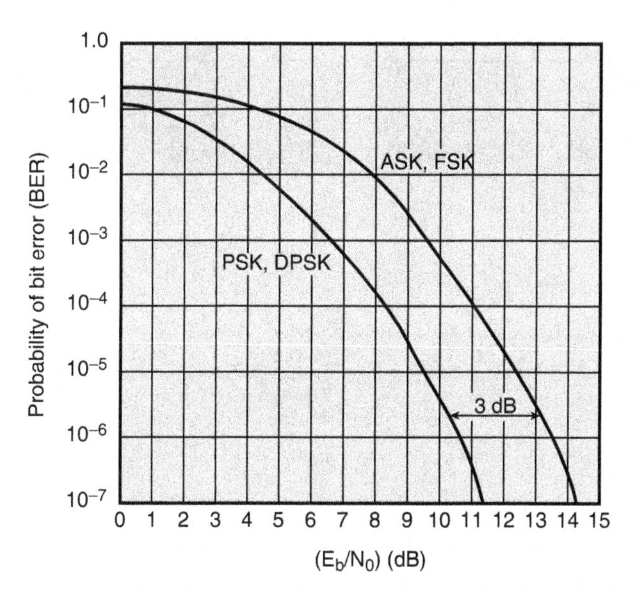

FIGURE 13.6 Theoretical Bit Error Rate for Various Encoding Schemes

Figure 13.7 shows the same information for various levels of M for MPSK. Note that the error probability for a given value of E_b/N_0 increases as M increases. On the other hand, Equation (13.7) shows that the bandwidth efficiency of MPSK increases as M increases. Thus, there is a trade-off between bandwidth efficiency and error performance: An increase in bandwidth efficiency results in an increase in error probability.

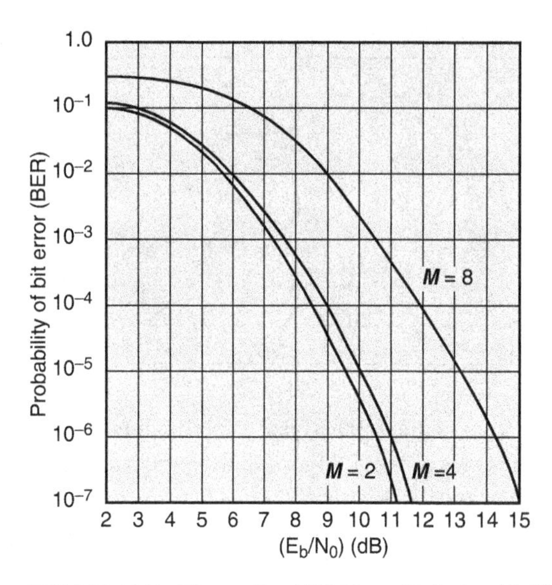

FIGURE 13.7 Theoretical Bit Error Rate for Multilevel PSK

Example

What is the bandwidth efficiency for FSK, ASK, PSK, and QPSK for a bit error rate of 10^{-7} on a channel with an SNR of 12 dB?

Using Equation (11.5), we have:

$$\left(\frac{E_b}{N_0}\right)_{dB} = 12\text{ dB} - \left(\frac{R}{B_T}\right)_{dB}$$

For FSK and ASK, from Figure 5.4:

$$\left(\frac{E_b}{N_0}\right)_{dB} = 14.2\text{ dB}$$

$$\left(\frac{R}{B_T}\right)_{dB} = -2.2\text{ dB}$$

$$\frac{R}{B_T} = 0.6$$

For PSK, from Figure 5.4:

$$\left(\frac{E_b}{N_0}\right)_{dB} = 11.2 \text{ dB}$$

$$\left(\frac{R}{B_T}\right)_{dB} = 0.8 \text{ dB}$$

$$\frac{R}{B_T} = 1.2$$

The result for QPSK must take into account that the baud rate $D = R/2$. Thus:

$$\frac{R}{B_T} = 2.4$$

As the preceding example shows, ASK and FSK exhibit the same bandwidth efficiency. PSK is better, and even greater improvement can be achieved with multilevel signaling.

Quadrature Amplitude Modulation

Quadrature amplitude modulation (QAM) is a popular modulation technique that is used in a number of wireless standards. This modulation technique is a combination of ASK and PSK. QAM can also be considered a logical extension of QPSK. QAM takes advantage of the fact that it is possible to send two different signals simultaneously on the same carrier frequency, by using two copies of the carrier frequency, one shifted by 90° with respect to the other. For QAM, each carrier is ASK modulated. The two independent signals are simultaneously transmitted over the same medium. At the receiver, the two signals are demodulated, and the results are combined to produce the original binary input.

Figure 13.8 shows the QAM modulation scheme in general terms. The input is a stream of binary digits arriving at a rate of R bps. The binary digits are represented by 1 and –1 for binary 1 and binary 0, respectively. This stream is converted into two separate bit streams of $R/2$ bps each, by taking alternate bits for the two streams. In Figure 13.8, the upper stream is ASK modulated on a carrier of frequency f_c by multiplying the bit stream by the carrier. Thus, a binary 0 is represented by $-\cos 2\pi f_c t$, and a binary 1 is represented by $+\cos 2\pi f_c t$, where f_c is the carrier frequency. This same carrier wave is shifted by 90° and used for ASK modulation of the lower binary stream. The two modulated signals are then added together and transmitted.

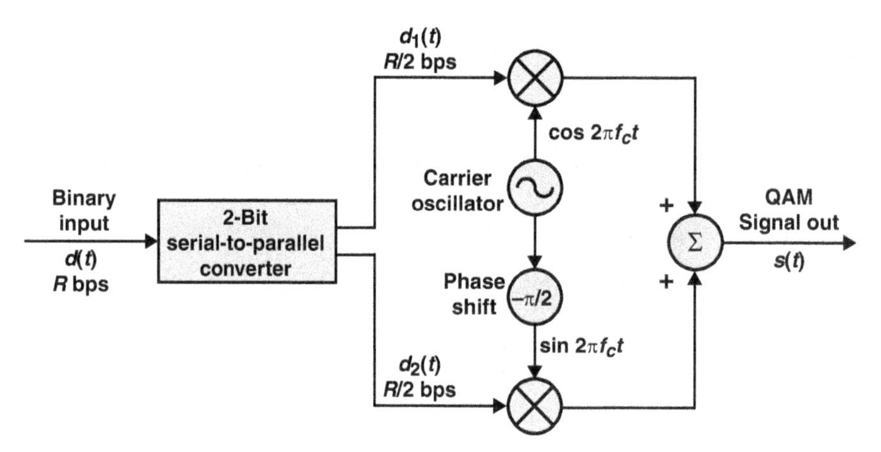

FIGURE 13.8 QAM Modulator

The transmitted signal can be expressed as follows:

$$\textbf{QAM:} \quad s(t) = d_1(t)\cos 2\pi f_c t + d_2(t)\sin 2\pi f_c t$$

Figure 13.9 shows a QAM demodulator that corresponds to the QAM modulator of Figure 13.8. It can be shown that this arrangement does recover the two signals $d_1(t)$ and $d_2(t)$, which can be combined to recover the original input, as illustrated next.

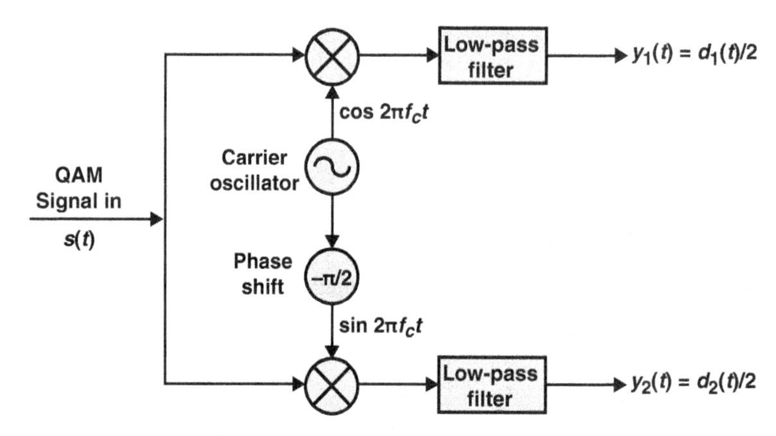

FIGURE 13.9 QAM Demodulator

Use the following identities:

$$\cos 2\alpha = 2\cos^2\alpha - 1$$

$$\sin 2\alpha = 2\sin\alpha \cos\alpha$$

Then:

$$s(t)\ \cos w_c t = d_1(t)\cos^2 w_c t + d_2(t)\sin w_c t\ \cos w_c t$$
$$= (1/2)d_1(t) + (1/2)d_1(t)\ \cos 2w_c t + (1/2)d_2(t)\ \sin 2w_c t$$

Use the following identities:

$$\cos 2\alpha = 1 - 2\ \sin^2\alpha$$

$$\sin 2\alpha = 2\sin\alpha\ \cos\alpha$$

Then:

$$s(t)\ \sin w_c t = d_1(t)\ \cos w_c t\ \sin w_c t + d_2(t)\sin^2 w_c t$$
$$= (1/2)d_1(t)\ \sin 2w_c t + (1/2)d_2(t) - (1/2)d_2(t)\ \cos 2w_c t$$

All terms at $2w_c$ are filtered out by the low-pass filter, yielding:

$$y_1(t) = (1/2)d_1(t);\ y_2(t) = (1/2)d_2(t)$$

If two-level ASK is used, then each of the two streams can be in one of two states, and the combined stream can be in one of 4 = 2 × 2 states. This is essentially QPSK. If four-level ASK is used (i.e., four different amplitude levels), the combined stream can be in one of 16 = 4 × 4 states. This is known as 16-QAM. Figure 13.10 shows the possible combinations of instantaneous values of the digital signals $d_1(t)$ and $d_2(t)$, in a format referred to as a *constellation*. For 16-QAM, each digital signal encodes 4 bits and takes on one of 16 values—4 positive and 4 negative. Figure 13.10 shows the usual bit assignment for the 16 constellation points.

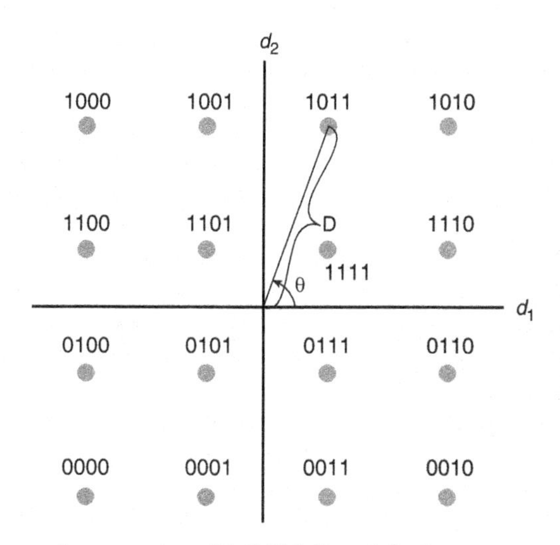

FIGURE 13.10 16-QAM Constellation

QAM can also be viewed as a combination of digital-amplitude and digital-phase modulation. Using trigonometric identities, we can rewrite the QAM equation in the form $s(t) = D(t)\cos(2\pi f_c t + \theta(t))$, where:

$$D(t)=\sqrt{d_1(t)^2+d_2(t)^2}, \quad \theta(t)=\tan^{-1}\left(\frac{d_2(t)}{d_1(t)}\right)$$

One of the points in Figure 13.10 is labeled with the magnitude and angle components.

Systems using 64 and 256 states have been implemented. The greater the number of states, the higher the data rate that is possible within a given bandwidth. Of course, as discussed previously, the greater the number of states, the higher the potential error rate due to noise and attenuation.

For comparison, Figure 13.11 shows diagrams for BPSK, QPSK, 16-QAM, and 64-QAM. The diagrams illustrate that as the modulation order increases, the distance between the points on the constellation decreases. Accordingly, small amounts of noise can cause greater problems. As the relative level of noise increases due to low signal strengths, so the area covered by a point on the constellation increases. If it becomes too large, the receiver is unable to determine in which position on the constellation the transmitted signal was meant to be, and this results in errors.

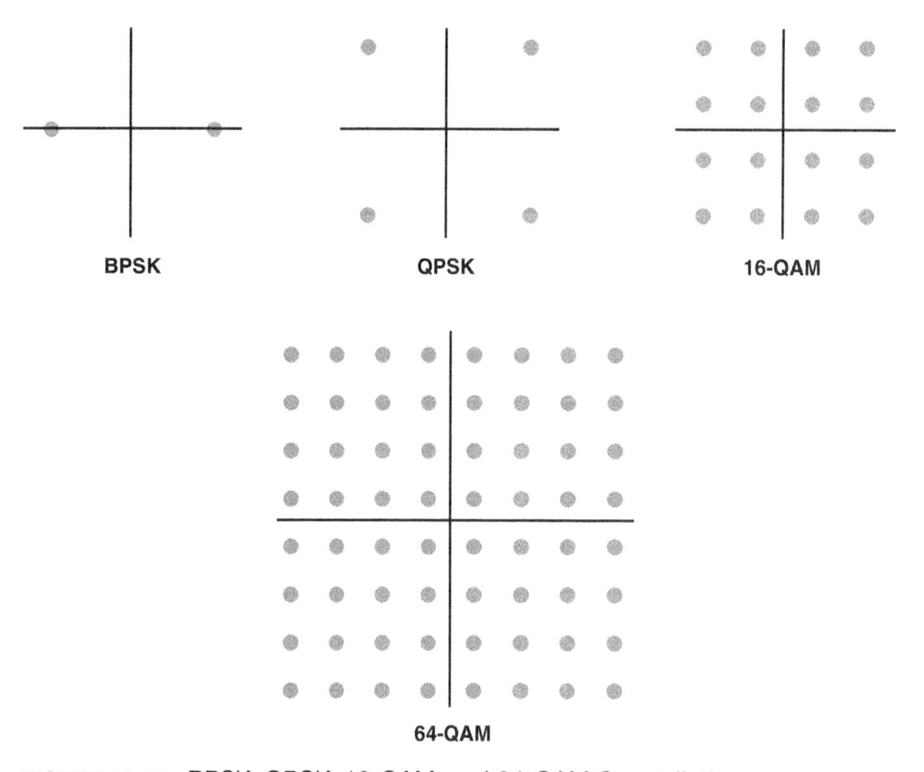

FIGURE 13.11 BPSK, QPSK, 16-QAM, and 64-QAM Constellations

Another challenge is that when a phase- or frequency-modulated signal is amplified in a radio transmitter, there is no need to use linear amplifiers, whereas when using QAM that contains an amplitude component, linearity must be maintained for reliable demodulation. Unfortunately, linear amplifiers are less efficient and consume more power than those that can be run in saturation, as is done for phase and frequency modulation.

Figure 13.12 shows the theoretical bit error rate (BER) performance for various levels of M for M-QAM. As expected, the BER for a given value of E_b/N_0 increases as M increases.

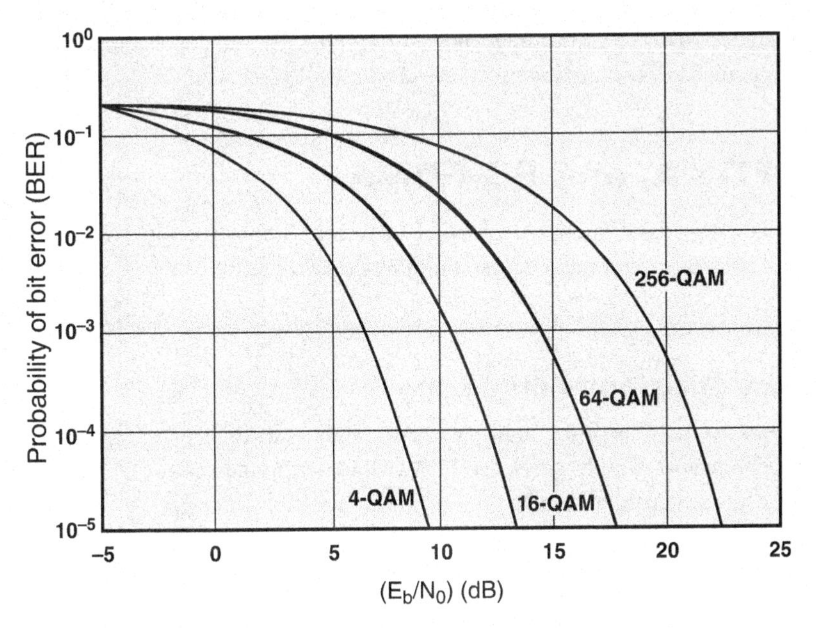

FIGURE 13.12 Theoretical Bit Error Rate for Multilevel QAM

π/2-BPSK

The coverage area of a cell or cell sector is limited by the transmission power in the uplink, which is considerably less than the transmission power available at the base station in the downlink. This is due to hardware limitations such as battery size in UE. To enhance cell coverage while limiting the cost of base station/antenna deployment, 3GPP introduced as an option a new modulation scheme, π/2-BPSK, for the uplink channel.

$\pi/2$-BPSK is a variation on BPSK that has been incorporated into several recent standards, including the 3GPP 5G specifications. As with BPSK, there is a phase shift of 180° between the two signal elements, and it can be represented by:

$$s(t) = \begin{cases} A\sin(2\pi ft + \pi/2) & \text{binary 1} \\ A\sin(2\pi ft - \pi/2) & \text{binary 2} \end{cases}$$

It can be shown that this modulation scheme, compared to the others listed in Figure 13.1, with appropriate spectrum shaping, produces low peak-to-average power ratio (PAPR) transmission without compromising error rate performance [KHAN20]. Section 13.2 discusses PAPR.

13.2 OFDM, OFDMA, and SC-FDMA

This section looks at some FDM-based techniques that are of increasing importance in broadband wireless networks. Annex 13A provides a summary of the multiplexing and multiple-access concepts used in this section.

Orthogonal Frequency-Division Multiplexing

OFDM, also called *multicarrier modulation*, uses multiple carrier signals at different frequencies, sending some of the bits on each channel. This is similar to FDM. However, in the case of OFDM, all of the subcarriers are dedicated to a single data source.

Figure 13.13 illustrates OFDM. Actual transmitter operation is simplified, but the basic concept can first be understood here. Suppose we have a data stream operating at R bps and an available bandwidth of Nf_b, centered at f_0. The entire bandwidth could be used to send the data stream, in which case each bit duration would be $1/R$. The alternative is to split the data stream into N substreams, using a serial-to-parallel converter. Each substream has a data rate of R/N bps and is transmitted on a separate subcarrier, with a spacing between adjacent subcarriers of f_b. Now the bit duration is N/R.

To gain a clearer understanding of OFDM, consider the scheme in terms of its base frequency, f_b. This is the lowest-frequency subcarrier. All of the other subcarriers are integer multiples of the base frequency: $2f_b$, $3f_b$, and so on, as shown in Figure 13.14a. The OFDM scheme uses advanced digital signal processing techniques to distribute the data over multiple carriers at precise frequencies. The relationship between the subcarriers is referred to as **orthogonality**.[1] The result is shown in Figure 13.4b, where because the signals overlap substantially, it appears that they are packed too closely together. However, one property of orthogonality is that the peaks of the power spectral density of each subcarrier occur at a point at which the power of other subcarriers is zero. Previous FDM

1. The term *orthogonal* is used in a number of contexts in technical literature. In general, an orthogonal set of primitives or capabilities spans the entire "capability space" of the system, and the primitives are in some sense non-overlapping or mutually independent. An example is a vector basis in geometry.

approaches are illustrated in Figure 13.4c, which assumes that signals should be spaced sufficiently apart in frequency to (1) avoid overlap in the frequency bands and (2) provide extra spacing known as *guard bands* to prevent the effects of adjacent carrier interference from out-of-band emissions. OFDM is able to drastically improve the use of frequency spectrum. The examples of Figure 8.2b and 8.2c show that the number of the signals that can be supported by OFDM compared to traditional FDM is greater by a factor of 6.

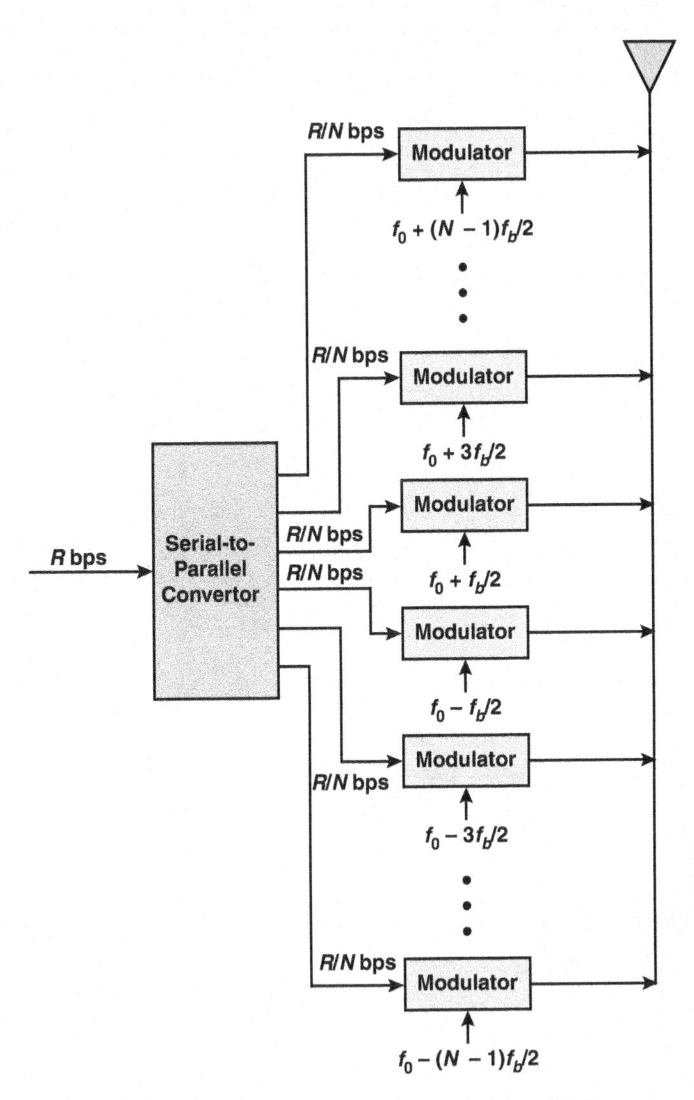

FIGURE 13.13 Orthogonal Frequency-Division Multiplexing

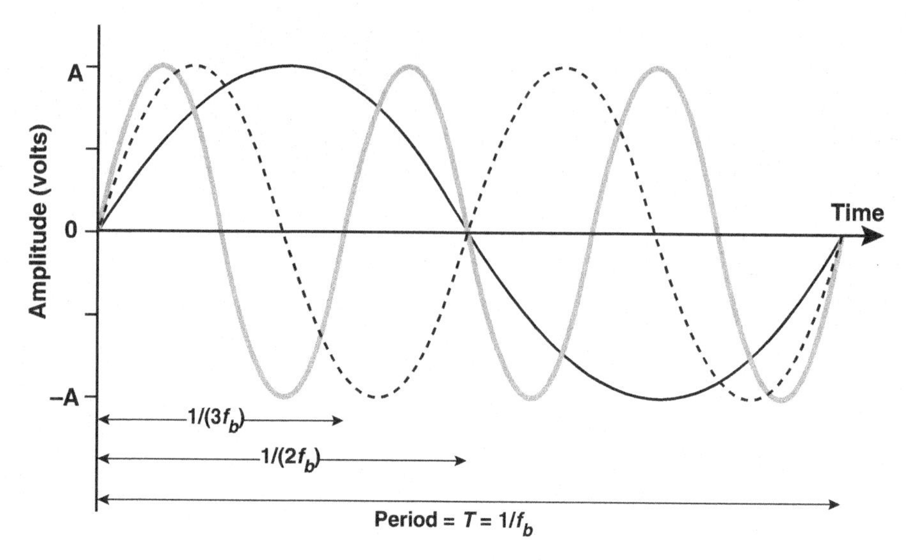

(a) Three subcarriers in time domain

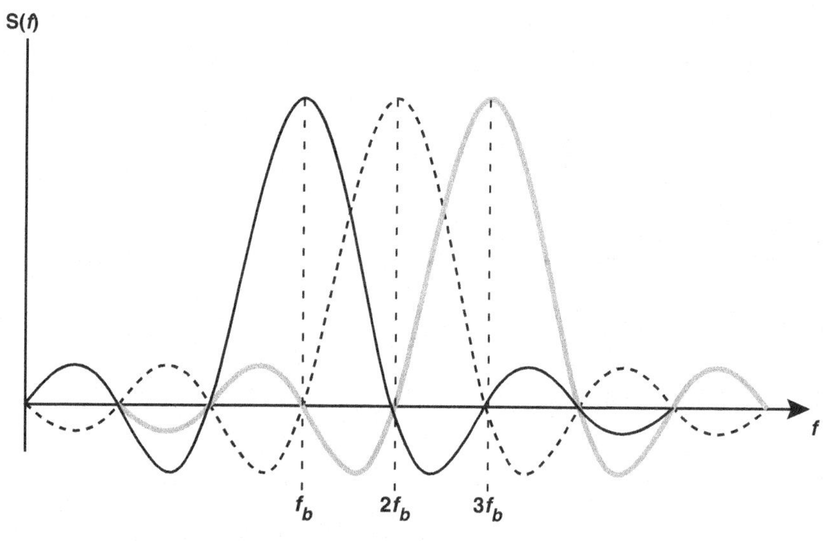

(b) Three subcarriers in frequency domain

FIGURE 13.14 Illustration of Orthogonality of OFDM

With OFDM, the subcarriers can be packed tightly together because there is minimal interference between adjacent subcarriers (and zero interference if the carrier spacing is not corrupted). In this context, orthogonality is defined as follows: Two signals, $s_1(t)$ and $s_2(t)$, are orthogonal if they meet this requirement:

Average over one bit time of $s_1(t)s_2(t) = 0$

Consider that a transmitter sends the sum of two orthogonal signals:

$$s(t) = s_1(t) + s_2(t)$$

The intended receiver can extract the $s_1(t)$ signal from the received signal by multiplying by $s_1(t)$ and averaging:

$$\text{Average of bit time of } s_1(t)s(t) = s_1(t)s_1(t) + s_1(t)s_2(t) = s_1^2(t) + 0$$

from which $s_1(t)$ can be recovered. If there are many signals that are all orthogonal, the receiver can remove all of the other signals and again recover $s_1(t)$.

The requirement for orthogonal digital signals that are subcarriers of OFDM is as follows: If the bit time of a subcarrier is T, then the base frequency f_b must be chosen to be a multiple of $1/T$. Every other signal will be a multiple of f_b such that $Mf_b = M/T$ for some integer M; all the signals will be orthogonal. One example of OFDM is that used for fourth-generation cellular LTE technology, which uses a subcarrier spacing of 15 kHz.

Note that Figure 13.14 depicts the set of OFDM subcarriers in a frequency band beginning with the base frequency. For transmission, the set of OFDM subcarriers is further modulated to a higher frequency band. For example, for the IEEE 802.11a LAN standard, the OFDM scheme consists of a set of 52 subcarriers with a base frequency of 0.3125 MHz. This set of subcarriers is then translated to the 5-GHz range for transmission.

OFDM has several advantages. First, frequency-selective fading affects only some subcarriers and not the whole signal. If the data stream is protected by a forward error-correcting code, this type of fading is easily handled. More importantly, OFDM overcomes intersymbol interference (ISI) in a multipath environment. As discussed in Section 11.4 in Chapter 11 (see Figure 11.9), ISI has a greater impact at higher bitrates because the distance between bits, or symbols, is smaller. With OFDM, the data rate is reduced by a factor of N, which increases the symbol time by a factor of N. Thus, if the symbol period is T_s for the source stream, the period for the OFDM signals is NT_s. This dramatically reduces the effect of ISI. As a design criterion, N is chosen so that NT_s is significantly greater than the root-mean-square (RMS) delay spread[2] of the channel.

As a result of these considerations, with the use of OFDM, it may not be necessary to deploy equalizers, which are complex devices whose complexity increases with the number of symbols over which ISI is present.

OFDM Implementation

OFDM implementation has two important operations that are required to create the benefits just described: the inverse fast Fourier transform (IFFT) and cyclic prefix.

2. In general terms, the delay spread is a measure of the difference between the time of arrival of the earliest significant multipath component (typically the line-of-sight component) and the time of arrival of the last multipath components. The RMS delay spread is the standard deviation (or root-mean-square) value of the delay of reflections, weighted proportionally to the energy in the reflected waves.

Inverse Fast Fourier Transform

Figure 13.13 shows a conceptual understanding of OFDM where a data stream is split into many lower-bitrate streams and then modulated on many different subcarriers. Such an approach, however, would result in a very expensive transmitter and receiver because it would involve many expensive oscillators.

Fortunately, OFDM can instead be implemented by taking advantage of the properties of the discrete Fourier transform (DFT). A DFT is an algorithm that generates a quantized Fourier transform, $X[k]$, of a discrete time-domain function, $x[n]$:

$$X[k] = \sum_{n=0}^{N-1} x[n] e^{-j2\pi kn/N}$$

for $0 \le k < N$, where $j = \sqrt{-1}$. The presence of an imaginary number in the equations is a matter of convenience. The imaginary component has a physical interpretation having to do with the phase of a waveform. (A discussion of this topic is beyond the scope of this book.) In the above equation, x is a sequence of N data points, with $x[i]$ the magnitude of the ith data point; X is a set of amplitude values, with $X[k]$ the amplitude of the kth subcarrier.

The inverse discrete Fourier transform, which converts the frequency values back to time domain values, is as follows:

$$x[n] = \sum_{k=0}^{N-1} X[k] e^{j2\pi kn/N}$$

for $0 \le n < N$. When this function is implemented using a number of data points N that is a power of two, the computational time is greatly reduced, and these transforms are then called the **fast Fourier transform (FFT)** and **inverse fast Fourier transform (IFFT)**.

The implementation of OFDM using the FFT and IFFT is illustrated in Figure 13.15. The data stream undergoes a serial-to-parallel (S/P) operation, which takes a sample from each carrier and makes a group of samples called an **OFDM symbol**. Each value in a sense gives a weight for each subcarrier. Then the IFFT (not the FFT) takes the values for these subcarriers and computes a time domain data stream to be transmitted that is a combination of these subcarriers. The IFFT operation has the effect of ensuring that the subcarriers do not interfere with each other. These values are put back into a serial stream with a P/S operation, and then the stream is modulated onto the carrier using one oscillator. At the receiver, the reverse operation is performed. An FFT module is used to map the incoming signal back to the M subcarriers, from which the data streams are recovered as the weights for each subcarrier are retrieved for each sample.

Note that in OFDM, the term *symbol* takes on a different meaning than is used in other contexts. An OFDM symbol is a group of samples, one from each subcarrier. This is the input to the IFFT operation. Thus, a transmission scheme using 16-QAM modulation and an OFDM block size of eight samples would be transmitting an OFDM symbol of eight 16-QAM symbols.

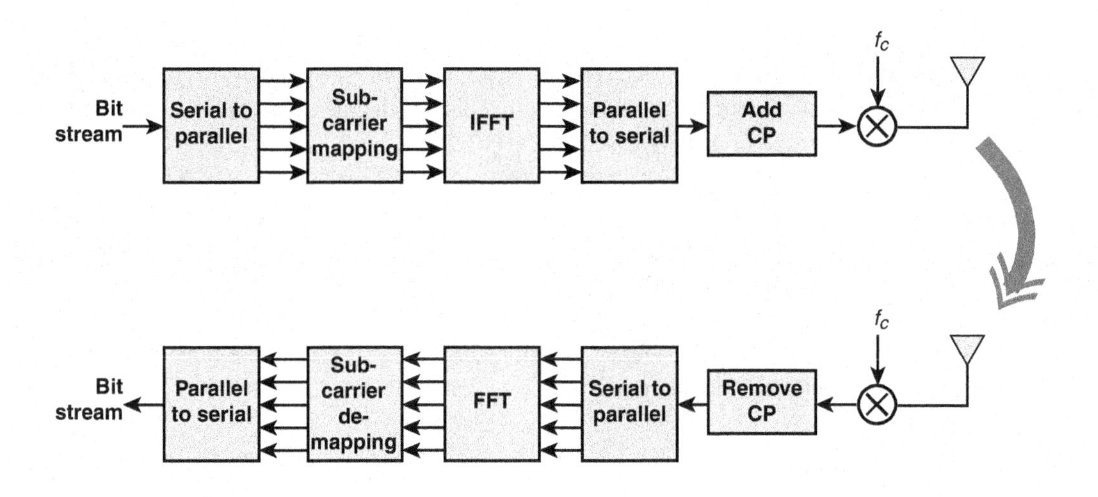

FFT = Fast Fourier transform
IFFT = Inverse fast Fourier transform
CP = Cyclic prefix

FIGURE 13.15 IFFT Implementation of OFDM

The Cyclic Prefix

Even though OFDM by definition limits ISI by using long symbol times, OFDM also uses a **cyclic prefix (CP)**, which goes another step further to combat ISI and completely eliminate the need for equalizers. The cyclic prefix is illustrated in Figure 13.16.

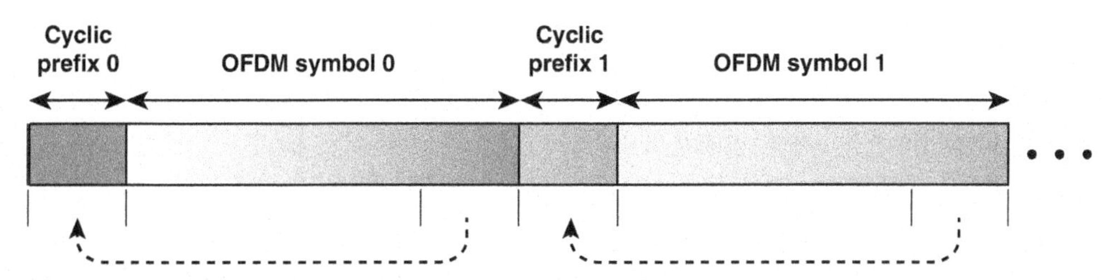

FIGURE 13.16 CP-OFDM Transmission Scheme

This accomplishes two functions:

- Additional time, known as a guard interval, is added to the beginning of the OFDM symbol before the actual data begins. This allows all residual ISI to diminish before it impacts the received data.

- This beginning time period is packed with data that is an actual copy of the data from the end of the OFDM symbol that is being sent. The effect of this is to isolate the parallel subchannels and allow for simple frequency-domain digital signal processing techniques. An explanation of this concept is beyond the scope of this book; see [SHAH10].

As an example of OFDM implementation, consider one implementation of the Long Term Evolution (LTE) cellular standard. LTE uses 15-kHz subcarriers and can use an OFDM symbol of 1024 subcarriers with the cyclic prefix accounting for a 7% guard time (for a nominal cyclic prefix and up to 25% for the extended cyclic prefix). The result would be 600 data subcarriers (the remainder as pilot or null subcarriers), 72 guard symbols using 4.7 μsec, an overall 142.7 μsec OFDM symbol time, and a data rate of 21.9 Mbps if using 16-QAM modulation.

Difficulties of OFDM

Even though OFDM has tremendous benefits and the implementation process has been highly simplified, there are still two key issues that must be addressed for successful OFDM implementation.

Peak-to-Average Power Ratio (PAPR)

An important consideration in the design of an OFDM-based transmission scheme is the peak-to-average-power ratio (PAPR). The PAPR equals the maximum power of a sample in a given OFDM transmit symbol divided by the average power of that OFDM symbol. In simple terms, PAPR is the ratio of peak power to the average power of a signal. It is expressed in dB.

PAPR occurs in a multicarrier system because the different subcarriers are out of phase with each other. So, at any point in time, the amplitudes of the different carriers differ. When all the points achieve the maximum value simultaneously, the output envelope suddenly shoots up, which causes a peak in the output envelope. Due to the presence of a large number of independently modulated subcarriers in an OFDM system, the peak value of the system can be significantly higher than the average. In LTE systems, OFDM signal PAPR is approximately 12 dB.

The significance of PAPR is its effect on the power amplifier (PA). The PA is the most power-consuming unit in a transceiver. For OFDM transmission, the PA must be selected for a desired output power, plus a margin to account for peak power. In other words, a higher-power PA is required than would be the case if there were no PAPR effect. This is especially important on uplink transmission from mobile devices and constrained Internet of Things (IoT) devices.

A number of approaches to mitigating the PAPR have been proposed ([RAHN13], [PARE15]). The modulation technique π/2-BPSK, described in Section 13.1, is a recently developed approach that is used in 5G.

Intercarrier Interference

In order to demodulate an OFDM signal, time and frequency synchronization is necessary. Because OFDM symbol times are long, the demands on time synchronization are somewhat relaxed compared to in other systems. Conversely, because OFDM frequencies are spaced as closely as possible, the frequency synchronization requirements are significantly more stringent. If they are not met, intercarrier interference (ICI) results. Timing and frequency synchronization algorithms are the responsibility of each equipment manufacturer, and these problems are some of the most challenging for OFDM implementation.

The CP provides an excellent way of ensuring orthogonality of carriers because it eliminates the effects of multipath. Because the CP causes a reduction in spectral efficiency, however, a certain level of ICI may be tolerated in an effort to reduce CP length. In addition, Doppler shift or mismatched oscillators of even one subcarrier can cause ICI in many adjacent subcarriers. The spacing between subcarriers has tight constraints and can be easily perturbed.

Because ICI can be a limiting factor for OFDM systems, implementations seek to find a balance between carrier spacing and OFDM symbol length. Short symbol duration reduces Doppler-induced ICI, but it also may cause the CP to be an unacceptably large part of the OFDM symbol time. Systems may also use different OFDM pulse shapes, use self-interference cancellation by modulating information across multiple carriers to reduce ICI, and implement frequency-domain equalizers.

Orthogonal Frequency-Division Multiple Access

Like orthogonal frequency-division multiplexing (OFDM), orthogonal frequency-division multiple access (OFDMA) employs multiple closely spaced subcarriers, but the subcarriers are divided into groups of subcarriers.[3] Each group is named a **subchannel**. The subcarriers that form a subchannel need not be adjacent. In the downlink, a subchannel may be intended for different receivers. In the uplink, a transmitter may be assigned one or more subchannels. Figure 13.17 contrasts OFDM and OFDMA; in the OFDMA case, adjacent subcarriers are used to form a subchannel in this figure.

Subchannelization defines subchannels that can be allocated to subscriber stations (SSs) depending on their channel conditions and data requirements. Using subchannelization, within the same time slot, a base station (BS) can allocate more transmit power to user devices (SSs) with lower SNR (signal-to-noise ratio), and less power to user devices with higher SNR. Subchannelization also enables the BS to allocate higher power to subchannels assigned to indoor SSs, resulting in better in-building coverage. Subchannels are further grouped into bursts, which can be allocated to wireless users. Each burst allocation can be changed from frame to frame as well as within the modulation order. This allows the base station to dynamically adjust the bandwidth usage according to the current system requirements.

Subchannelization in the uplink can save user device transmit power because it can concentrate power only on certain subchannels allocated to it. This power-saving feature is particularly useful for battery-powered user devices, the likely case in mobile 4G and 5G.

3. See Annex 13A for a discussion of the distinction between multiplexing and multiple access.

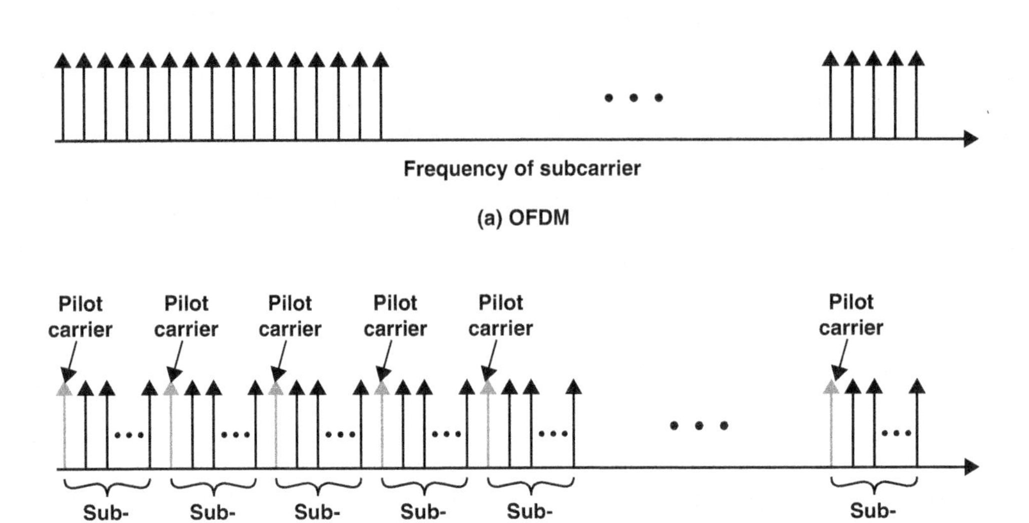

FIGURE 13.17 OFDM and OFDMA

Single-Carrier FDMA

Single-carrier FDMA (SC-FDMA) is a more recently developed multiple-access technique that is similar in structure and performance to OFDMA. One prominent advantage of SC-FDMA over OFDMA is the lower peak-to-average power ratio (PAPR) of the transmit waveform, which benefits the mobile user in terms of battery life and power efficiency. OFDM signals have a higher PAPR because, in the time domain, a multicarrier signal is the sum of many narrowband signals. At some time instances this sum is large, and at other times it is small, which means the peak value of the signal is substantially larger than the average value. Thus, SC-FDMA is superior to OFDMA. However, it is restricted to uplink use because the increased time-domain processing of SC-FDMA would entail considerable burden on the base station.

As shown in Figure 13.18, SC-FDMA performs a DFT prior to the IFFT operation, which spreads the data symbols over all the subcarriers carrying information and produces a virtual single-carrier structure. This then is passed through the OFDM processing modules to split the signal into subcarriers. Now, however, every data symbol is carried by every subcarrier.

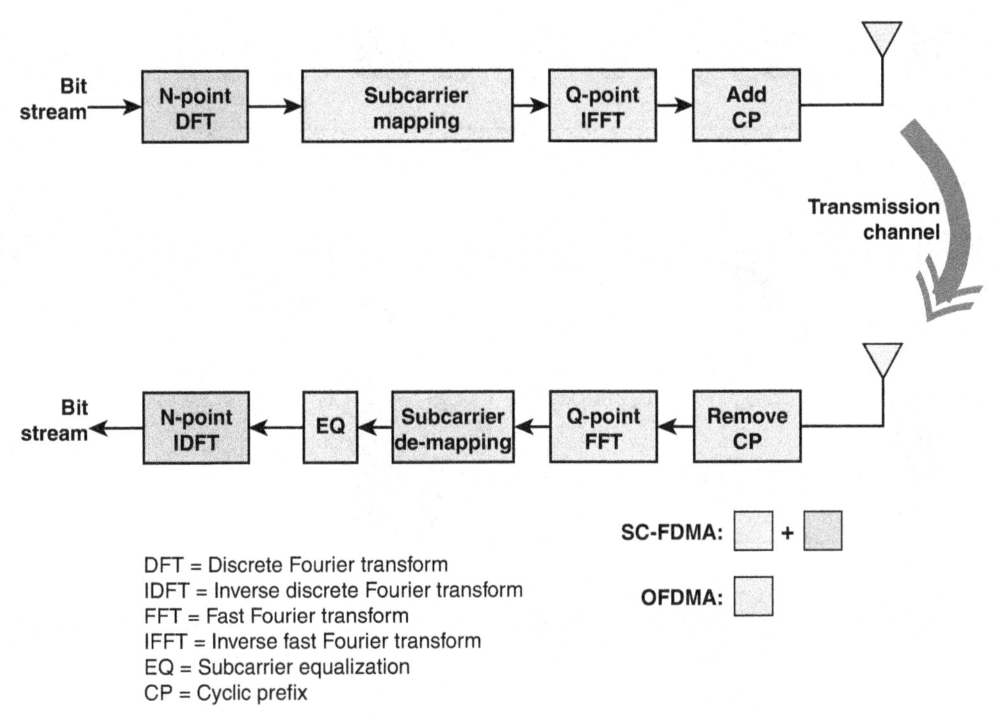

FIGURE 13.18 Simplified Block Diagram of OFDMA and SC-FDMA

Figure 13.19 provides an example of how the OFDM and SC-FDMA signals appear. As Figure 13.19 illustrates, with OFDM, a source data stream is divided into N separate data streams, and these streams are modulated and transmitted in parallel on N separate subcarriers, each with bandwidth f_b. The source data stream has a data rate of R bps, and the data rate on each subcarrier is R/N bps. For SC-FDMA, it appears from Figure 13.18 that the source data stream is modulated on a single carrier (hence the SC prefix to the name) of bandwidth $N \times f_b$ and transmitted at a data rate of R bps. The data is transmitted at a higher rate but over a wider bandwidth compared to the data rate on a single subcarrier of OFDM. However, because of the complex signal processing of SC-FDMA, the preceding description is not accurate. In effect, the source data stream is replicated N times, and each copy of the data stream is independently modulated and transmitted on a subcarrier, with a data rate on each subcarrier of R bps. Compared with OFDM, SC-FDMA transmits at a much higher data rate on each subcarrier, but because the same data stream is on each subcarrier, it is still possible to reliably recover the original data stream at the receiver.

A final observation concerns the term *multiple access*. With OFDMA, it is possible to simultaneously transmit either from or to different users by allocating the subcarriers during any one time interval to multiple users. This is not possible with SC-FDMA: At any given point in time, all the subcarriers are carrying the identical data stream and hence must be dedicated to one user. But over time, it is possible to provide multiple access. Thus, a better term for SC-FDMA might be SC-OFDM-TDMA, although that term is not used.

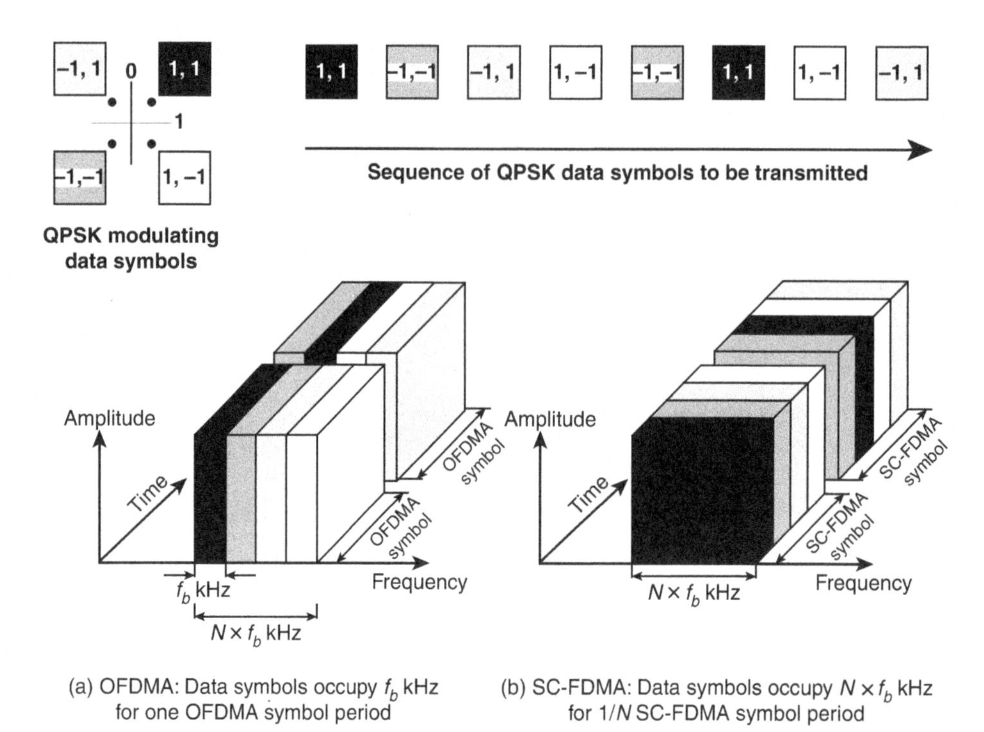

(a) OFDMA: Data symbols occupy f_b kHz for one OFDMA symbol period

(b) SC-FDMA: Data symbols occupy $N \times f_b$ kHz for $1/N$ SC-FDMA symbol period

FIGURE 13.19 Example of OFDMA and SC-FDMA

13.3 Waveforms and Numerologies

3GPP TR 38.802 (*Technical Specification Group Radio Access Network, Study on New Radio Access Technology Physical Layer Aspects [Release 14]*, September 2017) specifies the two waveforms supported for 5G: cyclic prefix orthogonal frequency-division multiplexing (CP-OFDM) for uplink and downlink use and discrete Fourier transform spread OFDM (DFT-S-OFDM) for uplink use only. This section provides an overview of each.

Numerology

In the context of the 3GPP specifications, the term *numerology* refers to the subcarrier spacing in an OFDM scheme. Table 13.2 shows the supported numerologies specified in 3GPP TS 38.211. Using the numbers 0 through 4, the subcarrier spacing scales as a power of two. This facilitates symbol boundary alignment between different subcarrier spacings.

TABLE 13.2 Supported Transmission Numerologies

μ	0	1	2	3	4
Δf = 2μ × 15 kHz	15	30	60	120	240
OFDM symbol duration	66.67 μs	33.33 μs	16.67 μs	8.33 μs	4.17 μs
Cyclic prefix duration	4.69 μs	2.34 μs	1.17 μs	0.59 μs	0.29 μs
OFDM symbol including CP	71.35 μs	35.68 μs	17.84 μs	8.91 μs	4.46 μs
Number of OFDM symbols per slot	7 or 14	7 or 14	7 or 14	14	14
Slot duration	500 μs or 1,000 μs	250 μs or 500 μs	125 μs or 250 μs	125 μs	62.5 μs

The use of multiple numerologies provides flexibility for optimal support based on the frequency band and the size of the cell. With respect to frequency band, phase noise increases with carrier frequency but decreases with subcarrier spacing. Thus, to operate in higher frequency ranges, wider subcarrier spacing is needed.

With respect to cell size, when a smaller subcarrier spacing is used, the symbol length is larger. With longer symbol length, the CP of the OFDM symbol can be longer at the same ratio of the CP duration-to-OFDM symbol duration (i.e., the CP overhead ratio). Thus, with smaller subcarrier spacing, a system is more tolerant of the effect of multipath delay spread at the same CP overhead ratio.

CP-OFDM

As the name indicates, CP-OFDM dictates the use of a cyclic prefix. [ZAID16] provides detail on CP-OFDM and lists the following reasons that CP-OFDM is among the most appropriate candidates for 5G:

- CP-OFDM ranks best on the performance indicators that matter most: compatibility with multi-antenna technologies, high spectral efficiency, and low implementation complexity.
- CP-OFDM is well localized in the time domain, which is important for latency-critical applications and time-division duplexing (TDD) deployments.
- CP-OFDM is also more robust to oscillator phase noise and Doppler than other multicarrier waveforms. Robustness to phase noise is crucial for operation at high carrier frequencies (e.g., mmWave band).

OFDM has two drawbacks: less frequency localization and high peak-to-average power ratio (PAPR), like all other multicarrier waveforms. However, there are well-established simple techniques for reducing PAPR (e.g., clipping, companding) and improving frequency localization (e.g., windowing). These techniques can easily be applied to CP-OFDM at the transmitter in a receiver-agnostic way.

Table 13.2 shows the CP durations for the various 5G numerologies.

DFT-S-OFDM

DFT-S-OFDM is a variant of SC-FDMA. Thus, it spreads each OFDM symbol across all of the subcarriers in one time slot. The elements of the method are shown in Figure 13.20.

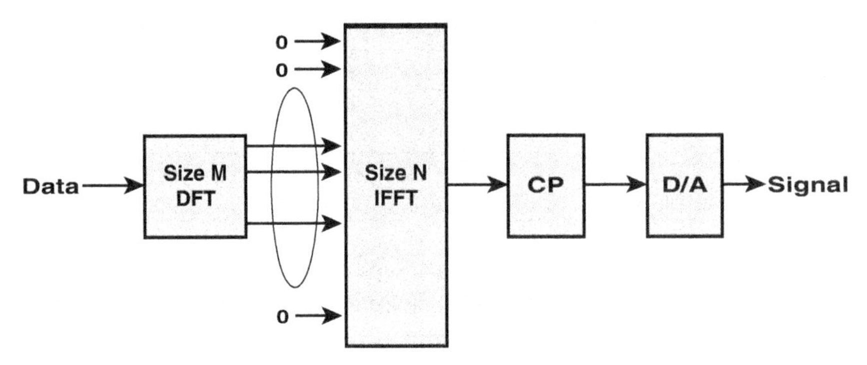

FIGURE 13.20 DFT-S-OFDM Simplified Block Diagram

DFT-S-OFDM works as follows:

- Discrete Fourier transform (DFT) converts a block of K QAM samples into the frequency domain.
- The resulting K frequency elements are mapped to selected subcarriers in an input of length M ($K < M$) to an inverse fast Fourier transform (IFFT).

An advantage of DFT-S-OFDM over CP-OFDM is reduced PAPR—and thus reduced terminal power consumption.

DFT-S-OFDM is similar to SC-FDMA but has the advantage of allowing noncontiguous (clustered) groups of subcarriers to be allocated for transmission by a single UE, thus increasing the flexibility available for frequency-selective scheduling. Figure 13.21 illustrates the two types of subcarrier mapping for DFT-S-OFDM. In localized mapping, the DFT outputs are mapped to a subset of consecutive subcarriers, thereby confining them to only a fraction of the system bandwidth. In distributed mapping, the DFT outputs of the input data are assigned to subcarriers over the entire bandwidth noncontinuously, resulting in zero amplitude for the remaining subcarriers. Localized mapping creates the opportunity to easily introduce multiple access, as different blocks of subcarriers can be assigned to different channels. Distributed mapping is more robust with respect to frequency-selective fading and offers additional frequency diversity gain because the information is spread across the entire system bandwidth.

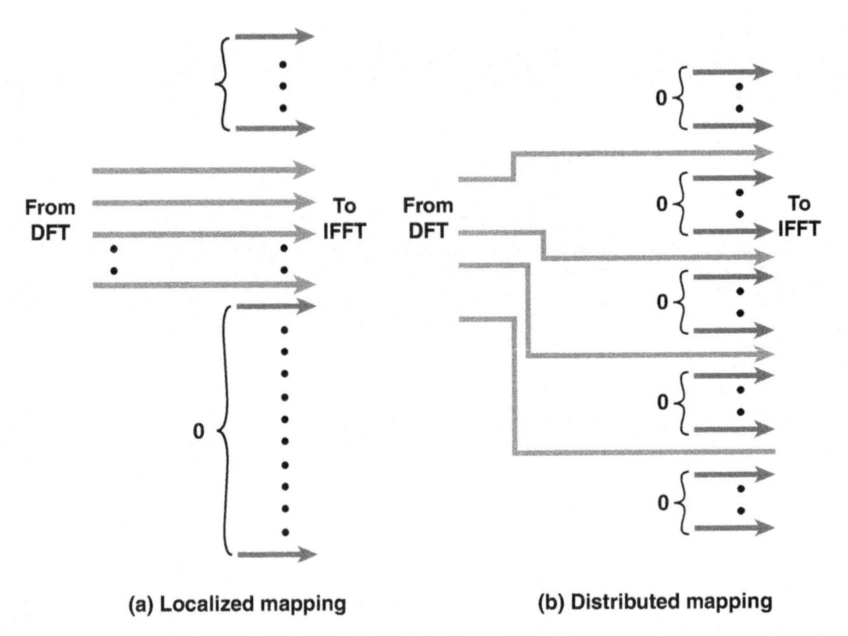

FIGURE 13.21 Localized and Distributed Mapping

13.4 Key Terms and Review Questions

Key Terms

amplitude-shift keying (ASK)	discrete Fourier transform (DFT)
baud	fast Fourier transform (FFT)
bandwidth efficiency	frequency-shift keying (FSK)
binary PSK (BPSK)	intercarrier interference (ICI)
carrier signal	inverse fast Fourier transform (IFFT)
CP-OFDM	phase-shift keying (PSK)
cyclic prefix (CP)	modulation
data rate	modulation rate
DFT-S-OFDM	multilevel PSK (MPSK)
differential PSK (DPSK)	numerology

orthogonal frequency-division modulation (OFDM)	single-carrier FDMA (SC-FDMA)
orthogonal frequency-division multiple access (OFDMA)	subcarrier
orthogonality	subchannel
peak-to-average power ratio (PAPR)	subchannelization
quadrature amplitude modulation (QAM)	waveform
quadrature phase-shift keying (QPSK)	π/2-BPSK

Review Questions

1. How are binary values represented in amplitude-shift keying, and what is the limitation of this approach?
2. How are binary values represented in frequency-shift keying?
3. How are binary values represented in phase-shift keying?
4. What is the difference between QPSK and offset QPSK?
5. What is the difference between data rate and modulation rate?
6. What is the trade-off between bandwidth efficiency and error performance for MPSK?
7. What is QAM?
8. What is π/2-BPSK?
9. Briefly explain OFDM.
10. Define orthogonality in the context of OFDM.
11. What roles do FFT and IFFT play in implementing OFDM?
12. What is the purpose of the cyclic prefix?
13. Why is PAPR an important consideration in the design of OFDM schemes?
14. Why is intercarrier interference an important consideration in the design of OFDM schemes?
15. What is the difference between OFDM and OFDMA?
16. What is the difference between OFDMA and SC-OFDM?

13.5 References and Documents

References

COUC13 Couch, L. *Digital and Analog Communication Systems.* Upper Saddle River, NJ: Pearson, 2013.

KHAN20 Khan, M., et al. "Low PAPR DMRS Sequence Design for 5G-NR Uplink." *2020 International Conference on Communication Systems & Networks*, 2020.

PARE15 Paredes, M., and García, J. "The Problem of Peak-to-Average Power Ratio in OFDM Systems." CoRR abs/1503.08271, 2015. https://arxiv.org/pdf/1503.08271.pdf

RAHN13 Rahmataliah, Y., and Mohan, S. "Peak-to-Average Power Ratio Reduction in OFDM Systems: A Survey and Taxonomy." *IEEE Communications Surveys & Tutorials*, Fourth Quarter 2013.

SHAH10 Shah, D., Rindhe, B., and Narayankhedkar, S. "Effects of Cyclic Prefix on OFDM System." *International Conference and Workshop on Emerging Trends in Technology*, February 2010.

ZAID16 Zaidi, A., et al. "Waveform and Numerology to Support 5G Services and Requirements." *IEEE Communications Magazine*, November 2016.

Documents

3GPP TR 38.802 *Technical Specification Group Radio Access Network, Study on New Radio Access Technology Physical Layer Aspects (Release 14).* September 2017.

3GPP TS 38.211 *Technical Specification Group Radio Access Network, NR, Physical channels and modulation (Release 16).* June 2020.

Annex 13A: Multiplexing and Multiple Access

This annex introduces and contrasts the concepts of multiplexing and multiple access.

Multiplexing

In both local and long-haul communications, it is almost always the case that the capacity of the transmission medium exceeds that required for the transmission of a single signal. To make cost-effective use of the transmission system, it is desirable to use the medium efficiently by sharing the capacity. This is referred to as multiplexing, and two techniques are in common use: frequency-division multiplexing (FDM) and time-division multiplexing (TDM).

FDM is possible when the useful bandwidth of the medium exceeds the required bandwidth of a given signal. A number of signals can be carried simultaneously if each signal is modulated onto a different carrier frequency and the carrier frequencies are sufficiently separated so that the bandwidths of the signals do not overlap. A general case of FDM is shown in Figure 13.22a. Six signal sources are fed into a multiplexer that modulates each signal onto a different frequency ($f_1, \ldots, f_6$). Each signal requires a certain bandwidth centered around its carrier frequency, referred to as a *channel*. To prevent interference, the channels are separated by *guard bands*, which are unused portions of the spectrum.

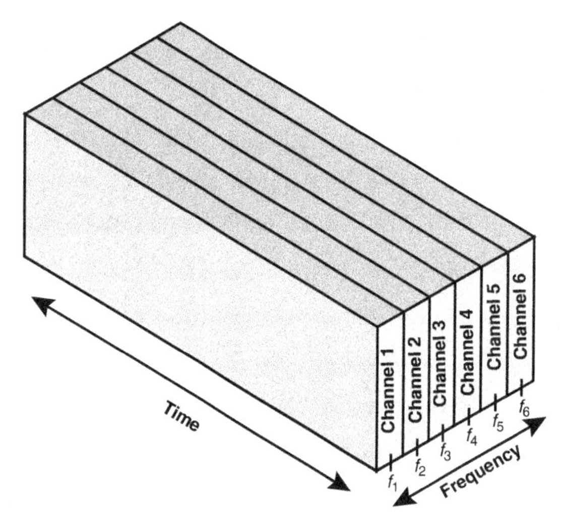

(a) Frequency-division multiplexing

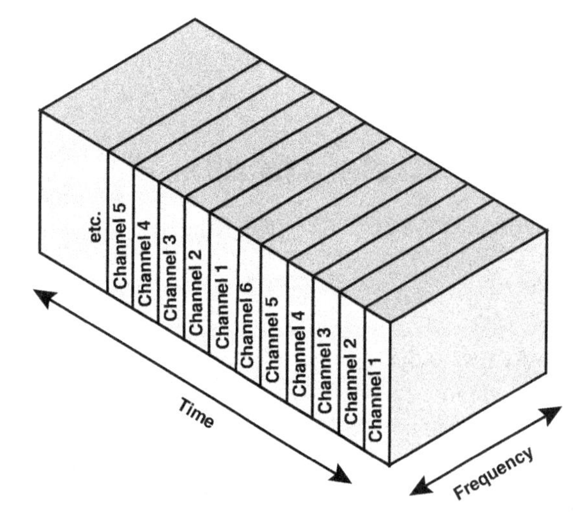

(b) Time division multiplexing

FIGURE 13.22 FDM and TDM

TDM is possible when the achievable bitrate (sometimes, unfortunately, called bandwidth) of the medium exceeds the required data rate of a digital signal. Multiple digital signals can be carried on a single transmission path by interleaving portions of each signal in time. The interleaving can be at the bit level or in blocks of bytes or larger quantities. For example, the multiplexer in Figure 13.22 has six inputs that might each be, say, 1 Mbps. A single line with a capacity of 6 Mbps (plus overhead capacity) could accommodate all six sources. Analogously to FDM, the sequence of time slots dedicated to a particular source is called a *channel*. One cycle of time slots (one per source) is called a *frame*.

TDM is not limited to digital signals. Analog signals can also be interleaved in time. Also, with analog signals, a combination of TDM and FDM is possible. A transmission system can be frequency divided into a number of channels, each of which is further divided via TDM.

Multiple Access

The terms *multiplexing* and *multiple access* are often confused with one another. In essence, multiplexing is a function that permits two or more data sources to share a common transmission medium such that each data source has its own channel. This is a physical layer function concerned with dividing a larger-capacity channel into multiple smaller-capacity subchannels. Multiple access is a link layer function for allowing multiple users to share a transmission channel by providing each with a subchannel.

Using the concepts of the preceding paragraph, the following definitions can be made:

- **Frequency-division multiplexing (FDM):** A physical layer technique in which multiple baseband signals are modulated on different frequency carrier waves and added together to create a composite signal. The effect of FDM is to divide transmission bandwidth into multiple subchannels, each of which is dedicated to a particular baseband signal.
- **Frequency-division multiple access (FDMA):** An access method at the data link layer based on FDM principles, in which different frequency bands are allocated to different data streams. The data link layer in each station tells its physical layer to make a bandpass signal from the data passed to it. The signal must be created in the allocated band. There is no multiplexer at the physical layer. The signals created at each station are automatically bandpass filtered. They are mixed when they are sent to the common channel. FDMA supports demand assignment, in which the assignment of frequency bands to users changes over time.
- **Time-division multiplexing (TDM):** A physical layer technique of transmitting and receiving independent signals over a common signal path by means of synchronized switches at each end of the transmission line so that each signal appears on the line only a fraction of time in an alternating pattern. Thus, multiple stations may share the same frequency channel but use only part of its capacity.
- **Time-division multiple access (TDMA):** An access method at the data link layer based on TDM principles. TDMA provides different time slots to different transmitters in a cyclically repetitive frame structure. For example, node 1 may use time slot 1, node 2 time slot 2, and so on until the last transmitter, when it starts over. TDMA supports demand assignment, in which the assignment of time slots to users changes over time.

Chapter 14

Air Interface Channel Coding

Learning Objectives

After studying this chapter, you should be able to:

- Provide the standard definition of error burst
- Explain the difference between a binary erasure channel and a binary symmetric channel
- Make a presentation on the basic concepts of forward error correction
- Explain the concept of Hamming distance and its significance for error correction
- Describe the operation of LDPC codes
- Describe the operation of polar codes
- Summarize the 3GPP specifications for channel coding
- Explain the operation of HARQ and the options for how it could be implemented

3GPP TS 38.212 (*Technical Specification Group Radio Access Network, NR, Multiplexing and Channel Coding (Release 16)*, December 2020) specifies two forward error correction techniques for the air interface: low-density parity-check coding and polar coding. TR 38.802 (*Technical Specification Group Radio Access Network, Study on New Radio Access Technology Physical Layer Aspects (Release 14)*, September 2017) contains the results of a study into NR physical layer aspects and is useful for understanding the reasoning behind the concepts.

The chapter begins with an introduction to important basic concepts related to transmission errors, in Section 14.1. Section 14.2 provides a general introduction to forward error correction. Sections 14.3 and 14.4 explain the basic concepts of parity-check matrix codes and the details of low-density parity-check codes, respectively. Section 14.6 covers polar coding. Section 14.7 looks at the specific use of the two techniques in 3GPP specifications.

Section 14.7 introduces the related error control technique hybrid automatic repeat request (HARQ).

14.1 Transmission Errors

This section provides an overview of basic concepts related to transmission errors.

Error Burst

In digital transmission systems, an error occurs when a bit is altered between transmission and reception—that is, when a binary 1 is transmitted and a binary 0 is received, or a binary 0 is transmitted and a binary 1 is received. Alternatively, as explained later in this chapter, a binary bit may be transmitted and then an error indication may be received. Two general types of errors can occur: single-bit errors and burst errors. A single-bit error is an isolated error condition that alters one bit but does not affect nearby bits. A burst error of length *B* is a contiguous sequence of *B* bits in which the first and last bits and any number of intermediate bits are received in error. More precisely, ITU-T Recommendation Q.9 (*Vocabulary of Switching and Signalling Terms*, November 1988) defines an error burst as follows:

A group of bits in which two successive erroneous bits are always separated by less than a given number x of correct bits. The number x should be specified when describing an error burst. The last erroneous bit in the burst and the first erroneous bit in the following burst are accordingly separated by x correct bits or more.

Thus, in an error burst, there is a cluster of bits in which a number of errors occur, although not necessarily all of the bits in the cluster suffer errors. Figure 14.1 provides an example of both types of errors.

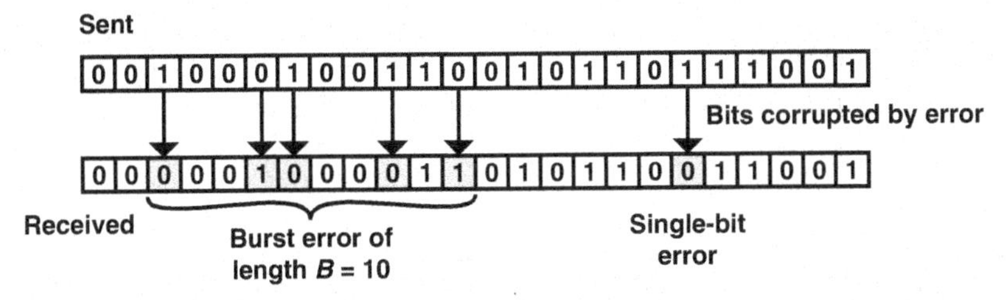

FIGURE 14.1 Burst and Single-Bit Errors

A single-bit error can occur in the presence of white noise, when a slight random deterioration of the signal-to-noise ratio is sufficient to cause the receiver to make an incorrect decision on a single bit. Burst errors are more common and more difficult to deal with. Burst errors can be caused by impulse noise or by fading in a mobile wireless environment.

Note that the effects of burst errors are greater at higher data rates.

Example

Say that an impulse noise event or a fading event of 1 μs occurs. At a data rate of 10 Mbps, there is a resulting error burst of 10 bits. At a data rate of 100 Mbps, there is an error burst of 100 bits.

Data Transmission Channels

A **discrete channel** comprises an input alphabet X (e.g., 0, 1), an output alphabet Y, and a likelihood function (probability of a particular value $y \in Y$, given an input value $x \in X$ for all x, y). A **discrete memoryless channel (DMC)** is a discrete channel for which the probability distribution of the output depends only on the current input and is independent of preceding inputs and outputs.

The **mutual information** $I(X; Y)$ of the input X and output Y of a communication channel is a measure of how much knowledge of the input reduces the uncertainty in the output. A result from information theory provides the following equation:

$$I(X;Y) = \sum_{x,y} p(x)p(y|x) \log \frac{p(y|x)}{p(y)}$$

where $p(x)$ is the probability of occurrence of x in X, and $p(y|x)$ is the probability of observing output value y given that input value x is sent. $I(X; Y)$ takes on values from 0 through 1. A value of 0 indicates that knowing x provides no information about the value of y. A value of 1 indicates that knowledge of x is sufficient to determine the value of y. For a communications channel W, $I(X; Y)$ is often expressed as $I(W)$.

The information capacity C of a channel is the maximum value of I(X; Y) taken over all possible input distributions p(x):

$$C = \max_{p(x)} I(X;Y)$$

Two important models used in the development of error-correcting codes are the binary symmetric channel and the binary erasure channel. A **binary symmetric channel (BSC)** has input and output alphabets of {0, 1}. The channel has a probability p of flipping a bit, as shown in Figure 14.2a.

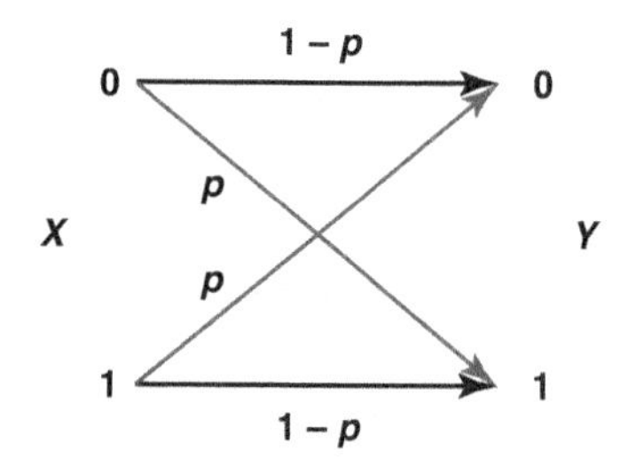

(a) Binary symmetric channel (BSC)

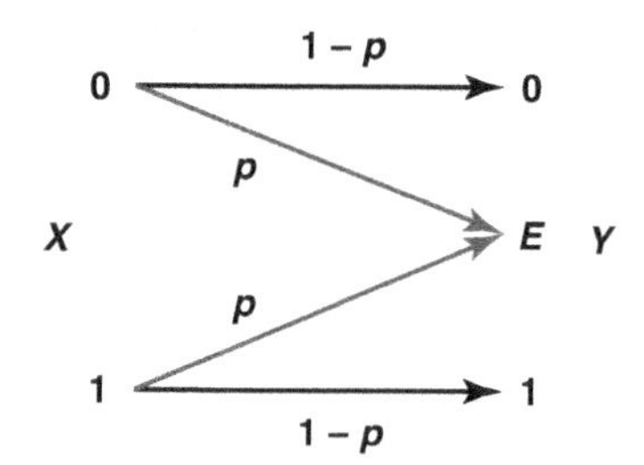

(b) Binary erasure channel (BEC)

FIGURE 14.2 Binary Symmetric Channel and Binary Erasure Channel

Thus:

$$\Pr[y=0 \mid x=0] = 1-p$$

$$\Pr[y=0 \mid x=1] = p$$

$$\Pr[y=1 \mid x=0] = p$$

$$\Pr[y=1 \mid x=1] = 1-p$$

The maximum value of I(X, Y) for a BSC occurs when Pr[x = 0] = Pr[x = 1] = 0.5. This yields an information channel capacity of a BSC of:

$$C = 1 - \mathrm{H}(p) = 1 + p\log(p) + (1-p)\log(1-p)$$

where H(p) is the binary entropy function. In essence, H(p) is the average amount of information for the random variable. If one of the two events is certain ($p = 1$ or $p = 0$), then the entropy is zero and $C = 1$. In terms of a communications channel, if p is 0 or 1, then the receiver is always certain of the transmitted bit value. The maximum value of H(X) = 1 is reached when the two outcomes are equally likely ($p = 0.5$), resulting in $C = 0$. If a communications channel preserves or flips a bit with equal probability, then there is no way to recover the transmitted bit value unless error correction coding, described later in this chapter, is applied.

BSC is a common communications channel model used in coding theory and information theory.

A **binary erasure channel (BEC)** has an input alphabet X of {0, 1} and an output alphabet Y of {0, 1, E}, where E is the erasure symbol. For an erasure probability of p, the channel exhibits the following conditional probabilities, as shown in Figure 14.2b:

$$\Pr[y=0 \mid x=0] = 1-p$$

$$\Pr[y=0 \mid x=1] = 0$$

$$\Pr[y=1 \mid x=0] = 0$$

$$\Pr[y=1 \mid x=1] = 1-p$$

$$\Pr[y=\mathrm{E} \mid x=0] = p$$

$$\Pr[y=\mathrm{E} \mid x=1] = p$$

The maximum value of I(X, Y) for a BEC occurs when Pr[x = 0] = Pr[x = 1] = 0.5. This yields an information channel capacity of a BEC of:

$$C = 1 - p$$

Note that this model excludes the probability of a bit being flipped (i.e., of a 0 being transmitted and a 1 received or vice versa). This limitation may appear unrealistic for an actual wireless link. However, the derivation of the polar code, discussed in Section 14.5, is easily done with the BEC assumption, but it is then applicable to an actual wireless channel.

14.2 Forward Error Correction

This section introduces block error correction and then examines the concept of Hamming distance.

Block Error Correction

A common technique used by link layer and transport protocols is error detection. In general, this technique involves the use of an error-detection code, typically a cyclic redundancy check (CRC), appended to the end of each transmitted block and checked for each received block. If an error is detected, the link or transport protocol includes procedures whereby the sending entity learns that an error is detected and retransmits the block that contained an error. For wireless applications, this approach is inadequate for two reasons:

- The bit error rate on a wireless link can be quite high, and a high rate would result in a large number of retransmissions.
- In some cases, especially with satellite links, the propagation delay is very long compared to the transmission time of a single frame. The result is a very inefficient system. The common approach to retransmission is to retransmit the frame in error plus all subsequent frames. With a long data link, an error in a single frame necessitates retransmitting many frames.

Instead, it would be desirable to enable the receiver to correct errors in an incoming transmission on the basis of the bits in that transmission. Figure 14.3 shows in general how this is done. On the transmission end, each k-bit block of data is mapped into an n-bit block ($n > k$) called a **codeword**, using a forward error correction (FEC) encoder. This is referred to as an (n, k) block error correction code. The codeword is then transmitted; in the case of wireless transmission, a modulator produces an analog signal for transmission. During transmission, the signal is subject to noise, which may produce bit errors in the signal. At the receiver, the incoming signal is demodulated to produce a bit string that is similar to the original codeword but that may contain errors.

This block is passed through an FEC decoder, and one of five possible outcomes results:

- **No errors:** If there are no bit errors, the input to the FEC decoder is identical to the original codeword, and the decoder produces the original data block as output.
- **Detectable, correctable errors:** For certain error patterns, the decoder is able to detect and correct those errors. Thus, even though the incoming data block differs from the transmitted codeword, the FEC decoder can map this block into the original data block.
- **Detectable, not correctable errors:** For certain error patterns, the decoder can detect but not correct the errors. In this case, the decoder simply reports an uncorrectable error.

- **Detectable, falsely correctable errors:** For certain, typically rare, error patterns, the decoder detects an error but does not correct it properly. It assumes that a certain block of data was sent when in reality a different one was sent that differs in at least one bit position. The receiver cannot distinguish this case from the case of detectable and correctable errors unless another layer of error correction is applied.
- **Undetectable errors:** For certain even more rare error patterns, the decoder does not detect that any errors have occurred and maps the incoming n-bit data block into a k-bit block that differs from the original k-bit block. The receiver cannot distinguish this case from the case of no errors unless another layer of error correction is applied.

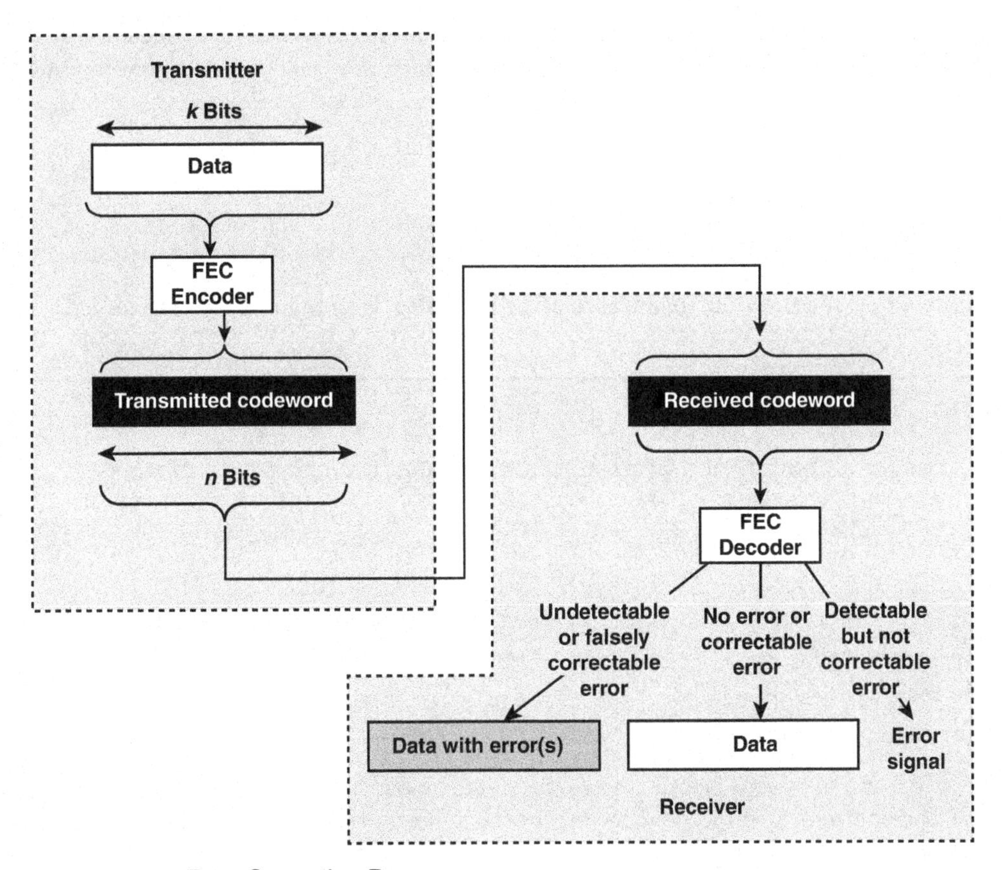

FIGURE 14.3 Error Correction Process

How is it possible for the decoder to correct bit errors? In essence, error correction works by adding redundancy to the transmitted message. Consider a scheme on a BSC with $p = 0.5$, in which a source binary 0 is transmitted as the codeword 0000 and binary 1 is transmitted as the codeword 1111. The redundancy makes it possible for the receiver to deduce what the original message was, even in the face of a high error rate. If a 0010 is received, the receiver recognizes that this is not a valid codeword (i.e., it is a **nonvalid codeword**). The receiver can assume that a 0000 was sent, corresponding to the original binary 0, because only one bit change would have occurred to make this happen. There is, however, a much more unlikely yet possible scenario in which a 1111 was sent. The decoder would then make a mistake by assuming that a 0 was sent. Consider if another received codeword were 0011. In this case, the decoder would not be able to decide because it would be equally likely that 0000 or 1111 was sent.

Hamming Distance

The Hamming distance is an important concept in assessing the effectiveness of error correction codes. The **Hamming distance** $\mathbf{d(v_1, v_2)}$ between two *n*-bit binary sequences $\mathbf{v}_1$ and $\mathbf{v}_2$ is the number of bits in which $\mathbf{v}_1$ and $\mathbf{v}_2$ disagree. For example, if:

$$\mathbf{v}_1 = 011011, \quad \mathbf{v}_2 = 110001$$

then:

$$d(\mathbf{v}_1, \mathbf{v}_2) = 3$$

The following example illustrates the significance of the Hamming distance for an (n, k) block code.

Example

For $k = 2$ and $n = 5$, consider the following mapping from a data block to a codeword, as performed by an FEC encoder:

Data Block	Codeword
00	00000
01	00111
10	11001
11	11110

Now suppose that a codeword block is received with the bit pattern 00100. This is not a valid codeword, and so the receiver has detected an error. Can the error be corrected? The receiver cannot be sure which data block was sent because 1, 2, 3, 4, or even all 5 of the bits that were transmitted may have been corrupted by noise. However, notice that it would require only a single bit change to transform the valid codeword 00000 into 00100. It would take two bit changes to transform 00111 to 00100, three bit changes to transform 11110 to 00100, and four bit changes to transform 11001 into 00100. Thus, the most likely codeword that was sent was 00000, and therefore the receiver's best hypothesis is to map the received block 00100 to data block 00, although the receiver does not know for certain that the transmitter sent 00. This is error correction. In terms of Hamming distances:

d(00000, 00100) = 1

d(00111, 00100) = 2

d(11001, 00100) = 4

d(11110, 00100) = 3

So the rule for error correction is that if an nonvalid codeword is received, then the valid codeword that is closest to it (i.e., the minimum Hamming distance) is selected. This works only if there is a unique valid codeword at a minimum distance from each nonvalid codeword.

For this example, it is not true that for every nonvalid codeword there is one and only one valid codeword at a minimum distance. There are $2^5 = 32$ possible codewords, of which 4 are valid, leaving 28 nonvalid codewords. For the nonvalid codewords:

Nonvalid Codeword	Minimum Distance	Valid Codeword	Nonvalid Codeword	Minimum Distance	Valid Codeword
00001	1	00000	10000	1	00000
00010	1	00000	10001	1	11001
00011	1	00111	10010	2	00000 or 11110
00100	1	00000	10011	2	00111 or 11001
00101	1	00111	10100	2	00000 or 11110
00110	1	00111	10101	2	00111 or 11001
01000	1	00000	10110	1	11110
01001	1	11001	10111	1	00111
01010	2	00000 or 11110	11000	1	11001
01011	2	00111 or 11001	11010	1	11110
01100	2	00000 or 11110	11011	1	11001
01101	2	00111 or 11001	11100	1	11110
01110	1	11110	11101	1	11001
01111	1	00111	11111	1	11110

There are eight cases in which a nonvalid codeword is a distance of 2 from two different valid codewords. Thus, if one such nonvalid codeword is received, an error in two bits could have caused it, and the receiver has no way to choose between the two alternatives. An error is detected but cannot be corrected. The only remedy is retransmission. However, in every case in which a single bit error occurs, the resulting codeword is of distance 1 from only one valid codeword, and the decision can be made. This code is therefore capable of correcting all single-bit errors but cannot correct double bit errors. Another way to see this is to look at the pairwise distances between valid codewords:

d(00000, 00111) = 3

d(00000, 11001) = 3

d(00000, 11110) = 4

d(00111, 11001) = 4

d(00111, 11110) = 3

d(11001, 11110) = 3

The minimum distance between valid codewords is 3. Therefore, a single bit error will result in a nonvalid codeword that is a distance of 1 from the original valid codeword but a distance of at least 2 from all other valid codewords. As a result, the code can always correct a single-bit error. Note that the code also will always detect a double-bit error.

The preceding example illustrates the essential properties of a block error-correcting code. An (n, k) block code encodes k data bits into n-bit codewords. Thus the design of a block code is equivalent to the design of a function of the form $\mathbf{v}_c = f(\mathbf{v}_d)$, where $\mathbf{v}_d$ is a vector of k data bits, and $\mathbf{v}_c$ is a vector of n codeword bits.

With an (n, k) block code, there are 2^k valid codewords out of a total of 2^n possible codewords. The ratio of redundant bits to data bits, $(n - k)/k$, is called the **redundancy** of the code, and the ratio of data bits to total bits, k/n, is called the **code rate**. The code rate is a measure of how much additional bandwidth is required to carry data at the same data rate as without the code. For example, a code rate of 1/2 requires double the bandwidth of an uncoded system to maintain the same data rate. The preceding example has a code rate of 2/5 and so requires a bandwidth 2.5 times the bandwidth for an uncoded system. If the data rate input to the encoder is 1 Mbps, then the output from the encoder must be at a rate of 2.5 Mbps to keep up.

For a code consisting of the codewords $\mathbf{w}_1, \mathbf{w}_2, \ldots, \mathbf{w}_s$, where $s = 2^k$, the minimum distance d_{min} of the code is defined as:

$$d_{\min} = \min_{i \neq j}\left[d\left(\mathbf{w}_i, \mathbf{w}_j\right)\right]$$

It can be shown (e.g., see [LIN04], [JOHN10]) that the following conditions hold: For a given positive integer t, if a code satisfies $d_{min} \geq 2t + 1$, then the code can correct all bit errors up to and including errors of t bits. If $d_{min} \geq 2t$, then all errors $\leq t - 1$ bits can be corrected, and errors of $2t$ bits can be detected but not, in general, corrected. Conversely, any code for which all errors of magnitude $\leq t$ are corrected must satisfy $d_{min} \geq 2t + 1$, and any code for which all errors of magnitude $\leq t - 1$ are corrected and all errors of magnitude t are detected must satisfy $d_{min} \geq 2t$.

Another way of putting the relationship between d_{min} and t is to say that the maximum number of guaranteed correctable errors per codeword satisfies:

$$t = \left\lfloor \frac{d_{min} - 1}{2} \right\rfloor$$

Furthermore, the maximum number of errors, t, that can be detected satisfies:

$$T = d_{min} - 1$$

To see this, consider that if d_{min} errors occur, this could change one valid codeword into another. Any number of errors less than d_{min} cannot result in another valid codeword.

The design of a block code involves a number of considerations:

- For given values of n and k, we would like the largest possible value of d_{min}.
- The code should be relatively easy to encode and decode, requiring minimal memory and processing time.
- We would like the number of extra bits, $(n - k)$, to be small to reduce bandwidth.
- We would like the number of extra bits, $(n - k)$, to be large to reduce error rate.

Clearly, the last two objectives are in conflict, and trade-offs must be made.

Before looking at specific codes, it will be useful to examine Figure 14.4. The literature on error-correcting codes frequently includes graphs of this sort to demonstrate the effectiveness of various encoding schemes. Recall from Chapter 13, "Air Interface Physical Layer," that coding can be used to reduce the required E_b/N_0 value to achieve a given bit error rate. The coding discussed in Chapter 13 has to do with the definition of signal elements to represent bits. The coding discussed in this chapter also has an effect on E_b/N_0. In Figure 14.4, the curve on the right is for an uncoded modulation system; the shaded region represents the area in which potential improvement can be achieved. In this region, a smaller BER (bit error rate) is achieved for a given E_b/N_0, and, conversely, for a given BER, a smaller E_b/N_0 is required. The other curve is a typical result of a code rate of one-half (i.e., an equal number of data and check bits). Note that at an error rate of 10^{-6}, the use of coding allows a reduction in E_b/N_0 of 2.77 dB. This reduction is referred to as the **coding gain**, which is defined as the reduction, in decibels, in the required E_b/N_0 to achieve a specified BER of an error correction coded system compared to an uncoded system using the same modulation.

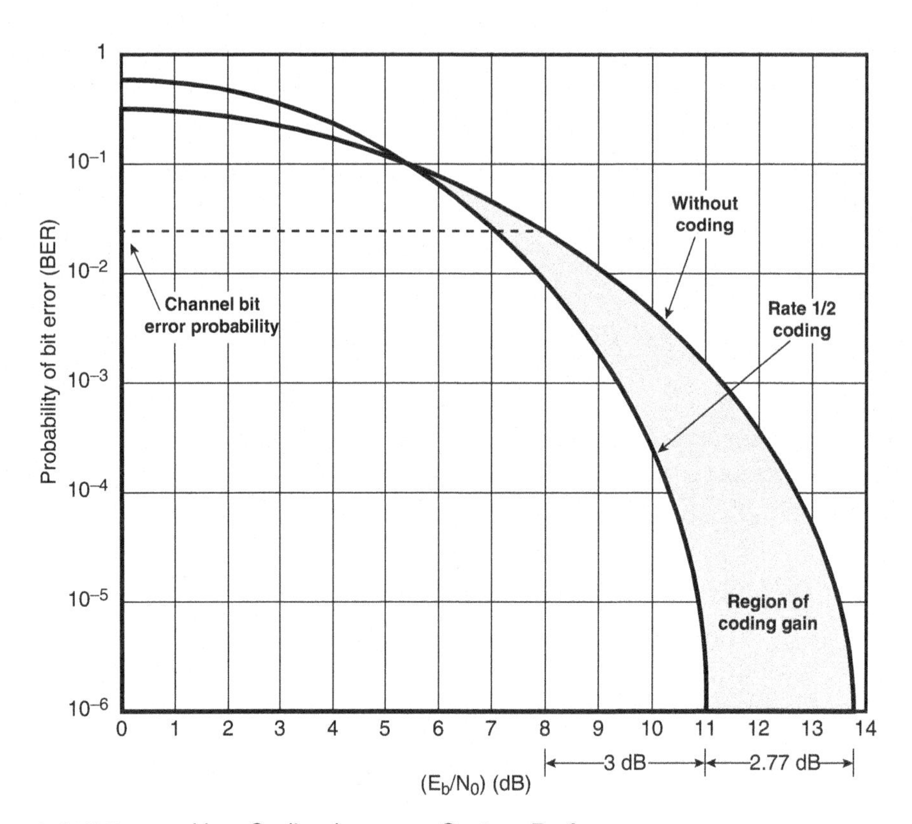

FIGURE 14.4 How Coding Improves System Performance

It is important to realize that the BER for the second rate 1/2 curve refers to the rate of uncorrected errors and that the E_b value refers to the energy per data bit. Because the rate is 1/2, there are two bits on the channel for each data bit, effectively reducing the data throughput by 1/2 (or requiring twice the raw data rate) as well. The energy per coded bit is half that of the energy per data bit, or a reduction of 3 dB. Based on the energy per coded bit for this system, the channel bit error rate is about 2.4×10^{-2}, or 0.024.

Finally, note that below a certain threshold of E_b/N_0, the coding scheme actually degrades performance. In the example in Figure 14.4, the threshold occurs at about 5.4 dB. Below this threshold, the extra check bits add to the system overhead that reduces the energy per data bit, causing increased errors. Above the threshold, the error-correcting power of the code more than compensates for the reduced E_b, resulting in a coding gain.

14.3 Parity-Check Matrix Codes

An important type of FEC code is the low-density parity check code (LDPC). LDPC codes are increasingly common in high-speed wireless specifications, including Wi-Fi, satellite, and cellular. LDPC codes exhibit performance in terms of bit error probability that is very close to the Shannon limit and can be efficiently implemented for high-speed use.

Before we discuss LDPC codes further, it is instructive to introduce the more general class of parity-check codes, of which the LDPC code is a specific example.

Consider a simple parity bit scheme used on blocks of n bits, consisting of $k = n - 1$ data bits and 1 parity-check bit, with even parity used.[1] Let c_1 through c_{n-1} be the data bits and c_n be the parity bit. Then the following condition holds:

$$c_1 \oplus c_2 \oplus \ldots \oplus c_n = 0 \qquad \textbf{(Equation 14.1)}$$

where addition is modulo 2 (equivalently, addition is the XOR function). There are 2^n possible codewords, of which 2^{n-1} are valid codewords. The valid codewords are those that satisfy Equation (14.1). If any of the valid codewords is received, the received block is accepted as being free of errors, and the first $n - 1$ bits are accepted as the valid data bits. This scheme can detect single-bit errors but cannot perform error correction.

Now generalize the parity check concept to consider codes whose words satisfy a set of $m = n - k$ simultaneous linear equations. A **parity-check code** that produces n-bit codewords is the set of solutions to the following set of equations:

$$\begin{aligned} h_{11}c_1 \oplus h_{12}c_2 \oplus \ldots + h_{1n}c_n &= 0 \\ h_{21}c_1 \oplus h_{22}c_2 \oplus \ldots \oplus h_{2n}c_n &= 0 \\ &\cdot \\ &\cdot \\ &\cdot \\ h_{m1}c_1 \oplus h_{m2}c_2 \oplus \ldots \oplus h_{mn}c_n &= 0 \end{aligned} \qquad \textbf{(Equation 14.2)}$$

where the coefficients h_{ij} take on the binary values 0 or 1. The specific set of values of h_{ij} define a specific code.

The $m \times n$ matrix $\mathbf{H} = [h_{ij}]$ is called the **parity-check matrix**. Each of the m rows of **H** corresponds to one of the individual equations in Equation (14.2). Each of the n columns of **H** corresponds to one bit of the codeword. If we represent the codeword by the row vector $\mathbf{c} = [c_j]$, then the equation set in Equation (14.2) can be represented as:

$$\mathrm{Hc^T = cH^T = 0} \qquad \textbf{(Equation 14.3)}$$

An (n, k) parity-check code encodes k data bits into an n-bit codeword. Typically, and without loss of generality, the convention used is that the leftmost k bits of the codeword reproduce the original k data bits, and the rightmost $(n - k)$ bits are the check bits (see Figure 14.5). This form is known as a **systematic code**. Thus, in the parity-check matrix **H**, the first k columns correspond to data bits, and the remaining columns correspond to check bits. With an (n, k) block code, there are 2^k valid codewords out of a total of 2^n possible codewords.

1. *Even parity* means that the total number of 1s in a block of data, including the parity bits, is even.

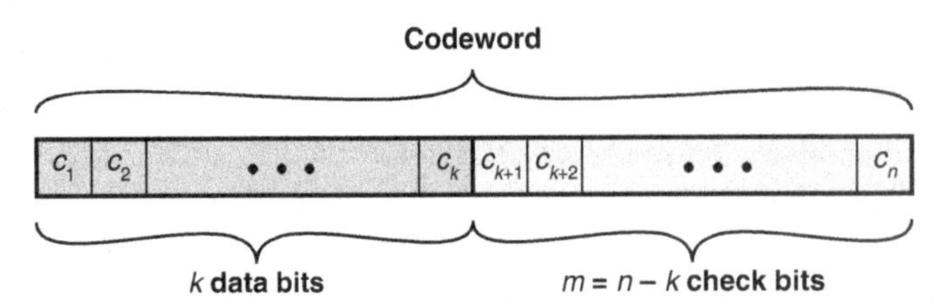

FIGURE 14.5 Structure of a Parity Check Codeword

The fundamental constraint on a parity-check code is that the code must have $(n - k)$ linearly independent equations. A code may have more equations, but only $(n - k)$ of them will be linearly independent. Without loss of generality, we can limit H to have the form:

$$\mathbf{H} = [\mathbf{A}\ \mathbf{I}_{n-k}]$$

where $\mathbf{I}_{n-k}$ is the $(n-k) \times (n-k)$ identity matrix, and $\mathbf{A}$ is a $(n-k) \times k$ matrix. The linear independence constraint is satisfied if and only if the determinant of $\mathbf{A}$ is nonzero. With this constraint, the k data bits may be specified arbitrarily in the equation set in Equation (14.2). The set of equations can then be solved for the values of the check bits. Put another way, for each of the 2^k possible sets of data bits, it is possible to uniquely solve the equation set in Equation (14.2) to determine the $(n - k)$ check bits.

Consider a (7, 4) check code defined by the constraint equations:

$$\begin{aligned} c_1 \oplus c_2 \oplus c_3 \oplus c_5 &= 0 \\ c_1 \oplus c_3 \oplus c_4 \oplus c_6 &= 0 \\ c_1 \oplus c_2 \oplus c_4 \oplus c_7 &= 0 \end{aligned} \qquad \textbf{(Equation 14.4)}$$

Using the parity-check matrix, we have:

$$\underbrace{\begin{bmatrix} 1 & 1 & 1 & 0 & 1 & 0 & 0 \\ 1 & 0 & 1 & 1 & 0 & 1 & 0 \\ 1 & 1 & 0 & 1 & 0 & 0 & 1 \end{bmatrix}}_{\mathbf{H}} \begin{bmatrix} c_1 \\ c_2 \\ c_3 \\ c_4 \\ c_5 \\ c_6 \\ c_7 \end{bmatrix} = \begin{bmatrix} 0 \\ 0 \\ 0 \end{bmatrix} \qquad \textbf{(Equation 14.5)}$$

For a parity-check code, as with any FEC code, three functions need to be performed:

- **Encoding:** For a given set of k data bits, generate the corresponding n-bit codeword.
- **Error detection:** For a given codeword, determine if there are one or more bits in error.
- **Error correction:** If an error is detected, perform error correction.

Encoding

For our example, to form a codeword, we first choose values for data bits c_1, c_2, c_3, and c_4; for example, $c_1 = 1$, $c_2 = 1$, $c_3 = 0$, $c_4 = 0$. We then solve the constraint equation set in Equation (14.4) by rewriting the equations so that we show each check bit as a function of data bits:

$$c_5 = c_1 \oplus c_2 \oplus c_3$$
$$c_6 = c_1 \oplus c_3 \oplus c_4 \qquad \textbf{(Equation 14.6)}$$
$$c_7 = c_1 \oplus c_2 \oplus c_4$$

Thus the codeword is 1100010. With 4 information bits, there are a total of 16 valid codewords out of the $2^7 = 128$ possible codewords. We can solve for each of the 16 possible combinations of data bits to calculate these codewords. The results are shown in Table 14.1.

TABLE 14.1 (7, 4) Parity Check Code Defined by Equation (14.6)

Data Bits				Check Bits		
c_1	c_2	c_3	c_4	c_5	c_6	c_7
0	0	0	0	0	0	0
0	0	0	1	0	1	1
0	0	1	0	1	1	0
0	0	1	1	1	0	1
0	1	0	0	1	0	1
0	1	0	1	1	1	0
0	1	1	0	0	1	1
0	1	1	1	0	0	0
1	0	0	0	1	1	1
1	0	0	1	1	0	0
1	0	1	0	0	0	1
1	0	1	1	0	1	0
1	1	0	0	0	1	0
1	1	0	1	0	0	1
1	1	1	0	1	0	0
1	1	1	1	1	1	1

A more general approach to encoding is to create a $k \times n$ **generator matrix** for the code. Using the equation set in Equation (14.4), we have:

$$\begin{bmatrix} c_1 & c_2 & c_3 & c_4 & c_5 & c_6 & c_7 \end{bmatrix} = \begin{bmatrix} c_1 & c_2 & c_3 & c_4 \end{bmatrix} \underbrace{\begin{bmatrix} 1 & 0 & 0 & 0 & 1 & 1 & 1 \\ 0 & 1 & 0 & 0 & 1 & 0 & 1 \\ 0 & 0 & 1 & 0 & 1 & 1 & 0 \\ 0 & 0 & 0 & 1 & 0 & 1 & 1 \end{bmatrix}}_{\mathbf{G}}$$

(Equation 14.7)

By convention, as described previously, the first k bits of **c** are the data bits. Label the data bits of **c** as $\mathbf{u} = [u_i]$, where $u_i = c_i$, for $i = 1$ to k. Then, the codeword corresponding to a data block is determined by:

$$c = uG$$ **(Equation 14.8)**

The first k columns of **G** are the identity matrix $\mathbf{I}_k$. **G** can be calculated from **H** as follows:

$$\mathbf{G} = [\mathbf{I}_k\ \mathbf{A}^T]$$

Error Detection

Doing error detection is simply a matter of applying Equation (14.3). If $\mathbf{Hc}^T$ yields a nonzero vector, then an error is detected. Using the example code, suppose that the codeword 1100010 is sent over a transmission channel, and 1101010 is received. We apply the parity-check matrix to the received codeword:

$$\underbrace{\begin{bmatrix} 1 & 1 & 1 & 0 & 1 & 0 & 0 \\ 1 & 0 & 1 & 1 & 0 & 1 & 0 \\ 1 & 1 & 0 & 1 & 0 & 0 & 1 \end{bmatrix}}_{\mathbf{H}} \begin{bmatrix} 1 \\ 1 \\ 0 \\ 1 \\ 0 \\ 1 \\ 0 \end{bmatrix} = \begin{bmatrix} 0 \\ 1 \\ 1 \end{bmatrix}$$

So the error is detected. Further, the resulting column vector is referred to as the **syndrome**. The syndrome indicates which of the individual parity-check equations do not equal 0. The result in this example indicates that the second and third parity-check equations in **H** are not satisfied. Thus at least one of the bits c_1, c_2, c_3, c_4, c_6, c_7 is in error.

Error Correction

One approach to error correction when a nonvalid codeword is received is to choose the valid codeword that is closest to the nonvalid codeword in terms of Hamming distance. This method works only in those instances in which there is a unique valid codeword at a minimum distance from the given nonvalid codeword. We can define the minimum distance of a code as the minimum Hamming distance d_{min} between any two valid codewords. The code can detect all patterns of $(d_{min} - 1)$ or fewer bit errors and can correct all patterns of $\left\lfloor \frac{d_{min} - 1}{2} \right\rfloor$ or fewer bit errors.

A parity-check code **H** has a minimum distance d_{min} if some XOR sum of d_{min} columns of **H** is equal to zero but no XOR sum of fewer than d_{min} columns of **H** is equal to zero. This property is of fundamental importance in the design of most parity-check codes. That is, the codes are designed so as to maximize d_{min}. An important exception to this approach is the low-density parity check code, discussed in the next section.

For the example **H** that we have been using, no single column is the **0** vector, and no two columns sum to **0**. But there are several instances in which three columns sum to **0** (e.g., the fourth, sixth, and seventh columns). Thus, $d_{min} = 3$.

The brute-force approach to error correction would be to compare a received nonvalid codeword to all 2^k valid codewords and pick the one with minimum distance. This approach is feasible only for small k. For codes with thousands of data bits per codeword, a variety of approaches have been developed. A discussion of these approaches is beyond the scope of this book.

14.4 Low-Density Parity-Check Codes

A **regular LDPC code** is a parity-check code with parity-check matrix **H** with the following properties:

1. Each code bit is involved with w_c parity constraints, and each parity constraint involves w_r bits.
2. Each row of **H** contains w_r 1s, where w_r is constant for every row.
3. Each column of **H** contains w_c 1s, where w_c is constant for every column.
4. The number of 1s in common between any two columns is zero or one.
5. Both w_r and w_c are small compared to the number of columns (i.e., the length of the codeword) and the number of rows (i.e., $w_c << n$ and $w_r << m$).

Because of property 5, **H** has a small density of 1s. That is, the elements of **H** are almost all equal to 0—hence the designation *low density*.

In practice, properties 2 and 3 are often violated in order to avoid having linearly dependent rows in **H**. For an **irregular LDPC code**, we can say that the average number of 1s per row and the average number of 1s per column are small compared to the number of columns and the number of rows.

Code Construction

A number of approaches have been developed for the construction of LDPC codes; these approaches differ in the definition of the LDPC parity-check matrix. One of the earliest approaches was proposed by Robert Gallager [GALL62]. The code is constructed as a stack of w_c submatrices. The topmost submatrix has n columns and $(n-k)/w_r$ rows. The first row of this submatrix has 1s in the first w_r positions and 0s elsewhere, the second row has 1s in the second group of w_r positions and 0s elsewhere, and so on. The other submatrices are random column permutations of the first submatrix. Figure 14.6a provides an example with $w_r = 4$.

$$\left[\begin{array}{cccccccccccc}
1&1&1&1&0&0&0&0&0&0&0&0\\
0&0&0&0&1&1&1&1&0&0&0&0\\
0&0&0&0&0&0&0&0&1&1&1&1\\
\hline
1&0&1&0&0&1&0&0&0&1&0&0\\
0&1&0&0&0&0&1&1&0&0&0&1\\
0&0&0&1&1&0&0&0&1&0&1&0\\
\hline
1&0&0&1&0&0&1&0&0&1&0&0\\
0&1&0&0&0&1&0&1&0&0&1&0\\
0&0&1&0&1&0&0&0&1&0&0&1
\end{array}\right]$$

(a) Gallager parity-check matrix with $w_c = 3, w_r = 4$

$$\left[\begin{array}{cccccccccccc}
1&0&0&0&0&1&0&1&0&1&0&0\\
1&0&0&1&1&0&0&0&0&0&1&0\\
0&1&0&0&1&0&1&0&1&0&0&0\\
0&0&1&0&0&1&0&0&0&0&1&1\\
0&0&1&0&0&0&1&1&0&0&0&1\\
0&1&0&0&1&0&0&0&1&0&1&0\\
1&0&0&1&0&0&1&0&0&1&0&0\\
0&1&0&0&0&1&0&1&0&1&0&0\\
0&0&1&1&0&0&0&0&1&0&0&1
\end{array}\right]$$

(b) MacKay-Neal parity-check matrix with $w_c = 3, w_r = 4$

$$\left[\begin{array}{ccc|ccccccccc}
1&0&0&1&0&0&0&0&0&0&0&0\\
1&0&0&1&1&0&0&0&0&0&0&0\\
0&1&0&0&1&1&0&0&0&0&0&0\\
0&0&1&0&0&1&1&0&0&0&0&0\\
0&0&1&0&0&0&1&1&0&0&0&0\\
0&1&0&0&0&0&0&1&1&0&0&0\\
1&0&0&0&0&0&0&0&1&1&0&0\\
0&1&0&0&0&0&0&0&0&1&1&0\\
0&0&1&0&0&0&0&0&0&0&1&1
\end{array}\right]$$

(c) Irregular repeat-accumulate parity-check matrix

FIGURE 14.6 Examples of LDPC Parity-Check Matrices

An alternative approach, proposed by MacKay and Neal [MACK99], is illustrated with an example in Figure 14.6b. For this scheme, a transmitted block length n and a source block length k are selected. The initial value of w_c is set to an integer greater than or equal to 3, and the initial value of w_r is set to $w_c n/m$. Columns of **H** are created one at a time from left to right. The first column is initially generated randomly, subject to the constraint that its weight must equal the initial value of w_c. The nonzero entries for each subsequent column are chosen randomly, subject to the constraint that each row weight does not exceed w_r. There may need to be some backtracking and relaxation of constraints to fill the entire matrix.

The two codes so far described are not systematic. That is, the data bits do not correspond to the first k columns of **H**. In general, with a nonsystematic code, it is more difficult to determine a technique for generating a codeword from a block of data bits. An LDPC construction technique that can be easily converted to a systematic equivalent is a **repeat-accumulate code** [DIVS98]. Construction begins with the rightmost column with a single 1 bit at the bottom of the column. The next column to the left has 1s in the lower two bit positions. For subsequent columns, the pair of 1s is shifted up one position, until the 1s reach the top of the column. The remaining k columns each have a single 1 bit per row. Figure 14.6c shows an example. The first parity bit can be computed as $c_4 = c_1$, the second as $c_5 = c_4 + c_1$, the third as $c_6 = c_5 + c_2$, and so on. Thus, after the first parity bit, each subsequent parity bit is a function of the preceding parity bit and one of the data bits.

Error Correction

As with any parity-check code, error detection of an LDPC code can be performed using the parity-check matrix. If $\mathbf{Hc}^\mathrm{T}$ yields a nonzero vector, then an error is detected. Because there are relatively few 1s in the parity-check matrix, this computation is reasonably efficient.

Error correction is a more complex process. Depending on the nature of the construction of the LDPC code, a number of methods have been developed to provide reasonably efficient computation. Here we give a brief overview.

Many error-detection techniques make use of a representation of an LDPC code known as a **Tanner graph**. The graph contains two kinds of nodes: check nodes, which correspond to rows of **H**, and bit nodes, which correspond to columns of **H** and hence to the bits of the codeword. Construction is as follows: Check node j is connected to the variable node I if and only if element h_{ij} in **H** is 1. Figure 14.7 shows an example.

Table 14.2a lists the constraint equations. Table 14.2b lists the messages sent and received by check nodes, and Table 14.2c lists the estimation of codeword bit values.

$$H = \begin{bmatrix} 0 & 1 & 0 & 1 & 1 & 0 & 0 & 1 \\ 1 & 1 & 1 & 0 & 0 & 1 & 0 & 0 \\ 0 & 0 & 1 & 0 & 0 & 1 & 1 & 1 \\ 1 & 0 & 0 & 1 & 1 & 0 & 1 & 0 \end{bmatrix}$$

(a) Parity-check matrix

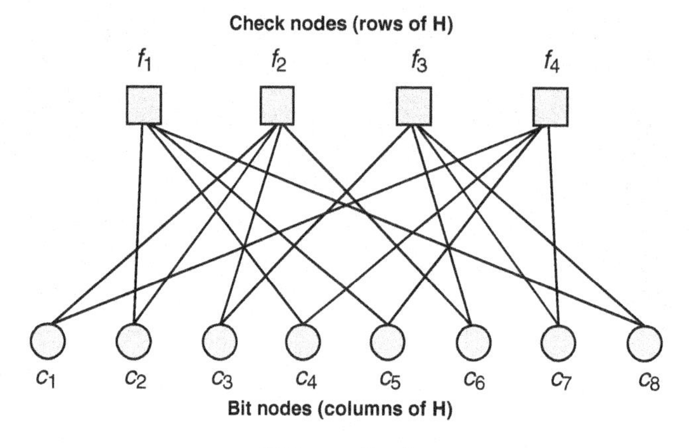

(b) Tanner Graph

FIGURE 14.7 Example of an LDPC Code

TABLE 14.2 Example Error Correction Technique for the LDPC Code of Figure 14.7

(a) Constraint Equations

Row of H	Check Node	Equation
1	f_1	$c_2 \oplus c_4 \oplus c_5 \oplus c_8 = 0$
2	f_2	$c_1 \oplus c_2 \oplus c_3 \oplus c_6 = 0$
3	f_3	$c_3 \oplus c_6 \oplus c_7 \oplus c_8 = 0$
4	f_4	$c_1 \oplus c_4 \oplus c_5 \oplus c_7 = 0$

(b) Messages Sent and Received by Check Nodes

Check Node	Messages				
f_1	Received:	$c_2 \to 1$	$c_4 \to 1$	$c_5 \to 0$	$c_8 \to 1$
	Sent:	$0 \to c_2$	$0 \to c_4$	$1 \to c_5$	$0 \to c_8$
f_2	Received:	$c_1 \to 1$	$c_2 \to 1$	$c_3 \to 0$	$c_6 \to 1$
	Sent:	$0 \to c_1$	$0 \to c_2$	$1 \to c_3$	$0 \to c_6$
f_3	Received:	$c_3 \to 0$	$c_6 \to 1$	$c_7 \to 0$	$c_8 \to 1$
	Sent:	$0 \to c_3$	$1 \to c_6$	$0 \to c_7$	$1 \to c_8$
f_4	Received:	$c_1 \to 1$	$c_4 \to 1$	$c_5 \to 0$	$c_7 \to 0$
	Sent:	$1 \to c_1$	$1 \to c_4$	$0 \to c_5$	$0 \to c_7$

(c) Estimation of Codeword Bit Values

Bit Node	Codeword Bit	Messages from Check Nodes	Decision
c_1	1	$f_2 \rightarrow 0$ $f_4 \rightarrow 1$	1
c_2	1	$f_1 \rightarrow 0$ $f_2 \rightarrow 0$	0
c_3	0	$f_2 \rightarrow 1$ $f_3 \rightarrow 0$	0
c_4	1	$f_1 \rightarrow 0$ $f_4 \rightarrow 1$	1
c_5	0	$f_1 \rightarrow 1$ $f_4 \rightarrow 0$	0
c_6	1	$f_2 \rightarrow 0$ $f_3 \rightarrow 1$	1
c_7	0	$f_3 \rightarrow 0$ $f_4 \rightarrow 0$	0
c_8	1	$f_1 \rightarrow 0$ $f_3 \rightarrow 1$	1

Now consider a very simple error-correction technique using the LDPC example of Figure 14.7. This technique is not efficient and is not used in practice. However, it is simple and gives some idea of the basic mechanisms involved in a typical error-correction algorithm for LDPC codes.

Assume for this example that the source codeword is 1001010. This is a valid codeword because $\mathbf{Hc}^{\mathbf{T}} = \mathbf{0}$. Assume that the second codeword bit is changed during transmission so that the received codeword is 11010101. The error-detection algorithm detects that one or more bit errors have occurred:

$$\begin{bmatrix} 0 & 1 & 0 & 1 & 1 & 0 & 0 & 1 \\ 1 & 1 & 1 & 0 & 0 & 1 & 0 & 0 \\ 0 & 0 & 1 & 0 & 0 & 1 & 1 & 1 \\ 1 & 0 & 0 & 1 & 1 & 0 & 1 & 0 \end{bmatrix} \begin{bmatrix} 1 \\ 1 \\ 0 \\ 1 \\ 0 \\ 1 \\ 0 \\ 1 \end{bmatrix} = \begin{bmatrix} 1 \\ 1 \\ 0 \\ 0 \end{bmatrix}$$

Error correction proceeds with the following steps:

Step 1. Using the Tanner graph of Figure 14.7b, each bit node sends its bit value to each linked check node (refer to Table 14.2b). Thus, c_1, which has the codeword bit value 1, sends this value to nodes f_2 and f_4.

Step 2. Each check node uses its corresponding constraint equation to calculate a bit value for each linked bit node and sends the value to the bit node. For example, the constraint equation for f_1, determined by the first row of **H**, is $c_2 \oplus c_4 \oplus c_5 \oplus c_8 = 0$. f_1 solves this equation four times, successively using each of its four inputs as the variable to be solved for. So, f_1 has received the values $c_2 = 1$, $c_4 = 1$, $c_5 = 0$, $c_8 = 1$. First, it solves $c_2 = c_4 \oplus c_5 \oplus c_8 = 0$ and sends the 0 to c_2. Then it solves $c_4 = c_2 \oplus c_5 \oplus c_8 = 0$ and sends the 0 to c_4—and so on for the remaining two variables.

Step 3. Each of the incoming bit values to each bit node is an estimate of the corrected value for that node. The original bit value of the node is another estimate. The bit node chooses the majority value as its final estimate of the value (refer to Table 14.2c). Thus, for c_1, two of the three estimates are 1, and 1 is chosen as the final value for the first bit of the codeword.

These three steps are repeated until the codeword stops changing or until all the constraint equations are satisfied. In this case, the codeword stabilizes after just one iteration.

The same concept of message passing between bit nodes and check nodes can be used with a probabilistic basis. In essence, the probability of a codeword bit being 1 or 0 is calculated in each iteration, and the loop stops when all of the probabilities pass a certain threshold.

There are a number of other approaches to error correction, many of which exploit the Tanner graph structure.

Encoding

A straightforward method of encoding data bits to form a codeword is to calculate the generator matrix **G** from the parity-check matrix **H**. Typically, for LDPC codes, this is a difficult and resource-intensive operation. Accordingly, a variety of techniques have been developed for solving the parity-check constraint equations to show the parity bits as a function of the data bits. For the repeat-accumulate code, as has been shown, this is a simple process. For other forms of LDPC codes, the encoding process is considerably more complex and often involves the use of the Tanner graph.

One other point worth noting is that, in general, with the exception of the repeat-accumulate code, the LDCP code is not systematic. That is, the data bits are not necessarily grouped together at the beginning of the codeword. In essence, the code designer selects codeword bits to be the data bits, and then the parity bits are calculated.

14.5 Polar Coding

In 2009, Erdal Arikan designed polar codes that also approach the Shannon limit but that are more structured than LDPC codes [ARIK09]. The mathematical structure of polar codes is such that their efficiency can be calculated and guaranteed. In practical applications, polar codes still perform worse than LDPC codes, but they are closing the gap, and 5G phones use both kinds of codes.

The polar code is a recursive FEC technique based on repeated use of the exclusive-OR (XOR) function. Polar codes can be constructed for any codeword length n that is a power of 2 and any number of data bits $k \leq n$.

Polar Encoder

This subsection first looks at a function G_N that takes N bits as input and produces N bits as output and then shows how this function is used to construct a (N, K) code. The most elemental form of G_N is G_2, shown in the upper left of Figure 14.8.

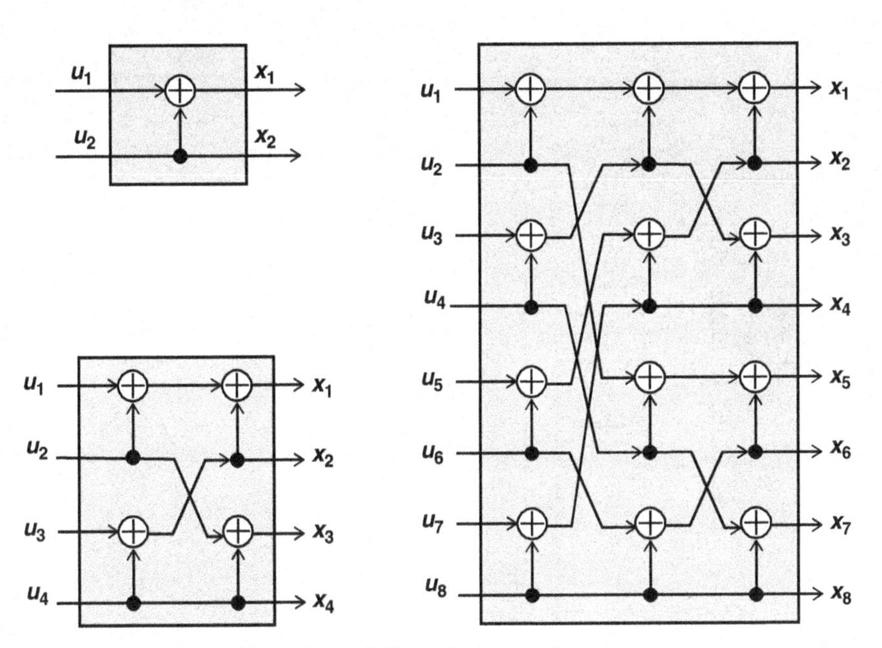

FIGURE 14.8 Polar Encoders of Sizes 2, 4, and 8

G_2 is defined by the equations:

$$x_1 = u_1 \oplus u_2$$

$$x_2 = u_2$$

Figure 14.9 indicates how higher levels of G are constructed by the recursive application of G_2. In particular, G_{2N} is constructed using N copies of G_2 and two copies of G_N. The box marked R_{2N} is a permutation called the *reverse shuffle*, which copies its odd-numbered inputs to its first N outputs and copies its even-numbered inputs to its last N outputs. Figure 14.8 shows the XOR implementation for G_2, G_4, and G_8.[2]

2. Various versions of G_4, G_8, and so on appear in the literature, using somewhat different recursive constructions. All are functionally equivalent.

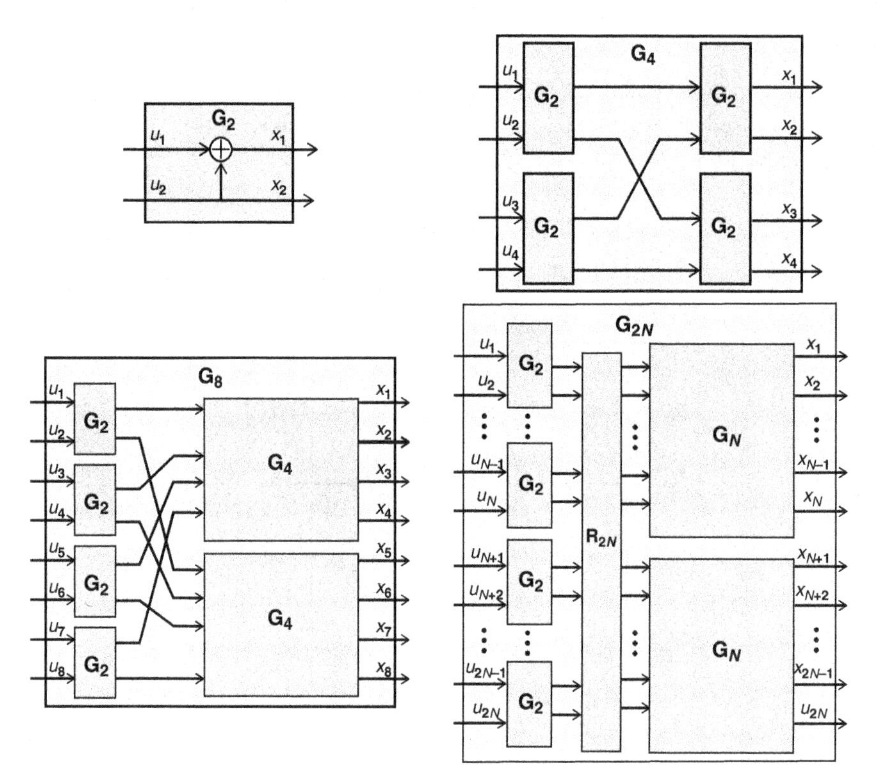

FIGURE 14.9 Recursive Construction of G_{2N}

Thus, G_N takes N binary inputs and produces N binary outputs. To construct a (N, K) code, the K data bits are each assigned to one of the input lines. The remaining input lines are frozen (i.e., have a fixed value of 0). The design challenge is to choose which K input bits will be the data bits to produce an efficient and effective code. A full mathematical development is beyond the scope of this book, but an intuitive explanation is provided in this section.

The essence of the technique for designing a polar code is as follows: Consider a general transmission model for a polar code. The N input bits $(u_1, u_2, \ldots)$ are encoded by G_N to produce an N-bit codeword $(x_1, x_2, \ldots)$. The bits of the codeword are then transmitted through a channel W that may introduce errors, so that the channel output $(y_1, y_2, \ldots)$ may differ from the codeword in some bit positions. Figure 14.10a represents this model. Figure 14.10b shows an (8, 4) encoder in which the data bits are assigned to inputs u_4, u_6, u_7, and u_8, and the remaining inputs are set to 0.

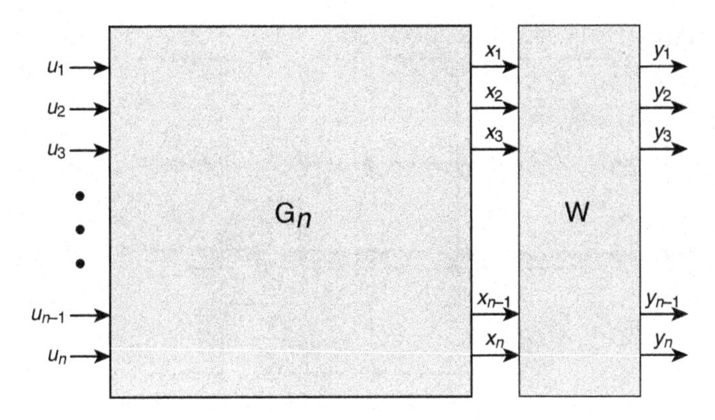

(a) General Transmission Model

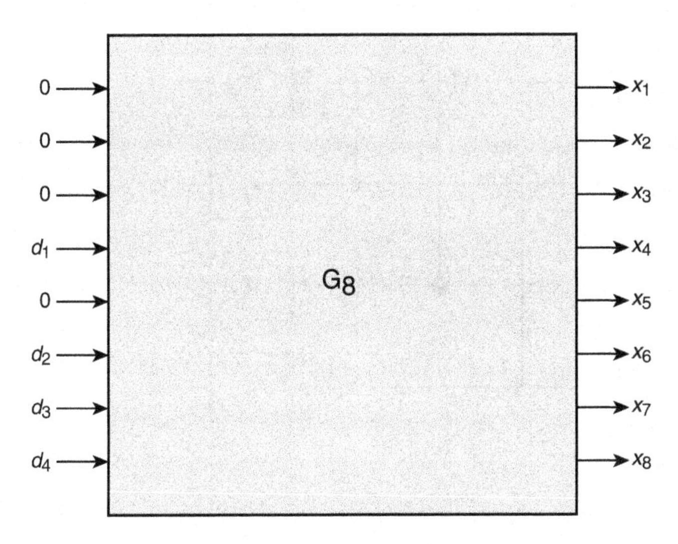

(b) An (8, 4) Polar Encoder

FIGURE 14.10 Polar Encoding

Synthetic Channels

Code design is based on the concept of a synthetic channel. This subsection provides an intuitive explanation of the concept. The derivation is based on the use of a binary erasure channel (BEC).

First, consider the transmission of two bits over two copies of a BEC W that has a successful transmission probability of $1 - p$, as shown in Figure 14.11a. Suppose the erasure probability is $p = 0.3$. Then the probability that a bit is successfully received is 0.7. Equivalently, $I(W) = 0.7$. This subsection uses the example of $I(W) = 0.7$ throughout.

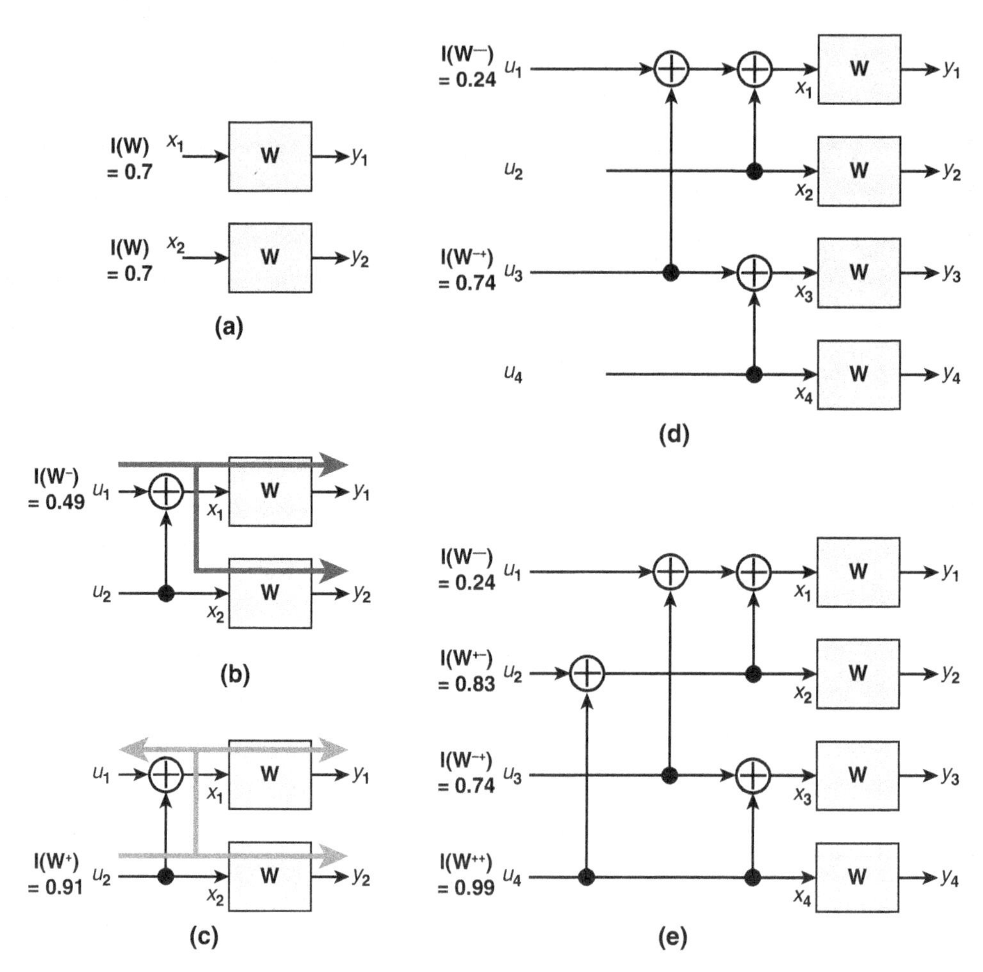

FIGURE 14.11 Channel Polarization

Consider the two-bit polar encoder shown in Figure 14.8, with the x_i passed through a BEC W to produce y_i. Now define two synthetic channels W^- and W^+. W^- has input u_1 and output $(y_1, y_2) = (u_1 \oplus u_2, u_2)$, shown in Figure 14.11b. The following are the possible outcomes of transmitting one bit on each W channel:

- Both channels are successfully received; this outcome has probability $(1 - p)^2$.
- The first channel yields an erase (E), and the second one is received successfully; this outcome has probability $p(1-p)$.
- The first channel is received successfully and the second one yields an erase (E); this outcome has probability $(1-p)p$.
- Both channels are erased; this outcome has probability p^2.

To summarize:

$$(y_1, y_2) = \begin{cases} (u_1 \oplus u_2, u_2) & (1-p)^2 \\ (\mathrm{E}, u_2) & p(1-p) \\ (u_1 \oplus u_2, \mathrm{E}) & (1-p)p \\ (\mathrm{E}, \mathrm{E}) & p^2 \end{cases}$$

Only in the first case can u_1 be recovered, by calculating:

$$u_1 = u_1 \oplus u_2 \oplus u_2$$

For W, the probability of success is 0.7. For W^-, the probability of success for $p = 0.3$ is $(1 - p)^2 = (1 - 0.3)^2 = 0.49$. So W^- is inferior to W in performance.

Now consider W^+ to be defined as having input u_2 and output $(y_1, y_2, u_1) = (u_1 \oplus u_2, u_2, u_1)$, shown in Figure 14.11c. This channel estimates the input received on the u_2 channel from (y_1, y_2, u_1). This is thus a synthetic or virtual channel because in fact only y_1 and y_2 are outputs of the transmission channel. The explanation is that a known value, such as 0, is sent on the W^- channel. This improves the performance on the W^+ channel. In essence, the scheme is to send data bits on "good" bit channels and to set each of the bad-bit channels to the same known value.

The two W channels yield the same probabilities as before, so the possible outcomes can be summarized as follows:

$$(y_1, y_2, u_1) = \begin{cases} (u_1 \oplus u_2, u_2, u_1) & (1-p)^2 \\ (\mathrm{E}, u_2, u_1) & p(1-p) \\ (u_1 \oplus u_2, \mathrm{E}, u_1) & (1-p)p \\ (\mathrm{E}, \mathrm{E}, u_1) & p^2 \end{cases}$$

In the first and second cases, u_2 is successfully decoded as it was received successfully. In the third case, $u_2 = u_1 \oplus u_2 \oplus u_2$. Thus the probability of successful decoding is:

$$(1 - p)^2 + p(1 - p) + (1 - p)p = 1 - p^2$$

For the example of $p = 0.3$, the comparable results then are:

W success probability $= 1 - p = 0.7$

W^- success probability $= (1 - p)^2 = 0.49$

W^+ success probability $= 1 - p^2 = 0.91$

The following relationships apply:

$I(W^-) = I(W)^2$

$I(W^+) = 2I(W) - I(W)^2$

The derivation of the two synthetic channels yields the following important conditions:

- **Polarization:** $I(W^-) \leq I(W) \leq I(W^+)$
- **Conservation of information:** $I(W^-) + I(W^+) = 2 \times I(W)$

To summarize, the W channel has been duplicated to produce two channels, W^- and W^+. This is done by making use of the G_2 encoder. This process can be applied recursively. Using G_2 in a cascade, W^- is split into W^{--} and W^{-+}; W^+ is split into W^{+-} and W^{++}. It is easier to see this process in stages. Figure 14.11d shows that W^- has been duplicated, and the G_2 transformation is applied to get W^{--} and W^{-+}. Figure 14.11e completes the duplication by applying G_2 to the W^+ channels.

The following relationships apply:

$I(W^{--}) = I(W^-)^2$

$I(W^{-+}) = 2I(W^-) - I(W^-)^2$

$I(W^{+-}) = I(W^+)^2$

$I(W^{++}) = 2I(W^+) - I(W^+)^2$

Returning to the example in which the probability of erasure is $p = 0.3$, the values for the channels are:

$I(W^{--}) = 0.24$

$I(W^{-+}) = 0.74$

$I(W^{+-}) = 0.83$

$I(W^{++}) = 0.99$

The result is one very good channel, one very bad channel, and two fairly good channels. Through repeated application of channel splitting, a point is reached at which all channels are either very good or very bad. The very good channels can then be used to transmit data bits. It can be shown that for an n-bit codeword and n synthetic channels, as n becomes large, $nI(W)$ channels will have a capacity of very close to 1, and $n(1 - I(W))$ channels will have a capacity close to 0. The resulting coding strategy is to encode $nI(W)$ data bits by assigning them to the error-free channels and to input 0 to the useless channels.

To generalize, n levels into this process transforms n uses of channel W into one use each of n channels. Channel W_j ($j = 1,\ldots, n$) has input u_j and output (y_i, ($i = 1,\ldots, j$); u_i, ($i = 1,\ldots, j - 1$)). The y_i are taken directly from the W channels. For the u_i, the following holds: If W_j is a bad channel, then u_i is known to be 0; if W_j is a good channel, then u_i is transmitted on an extremely reliable channel, and the decoded output for u_i is taken as valid.

Decoding

Like any other error-correcting code, a polar code produces a codeword that is subject to error on transmission. The receiver must take the received codeword and estimate the values of the data bits encoded at the transmission's end. In general terms:

1. The bits of a k-bit data block are assigned to k of the n bits ($u_1,\ldots, u_n$) input to a polar encoder.
2. The output of the decoder consists of n bits $X = (x_1,\ldots, x_n)$ which are transmitted through n instances of a W channel.
3. The receiver passes the received bits $Y = (y_1, \ldots, y_n)$ through a polar decoder to produce estimates of the ui bits $\left(\hat{u}_1,\ldots,\hat{u}_n\right)$.

The challenge with polar coding is to find an efficient decoder. This subsection summarizes the successive cancellation (SC) approach introduced in [ARIK09]. A number of more efficient but more complex methods have been proposed more recently (e.g., [ELKE18], [BALA15]).

The SC approach is suggested by the way in which the good and bad channels were derived in the preceding subsection. The SC decoder estimates the value of one bit at a time. To estimate u_i, the decoder takes as input the channel output Y and the previous bit estimates $\left(\hat{u}_1,\ldots,\hat{u}_{i-1}\right)$, which are assumed to be correct. Then:

- If u_i is frozen (preset to 0), set $\hat{u}_i = 0$.
- Otherwise, choose $\hat{u}_i$ to be the bit value that would have the higher probability of producing the channel output Y, taking into account the estimates that have already been produced for ($u_1,\ldots, u_{i-1}$). Express the probability as:

$$PW_{i,n}\left(y_1,\ldots,y_n,\hat{u}_1,\ldots,\hat{u}_{i-1}|u_i\right)$$

In other words, given that the decoder is presented with Y and has produced the estimates $\left(\hat{u}_1,\ldots,\hat{u}_{i-1}\right)$, which are presumably reliable, then is it more likely that the value of u_i is 0 or 1?

To compute $PW_{i,n}(y_1,\ldots, y_n, \hat{u}_1,\ldots,\hat{u}_{i-1} | u_i = 0)$, choose specific bit values for $u_i + 1$ through u_n and compute:

$$X = G_n(\hat{u}_1,\ldots,\hat{u}_{i-1}, 0, u_{i+1},\ldots, u_n)$$

The probability of sending this particular value of X through channel W and receiving Y is:

$$\Pr\left[Y|X\right] = \prod_{i=1}^{n} \Pr\left[y_i|x_i\right]$$

Calculate the expected value of this probability as the average over all possible values of u_{i+1} through u_n, keeping in mind that some inputs are fixed at 0 and some have an equal probability of being 0 or 1. Finally, if

$$PW_{i,n}(y_1,\ldots, y_n, \hat{u}_1,\ldots,\hat{u}_{i-1} | u_i = 0) \geq PW_{i,n}(y_1,\ldots, y_n, \hat{u}_1,\ldots,\hat{u}_{i-1} | u_i = 1)$$

set $\hat{u}_i = 0$; otherwise, set $\hat{u}_i = 1$. This rule, as described, has a complexity of $O(2^n)$. However, [ARIK09] describes a recursive method that reduces the complexity to $O(n \log n)$.

14.6 3GPP Channel Coding Specification

This section provides a brief overview of the variants of LDPC and polar coding specified by 3GPP. Table 14.3 indicates the channel coding used on the various types of 5G NR physical channels.

TABLE 14.3 Channel Coding for 5G NR Physical Channels

Channel	Description	Channel Coding
Uplink		
Physical Uplink Control Channel (PUCCH)	Carries uplink control information (UCI), which includes scheduling requests, channel quality indicator (CQI) information, and hybrid automatic repeat request (HARQ) information	Polar code
Physical Uplink Shared Channel (PUSCH)	Carries application data and also carries radio resource control (RRC) signaling messages and UCI	LDPC for application data Polar code for UCI
Downlink		
Physical Downlink Control Channel (PDCCH)	Conveys downlink control information (DCI), which includes scheduling decisions for PDSCH reception and for scheduling grants for transmission on the PUSCH	Polar code
Physical Downlink Shared Channel (PDSCH)	Carries application data and paging information	LDPC
Physical Broadcast Channel (PBCH)	Carries part of the system information required for UE to connect to a NR base station (gNodeB)	Polar code

According to 3GPP, a physical channel corresponds to a set of resource elements carrying information originating from higher layers. In frequency-division duplexing (FDD) mode, a physical channel is defined by code, frequency and, in the uplink, relative phase. In time-division duplexing (TDD) mode, a physical channel is defined by code, frequency, and time slot.

Quasi-Cyclic Low-Density Parity-Check Codes

3GPP specifies a structured form of LDPC codes known as quasi-cyclic low-density parity-check (QC-LDPC) codes, which exhibit advantages over other types of LDPC codes with respect to the hardware implementations of encoding and decoding using simple shift registers and logic circuits.

This subsection provides a brief overview; for a more detailed description, see [RICH18] and [HUI18].

The QC-LDPC code uses a rate-compatible matrix, called a base graph (see Figure 14.12).

A B C
D E

A = information bits
B = square matrix
C = zero matrix
D = extension bits
E = identity matrix

(a) Base matrix used to construct parity check matrix

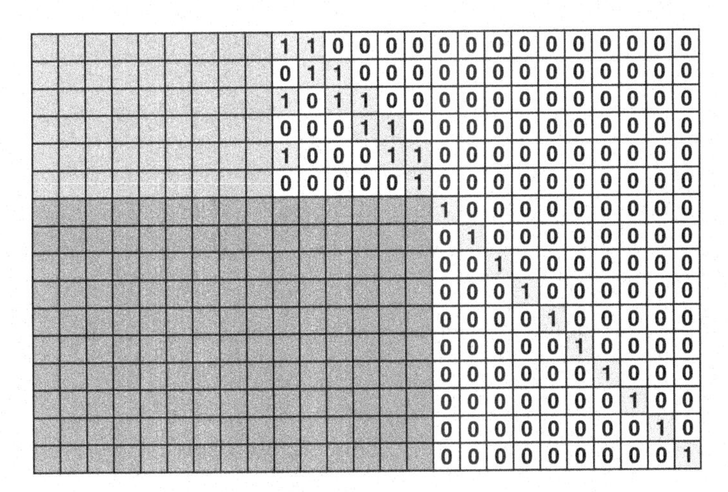

(b) Matrix example

FIGURE 14.12 Quasi-Cyclic Low-Density Parity-Check (QC-LDPC) Scheme

The base graph consists of the following submatrices:

- **A** corresponds to information bits.
- **B** is a square matrix that corresponds to parity bits.
- **C** is a matrix of all zeros.
- **D** is an extension matrix that relates to data link layer error control, which is beyond the scope of this book.
- **E** is the identity matrix.

TS 38.212 defines two base matrices, denoted by BG1 and BG2, respectively. These two matrices have the same structure, but BG1 is designed for information lengths up to 8448 bits and code rates from 2/3 to 8/9, while BG2 is designed for smaller information lengths no more than 3840 bits and code rates from 1/5 to 2/3. Because the parity-check matrix is smaller for the higher code rates, decoding latency and complexity is less for these rates. This enables high peak throughput and low latency. The expanded matrix for lower rates, which yields higher latency and complexity, is suitable for use cases requiring high reliability.

To create a parity check matrix **H**, each element of the base graph is replaced by a $Z \times Z$ matrix. Z is determined by the code block size and the base graph used. Z ranges from 2 to 384. Each 0 bit in the base graph is replaced by a matrix of all zeros. Each 1 bit is replaced by a row-wise circular right shift of an identity matrix, in which the amount of the shift is determined by the position of the bit in the base matrix.

Polar Coding with CRC

3GPP specifies the augmenting of the error-correcting polar code with the error-detecting cyclic redundancy check (CRC). The purpose of the addition of the CRC is to reduce the complexity and latency of the decoding process and to improve error performance. In essence, the form of SC decoding typically used generates a list of potential estimated codewords. The inclusion of a CRC enables the decoder to pick the most likely codeword.

This subsection provides a brief overview; for a more detailed description, see [BIOG18], [EGIL19], and [HUI18].

Figure 14.13 provides a simplified flowchart for the encoding process for CRC-aided polar codes. There are slight differences in the uplink and downlink versions. The downlink version is designed to minimize the complexity and resource consumption of the decoding process, which must often be done on constrained devices. The uplink version is designed to compensate as much as possible for the weaker signal carried on the uplink channel.

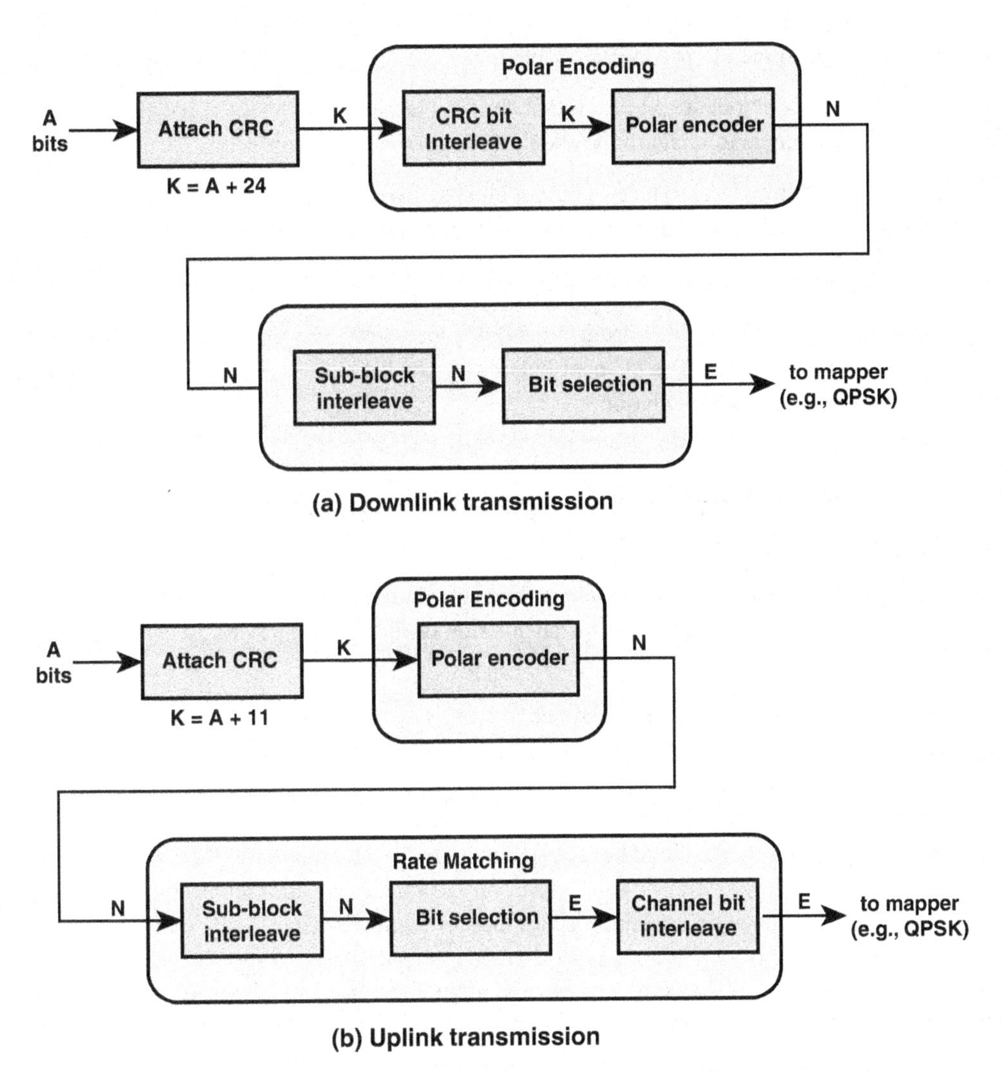

FIGURE 14.13 Polar Encoding with CRC

The most important functions for the downlink encoding process are as follows:

1. Begin with an A-bit block of data bits to be encoded. Compute a 24-bit CRC to yield a block length of $K = A + 24$ bits. This expanded block is presented to the polar encoder.

2. The K input bits are interleaved prior to polar encoding. Thus, the CRC bits are scattered throughout the input block rather than being placed as a unit at the end of the block. This interleaving is not performed for the uplink. The bit interleaver is designed in such a way that each CRC bit only depends on the previous information bits during SC decoding so that the CRC bit value can be calculated without waiting for information bits that come after the given CRC bit. This enables the decoder to discard possible erroneous candidates early, speeding up decoding.

3. Polar encoding is performed to produce an N-bit codeword.
4. Subblocks of the coder output are interleaved and placed in a circular buffer of length N. The effect of the interleaving is to increase the code's error correction capability.
5. Rate matching is performed to produce as output blocks of size E to yield the desired rate. The value E may be greater than, less than, or equal to N. In any case, E bits are taken from the circular buffer for output.

Uplink processing differs in several ways from downlink processing. The uplink encoder does not include interleaving of the CRC bits. The channel interleaver is a bit interleaver designed to improve coding performance for high-order modulation.

14.7 Hybrid Automatic Repeat Request

Automatic repeat request (ARQ) is a mechanism used in data link control and transport protocols that relies on the use of an error-detection code, such as the cyclic redundancy check (CRC). In essence, a receiver automatically initiates a request for retransmission of a data block when an error in transmission is detected.

In practical wireless system implementation, neither forward error correction (FEC) nor ARQ is an adequate one-size-fits-all solution. FEC may add unnecessary redundancy (i.e., use extra bandwidth) if channel conditions are good, and ARQ with error detection may cause excessive delays from retransmissions in poor channel conditions. Therefore, a solution known as *hybrid automatic repeat request (HARQ)* has been widely implemented in today's systems; it uses a combination of FEC to correct the most common errors and ARQ for retransmission when FEC cannot make corrections. Going beyond this basic concept, the following additional approaches may be implemented:

- **Soft decision decoding:** The decoding process can provide not just an assessment of a bit being 0 or 1 but also levels of confidence in those results.
- **Chase combining:** Previous frames that were not corrected by FEC need not be discarded. The soft decision information can be stored and then combined with soft decision information from retransmissions. The decoders in the receivers then use information from multiple frames, not just the current frame. This results in stronger FEC capabilities. In *chase combining*, exactly the same frames are retransmitted each time and soft combined.
- **Incremental redundancy:** Each time a sender retransmits, different coding information can be provided. This can accomplish two goals:
  - **Lower overhead:** The initial packets can include less coding, reducing overhead. For example, the first frame may only include a few bytes of an error detection code like CRC, with later frames then including FEC after the first frame has errors.

- **Stronger correction:** The retransmissions can provide stronger coding at lower coding rates. If adapted to the current wireless channel environment, this increases the probability of a successful transmission by the second or third frames.

- **Puncturing:** To provide the various coding rates for incremental redundancy, a different FEC coding algorithm could be used each time. A simpler approach, however, is puncturing, which removes bits to increase the coding rate.

Example

Consider an FEC coder that produces a 1/3 rate code that is punctured to become a 1/2 rate code. Say that there are 100 bits of data that become a 300-bit FEC codeword. To become a 1/2 rate FEC codeword, there need to be 2 bits of codeword for every 1 bit of data—hence a 200-bit codeword. This means 100 bits (i.e., 1 of every 3 bits of the original FEC code) need to be punctured. At the receiver, the missing 100 bits would be replaced before decoding. These could just be replaced with random numbers, which would mean that roughly 50 of those would coincidentally be correct and the other 50 incorrect. The original FEC code may actually still be sufficiently effective to correct those errors if the received signal-to-noise ratio is relatively good. If the received SNR is not high enough, a later retransmission might use less puncturing or puncture different bits.

In general, a punctured code is weaker than an unpunctured code at the same rate. However, simply puncturing the same code to achieve different coding rates allows the decoder structure to remain the same, so you don't have multiple decoders for different code rates. The benefits of this reduction in complexity can outweigh the reduction in performance due to puncturing. Used with HARQ incremental redundancy, puncturing takes a single output from an FEC coder and removes more or different bits each time.

- **Adaptive modulation and coding:** Systems use channel quality information (CQI) to estimate the best modulation and coding to work with HARQ. For example, LTE uses CQI to determine the highest modulation and coding rate that would provide a 10% block error rate for the first HARQ transmission. Also, if the CQI changes in the middle of an HARQ process, the modulation and coding might be adapted.

- **Parallel HARQ processes:** Some systems wait until the HARQ process finishes for one frame before sending the next frame; this is known as a stop-and-wait protocol. The process of waiting for an acknowledgment (ACK) or negative acknowledgment (NACK), followed by possible multiple retransmissions can be time-consuming, however. Therefore, some HARQ implementations allow for multiple open HARQ operations to occur at the same time. This is known as an *N-Channel Stop-and-Wait* protocol.

14.8 Key Terms and Review Questions

Key Terms

binary erasure channel (BEC)	hybrid automatic repeat request (HARQ)
binary symmetric channel (BSC)	low-density parity-check (LDPC) code
block error correction	mutual information
code rate	parity-check matrix code
coding gain	polar code
discrete channel	quasi-cyclic low-density parity-check (QC-LDPC)
discrete memoryless channel	redundancy
error burst	repeat-accumulate code
error correction	synthetic channel
error detection	systematic code
forward error correction	Tanner graph
generator matrix	
Hamming distance	

Review Questions

1. What is an error burst?
2. What is the difference between a BEC and a BSC?
3. In an (n, k) block error-correcting code, what do n and k represent?
4. What are the five possible outcomes of an FEC decoder?
5. What is a parity-check code?
6. How many simultaneous equations define an (n, k) parity-check code?
7. How do the equations mentioned in the preceding question relate to the parity-check matrix?
8. What is the relationship between the redundancy and code rate of an (n, k) code?
9. What is the purpose of the syndrome in a parity-check code?
10. What properties determine a regular LDPC code?
11. What elementary function is the foundation of polar codes?

12. What is a synthetic channel?
13. Explain the concept of polarization.
14. How does soft decision decoding improve HARQ?

14.9 References and Documents

References

ARIK09 Arikan, J. "Channel Polarization: A Method for Constructing Capacity-Achieving Codes for Symmetric Binary-Input Memoryless Channels." *IEEE Transactions on Information Theory*, July 2009.

BALA15 Balatsoukas-Stimming, A. "LLR-Based Successive Cancellation List Decoding of Polar Codes." *IEEE Transactions on Signal Processing*, June 2015.

BIOG18 Bioglio, V., Condo, C., and Land, I. "Design of Polar Codes in 5G New Radio." *IEEE Communications Surveys & Tutorials*, April 2018.

DIVS98 Divsalar, D., Jin, H., and McEliece, J. "Coding Theorems for 'Turbo-Like' Codes." *Proceedings, 36th Allerton Conference on Communication, Control, and Computing*, September 1998.

EGIL19 Egilmez, Z., et al. "The Development, Operation and Performance of the 5G Polar Codes." *IEEE Communications Surveys & Tutorials*, December 2019.

ELKE18 Elkelesh, A., et al. "Belief Propagation List Decoding of Polar Codes." *IEEE Communications Letters*, June 2018.

GALL62 Gallager, R. "Low-Density Parity-Check Codes." *IRE Transactions on Information Theory*, January 1962.

HUI18 Hui, D., et al. "Channel Coding in 5G New Radio." *IEEE Vehicular Technology Magazine*, December 2018.

JOHN10 Johnson, S. *Iterative Error Correction: Turbo, Low-Density Parity-Check, and Repeat-Accumulate Codes*. Cambridge, UK: Cambridge University Press, 2010.

LIN04 Lin, S., and Costello, D. *Error Control Coding*. Upper Saddle River, NJ: Pearson, 2004.

MACK99 Mackay, D., and Neal, R. "Good Error-Correcting Codes Based on Very Sparse Matrices." *IEEE Transactions on Information Theory*, May 1999.

RICH18 Richardson, T., and Kudekar, S. "Design of Low-Density Parity Check Codes for 5G New Radio." *IEEE Communications Magazine*, March 2018.

Documents

3GPP TR 38.802 *Technical Specification Group Radio Access Network, Study on New Radio Access Technology Physical Layer Aspects (Release 14).* September 2017.

3GPP TS 38.212 *Technical Specification Group Radio Access Network, NR, Multiplexing and Channel Coding (Release 16).* December 2020.

ITU-T Q.9 *Vocabulary of Switching and Signalling Terms*. November 1988.

Chapter 15

5G Radio Access Network

Learning Objectives

After studying this chapter, you should be able to:

- Explain the roles of gNB, ng-eNB, en-gNB, and eNB nodes in a RAN architecture
- Discuss the various interfaces that are prominent parts of a RAN
- Present an overview of the fronthaul, midhaul, and backhaul elements of a RAN transport network
- Make a presentation on the need for and the architecture of integrated access and backhaul

This chapter provides an overview of key aspects of the radio access network portion of a 5G system. Section 15.1 provides an overview of radio access network (RAN) architecture, highlighting the roles of various types of RAN nodes. Section 15.2 details the functional split between the RAN and the core network. Section 15.3 describes the RAN channel structure and defines the logical, transport, and physical channels, as well as the relationship between them. Section 15.4 discusses the protocol architecture and key RAN interfaces. Section 15.5 provides an overview of the RAN transport network. Finally, Section 15.6 discusses concepts related to integrated architecture and backhaul.

15.1 Overall RAN Architecture

The overall RAN architecture, in terms of the deployment of base station RAN nodes, is dictated by the need to coexist with 4G user equipment (UE) and 4G core networks for an extended period. Figure 15.1, from [GSMA20], shows the relative market share of 2G through 5G offerings over time. In 2019, 4G became the dominant mobile technology across the world, with more than 4 billion connections, accounting for 52% of total connections (excluding licensed cellular IoT). 4G connections will continue to grow for the next few years, peaking at just under 60% of global connections by 2023. It is clear that 4G UE will form a major portion of the cellular demand for quite a few years to come.

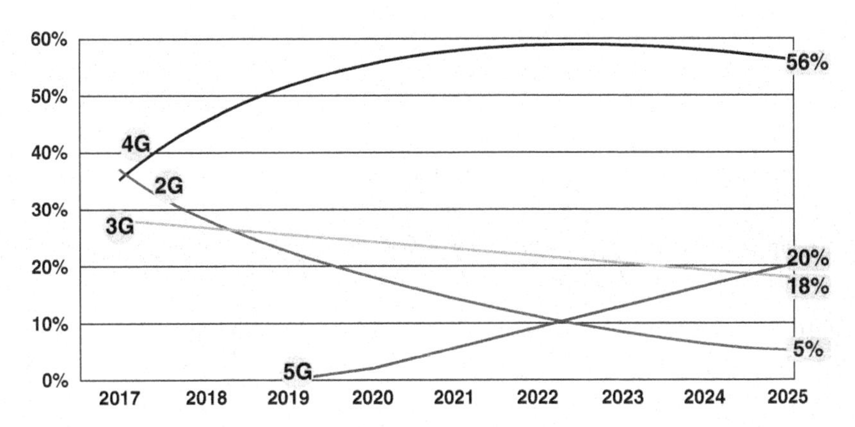

FIGURE 15.1 Global Cellular Market Share

The various options and mechanisms involved in supporting this 4G/5G transition form a complex subject. This section provides an overview of the general architectural considerations; [GSMA18] and [GSMA19] provide more details.

The most important requirement for 5G carriers is to provide full support to both 4G and 5G UE. Figure 15.2, from TS 38.300 (*Technical Specification Group Radio Access Network, NR, NR and NG-RAN Overall Description, Stage 2 [Release 16]*, September 2020), is a simplified view of the overall RAN architecture and its interface to the 5G core network for providing that support.

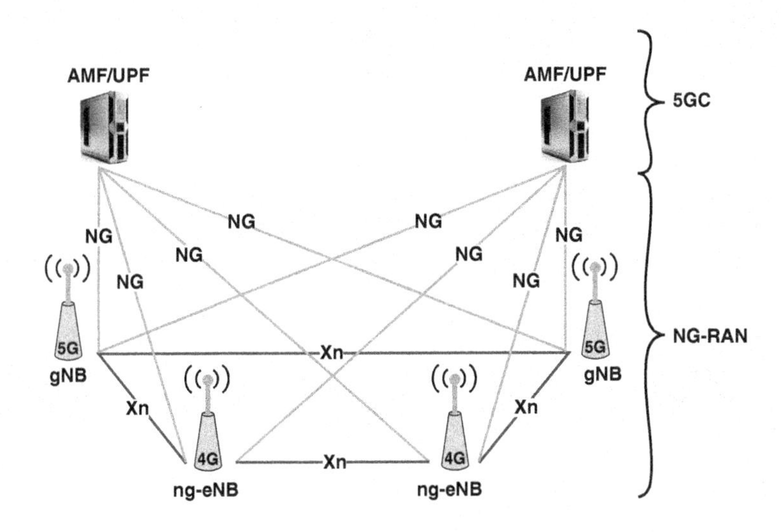

AMF = Access and Mobility Management Function
UPF = User Plane Function
5GC = 5th-Generation Core Network
NG-RAN = 5th-Generation (Next Generation) Radio Access Network

FIGURE 15.2 Overall Radio Access Network Architecture

Figure 15.2 depicts two types of base stations. The gNB node provides 5G NR (New Radio) user plane and control plane protocol terminations toward the UE and connects via the NG interface to the 5GC (5G core). The ng-eNB node provides 4G, referred to as E-UTRA (Evolved Universal Terrestrial Radio Access), user plane and control plane protocol terminations toward the UE and connects via the NG interface to the 5GC. The base stations interconnect with each other by means of the Xn interface. Base stations connect to the core network through the NG interfaces, more specifically to the AMF (access and mobility management function) by means of the NG-C interface and to the UPF (user plane function) by means of the NG-U interface.

Figure 15.3, from TS 38.401 (*Technical Specification Group Radio Access Network, NG-RAN, Architecture Description [Release 16]*, September 2020), provides a different perspective on key 5G RAN interfaces. A gNB may be a single integrated system, referred to as a monolithic or non-split node. Or a gNB may be organized as a split node, consisting of a gNB-central unit (gNB-CU) and one or more gNB-distributed units (gNB-DUs). The CU processes non-real-time protocols and services, and the DU processes physical layer protocol and real-time services. One gNB-DU supports one or multiple cells. One cell is supported by only one gNB-DU. A gNB-CU and the gNB-DU units are connected via the F1 logical interface. One gNB-DU is connected to only one gNB-CU. For resiliency, a gNB-DU may also be able to connect to another gNB-CU (if the primary gNB-CU fails) through appropriate implementation. NG, Xn, and F1 are logical interfaces.

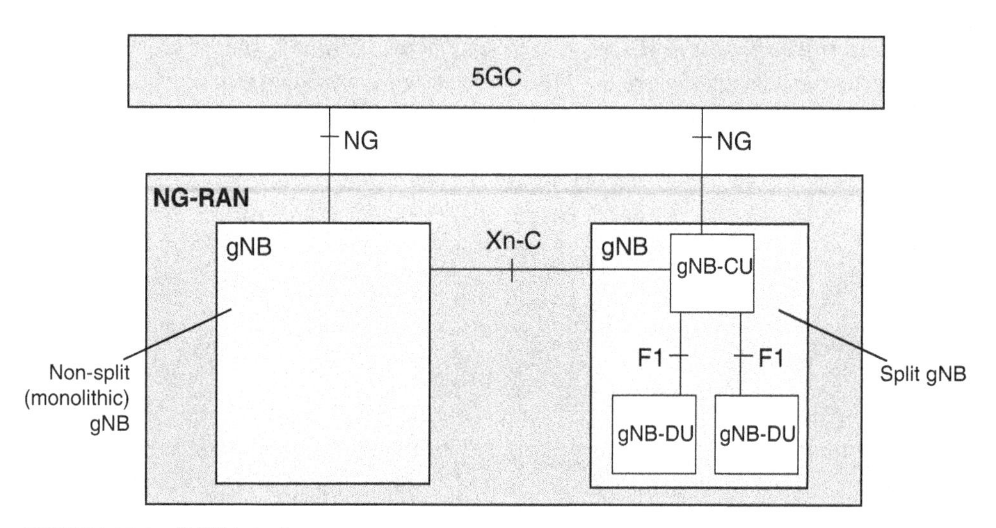

FIGURE 15.3 RAN Interfaces

Figure 15.4 shows the logical structure of a split RAN node. The figure shows logical nodes (i.e., CU-C, CU-U, and DU) internal to a logical gNB/en-gNB. Protocol terminations of the NG and Xn interfaces are depicted as ellipses in the figure. The terms *central entity* and *distributed entity* refer to physical network nodes.

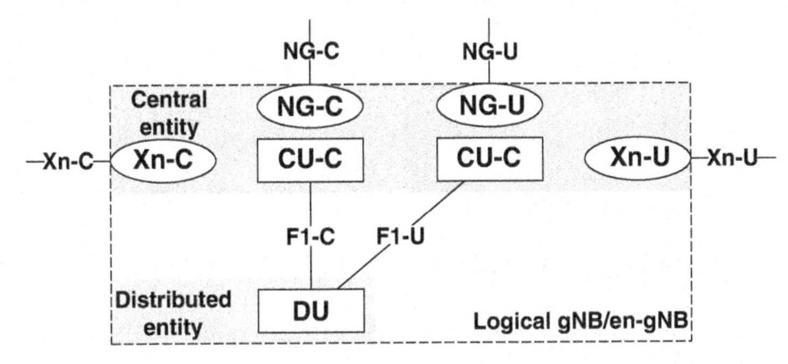

FIGURE 15.4 Example of the Deployment of a Logical gNB

Although the primary emphasis in the development of transition schemes is the support of 4G devices on 5G networks, there is also a need to consider support for 5G devices that are in use in locations that result in routing through 4G core networks. 5G devices also include the capability of operating on a 4G RAN and core network. However, a technique known as **dual connectivity** enables provision of better performance for 5G devices on a 4G core network.

Figure 15.5, from TS 37.340 (*Technical Specification Group Radio Access Network, NR, Evolved Universal Terrestrial Radio Access [E-UTRA] and NR, Multi-connectivity, Stage 2 [Release 16]*, September 2020), illustrates the architecture. This architecture supports the EN-DC or E-UTRA New Radio dual connectivity option for rapid deployment of 5G for current operators of 4G networks. The en-gNB provides a 5G radio interface for UEs to connect to the 4G core network. A 4G eNB acts as a primary, or controlling, node that is in control of the radio connection with the UE, and the en-gNB is used as a secondary, or controlled, node. This arrangement provides very high bitrates to the UEs that support dual connectivity without much impact on the core infrastructure. A typical application is the deployment of small cells (such as secondary millimeter wave cells) to enhance the capacity of a macro cell, where the UEs can communicate with both cells simultaneously (dual connectivity). Note that the en-gNB and the eNB can communicate through the X2 interface.

ITU-R Report M.2320 (*Future Technology Trends of Terrestrial IMT Systems*, November 2014) lists the following benefits of the dual connectivity option:

- **Mobility robustness:** Vital control plane information such as handover commands can be transmitted from an overlaid macro node even if the user data is provided by a low-power node. A terminal being connected to a single node only may lose the connection if it has moved outside the coverage area of the low-power node before the handover procedure is completed; this problem can be avoided if the overlaid wide-area macro layer is responsible for transmitting handover commands. Alternatively, the handover commands can be transmitted from both the source and the destination nodes.
- **User throughput enhancement:** By aggregating data streams from different sites, a higher user throughput can be achieved. This can be seen as an extension of carrier aggregation. Different nodes can also be used for different data flows, depending on the quality of service (QoS) needs.

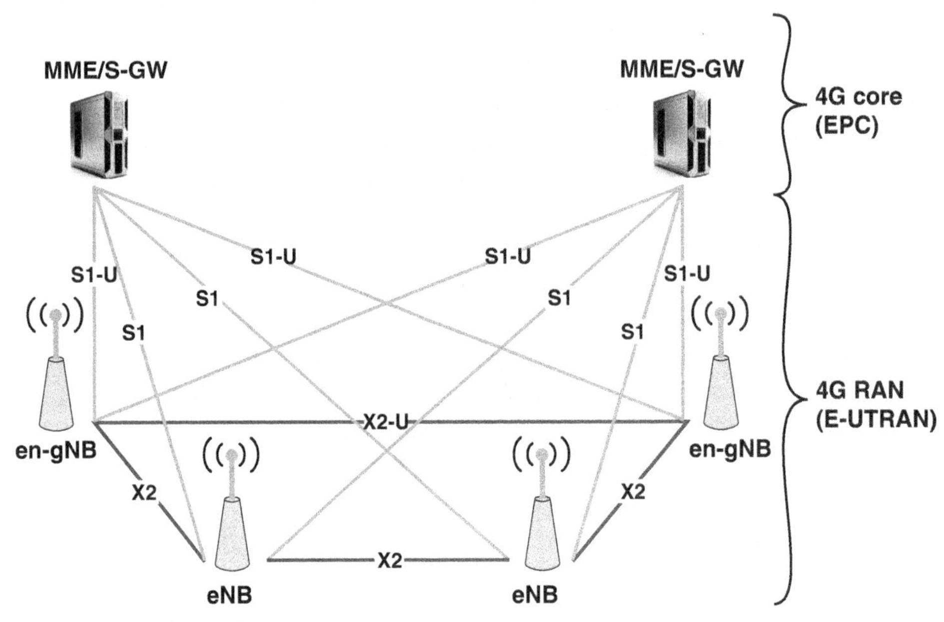

EN-DC = E-UTRA-NR Dual Connectivity
EPC = Evolved Packet Core
E-UTRAN = Evolved Universal Terrestrial Radio Access Network
MME = Mobile Management Entity
S-GW = Serving Gateway

FIGURE 15.5 EN-DC Overall Architecture

- **Signaling reduction:** Depending on realization of inter-node radio resource aggregation, signaling overhead toward the core network can potentially be saved by keeping the mobility anchor in an overlaid macro node.
- **Uplink–downlink separation:** Uplink reception may occur in a network node different from that used for downlink transmissions. Uplink–downlink separation is particularly advantageous in a heterogeneous deployment where the node with the strongest received signal at the terminal (and thus typically used for downlink transmissions) may not be the node with the lowest path loss to the terminal (and thus not the best node for uplink reception) as the difference in transmission power and/or load between a macro node and a low-power node can be substantial.
- **Interference reduction:** Interference reduction refers to minimizing unnecessary transmissions from the low-power nodes. As terminals can obtain essential system information from the overlaid macro layer, there is no need to also broadcast this information from the low-power nodes. This can help in reducing the overall energy consumption as a low-power node only needs to transmit when there is a terminal to serve in its coverage area.

Table 15.1 summarizes types of RAN nodes.

TABLE 15.1 Types of RAN Nodes

Node Name	User Equipment (UE)	Core Network
gNB	5G	5G
ng-eNB	4G	5G
en-gNB	5G	4G
eNB	4G	4G

15.2 RAN–Core Functional Split

Figure 15.6, from TS 38.300, shows the major functional elements performed by the RAN, together with functions within the core network that specifically relate to the RAN. The outer shaded boxes depict the logical nodes, and the inner white boxes depict the main functions at each node.

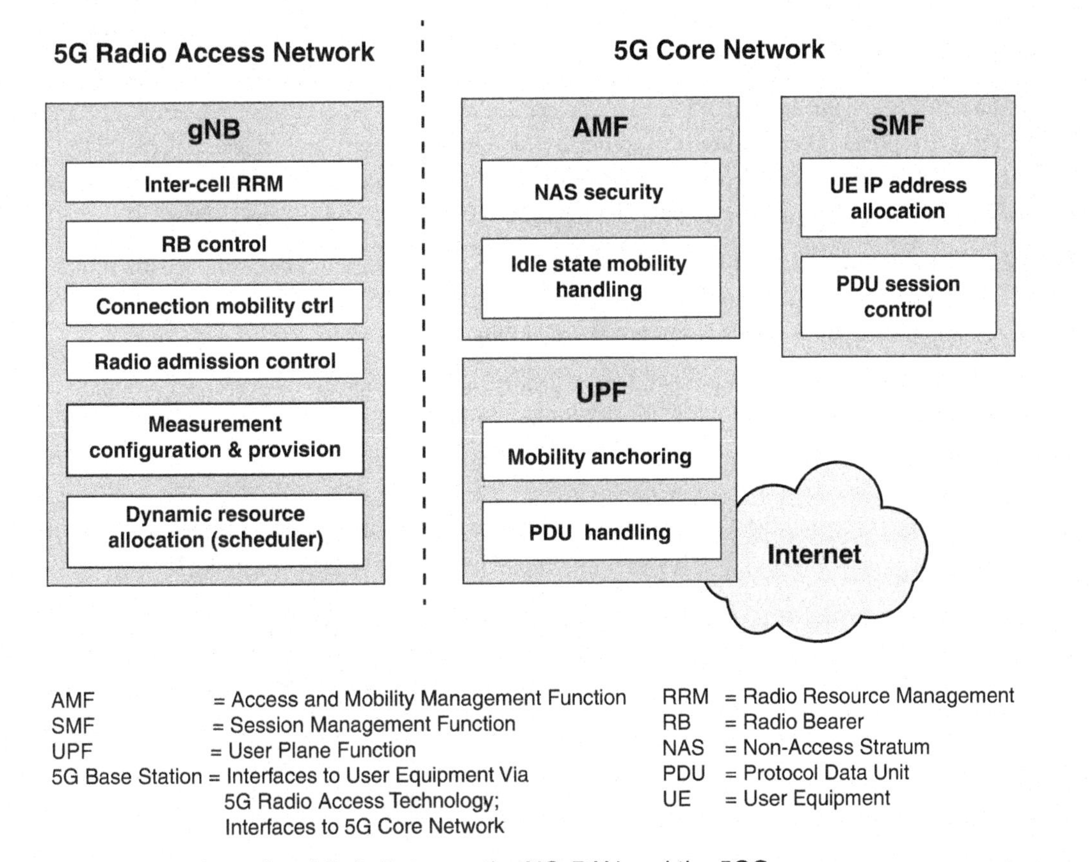

FIGURE 15.6 Functional Split Between the NG-RAN and the 5GC

RAN Functional Areas

Figure 15.6 depicts the following functional areas in the NG-RAN (next-generation radio access network):

- **Inter-cell radio resource management:** Allows the UE to detect neighbor cells, query about the best serving cell, and support the network during handover decisions by providing measurement feedback.
- **Radio bearer control (RBC):** Consists of the procedure for configuration (e.g., security), establishment, and maintenance of the radio bearer (RB) on both the uplink and downlink with different levels of quality of service (QoS). The term *radio bearer* refers to an information transmission path of defined capacity, delay, bit error rate, and other parameters.
- **Connection mobility control (CMC):** Functions both in UE idle mode and connected mode. Idle mode is a state in which the UE is switched on but does not have any established connection. Connected mode is the state in which the UE is switched on and has an established connection. In idle mode, the CMC performs cell selection and reselection. The connected mode involves handover procedures triggered on the basis of the outcome of CMC algorithms.
- **Radio admission control (RAC):** Decides whether a new radio bearer admission request is admitted or rejected. The objective is to optimize radio resource usage while maintaining the QoS of existing user connections. Note that the RAC decides on admission or rejection for a new radio bearer, whereas the RBC takes care of bearer maintenance and bearer release operations.
- **Measurement configuration and provision:** Consists of provisioning the configuration of the UE for radio resource management procedures such as cell selection and reselection and for requesting measurement reports to improve scheduling.
- **Dynamic resource allocation (scheduler):** Consists of scheduling RF resources according to their availability on the uplink and downlink for multiple UEs according to the QoS profiles of a radio bearer.

The functions in the preceding list all concern radio resource management and constitute only a small portion of the total functionality of RAN nodes. TS 38.300 lists the following additional functions:

- IP and Ethernet header compression, encryption, and integrity protection of data
- Selection of an AMF at UE attachment when no routing to an AMF can be determined from the information provided by the UE
- Routing of the user plane data toward the UPF(s)
- Routing of control plane information toward the AMF
- Connection setup and release
- Scheduling and transmission of paging messages

- Scheduling and transmission of system broadcast information, which originates from the AMF or operations and maintenance (O&M)
- Measurement and measurement reporting configuration for mobility and scheduling
- Transport layer packet marking in the uplink
- Session management
- Support of network slicing
- QoS flow management and mapping to data radio bearers
- Support of UEs in the radio resource control inactive (RRC_INACTIVE) state
- Distribution function for non-access stratum (NAS) messages
- Radio access network sharing
- Dual connectivity
- Tight interworking between NR and E-UTRA
- Maintenance of security and radio configuration for user plane cellular IoT optimization

Core Functional Areas

On the core network side, the NG-RAN nodes interact primarily with the following three functions:

- **Access and mobility management function (AMF):** The AMF provides UE authentication, authorization, and mobility management services. Two key functional components of AMF are the following:
  - Non-access stratum (NAS) is the highest protocol layer of the control plane between a UE and the mobility management entity (MME) in the core network. Main functions of the protocols that are part of the NAS are the support of mobility of the UE and the support of session management procedures to establish and maintain IP connectivity between the UE and a packet data network gateway (PDN GW). The NAS is used to maintain continuous communications with the UE as it moves. In contrast, the access stratum carries information over the wireless portion of a connection. NAS security involves IP header compression, encryption, and integrity protection of data based on the NAS security keys derived during the registration and authentication procedure.
  - Idle state mobility handling deals with cell selection and reselection while the UE is in idle mode, as well as reachability determination.
- **Session management function (SMF):** The UE IP address allocation process assigns an IP address to the UE at session establishment. This ensures the ability to route data packets within the 5G system and also supports data reception and forwarding to outside networks

and provides interconnectivity to external packet data networks (PDNs). In cooperation with the UPF, the SMF establishes, maintains, and releases a protocol data unit (PDU) session for user data transfer, which is defined as an association between the UE and a data network that provides a PDU connectivity.

- **User plane function (UPF):** UE mobility handling deals with ensuring that there is no data loss when there is a connection transfer due to handover that involves changing anchor points. Once a session is established, the UPF has a responsibility for PDU handling. This includes the basic functions of packet routing, forwarding, and QoS handling.

The 5GC functions depicted in Figure 15.6 constitute only a small portion of the total 5GC functionality, emphasizing interfacing with the RAN. Chapter 9, "Core Network Functionality, QoS, and Network Slicing," provides a more comprehensive discussion of core network functionality.

15.3 RAN Channel Structure

3GPP has defined a channel structure for conveying uplink and downlink data and control information across the 5G air interface and 5G RAN. The structure consists of a hierarchy of channels, as shown in Figure 15.7. The figure also depicts the mapping between channels at adjacent layers. This section provides an overview of the three layers of the architecture and defines the individual channel types.

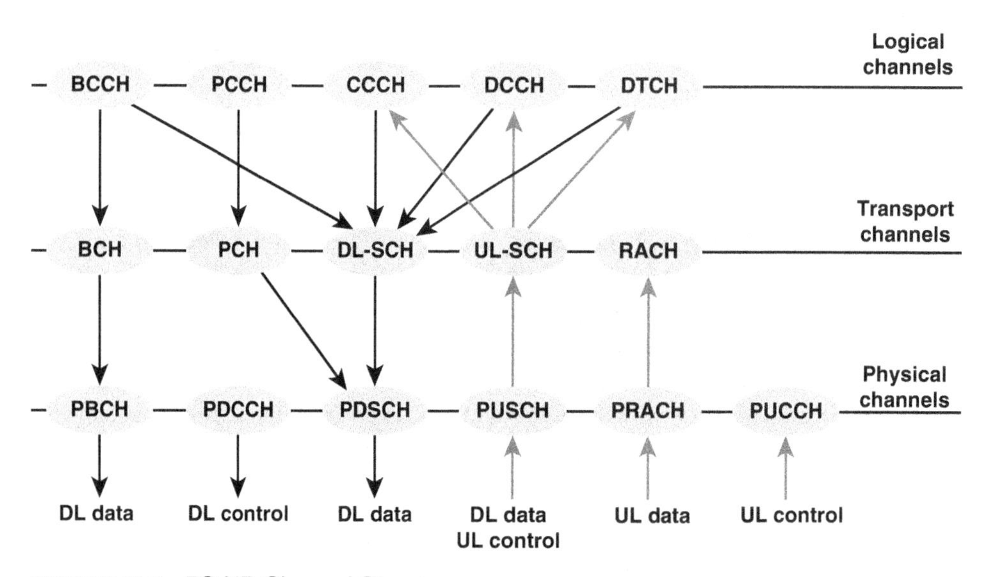

FIGURE 15.7 5G NR Channel Structure

Logical Channel

A logical channel is an information stream dedicated to the transfer of a specific type of information over the radio interface. Multiple logical channels can be mapped onto a single transport channel. One logical channel can also be mapped onto or duplicated on multiple transport channels. Logical channels are classified into two groups: control channels and traffic channels. Control channels are used for the transfer of control plane information only. Traffic channels are used for the transfer of user plane information only.

The logical control channel types are as follows:

- **Broadcast Control Channel (BCCH):** A channel used within the downlink for sending broadcast information to the UEs within that cell. The system information transmitted by the 5G NR BCCH is divided into different blocks:
  - **Master Information Block (MIB):** There is one MIB, and it is mapped onto the BCH transport channel and then onto the PBCH physical channel.
  - **System Information Block (SIB):** There are several SIBs, and they are mapped onto the DL-SCH transport channel and then onto the PDSCH physical channel.
- **Paging Control Channel (PCCH):** A downlink channel used to page the UEs whose locations at cell levels are not known to the network. As a result, the paging message needs to be transmitted in multiple cells. The PCCH is mapped to the PCH transport channel and then to the PDSCH physical channel.
- **Common Control Channel (CCCH):** A channel used on both the downlink and uplink for transmitting control information to and from the UEs. The channel is used for initial access for UEs that do not have a radio resource control (RRC) connection.
- **Dedicated Control Channel (DCCH):** A bidirectional channel to carry dedicated control information between the UE and the network. It is used by the UE and the network after an RRC connection has been established.

There is one logical channel for user plane information:

- **Dedicated Traffic Channel (DTCH):** A bidirectional channel dedicated to one UE and used for carrying user information to and from a specific UE and the network.

Transport Channels

3GPP views physical layer transport services as being provided by a transport channel that is in a one-to-one relationship with a physical channel. Transport services are described based on how and with what characteristics data is transferred over the radio interface. This should be clearly separated from the classification of what is transported, which relates to the concept of logical channels at the medium access control (MAC) sublayer.

The downlink transport channel types are as follows:

- **Broadcast Channel (BCH):** The BCH broadcasts BCCH system information—specifically, the MIB information—in the entire coverage area of the cell. Broadcast occurs as a single message to the entire coverage area or using multiple BCH instances via beamforming. A fixed predefined transport format is used.
- **Paging Channel (PCH):** The PCH is used for carrying paging information from the PCCH logical channel. The PCH supports discontinuous reception (DRX) to enable the UE to save battery power by waking up at a specific time to receive the PCH. A paging message is broadcast as a single message or by beamforming.
- **Downlink Shared Channel (DL-SCH):** This is the main transport channel used for transmitting downlink data. The channel supports all the key 5G NR features, including dynamic rate adaptation, HARQ (hybrid automatic repeat request), channel-aware scheduling, and spatial multiplexing. The DL-SCH is also used for transmitting some parts of the BCCH system information. Each UE has a DL-SCH for each cell to which it is connected.

The uplink transport channel types are as follows:

- **Uplink Shared Channel (UL-SCH):** This uplink counterpart to the DLSCH is used for transmission of uplink data.
- **Random-Access Channel (RACH):** This channel carries the random access preamble, which is used to overcome the message collisions that can occur when UEs access the system simultaneously.

Physical Channels

A physical channel corresponds to a set of resource elements carrying information originating from higher layers. In FDD (frequency-division duplexing) mode, a physical channel is defined by code, frequency, and, in the uplink, relative phase. In TDD (time-division duplexing) mode, a physical channel is defined by code, frequency, and time slot.

The downlink physical channel types are as follows:

- **Physical Downlink Shared Channel (PDSCH):** Carries data by sharing the capacity on a time and frequency basis. The physical channel carries a variety of data, including user data, UE-specific higher-layer control messages mapped down from higher channels, system information blocks (SIBs), and paging data. The PDSCH uses an adaptive modulation format that depends on the link's signal-to-noise ratio. It also uses a flexible coding scheme. This combination yields flexible coding and a flexible data rate.

- **Physical Downlink Control Channel (PDCCH):** Carries downlink control data, which includes scheduling decisions for PDSCH reception and data for scheduling grants for transmission on the PUSCH. The PDCCH uses QPSK as its modulation format and polar coding as the coding scheme, except in the case of small packets of data.
- **Physical Broadcast Channel (PBCH):** Carries part of the system information required for a UE to connect to an NR base station (gNodeB). Its function is to provide UEs with the master information block (MIB). A further function of the PBCH in conjunction with the control channel is to support the synchronization of time and frequency. This aids with cell acquisition, selection, and reselection. The PBCH uses a fixed data format, and there is one block that extends over a TTI of 80 ms. The PBCH uses QPSK modulation, and it transmits a cell-specific demodulation reference signal that can be used to aid with beamforming.

The uplink transport channel types are as follows:

- **Physical Random Access Channel (PRACH):** Used for channel access. It transmits an initial random access preamble consisting of sequences that may be of two different lengths. A long sequence is 839, which is applied to the subcarrier spacings of 1.25 kHz and 5 kHz. Short sequence lengths of 139 are applied to subcarrier spacings of 15 kHz and 30 kHz for FR1 bands and 60 kHz and 120 kHz for FR2 bands.
- **Physical Uplink Shared Channel (PUSCH):** Carries application data and also carries radio resource control (RRC) signaling messages and uplink control information (UCI). Like the PDSCH, the PUSCH has a flexible format. The allocation of frequency resources is undertaken using resource blocks along with a flexible modulation and coding scheme that is dependent upon the link's signal-to-noise ratio.
- **Physical Uplink Control Channel (PUCCH):** Carries UCI, which includes scheduling requests, channel quality indicator (CQI) information, and hybrid automatic repeat request (HARQ) information. It is also possible that, depending on the resource allocation, the uplink control information or data may also be sent on the PUSCH, even though in the downlink direction control information is always sent on the PDCCH.

15.4 RAN Protocol Architecture

This section provides an overview of the RAN protocol architecture at three key interfaces: the air interface to the UE, the NG interface to the core network, and the Xn interface between RAN nodes.

Air Interface Protocol Architecture

Figure 15.8 shows the protocol stack at the air interface, which consists of separate protocol interfaces for user plane and control plane traffic.

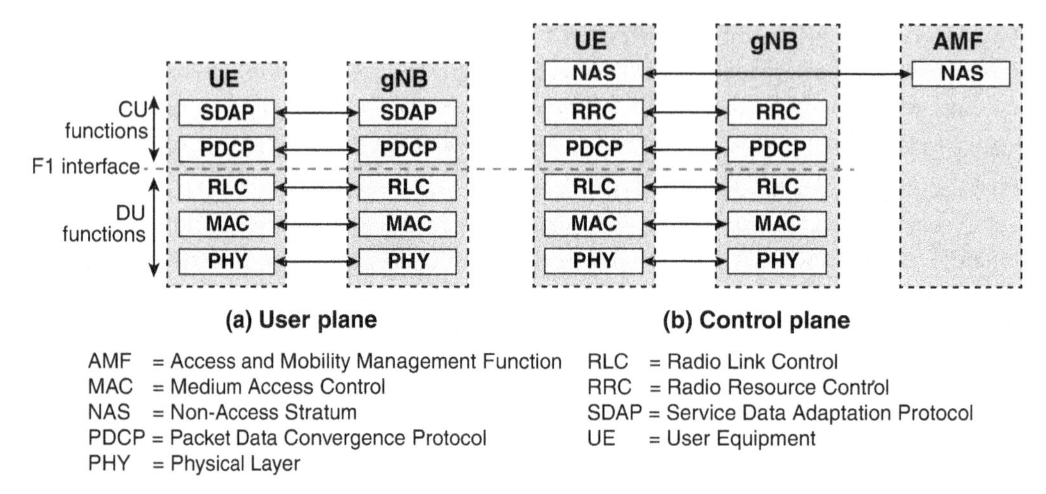

FIGURE 15.8 Air Interface Protocol Stack

The bottom four layers are the same for both the user and control planes:

- **Physical layer (PHY):** Covers the physical radio interface between data transmission devices. This layer is concerned with specifying the characteristics of the transmission medium, including frequency bands and data rates. This layer also covers modulation schemes, such as QAM (quadrature amplitude modulation). Channel coding, such as polar and low-density parity check (LDPC) coding, is performed at this layer.
- **Medium access control (MAC):** This layer is concerned with regulating access to a shared medium. It includes access control mechanisms such as orthogonal frequency-division multiple access (OFDMA) and single-carrier FDMA. This layer provides priority handling between data packets from different radio bearers and error correction through hybrid automatic repeat request (HARQ).
- **Radio link control (RLC):** RLC functions include error correction with automatic repeat request (ARQ), concatenation and segmentation in order to match the transmitted protocol data unit (PDU) size to the available radio resources, in-sequence delivery, and protocol error handing.
- **Packet data convergence protocol (PDCP):** Functions of PDCP include IP header encryption and decryption, integrity protection for the user plane data using error detection codes, duplication of transmitted PDCP PDUs over different data paths, and reordering and duplicate detection of received PDCP PDUs.

The top layer of the user plane protocol architecture is the **Service Data Adaptation Protocol (SDAP)**. SDAP supports the flow-based QoS model of the 5G core network. With this QoS model, the core network can configure different QoS requirements for different QoS flows of a PDU session. The SDAP layer provides mapping of IP flows with different QoS requirements to radio bearers that are configured appropriately to deliver that required QoS. The mapping between QoS flows and radio bearers may be configured and reconfigured by RRC signaling, but it can also be changed more dynamically without the involvement of RRC signaling through a reflective mapping process. (Reflective QoS is discussed in Section 9.3 of Chapter 9.)

The control plane protocol stack includes the same four lower layers as the user plane (i.e., PHY, MAC, RLC, PDCP). On top of these four layers in the control plane are the following layers:

- **Radio resource control (RRC):** RRC is responsible for control and configuration of the radio-related functions in the UE. For each connection to UE, RRC operates using a three-state model, as shown in Figure 15.9. The RRC inactive state provides battery efficiency similar to RRC idle but with a UE context remaining stored within the NG-RAN so that transitions to/from RRC connected are faster and incur less signaling overhead. RRC supports an on-demand system information mechanism that enables the UE to request when specific system information is required instead of allowing the NG-RAN to consume radio resources to provide frequent periodic system information broadcasts.

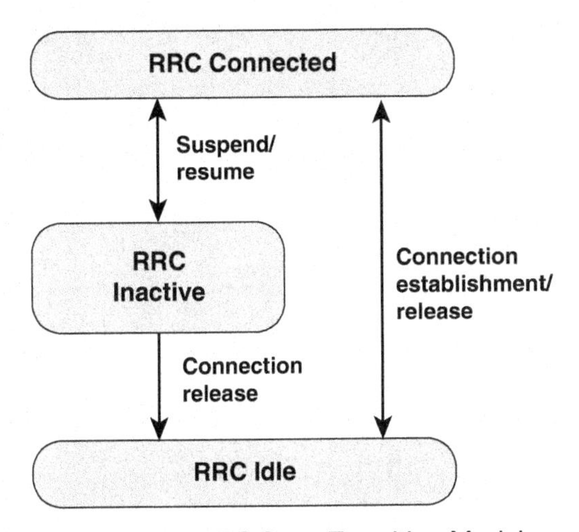

FIGURE 15.9 RRC State Transition Model

- **Non-access stratum (NAS):** The NAS consists of protocols between UE and the core network that are not terminated in the RAN. Specifically, NAS protocols terminate in the UE and the AMF of the 5G core network and are used for core network–related functions such as registration, authentication, location updating, and session management.

Figure 15.8 also indicates the functional split between central unit (CU) and distributed unit (DU) for a distributed architecture. The RLC, MAC, and physical layers reside in the DU. The CU contains the PDCP and SDAP layers in the user plane and the PDCP and RRC layers in the control plane.

Channel Structure

Figure 15.10, based on a figure in TS38.300, shows the uplink channels and bearers defined across the air interface. This provides insight into how QoS flows are handled at this interface. The terms *physical channel*, *transport channel*, and *logical channel* are defined in the preceding section. The remaining terms on the right side of Figure 15.10 are defined as follows:

- **QoS flow:** The lowest granularity of a traffic flow where QoS and charging can be applied
- **Radio bearer:** An information transmission path across the air interface of defined capacity, delay, bit error rate, and other parameters
- **RLC channel:** A logical channel between two RLC protocol entities through which RLC data and control information are exchanged

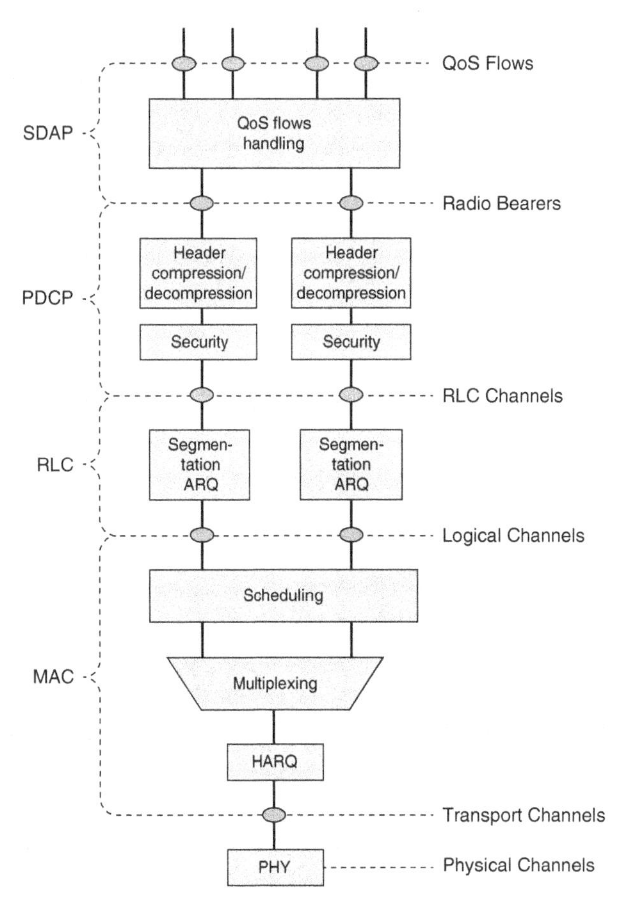

FIGURE 15.10 Channels/Bearer Architecture

There is no strict one-to-one relationship between QoS flows and radio bearers. It is up to the RAN to determine the resources that QoS flows can be mapped to and to release them. The functions of the PDCP in effect transform a radio bearer transmission path into an RLC channel. There is a one-to-one mapping from the RLC channel to the underlying logical channel that carries the RLC data and control information. There may be multiple logical channels multiplexed onto a single transport channel, which maps to a single physical channel.

The structure of the downlink flows is essentially the same as shown in Figure 15.10, except that QoS flows from multiple UEs may be scheduled as a group.

RAN–Core Network Interface Protocol Architecture

The interface between a gNB node in a 5G RAN and the 5G core network is labeled the NG interface, as shown in Figure 15.2. Figure 15.11 shows the protocol stack at the NG interface, which consists of separate protocol interfaces for user plane and control plane traffic. The NG user plane interface is defined between the NG-RAN node and the UPF. It provides non-guaranteed delivery of PDU session user plane PDUs between the NG-RAN node and the UPF. The NG control plane interface is defined between the NG-RAN node and the AMF.

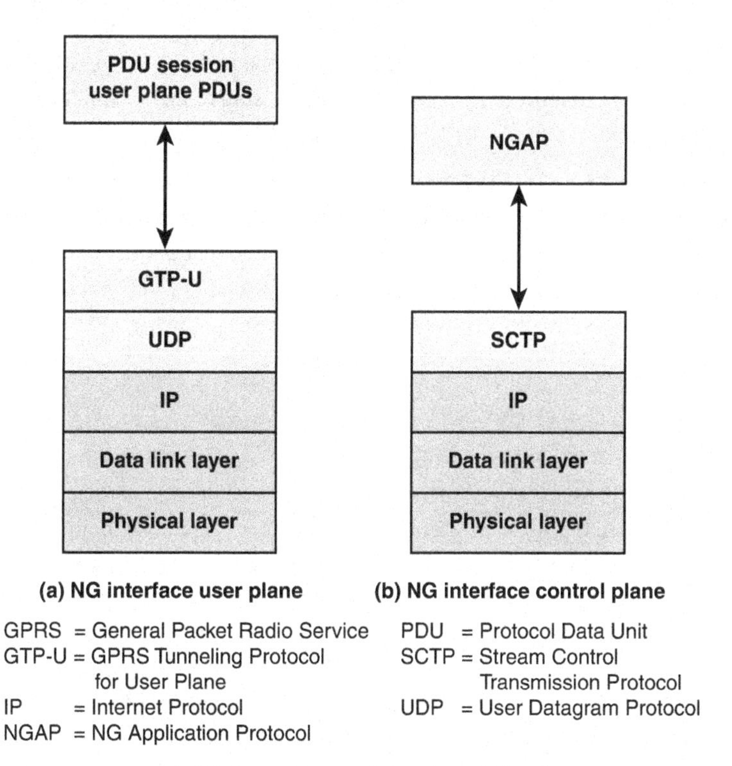

FIGURE 15.11 RAN–Core Network Protocol Stack

The bottom three layers are the same for both user and control planes:

- **Physical layer:** This layer covers the physical interface between data transmission devices. This may be a wired link, such as optical fiber, or a fixed wireless link. This layer is concerned with specifying the characteristics of the transmission medium, the nature of the signals, the data rate, and related matters.
- **Data link layer:** This layer provides for the reliable transfer of information across the physical link. This layer sends data as link layer protocol data units (PDUs), often called frames, with synchronization, flow control, and error control capabilities.
- **Internet Protocol (IP):** IP is a standard protocol designed for use over interconnected systems of data networks, as well as over a single network or data link. IP transmits blocks of data called IP datagrams from source to destination, where sources and destinations are hosts identified by fixed-length IP addresses. IP provides for fragmentation and reassembly of long datagrams. IP is connectionless.

The following layers are on top of IP in the user plane:

- **User Datagram Protocol (UDP):** UDP is a connectionless transport layer protocol that provides for the exchange of transport layer datagrams without acknowledgments or guaranteed delivery. UDP adds a port-addressing capability to IP so that a specific source and destination application or service is designated.
- **GPRS Tunneling Protocol for User Plane (GTP-U):** This layer supports multiplexing of traffic of different PDU sessions (via N3) and carries QoS marking. GTP-U tunnels are used to carry encapsulated user plane PDUs between a given pair of GTP-U tunnel endpoints.

The control plane protocol stack includes the same three lower layers as the user plane (i.e., PHY, data link, IP). On top of these three layers in the control plane are the following layers:

- **Stream Control Transmission Protocol (SCTP):** This is a reliable transport layer protocol. Like Transmission Control Protocol (TCP), SCTP ensures reliable transport of PDUs with congestion control. In contrast to TCP, SCTP allows delivery of out-of-order packets to applications; this type of delivery is more suitable for delay-sensitive applications. The multihoming feature of SCTP enables transparent handover over several heterogeneous overlapping wireless networks. One path with specified destination and source addresses plays the role of primary path. The remaining paths are secondary paths. SCTP can monitor, at runtime, delay and jitter on all active paths, and it makes the paths available to the application. Heartbeat messages are sent over secondary paths to collect required measurements. The collected key performance indicators (KPIs) are mapped to quality of experience (QoE) values using a suitable QoE/QoS mapping model. The path quality is compared at regular intervals, and the client decides whether a network switch is necessary according to its customized and internal policy.

- **NG Application Protocol (NGAP):** This is an application signaling protocol. NGAP provides the control plane signaling between an NG-RAN node and the access and mobility management function (AMF). It is transferred over SCTP. NGAP provides a transport function between the UE and the AMF by offering NAS signaling transport. The NAS protocol provides mobility management and session management between the UE and the AMF.

The NG control interface provides the following functions:

- NG interface management
- UE context management
- UE mobility management
- Transport of NAS messages
- Paging
- PDU session management
- Configuration transfer
- Warning message transmission

Xn Interface Protocol Architecture

Figure 15.12 shows the protocol stack at the Xn interface between RAN nodes, which consists of separate protocol interfaces for user plane and control plane traffic. It is identical to the NG interface protocol architecture except at the top layer.

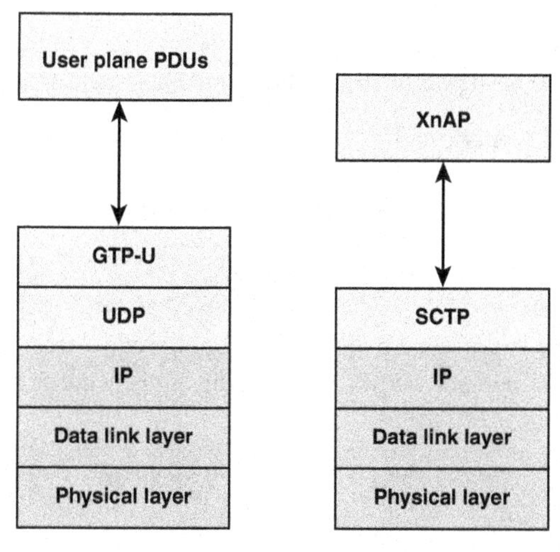

FIGURE 15.12 Xn Interface Protocol Stack

The Xn user interface supports data forwarding and flow control of user plane PDUs. The Xn control interface supports interface management, UE mobility management (including context transfer and RAN paging), and dual connectivity.

15.5 NG RAN Transport Network

The NG RAN transport network is a collection of communication links that interconnect nodes of the RAN and communication links that connect RAN elements to the 5G core networks. Links between RAN elements and UEs are generally not considered part of the RAN transport network.

ITU-T Technical Report GSTR-TN5G (*Transport Network Support of IMT-2020/5G*, October 2018) lists four possible RAN deployment scenarios that dictate the transport network architecture. The scenarios concern the manner in which the three main elements of a RAN—the CU, DU, and radio unit (RU)—are grouped. The RU is equipment that contains the wireless transmitter and receiver. It generates and receives radio signals transmitted over the airwaves via tower-mounted antennas. When the RU is physically distant from the base station, it is referred to as a remote radio unit (RRU). The following terminology is used for the communication links or networks between elements:

- **Backhaul:** A network path between base station systems and a core network. The distance covered by a backhaul network between the core network and a base station could range from 1 km up to hundreds of kilometers.
- **Midhaul:** A network path between CUs and DUs that are physically separated. The typical range is 20 to 40 km.
- **Fronthaul:** A network path between centralized radio controllers and remote radio units of a base station function. The distance is in the range of less than 20 km.

Note that fronthaul connections are wireless, while midhaul and backhaul are primarily wireline (optical fiber or coaxial cable), thus enabling longer distances to be supported on midhaul and backhaul.

Figure 15.13 illustrates the following CU, DU, and RU scenarios:

- **Independent RRU, CU, and DU locations (part a):** In this scenario, there are fronthaul, midhaul, and backhaul networks.
- **RU and DU integration (part b):** In this scenario, an RU and a DU are deployed close to each other—perhaps hundreds of meters apart, such as in the same building. In order to reduce cost, an RU is connected to a DU just through straight fiber, and no transport equipment is needed. In this case, there are midhaul and backhaul networks.
- **Collocated CU and DU (part c):** In this scenario, the CU and DU are located together; consequently, there is no midhaul.
- **RRU, DU, and CU integration (part d):** This network structure may be used for small cell and hotspot deployments. There is only backhaul in this case.

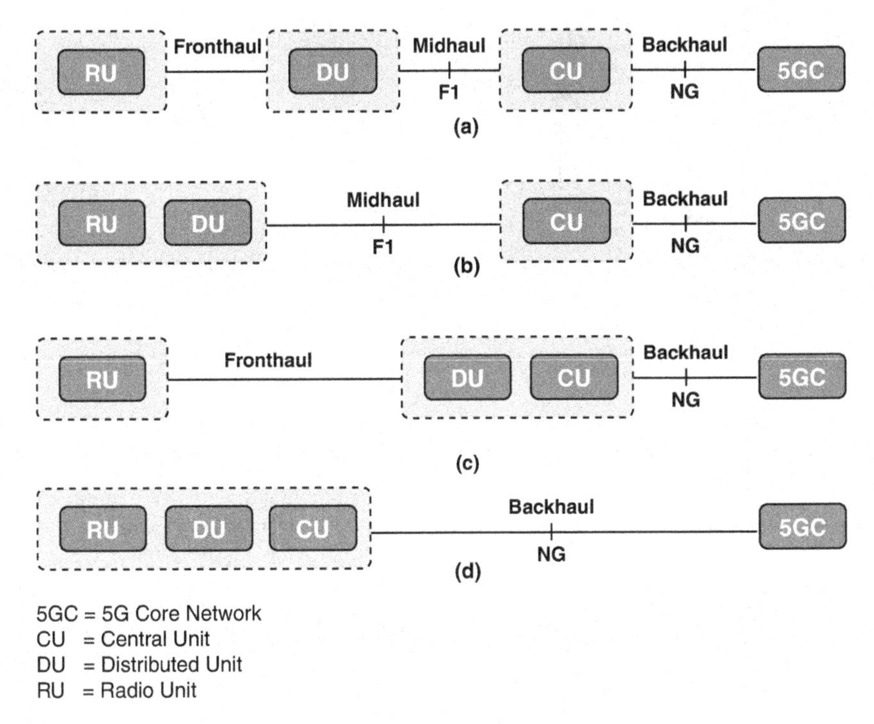

FIGURE 15.13 Possible CU, DU, and RU Combinations

At this point, it will be useful to clarify some terms. In a wireless system, **baseband** refers to the original frequency range of a transmission signal before it is modulated. A **baseband unit (BBU)** is a unit that processes baseband in telecommunication systems. A common arrangement is a BBU containing both CU and DU functionality, and one or more RUs, either collocated or remote (RRUs).

Figure 15.14, based on a figure in ITU-T GSTR-TN5G, provides further insight into the nature of RAN transport networks. The transport network deployment between RAN elements may be a single transmission link, such as a fiber-optic link. In many cases, particularly to cover long distances and to support many RAN elements, the transport network topology for fronthaul, midhaul, and backhaul segments is a switched network topology. The switches may be passive or active optical switches, packet-switched nodes using some form of wired transmission medium, microwave relay towers, or satellite links.

Figure 15.14 shows a correspondence between the fronthaul/midhaul/backhaul paradigm and the interfaces defined in 3GPP documents. Table 15.2, from ITU-T G.8300 (*Characteristics of Transport Networks to Support IMT 2020/5G*, May 2020), specifies this correspondence in greater detail.

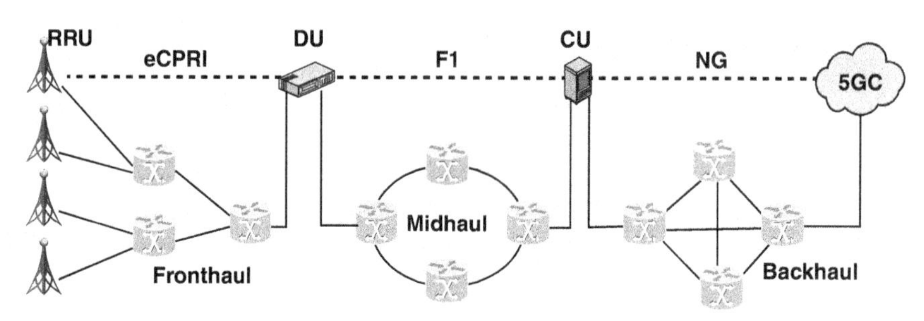

5GC = 5G Core Network
eCPRI = Enhanced Common Public Radio Interface
DU = Distributed Unit
F1 = Interface Label
CU = Central Unit
NG = Interface Label
RRU = Remote Radio Unit

FIGURE 15.14 Transport Network Architecture for Independent CU, DU, and RRU Deployment

TABLE 15.2 ITU-T and 3GPP Terminology for RAN Interfaces

3GPP Logical Interface or CPRI Name	ITU-T Transport Network Name	Interface Description	Transport Network Domains
CPRI/eCPRI	Fronthaul	Interface between RRU and DU	Metro access
F1	Midhaul	Logical interface between CU and DU	Metro access or metro aggregation
NG	Backhaul	Logical interface between gNB (CU) and 5GC	Metro access, metro aggregation, metro core, or backbone
Xn	Midhaul or backhaul	Logical interface between gNB nodes	Metro access, metro aggregation, metro core, or backbone

The final column of Table 15.2 includes the following important terms:

- **Metro access:** The access portion of a cellular network provides the last kilometer or last several kilometers of connectivity between UE and the radio unit. This corresponds to the air interface. For configurations with a remote RU, the term *metro access* roughly corresponds to the fronthaul network that connects the RUs to base stations or at least to the DU portion of a base station.
- **Metro aggregation:** The metro aggregation network aggregates the traffic of progressively larger sets of different uses and transmits this aggregated traffic over increasingly higher-capacity facilities.
- **Metro core:** The metro core acts as a regional network providing connectivity between the various access and aggregation networks within a given metropolitan area and connecting to a larger backbone network.

- **Backbone:** This term generally refers to high-speed long-haul transmission links and networks that connect metropolitan area networks to more distant resources. The backbone encompasses the switched core network as well as the backhaul network.

The boundaries between access facilities, aggregation facilities, and backbone facilities vary from network to network and are not always easy to identify with precision.

Another term used in Figure 15.14 is eCPRI (enhanced Common Public Radio Interface). CPRI and eCPRI are specifications for the interface between the RRU and the DU or base station. The eCPRI specification uses either Ethernet transport or optical fiber transport.

Figure 15.15, from [ERIK19] (and reprinted in ITU-T G.8300), shows typical deployment options for the RAN transport network. The specific type of deployment depends on the geographic location (urban, suburban, or rural), the application areas that predominate, and the availability of various types of transmission media infrastructure.

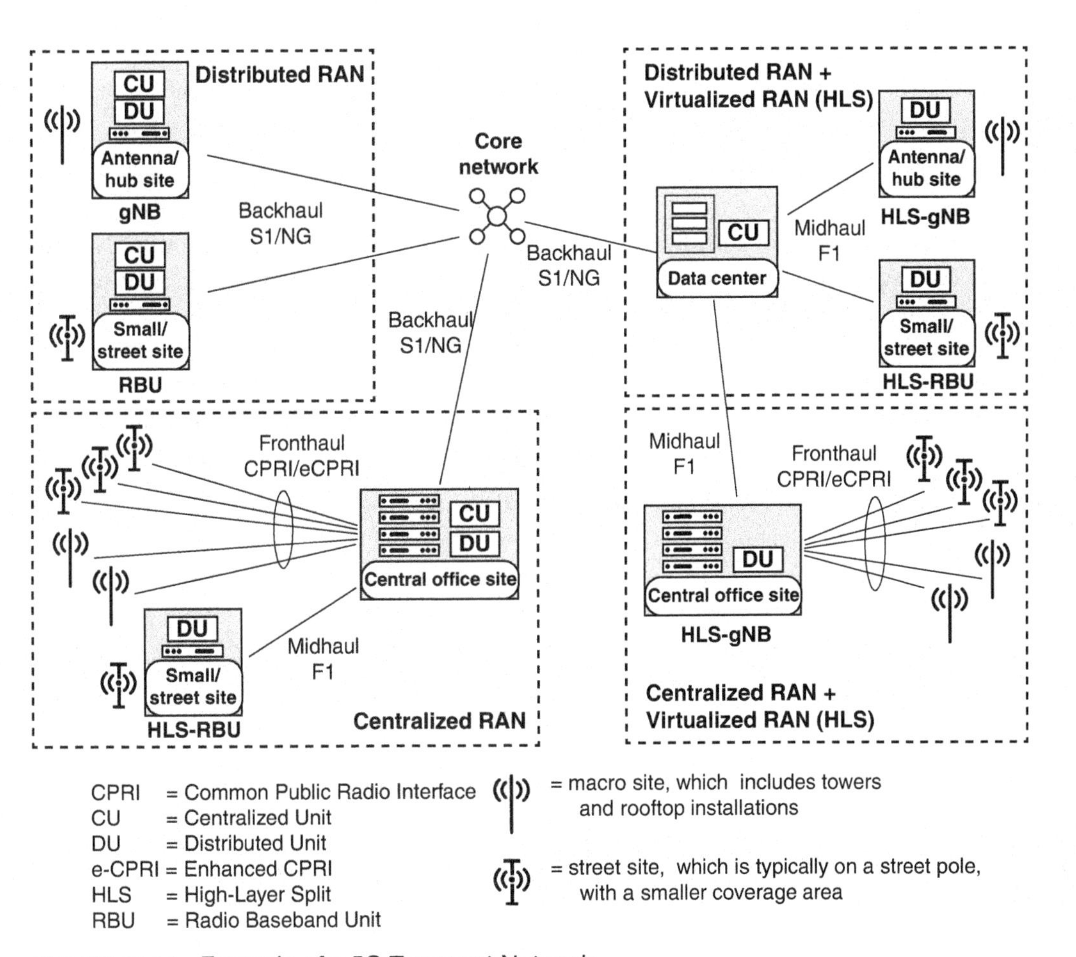

FIGURE 15.15 Example of a 5G Transport Network

The upper-left portion of Figure 15.15 depicts a **distributed RAN (DRAN)**, in which each base station contains all the network functionality (i.e., CU, DU, RU). 4G networks typically deploy the DRAN configuration. This is a flexible solution that has a number of advantages. It allows 5G deployments to take advantage of legacy 4G infrastructure. For areas that need a high antenna density, such as urban areas using millimeter wave (mmWave) antenna placement, DRAN street sites work well with nearby DRAN macro sites. Further, a baseband unit can be tailored for use in a small cell site, using integrated baseband functions, simple installation, and reduced site space; these are designated as radio baseband units (RBUs) in the figure.

The lower-left portion of Figure 15.15 depicts a **centralized RAN** configuration. In this configuration, the RU is remote from the gNB base station, which in turn may be in a non-split or split configuration. In the former case, a fronthaul transport network connects the remote RUs to the base station. In the latter case, the RU may be collocated with an element that includes the DU. In the figure, this is referred to as a high-layer split RBU (HLS-RBU).

The term *high-layer split* refers to the division of functions between the CU and the DU defined as the F1 interface in 3GPP documents. The HLS places most of the functions inside the DU, minimizing the functionality of the CU. Using an HLS reduces the bandwidth required between the CU and DU compared to with a lower-layer split.

A **virtualized RAN (vRAN)**, formerly referred to as a cloud RAN,[1] decouples hardware and software, allowing RAN functions typically run on a proprietary technology stack to exist as software workloads using commodity or custom hardware. Figure 15.16, from [UITT20], indicates how RAN virtualization is evolving. In earlier versions of vRAN, CU functions are implemented in software and deployed as virtual network functions (VNFs) at the edge of the network on standard commercial servers in a cloud data center. The DU functions may be collocated or distributed. In more recent vRAN implementations, both DU and CU functions are deployed as VNFs. The use of VNFs brings the benefits discussed in Section 8.4 of Chapter 8, "Network Functions Virtualization," including flexible allocation of computing and storage resources based on the traffic handled by each base station.

The right-hand side of Figure 15.14 gives examples of adding vRAN implementation to both distributed and centralized RAN configurations.

1. The term cloud RAN has been deprecated because of the potential confusion with centralized RAN as both of these terms have been referred to as CRAN or C-RAN.

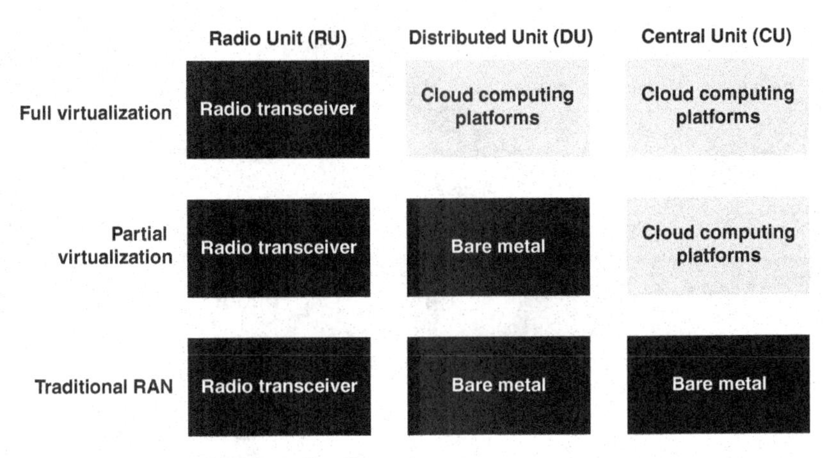

FIGURE 15.16 RAN Virtualization

15.6 Integrated Access and Backhaul

Figure 15.2 implies that all base stations have direct backhaul links to the core network. Typically, such links are optical fiber. But backhaul presents a significant challenge for dense small cell deployment. Requiring a direct link to the core, either wired or wireless, can limit where cells can be placed. Limited placement options results in suboptimal radio performance. Further, if backhaul transmission quality from a particular base station is poor, performance suffers for the entire network.

The approach taken by 3GPP to address this issue is integrated access and backhaul (IAB). The key concept is to use the same wireless access technology that provides an air interface to user equipment (UE) for creating a backhaul link between 5G RAN nodes. Figure 15.17 illustrates this concept. The base station labeled IAB-donor is shown with a fiber backhaul link to the core network. This node functions as an ordinary gNB node, providing access to the 5G core for UEs. It also acts as a relay providing core access to gNB nodes that do not have a direct link to the core network. Thus, the node labeled IAB-node 1 does not have a link to the core network but is close enough to IAB-donor to establish a backhaul link using the same air interface frequency bands used for UE access. Figure 15.17 shows that IAB support from an IAB-donor node is not limited to nearby base stations. Rather, it is possible to establish a chain of backhaul links so that more remote nodes can provide core access for local UEs. In the figure, IAB-node 2 establishes a backhaul link to IAB-node 1 and thus reaches IAB-donor through two backhaul links.

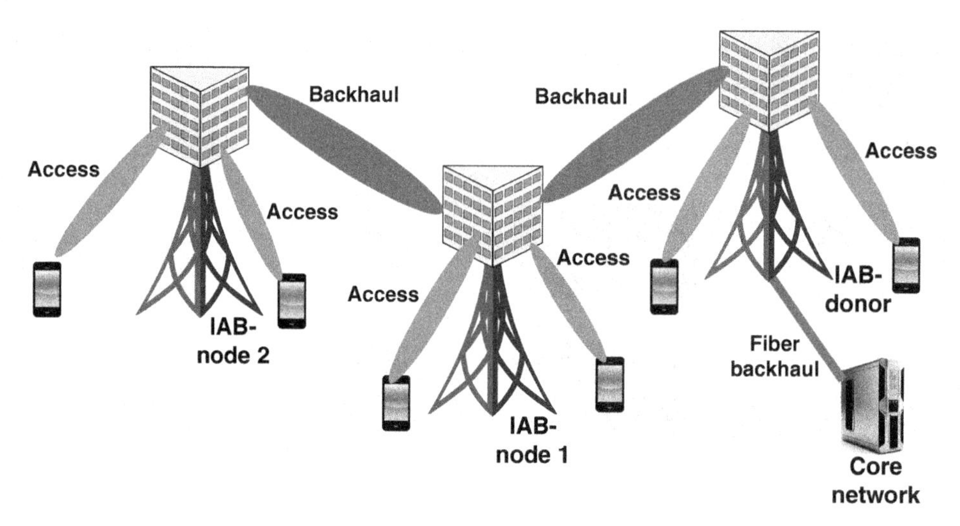

FIGURE 15.17 IAB Concept

For two nodes that share an IAB backhaul link, the upstream node is referred to as the *parent node*, and the downstream node is referred to as the *child node*, as discussed later in this chapter.

Table 15.3 defines key IAB terms.

TABLE 15.3 IAB Terminology

Term	Definition
IAB node	A RAN node that supports wireless relaying of NR access traffic from UEs via NR Uu backhaul links. It supports the UE function and the DU function of the CU/DU architecture.
IAB donor	A RAN node that provides connection to the core network for IAB nodes. It supports the CU function of the CU/DU architecture.
Access link	A link between UE and an IAB node or IAB donor.
Backhaul link	A link between an IAB node and an IAB child node or an IAB parent node.
IAB parent node	The node closer to the core network of two nodes that are adjacent in an IAB backhaul transmission path from UE to the core network. The parent schedules the backhaul downstream and upstream traffic to/from the child node.
IAB child node	The node farther from the core network of two nodes that are adjacent in an IAB backhaul transmission path from UE to the core network.
Uu interface	The NR air interface. This interface exists between a RAN node and UE as well on IAB backhaul links other than the link between the IAB donor and the core network.
Downstream	IAB backhaul traffic going away from the core network.
Upstream	IAB backhaul traffic going toward the core network.

IAB Architecture

Figure 15.18, based on a figure in 3GPP TS 23.501 (*Technical Specification Group Services and System Aspects, System Architecture for the 5G System [5GS], Stage 2 [Release 16]*, September 2020), provides a top-level view of the IAB architecture.

TS 23.501 summarizes the following key characteristics of the IAB functionality:

- IAB uses the CU/DU architecture illustrated in Figure 15.13, and the IAB operation via F1 (between IAB-donor and IAB-node) is invisible to the 5GC.
- IAB performs relaying at Layer 2 and, therefore, does not require a local user plane function (UPF).
- IAB supports multi-hop backhauling, as illustrated in Figure 15.15.
- IAB supports dynamic topology updates. That is, an IAB node can change the parent node (another IAB node or the IAB donor) during operation—for example, in response to backhaul link failure or blockage.

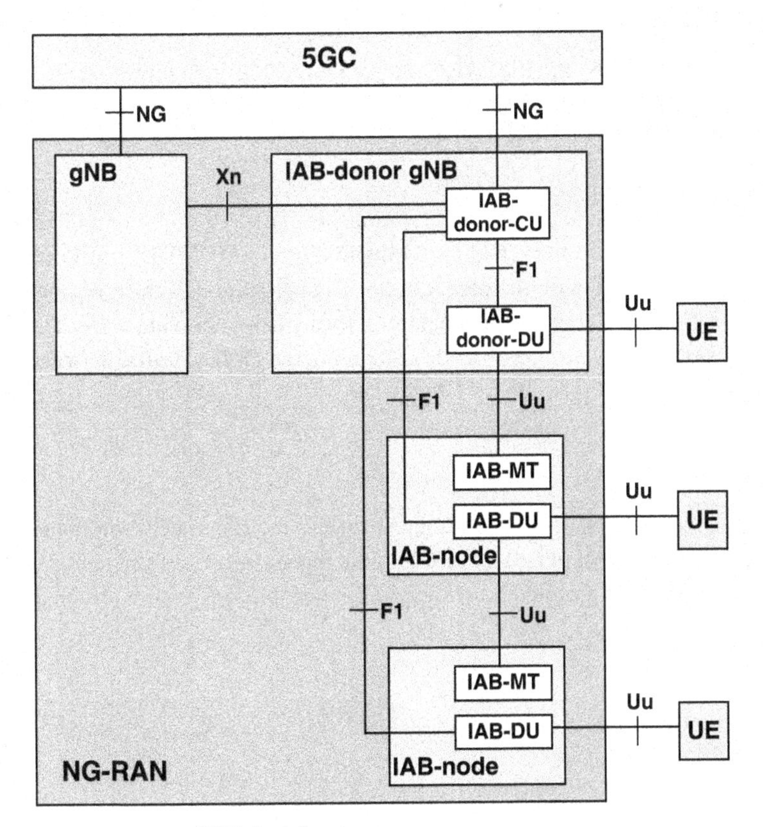

FIGURE 15.18 IAB Architecture

Two principal functions are implemented in each IAB node:

- **IAB-DU:** Provides NR Uu access to UEs that are in the cell supported by the IAB node. It also provides Uu access to any child node that is sufficiently close. The corresponding gNB-CU function exists in the IAB donor gNB and controls the gNB-DU via the F1 interface. Note that F1 is a logical interface and that communication between the IAB donor CU and an IAB node's gNB-DU involves one or more backhaul hops.
- **IAB-MT (mobile termination):** The logical functional part of an IAB node that supports the Uu interface toward the IAB donor or to an adjacent upstream IAB node. The IAB-MT function behaves as UE and reuses UE procedures to connect to (1) the gNB-DU on a parent IAB node or IAB donor for access and backhauling and (2) the gNB-CU on the IAB donor via RRC for control of the access and backhaul link. (Note that the term IAB-MT used in the 38 series of 3GPP specifications is referred to as IAB-UE in TS 23.501.)

The IAB donor performs centralized resource, topology, and route management for the IAB topology.

Parent/Child Relationship

Figure 15.19a, from TS 38.300, shows that the topology of a set of cooperating IAB nodes forms a directed acyclic graph with the IAB donor at its root. In this topology, a downstream neighbor node of an IAB-DU or the IAB-donor-DU is referred to as a child node, and an upstream neighbor node of the IAB-MT is referred to as a parent node.

Figure 15.19b provides another perspective on the parent/child relationship. The IAB donor or an IAB node in its role as a parent node is responsible for scheduling downlink/uplink (DL/UL) traffic on access links for UEs that communicate directly with the parent node. The parent node is also responsible for scheduling downstream/upstream traffic on the backhaul link to any neighbor child node. The IAB node at the end of the transmission chain is responsible for scheduling the DL/UL traffic between itself and its UEs.

IAB Protocol Architecture

Figure 15.20 shows the protocol architecture of the backhaul interface (F1 interface) between nodes in the IAB topology. It depicts the F1-C (control plane) and F1-U (user plane) interfaces carried over two backhaul hops. It should be clear what the architecture would be for a single hop or for more than two hops.

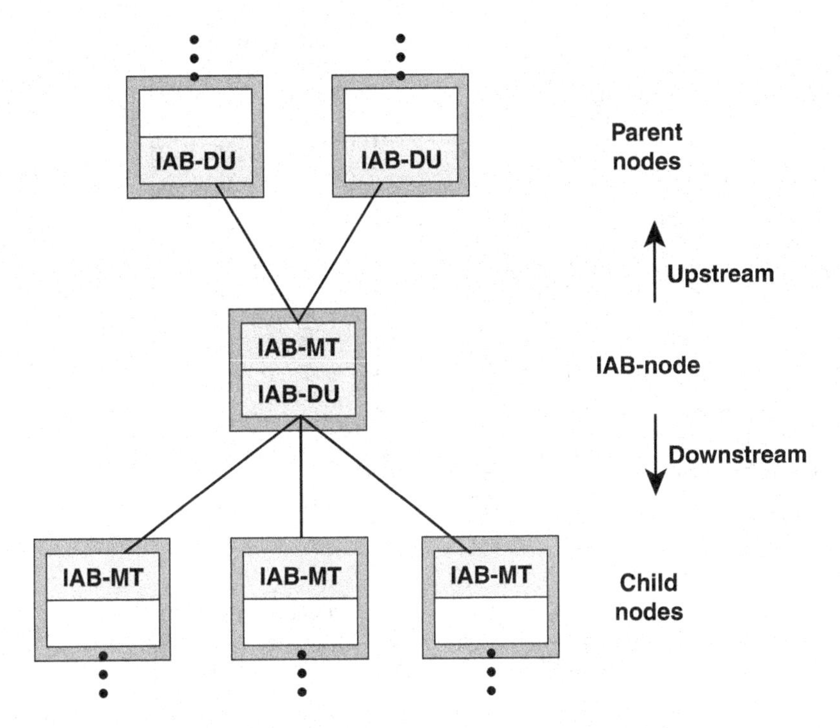

(a) Parent- and child-node relationship for IAB node

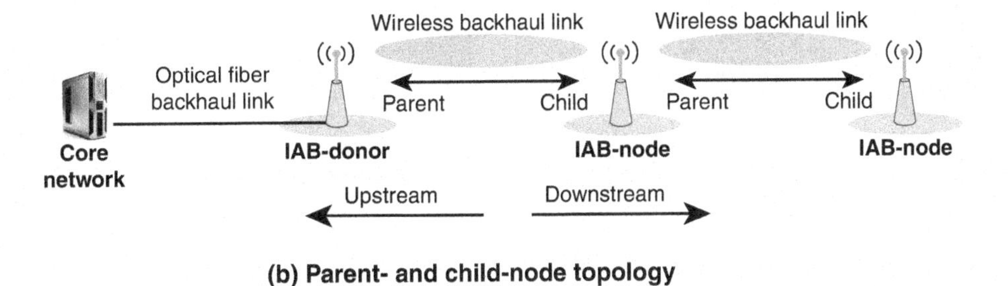

(b) Parent- and child-node topology

FIGURE 15.19 Parent/Child Relationship

In both the user and control planes, the lowest three layers are the same as are used in the air interface protocol stack (refer to Figure 15.8). This makes sense because the IAB backhaul links are using the air interface frequency band. Thus, the physical, medium access, and radio link control protocols are the same as for the air interface.

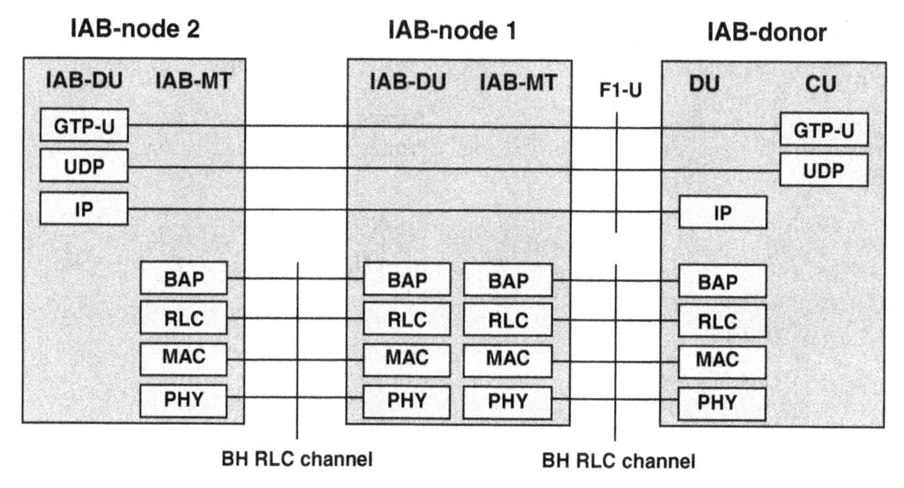

(a) Protocol stack for support of F1-U protocol

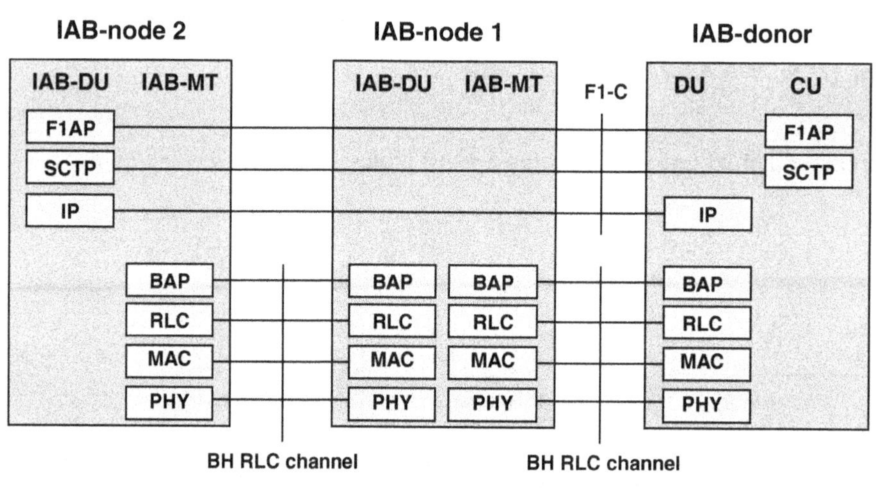

(b) Protocol stack for support of F1-C protocol

FIGURE 15.20 IAB Protocol Stacks

In both the user and control planes, the next layer is the Backhaul Adaptation Protocol (BAP). BAP services and functions include:

- Transfer of data
- Routing of packets to next hop
- Determination of BAP destination and BAP path for packets from upper layers

- Determination of egress backhaul radio link control (BH RLC) channels for packets routed to next hop
- Differentiation of traffic to be delivered to upper layers from traffic to be delivered to egress link
- Flow control feedback signaling
- Backhaul radio link failure (RLF) indication

The lower four layers operate hop by hop over individual backhaul links. The upper three layers are end to end between the IAB node that provides access to a UE and the IAB donor. In the user plane, IP and UDP provide connectionless transport service. The top layer is GTP-U (described in Section 15.3).

In the control plane, IP and SCTP provide a reliable, connection-oriented transport service. The top layer is the F1 Application Protocol (F1AP). F1AP provides the signaling messages on F1 that support the procedures for F1 management, UE context management, and RRC message transfer. The signaling messages/services can be non-UE associated and UE associated.

15.7 Key Terms and Review Questions

Key Terms

access and mobility management function (AMF)	distributed entity
air interface	distributed RAN
backbone	downlink
backhaul	downstream
base station	dual connectivity
baseband	fronthaul
baseband unit (BBU)	GPRS Tunneling Protocol for User Plane (GTP-U)
central entity	IAB donor
centralized RAN	IAB node
child node	integrated access and backhaul (IAB)
core	Internet Protocol (IP)
data link layer	logical channel

medium access control (MAC)	radio bearer
metro access	radio link control (RLC)
metro aggregation	RLC channel
metro core	radio resource control (RRC)
midhaul	Service Data Adaptation Protocol (SDAP)
NG Application Protocol (NGAP)	session management function (SMF)
non-access stratum (NAS)	Stream Control Transmission Protocol (SCTP)
packet data convergence protocol (PDCP)	transport channel
parent node	transport network
physical channel	user equipment (UE)
physical layer	uplink
protocol data unit (PDU)	upstream
QoS flow	User Datagram Protocol (UDP)
radio access network (RAN)	user plane function (UPF)
radio baseband unit (RBU)	virtualized RAN

Review Questions

1. List and define four types of RAN nodes.
2. What is dual connectivity?
3. List and define three core functional areas that are the primary means of interaction with a RAN.
4. What functions are performed by Service Data Adaptation Protocol?
5. What functions are performed by the radio resource control?
6. What is the non-access stratum?
7. What functions are performed by GTP-U?
8. What are the chief differences between UDP and SCTP?
9. What is the RAN transport network?

10. Explain the differences between fronthaul, midhaul, and backhaul.
11. Describe the air interface protocol architecture.
12. Describe the RAN–core interface protocol architecture.
13. Describe the Xn interface protocol architecture.
14. Describe four possible deployment scenarios for transport networks.
15. Explain the difference between metro access, metro aggregation, metro core, and backbone.
16. What is a virtualized RAN?
17. Explain the basic concept of IAB.
18. Explain the concepts of IAB donor, IAB node, parent node, and child node.

15.8 References and Documents

References

ERIK19 Eriksson, A., et al. "5G New Radio RAN and Transport Choices That Minimize TCO." *Ericsson Technology Review*, November 7, 2019.

GSMA18 GSM Association. *Road to 5G: Introduction and Migration.* April 2018.

GSMA19 GSM Association. *The 5G Guide: A Reference for Operators.* April 2019.

GSMA20 GSM Association. *The Mobile Economy.* 2020.

UITT20 Uitto, T. "Making Sense of ORAN and vRAN." Nokia blog, September 10, 2020. https://www.nokia.com/blog/making-sense-of-oran-and-vran-part-one/

Documents

3GPP TS 23.501 *Technical Specification Group Services and System Aspects, System Architecture for the 5G System (5GS), Stage 2 (Release 16).* September 2020.

3GPP TS 37.340 *Technical Specification Group Radio Access Network, NR, Evolved Universal Terrestrial Radio Access (E-UTRA) and NR, Multi-connectivity, Stage 2 (Release 16).* September 2020.

3GPP TS 38.300 *Technical Specification Group Radio Access Network, NR, NR and NG-RAN Overall Description, Stage 2 (Release 16).* September 2020.

3GPP TS 38.401 *Technical Specification Group Radio Access Network, NG-RAN, Architecture Description (Release 16).* September 2020.

ITU-R Report M.2320 *Future Technology Trends of Terrestrial IMT Systems*. November 2014.

ITU-T G.8300 *Characteristics of Transport Networks to Support IMT 2020/5G*. May 2020.

ITU-T Technical Report GSTR-TN5G *Transport Network Support of IMT-2020/5G*. October 19, 2018.

Appendix A

Review Questions and Solutions

Chapter 1: Cellular Networks: Concepts and Evolution

1. What geometric shape is used in cellular system design?

   Hexagonal

2. What is the principle of frequency reuse in the context of a cellular network?

   Each cellular base station is an assigned set of frequencies, with adjacent cells using non-overlapping sets of frequencies. The set of frequencies used in one cell may be reused in other nearby (but not adjacent) cells, thus allowing a frequency channel to be used for multiple simultaneous connections.

3. List five ways of increasing the capacity of a cellular system.

   - **Addition of new channels:** Typically, when a system is set up in a region, not all of the channels are used, and growth and expansion can be managed in an orderly fashion by adding new channels.
   - **Frequency borrowing:** In the simplest case, frequencies are taken from adjacent cells by congested cells. The frequencies can also be assigned to cells dynamically.
   - **Cell splitting:** In practice, the distribution of traffic and topographic features is not uniform, and this presents opportunities for capacity increase. Cells in areas of high usage can be split into smaller cells. Generally, the original cells are about 6.5 to 13 km in size. The smaller cells can themselves be split.
   - **Cell sectoring:** With cell sectoring, a cell is divided into a number of wedge-shaped sectors, each with its own set of channels—typically three sectors per cell. Each sector

is assigned a separate subset of the cell's channels, and directional antennas at the base station are used to focus on each sector. This can be seen in the triangular shape of typical cellular antenna configurations, where the antennas mounted on each side of the triangle are directed toward their respective one of the three sectors.

- **Small cells, or micro cells:** As cells become smaller, antennas move from the tops of tall buildings or hills, to the tops of small buildings or the sides of large buildings, and finally to lamp posts, where they form *picocells*. Each decrease in cell size is accompanied by a reduction in the radiated power levels from the base stations and the mobile units. Picocells are useful in city streets in congested areas, along highways, and inside large public buildings. If placed inside buildings, these are called *femtocells*, and they might be open to all users or only to authorized users (e.g., only those who work in the building).

4. What are the five principal elements of a cellular system?

   - **Base station:** A network element in a radio access network responsible for radio transmission and reception in one or more cells to or from the user equipment. A base station can have an integrated antenna or can be connected to an antenna by feeder cables. The base station interfaces the user terminal (through an air interface) to a radio access network infrastructure.
   - **Air interface:** Wireless interface between user equipment and the base station, also called a *radio interface*. The air interface specifies the method for transmitting information over the air between base stations and mobile units, including protocols, frequency, channel bandwidth, and the modulation scheme.
   - **Mobile telecommunications switching office (MTSO):** Used by a cellular service provider for originating and terminating functions for calls to or from end user customers of the cellular provider. Also known as *mobile switching center* (*MSC*).
   - **Radio access network (RAN):** The network that connects radio base stations to the core network. The RAN provides and maintains radio-specific functions, which may be unique to a given radio access technology, that allow users to access the core network. RAN components include base stations and antennas, MTSOs, and other management and transmission elements.
   - **Core network:** A central network that provides networking services to attached distribution and access networks.

5. What is the difference between a control channel and a traffic channel?

   Control channels are used to exchange information having to do with setting up and maintaining calls and with establishing a relationship between a mobile unit and the nearest BS. Traffic channels carry a voice or data connection between users.

6. What is a handoff?

   If a mobile unit moves out of range of one cell and into the range of another during a connection, the traffic channel has to change to one assigned to the base station in the new cell. The system makes this change without either interrupting the call or alerting the user.

7. Explain the paging function of a cellular system.

   Paging is a function of a base station that attempts to complete a connection to a called user by broadcasting a paging signal to locate the called unit.

8. What are the key differences between first- and second-generation cellular systems?

   - **Digital traffic channels:** The most notable difference between the two generations is that 1G systems are almost purely analog, whereas 2G systems are digital. In particular, 1G systems are designed to support voice channels using FM; digital traffic is supported only by the use of a modem that converts the digital data into analog form. 2G systems provide digital traffic channels. These systems readily support digital data; voice traffic is first encoded in digital form before transmission.
   - **Encryption:** Because all of the user traffic, as well as control traffic, is digitized in 2G systems, it is a relatively simple matter to encrypt all of the traffic to prevent eavesdropping. All 2G systems provide this capability, whereas 1G systems send user traffic in the clear, providing no security.
   - **Error detection and correction:** The digital traffic stream of 2G systems also lends itself to the use of error detection and correction techniques, such as those discussed in Chapter 14, "Air Interface Channel Coding." The result can be very clear voice reception.
   - **Channel access:** In 1G systems, each cell supports a number of channels. At any given time, a channel is allocated to only one user. 2G systems also provide multiple channels per cell, but each channel is dynamically shared by a number of users using time-division multiple access (TDMA) or code-division multiple access (CDMA).

9. What is the difference between TDMA and FDMA?

   - **TDMA:** A transmission channel is divided into a sequence of time slots. Repetitive time slots are assigned to an individual user to form a logical subchannel.
   - **FDMA:** A transmission channel is divided into a number of frequency bands. One or more frequency bands are assigned to an individual user to form a logical subchannel.

10. What are some key characteristics that distinguish third-generation cellular systems from second-generation cellular systems?

    - Voice quality comparable to that of the public switched telephone network
    - 144 kbps data rate available to users in high-speed motor vehicles over large areas

- 384 kbps available to pedestrians standing or moving slowly over small areas
- Support (to be phased in) for 2.048 Mbps for office use
- Symmetrical and asymmetrical data transmission rates
- Support for both packet-switched and circuit-switched data services
- An adaptive interface to the Internet to reflect efficiently the common asymmetry between inbound and outbound traffic
- More efficient use of the available spectrum in general
- Support for a wide variety of mobile equipment
- Flexibility to allow the introduction of new services and technologies

11. What are NodeB and eNodeB?

    The NodeB station interface with subscriber stations (referred to as user equipment [UE]) is based on CDMA, whereas the eNodeB air interface is based on OFDMA. eNodeB embeds its own control functionality rather than using a radio network controller (RNC), as does a NodeB. This means that the eNodeB now supports radio resource control, admission control, and mobility management, which were originally the responsibility of the RNC.

12. Briefly explain the X2 interface in LTE-Advanced.

    The X2 interface enables eNodeBs to interact with each other. The architecture is open so that there can be interconnections between different manufacturers. There is a control plane X2-C interface that supports mobility management, handover preparation, status transfer, UE context release, handover cancel, inter-cell interference coordination, and load management. The X2-U interface is the user plane interface used to transport data during X2 initiated handover.

Chapter 2: 5G Standards and Specifications

1. What are the main activities of ITU-R?

    - Develop draft ITU-R recommendations on the technical characteristics of, and operational procedures for, radiocommunication services and systems
    - Compile handbooks on spectrum management and emerging radiocommunication services and systems
    - Ensure optimal, fair, and rational use of the radio-frequency spectrum and satellite-orbit resources and coordinate matters related to radiocommunication services and wireless services

2. What is ITU-R's role in the development of 5G via IMT-2020 documents?

   ITU-R's role in the development of 5G via IMT-2020 includes developing and adopting the following:

   - International regulations on the use of the radio-frequency spectrum, referred to as the Radio Regulations (RR). To take into account the progress of technologies and the changes in spectrum uses, the RR are updated every four years by the ITU World Radiocommunication Conference (WRC). The RR are an international treaty that is binding on the 193 member states of the ITU. They are the basis for the harmonization of IMT spectrum worldwide.
   - Global standards for the overall requirements of IMT and for its radio interface (ITU-R recommendations).
   - Best practices in the implementation of these standards and regulations (ITU-R reports and handbooks).
   - Evaluation criteria and procedures to evaluate technology submissions for IMT-2020, as well as submission templates that proponents must utilize to organize the information that is required in a submission of a candidate technology for evaluation.

3. Briefly describe eMBB, mMTC, and URLLC.

   - **Enhanced mobile broadband (eMBB):** A 5G usage scenario characterized by high data rates for mobile devices.
   - **Massive machine type communication (mMTC):** A 5G usage scenario characterized by the ability to support huge numbers of devices, such as in a large IoT deployment.
   - **Ultra-reliable and low-latency communications (URLLC):** A form of machine-to-machine communications that enables delay-sensitive and mission-critical services that require very low end-to-end delay, such as tactile Internet, remote control of medical or industrial robots, driverless cars, and real-time traffic control.

4. List some examples of emerging use cases that can be supported by IMT-2020.

   - **Enhanced mobile broadband (eMBB):** Use cases include cloud access apps for commuters and other off-site employees, remote workers needing to communicate with the back office, or indeed an entire smart office where all devices are wirelessly and seamlessly connected.
   - **Massive machine type communication (mMTC):** Internet of Things (IoT) applications, such as smart agriculture and smart cities.
   - **Ultra-reliable and low-latency communications (URLLC):** Use cases such as tactile Internet, remote control of medical or industrial robots, driverless cars, and real-time traffic control.

5. What is ITU-R's role in the development of technical specifications for radio interface technologies for 5G?

   ITU-R itself is not developing the technical specifications for radio interface technologies (RITs). Rather, it has used the following process:

   1. ITU-R defined five test environments that together are representative of the real-world environments for 5G deployment.
   2. ITU-R developed a set of evaluation criteria for assessing the compliance of any proposed RIT with IMT-2020 minimum requirements. These requirements consist of:
      a. The ability to function in one or more of the test environments
      b. The use of approved 5G frequency bands
      c. Technical performance requirements
   3. ITU-R invited organizations to submit proposals for RITs or sets of radio interface technologies (SRITs) covering one or more of the test environments.

6. What is the difference between peak data rate and user-experienced data rate?

   - Peak data rate is the maximum achievable data rate under ideal conditions per user/device (in Gbit/s). The minimum target values are downlink peak data rate of 20 Gbit/s and uplink peak data rate of 10 Gbit/s.
   - User-experienced data rate is the achievable data rate that is available ubiquitously across the coverage area to a mobile user/device (in Mbit/s or Gbit/s). This rate will depend the type of environment. The target value for dense urban-eMBB is 100 Mbit/s downlink and 50 Mbit/s uplink.

7. Describe two types of latency that are relevant to IMT-2020.

   - **User plane latency:** The contribution by the radio network to the time from when the source sends a packet to when the destination receives it (in ms). The minimum requirements for user plane latency are 4 ms for eMBB and 1 ms for URLLC.
   - **Control plane latency:** The transition time from the most "battery-efficient" state (e.g., Idle state) to the start of continuous data transfer (e.g., Active state). The minimum requirement is 20 ms.

8. What is ITU-T's role in the development of 5G via IMT-2020 documents?

   With respect to IMT-2020, the role of ITU-T is complementary to that of ITU-R. ITU-R develops and adopts the international regulations on the use of the radio-frequency spectrum. ITU-R also develops and adopts global standards for the overall requirements of IMT and for its radio interface, as well as best practices in the implementation of these standards and regulations. ITU-T specifies requirements for overall non-radio aspects of the IMT-2020 network, especially with respect to network operations and support of service requirements. Whereas the

ITU-R recommendations and reports emphasize the air interface performance characteristics, the ITU-T's focus is on increased end-to-end flexibility, taking advantage of software-defined networking (SDN), network functions virtualization (NFV), and cloud computing.

9. What are the core network requirements from the network operation point of view defined by ITU-T?

   - **Network flexibility and programmability:** This is a major requirement for IMT-2020 networks. The network should be able to support a wide range of devices, users, and applications, with evolving requirements for each. Significant concepts in this regard are network functions virtualization(NFV), separation of user and control planes, and network slicing.
   - **Fixed mobile convergence:** The goal of this requirement is to enable subscriber access through multiple-access networks in seamless, integrated fashion.
   - **Enhanced mobility management:** The network should support a wide variety of mobility options.
   - **Network capability exposure:** The IMT-2020 network should provide suitable ways (e.g., via APIs) to expose network capabilities and relevant information (e.g., information for connectivity, QoS, and mobility) to third parties. This enables third parties to dynamically customize the network capabilities for diverse use cases within the limits set by the IMT-2020 network operator.
   - **Identification and authentication:** There should be a unified approach to user and device identification and authentication mechanisms.
   - **Security and personal data protection:** The IMT-2020 network must provide effective mechanisms to preserve security and personal data protection for different types of devices, users, and services, including rapid adaptation to dynamic network changes.
   - **Efficient signaling:** There are two aspects to this requirement. The signaling mechanisms should be designed to mitigate risks of control and data traffic bottlenecks. Also, the network should provide lightweight signaling protocols and mechanisms to accommodate limited-resource devices.
   - **Quality of service control:** The network should support different QoS levels for different services and applications.
   - **Network management:** The network should provide a unified network management framework to support interworking of different providers and management of legacy networks.
   - **Charging:** The IMT-2020 network needs to support different charging policies and requirements of network operators and service providers, including third parties that may be involved in a given IMT-2020 network deployment. The charging models to be

supported include, but are not limited to, charging based on volume, time, session, and application.

- **Interworking with non-IMT-2020 networks:** IMT-2020 networks should support user-transparent interworking with legacy networks.
- **IMT-2020 network deployment and migration:** The network design should accommodate incremental deployment with migration capabilities for services and related users.

10. Briefly explain network slicing.

    Network slicing permits a physical network to be separated into multiple virtual networks (i.e., logical segments) that can support different radio access networks or several types of services for certain customer segments, greatly reducing network construction costs through more efficient use of communication channels. In essence, network slicing allows the creation of multiple virtual networks atop a shared physical infrastructure. This virtualized network scenario devotes capacity to certain purposes dynamically, according to need. As needs change, so can the devoted resources. Using common resources such as storage and processors, network slicing permits the creation of slices devoted to logical, self-contained, and partitioned network functions. Network slicing supports the creation of virtual networks to provide a given QoS level, such as guaranteed delay, throughput, reliability, and/or priority.

11. What is fixed mobile convergence?

    ITU-T defines fixed mobile convergence as the capabilities that provide services and applications to end users, regardless of the fixed or mobile access technologies being used and independently of users' locations. This capability requires a unified core network for the new radio access technologies, as well as existing fixed and wireless networks.

12. List the main functional elements of a 5G core network.

    - **Network access control function (NACF):** Provides access to the CN services for the AN and for UE. NACF includes:
        - **Registration management:** Enables a UE to register for network access. NACF performs, among other actions, network slice instance selection, UE authentication, authorization of network access and network services, and network access policy control.
        - **Connection management:** Establishes and releases a signaling connection between the UE and the CN.
        - **SMF selection:** Determines the session management function that is most appropriate to establish and manage a session. In the context of IMT-2020, a session is an association between a UE and a data network that provides a protocol data unit (PDU) connectivity service.

- **Session management function (SMF):** Sets up and manages one or more sessions that provide connectivity between the local UE and a remote UE. This function deals with user path selection and enforcement of policies, including QoS policy and charging policy.
- **Policy control function (PCF):** Provides for control and management of policy rules.
- **Capability exposure function (CEF):** Enables the exposure of network functions and network slices as a service to third parties.
- **Network function registry (NFR) function:** Assists in the discovery and selection of required network functions.
- **Unified subscription management (USM) function:** Stores and manages UE context and subscription information, including, but not limited to, information on the UE's registration and mobility management, information on network functions that serve the UE, and information on session management. The USM function also provides authentication information of the UE to the ASF.
- **Network slice selection function (NSSF):** When a UE requests registration with the network, the NACF sends a network slice selection request to the NSSF with preferred network slice selection information. The NSSF responds with a message including a list of appropriate network slice instances for the UE.
- **Authentication server function (ASF):** Performs authentication between a UE and the network.
- **Application function (AF):** Interacts with application services that require dynamic policy control. The AF extracts session-related information (e.g., QoS requirements) from application signaling and provides it to the PCF in support of its rule generation.

13. What is 3GPP's role in the development of 5G technical specifications?

    3GPP began work in 2016 on defining 5G technical specifications for a new radio access technology known as 5G NR (i.e., 5G New Radio) and a next-generation network architecture known as 5G NGN (i.e., 5G NextGen). Unlike with previous generations, there are no longer competing standard bodies working on potential solutions for 5G.

14. List and briefly describe the performance requirements defined by 3GPP.

    - **High data rates and traffic densities:** Several 5G scenarios require the support of very high data rates or traffic densities, including urban and rural areas, office and home, and special deployments (e.g., massive gatherings, broadcast, residential, high-speed vehicles).
    - **Low latency and high reliability:** Some scenarios require the support of very low-latency and very high communications service availability, which implies very high reliability. The overall service latency depends on the delay on the radio interface,

transmission within the 5G system, transmission to a server that may be outside the 5G system, and data processing. Some of these factors depend directly on the 5G system itself, whereas for others the impact can be reduced with suitable interconnections between the 5G system and services or servers outside the 5G system, such as to allow local hosting of the services. TS 22.261 provides an overview of potential scenarios and references other technical specifications for specific requirements.

- **High accuracy positioning:** The 5G system needs to provide different 5G positioning services with configurable performance working points (e.g., accuracy, positioning service availability, positioning service latency, energy consumption, update rate, time to first fix) according to the needs of users, operators, and third parties. TS 22.261 lists quantitative requirements for a number of indoor and outdoor scenarios.
- **Key performance indicators (KPIs) for a 5G system with satellite access:** In some contexts, a 5G access network must use at least one satellite link. KPIs defined in TS 22.261 include minimum and maximum UE-to-satellite delay for various earth orbits, as well as maximum propagation delay.
- **High-availability IoT traffic:** This requirement is concerned specifically with medical monitoring but is applicable to other scenarios that require highly reliable machine type communication in both stationary and highly mobile settings.
- **High data rate and low latency:** This requirement defines data and latency requirements for scenarios such as audiovisual interaction, gaming, and virtual reality.
- **KPIs for UE to network relaying in the 5G system:** In several scenarios, it can be beneficial to relay communication between one UE and the network via one or more other UEs. This category includes performance requirements for various scenarios.

Chapter 3: Overview of 5G Use Cases and Architecture

1. List the emerging 5G use cases defined by ITU-R.

   Machine-type communication, broadband PPDR, transportation applications, utilities, industrial automation, remote control, surveying and inspection, healthcare, sustainability/environmental, smart city, wearables, smart homes, agriculture, media and entertainment, enhanced personal experience, and commercial airspace UAS

2. Explain the use of two dimensions in characterizing 5G uses cases by 5G Americas.

   One dimension consists of three usage scenarios similar to those defined by ITU-R: enhanced mobile broadband, massive IoT, and critical communications. The other dimension characterizes use cases based on whether they involve human-to-human, human-to-machine, or machine-to-machine communication.

3. Explain the concepts of families and categories used by NGMN to classify 5G use cases.

   At a high level, use cases are grouped into eight use case families. These families are roughly similar to the three usage scenarios defined by ITU-R but at a greater granularity. Each family reflects the dominant characteristic of the use cases in that family.

   Each family is in turn divided into a number of categories. The categories represent distinct types of demands on the 5G network in terms of user experience requirements and system performance requirements. For each use case category, one set of requirement values is given, which is representative of the extreme use cases(s) in the category. As a result, satisfying the requirements of a category leads to satisfying the requirements of all the use cases in this category.

4. What is softwarization?

   Softwarization is an overall approach for designing, implementing, deploying, managing, and maintaining network equipment and/or network components through software programming.

5. What are the main approaches to softwarization?

   - **Network functions virtualization (NFV):** The virtualization of compute, storage, and network functions by implementing these functions in software and running them on virtual machines.

   - **Software-defined networking (SDN):** An approach to designing, building, and operating large-scale networks based on programming the forwarding decisions in routers and switches via software from a central server. SDN differs from traditional networking, which requires configuration of each device separately and relies on protocols that cannot be altered.

   - **Edge computing:** A distributed information technology (IT) architecture in which client data is processed at the periphery of the network, as close to the originating source as possible.

   - **Cloud-edge computing:** A form of edge computing that offers application developers and service providers cloud computing capabilities as well as an IT service environment at the edge of a network. The aim is to deliver compute, storage, and bandwidth much closer to data inputs and/or end users.

6. Describe the three layers of the NGMN architecture framework.

   The infrastructure resource layer consists of the physical resources and system software of a fixed-mobile converged (FMC) network. The business enablement layer is a library of all functions required within a converged network in the form of modular architecture building blocks, including functions realized by software modules that can be retrieved from the repository to the desired location, and a set of configuration parameters for certain parts of the network (e.g., radio access). The business application layer contains specific applications and services that support users of the 5G network.

7. Define the three types of nodes that are used in the NGMN core network model.

   - **Cloud nodes:** These nodes provide cloud services, software, and storage resources. There are likely to be one or more central cloud nodes that provide traditional cloud computing service. In addition, cloud-edge nodes provide low latency and higher-security access to client devices at the edge of the network. All of these nodes include virtualization system software to support virtual machines and containers. NFV enables effective deployment of cloud resources to the appropriate edge node for a given application and given fixed or mobile user. The combination of SDN and NFV enables the movement of edge resources and services to dynamically accommodate mobile users.

   - **Networking nodes:** These are IP routers and other types of switches for implementing a physical path through the network for a 5G connection. SDN provides for flexible and dynamic creation and management of these paths.

   - **Access nodes:** These provide an interface to radio access networks (RANs), which in turn provide access to mobile user equipment (UE). SDN creates paths that use an access node for one or both ends of a connection involving a wireless device.

8. In the NGMN model, who are the key 5G users?

   - **Mobile network operator:** A mobile network operator is a wireless telecommunications organization that provides wireless voice and data communication for its subscribed mobile users. The operator owns or controls the complete telecom infrastructure for hosting and managing mobile communications between the subscribed mobile users with users in the same and external wireless and wired telecom networks, including radio spectrum allocation, wireless network infrastructure, backhaul infrastructure, billing, customer care, provisioning computer systems, and marketing and repair organizations. Mobile network operators are also known as wireless service providers, wireless carriers, and cellular companies.

   - **Enterprise:** An enterprise is a business that offers services over the mobile network. These services include applications that run on mobile devices and cloud-based services that enable application portability across multiple devices.

   - **Verticals:** An industry vertical is an organization that provides products and/or services targeted to a specific industry, trade, profession, or other group of customers with specialized needs. A vertical might provide a range of products or services useful in the banking industry or healthcare. In contrast, a horizontal provides products or services that address a specific need across multiple industries, such as accounting or billing products and services. Some 5G use cases are realized by standalone private networks managed by the vertical industry itself rather than the network service provider (NSP). A good example is factory automation. In such cases, the vertical can own and control its own application packages and business application layer.

- **Over-the-top (OTT) and third parties:** OTT or third-party services can be defined as any services provided over the Internet and mobile networks that bypass traditional operators' distribution channel. Cooperation between the mobile network operator and the OTT involves providing quality of service (QoS) and latency attributes in network slices.

9. What is the relationship between network slicing and the types of nodes in the core network?

   Slices effectively partition networks in such a way that different classes of user equipment, utilizing their respective sets of radio access technologies, would perceive quite different infrastructure configurations, even though they'd be accessing resources from the same pools.

10. Explain the concept of roaming.

   Roaming is the ability for a user to function in a serving network different from the home network, called the visited network.

11. What is the difference between a service-based representation and a reference point representation?

   - **Service-based representation:** NFs within the control plane enable other authorized NFs to access their services. This representation illustrates how a set of services is provided/exposed by a given network interface. This interface defines how one network function within the control plane allows other network functions that have been authorized to access its services. This representation also includes point-to-point reference points where necessary.

   - **Reference point representation:** This representation uses labeled point-to-point links to show the interaction that exists between two NFs or between an NF and an external functional module or network. The reference point representation is beneficial when showing message sequence charts. It shows the relationships between NFs that are used in the message sequence charts.

12. Summarize the major functions encompassed by the user plane function.

   - Packet routing and forwarding.

   - Anchor point for intra-/inter-RAT mobility (when applicable). Anchor points are transit nodes in the network used for forwarding PDUs along a session from a UE to the destination.

   - External PDU session point of interconnect to data network.

   - Packet inspection (e.g., application detection based on a service data flow [SDF] template). An SDF provides end-to-end packet flow between an end user and an application.

- User plane part of policy rule enforcement (e.g., gating, redirection, traffic steering).
- Traffic usage reporting.
- QoS handling for the user plane, such as uplink/downlink rate enforcement.
- Uplink traffic verification (SDFs to QoS flow mapping). A QoS flow is the lowest level of granularity for defining end-to-end QoS policies. A QoS flow may contain multiple SDFs; this is discussed in Chapter 9, "Core Network Functionality, QoS, and Network Slicing."
- Transport-level packet marking in the uplink and downlink.
- Downlink packet buffering and downlink data notification triggering.
- Sending and forwarding of one or more end markers to the source NG-RAN node.

13. What are the two primary types of nodes in a radio access network?
    - **gNB:** Provides 5G user plane and control plane protocol terminations toward the UE.
    - **ng-eNB:** Provides 4G (E-UTRA) user plane and control plane protocol terminations toward the UE and connects via the NG interface to the 5G core. This enables 5G networks to support UE that use the 4G air interface. However, the UE must still implement the 5G protocols to interact with the 5G core network.

14. List and briefly define the RAN functional areas.
    - **Inter-cell radio resource management:** Allows the UE to detect neighbor cells, query about the best serving cell, and support the network during handover decisions by providing measurement feedback.
    - **Radio bearer control (RBC):** Consists of the procedure for configuration (such as security), establishment, and maintenance of the radio bearer (RB) on both the uplink and downlink with different quality of service (QoS). The term *radio bearer* refers to an information transmission path of defined capacity, delay, bit error rate, and other parameters.
    - **Connection mobility control (CMC):** Functions both in UE idle mode and connected mode. In idle mode, UE is switched on but does not have an established connection. In connected mode, UE is switched on and has an established connection. In idle mode, CMC performs cell selection and reselection. The connected mode involves handover procedures triggered on the basis of the outcome of CMC algorithms.
    - **Radio admission control (RAC):** Decides whether a new radio bearer admission request is admitted or rejected. The objective is to optimize radio resource usage while maintaining the QoS of existing user connections. Note that RAC decides on admission or rejection for a new radio bearer, while RBC takes care of bearer maintenance and bearer release operations.

- **Measurement configuration and provision:** Consists of provisioning the configuration of the UE for radio resource management procedures such as cell selection and reselection and for requesting measurement reports to improve scheduling.
- **Dynamic resource allocation (scheduler):** Consists of scheduling RF resources according to their availability on the uplink and downlink for multiple pieces of UE, according to the QoS profiles of a radio bearer.

15. Which core network functions interact directly with the RAN?

    Access and mobility management, session management, and user plane

Chapter 4: Enhanced Mobile Broadband

1. What are the key drivers for the deployment of eMBB?
   - **Traffic demand:** Mobile traffic is expected to grow by 27% annually between 2019 and 2025, with most of the increase from video traffic. By 2025, video traffic will account for about 75% of all mobile traffic. It has been estimated that over 80% of cellular Internet traffic is consumed indoors. Users are demanding higher throughputs (e.g., to meet various video needs, such as HD video streaming).
   - **Operator competition:** Operators are endeavoring to provide more competitive, attractive data plans, and service offerings are being eroded by over-the-top (OTT) alternatives. Competition is mainly in two directions: to offer cheaper rates, often focusing on particular customer segments as pursued by many mobile virtual network operators (MVNOs), or to offer added value, exemplified by larger data caps, faster data rates, and bundling with, for example, zero-rated videos. Zero-rating is the practice of exempting an app from a user's monthly data plans; an application that is zero-rated does not count against a user's data cap, while all other applications do count against the cap.
   - **Incentives to improve attractiveness of countries, cities, and premises:** Many governments and municipalities see eMBB availability as a key to future productivity and economic growth (enriching the lives of citizens, attracting more tourists, and facilitating businesses) and are leading initiatives to improve broadband environments. Similarly, building owners have incentives to invest in connectivity provisioning to sustain property value. The same applies to any kind of space, including airports, hotels, shopping malls, coffee shops, public transport, entertainment venues (e.g., sports, music), and even cars and airplanes. The cost for connectivity provisioning is often paid by indirect sources, making the cost invisible to the end user.

2. What challenge is presented in the use of high-frequency bands in the indoor hotspot deployment scenario?

   Higher bands experience greater link losses.

3. Why does the dense urban environment require the use of a dense collection of small cells to supplement macro cells?

   - The concentrated collection of stationary, pedestrian, and vehicular users, with 5G use cases, generates a tremendous traffic load.
   - 5G mmWave networks are predominantly noise limited. As a result, only small cell sizes can be supported.

4. Summarize the eMBB data rate requirements.

Parameter	Downlink Value	Uplink Value
Peak data rate	20 Gbit/s	10 Gbit/s
User-experienced data rate (dense urban test environment)	100 Mbit/s	50 Mbit/s
Area traffic capacity (indoor hotspot test environment)	10 Mbit/s/m^2	—

5. Summarize the eMBB spectral efficiency requirements.

Parameter	Test Environment	Downlink Value (bit/s/Hz)	Uplink Value (bit/s/Hz)
Peak spectral efficiency	All	30	15
5th percentile user spectral efficiency	Indoor hotspot	0.3	0.21
	Dense urban	0.225	0.15
	Rural	0.12	0.045
Average spectral efficiency	Indoor hotspot	9	6.75
	Dense urban	7.8	5.4
	Rural	3.3	1.6

6. Summarize the eMBB latency requirements.

Parameter	Value	4G Requirement
User plane latency	4 ms	10 ms
Control plane latency	20 ms	100 ms

7. Summarize the eMBB mobility requirements.

Test Environment	Indoor Hotspot	Dense Urban	Rural	
Mobility classes supported	Stationary, pedestrian	Stationary, pedestrian, vehicular	Pedestrian, vehicular	High-speed vehicular
Traffic channel link data rate (bit/s/Hz)	1.5	1.12	0.8	0.45
4G traffic channel link data rate (bit/s/Hz)	1.0	0.75	0.55	0.25
Mobility (km/h)	10	30	120	500
4G mobility (km/h)	10	30	120	350

8. What are the key performance indicators for a smart office indoor hotspot use case?

KPI	Smart Office (NGMN)	Virtual Reality Office (METIS)
DL user-experienced data rate	1 Gbps Average load: 0.2 Gbps/user	At least 1 Gbps with 95% location availability and 5 Gbps with 20% location
UL user-experienced data rate	500 Mbps Average load: 27 Mbps/user	Same as downlink
Connection density	75,000/km^2 (0.75/10m^2)	200,000/km^2 (2/10 m^2)
Traffic density	DL: 15 Tbps/km^2 UL: 2 Tbps/km^2	100 Tbps/km^2 in DL and UL
Mobility	Pedestrian	Static or low mobility nomadic (less than 6 km/h)
Availability	User-experienced data rate should be available in at least 95% of the locations (including at the cell-edge) for at least 95% of the time	95% for 1 Gbps and 20% for 5 Gbps
Reliability	95%	99% during working hours
Latency	10 ms end-to-end	10 ms round-trip time

9. What are the key performance indicators for a dense urban information society use case?

KPI	Value
User-experienced throughput	DL: 300 Mbps UL: 60 Mbps
Traffic volume density	700 Gbps/km^2 (0.7 Gbps/m^2)
Latency	Web browsing: < 0.5 s for download of an average-size web page Video streaming: < 0.5 s Augmented reality processed in the cloud and locally: < 2–5 ms Device-to-device feedback: ≤ 1 ms
User/device density	200,000/km^2 (2/10 m^2)
User mobility	Most of the users and devices have velocities up to 3 km/h and, in some cases, up to 50 km/h

10. List some key requirements for supporting RSTT.

    - Railways require connectivity at speeds up to 500 km/hr, thus involving numerous rapid handovers.
    - Trains often travel in cuttings and tunnels that typically have poor radio-frequency (RF) coverage.
    - Trains require very high availability, reaching or exceeding 99.999%, due to the need to control driverless trains.
    - To improve security and safety, real-time passenger surveillance and front-looking obstacle detection cameras add the requirement of high uplink throughput capacity.

11. List and briefly describe four main application areas of RSTT.

    - **Train radio:** Train radio is used for communication between the train and the track side for signaling and traffic management to promote safe train operation. Train radio provides mobile interconnection to landline and mobile-to-mobile voice communication and also serves as the data transmission channel within various bearer services.
    - **Train positioning:** Train positioning systems gather all kind of train positioning information (e.g., exact locations of all units on trackside) relevant to train operations. This includes line- and location-oriented information.
    - **Train remote:** Train remote refers to the ability for a ground-based operator to remotely control the movements of a train. This feature is typically used for shunting operations in depots.
    - **Train surveillance:** Train surveillance systems enable the capture and transmission of video of the public and trackside areas, driver cabs, passenger compartments, platforms, and device monitoring.

Chapter 5: Massive Machine Type Communications

1. What are the mMTC performance requirements listed in M.2410?

   For mMTC, the only defined performance requirement is for connection density. M.2410 sets the minimum requirement as 10^6 devices per km^2.

2. Define *Internet of Things*.

   The expanding connectivity, particularly via the Internet, of a wide range of sensors, actuators, and other embedded systems. In almost all cases, there is no human user, and interaction is fully automated.

3. List and briefly define the principal components of an IoT-enabled thing.

   - **Sensor:** A sensor measures some parameter or set of parameters of a physical, chemical, or biological entity and delivers an electronic signal proportional to the observed characteristic, either in the form of an analog voltage level or a digital signal. In both cases, the sensor output is typically input to a microcontroller or other management element. Examples include temperature measurement, radiographic imaging, optical sensing, and audio sensing.
   - **Actuator:** An actuator receives an electronic signal from a controller and responds by interacting with its environment to produce an effect on some parameter or set of parameters of a physical, chemical, or biological entity. Examples include heating coils, cardiac electric shock delivery, electronic door locks, unmanned aerial vehicle operation, servo motors, and robotic arms.
   - **Microcontroller:** The "smart" in a smart device is provided by a deeply embedded microcontroller.
   - **Transceiver:** A transceiver contains the electronics needed to transmit and receive data. An IoT device typically contains a wireless transceiver that is capable of communication using Wi-Fi, ZigBee, or some other wireless protocol. By means of the transceiver, IoT devices can interconnect with other IoT devices, with the Internet, and with gateway devices to cloud systems.
   - **Power supply:** Typically, this is a battery.

4. Describe the three classes of constrained devices.

   - **Class 0:** These devices are very constrained devices, typically sensors, called *motes*, or *smart dust*. Motes can be implanted or scattered over a region to collect data and pass it on from one device to another to some central collection point. For example, a farmer, a vineyard owner, or an ecologist could equip motes with sensors that detect temperature, humidity, and so on, making each mote a mini weather station. Scattered throughout a field, vineyard, or forest, these motes would allow the tracking of microclimates. Class 0 devices generally cannot be secured or managed comprehensively in the traditional sense. They are most likely to be preconfigured with a very small data set, and they are reconfigured rarely, if at all.
   - **Class 1:** These devices are quite constrained in code space and processing capabilities, such that they cannot easily talk to other Internet nodes employing a full protocol stack. However, they are capable enough to use a protocol stack specifically designed for constrained nodes (such as the Constrained Application Protocol [CoAP]) and participate in meaningful conversations without the help of a gateway node.
   - **Class 2:** These devices are less constrained and are fundamentally capable of supporting most of the same protocol stacks as used on notebooks or servers. However, they are still

very constrained compared to high-end IoT devices. Thus, they require lightweight and energy-efficient protocols and low-transmission traffic.

5. Explain the relationship between edge, fog, core, and cloud.

   - **Edge:** At the edge of a typical enterprise network is a network of IoT-enabled devices, consisting of sensors and perhaps actuators.
   - **Fog:** The purpose of what is sometimes referred to as the fog computing level is to convert network data flows into information that is suitable for storage and higher-level processing. Processing elements at this level may deal with high volumes of data and perform data transformation operations, resulting in the storage of much lower volumes of data.
   - **Core:** The core network, also referred to as a *backbone network*, connects geographically dispersed fog networks as well as providing access to other networks that are not part of the enterprise network. Typically, the core network uses very high-performance routers, high-capacity transmission lines, and multiple interconnected routers for increased redundancy and capacity.
   - **Cloud:** The cloud network provides storage and processing capabilities for the massive amounts of aggregated data that originate in IoT-enabled devices at the edge. Cloud servers also host the applications that interact with and manage the IoT devices and that analyze the IoT-generated data.

6. What is the difference between mMTC and IoT?

   The concept of IoT is broader. An mMTC network is characterized by huge volumes of constrained devices that send and/or receive messages infrequently. The traffic is often tolerant of delay. Examples of use cases include low-cost sensors, meters, wearables, and trackers. Such devices are often deployed in challenging radio conditions, such as the basement of a building. Therefore, they require extended coverage and may rely solely on a battery power supply, which puts extreme requirements on the device's battery life.

7. What is the difference between NB-IoT and eMTC?

   NB-IoT is optimized for low-data, low-complexity, delay-tolerant, and non-critical applications that need greater connection density, such as utility meters, industrial and environmental sensors, agricultural monitors, and low-precision mobile trackers. The eMTC network is primarily designed for high-data-rate (compared to NB-IoT), low-latency, mobile, and voice-supporting applications. These include industrial handhelds, health monitors, wearables, and high-precision mobile trackers. It is able to leverage existing 4G base stations. Thus, eMTC is optimized for the broadest range of IoT applications.

8. What is the relationship between mMTC, NB-IoT, eMTC, and LPWA?

   Both NB-IoT and eMTC were developed by 3GPP late within the 4G time span; both support mMTC and can be considered categories or subsets of mMTC.

9. Provide an overview of the five use case categories defined by IoF2020.

   - **Arable farming:** The Internet of Arable Farming trial integrates IoT technologies, data acquisition (soil, crop, climate) using sensors and earth observation systems, crop growth models, and yield gap analysis tools.
   - **Dairy farming:** This use case involves real-time sensor data (e.g., neck collars and movement sensors for livestock) combined with GPS location data to create value in the dairy chain from "grass to glass"— to more efficiently use resources and produce quality foods and provide a better animal health, welfare, and environment implementation.
   - **Internet of fruits:** This use case integrates IoT technology throughout the whole supply chain from the field, logistics, and processing to the retailer.
   - **Internet of vegetables:** This is a combination of environmental control levels: fully controlled indoor growing with an artificial lighting system, semi-controlled greenhouse production, and nonregulated ambient conditions in open-air cultivation of vegetables.
   - **Internet of meat:** This use case aims to demonstrate how the growth of animals can be optimized and how communication in the whole supply chain can be improved using automated monitoring and control of advanced sensor-actuator systems. The data generated by events provides early warning (e.g., on health status) and helps improve the transparency and traceability of meat.

10. List and briefly describe the layers of a business process view model.

   - **Physical object layer:** Describes the physical environment. It depicts the elements used to manage the supply of water and nitrogen in the crop, including the machinery. The IoF2020 document does not explicitly show the IoT-enabled devices.
   - **Production control layer:** Represents the link between the physical world and the virtual world. So in this part, it can find processes that on one hand measure different kinds of information in the field and on the other hand act on crops (irrigation or fertilization). These processes have a short time horizon (e.g., minutes, seconds, milliseconds).
   - **Operations execution layer:** Groups all computational tasks using collected information as well as processes that take into account results of the management information layer. An example is the creation of maps to plan irrigation or apply fertilizer. This layer contains processes related to the definition, control, and performance of tasks with an intermediate time horizon (e.g., days, hours, minutes).

- **Management information layer:** Includes processes that take into account data from the operations layer and apply business rules to obtain several indicators. With some additional analysis, this results in the development of an overall plan related to the control of an entire enterprise. These processes have the longest time horizon (e.g., months, weeks, days).

11. What are the chief aspects of sustainability for a smart city?

    - **Economic:** The ability to generate income and employment for the livelihood of the inhabitants
    - **Social:** The ability to ensure that the welfare (e.g., safety, health, education) of the citizens can be equally delivered, despite differences in class, race, or gender
    - **Environmental:** The ability to protect future quality and reproducibility of natural resources
    - **Governance:** The ability to maintain social conditions of stability, democracy, participation, and justice

12. What are the key components that might be expected in a smart city integrated command and control center?

    - **City communication network:** This involves a combination of optical fiber cable and cellular network capability.
    - **Integrated command and control center (ICCC):** The ICCC is the central repository for management and monitoring of all ICT-based smart city components, such as the solid waste management system, the smart street lighting control system, Wi-Fi, smart transport, smart bus stops, CCTV surveillance, digital signages, IoT sensors (environment, etc.), and the public information system (PIS).
    - **Data center and disaster recovery:** Cloud-based data storage has backup.
    - **City and enterprise geographic information system (GIS):** A comprehensive GIS application is used for planning, management, and governance in the context of the entire functioning of the organization.
    - **CCTV-based real-time public safety system:** CCTV cameras are installed at various locations across the city for safety along with public address system and variable message signboard (VaMS), emergency/panic box system, and other services.
    - **Intelligent traffic management:** CCTV cameras and traffic violation sensors are installed at various locations across the city for traffic management and enforcement system, such as red light violation detection (RLVD), automatic number plate recognition (ANPR), and speed detection.

- **Environmental sensors:** Smart environmental sensors gather data about pollution, ambient conditions (light, noise, temperature, humidity, and barometric pressure), weather conditions (rain), levels of gases in the city (pollution), and other events on an hourly and subsequently on a daily basis.
- **GIS-/GPS-enabled solid waste management:** A GIS-/GPS-enabled solid waste management system provides end-to-end management and monitoring of garbage collection and processing.
- **Adaptive traffic management system:** A system for control and management of traffic controls the traffic signals on certain stretches of road with sensor-based automation of signals.
- **Integration components:** Present and future systems are integrated with the ICC, including an e-Governance system, smart LED lighting, smart bus stops, a SCADA system, a sewage system, Wi-Fi hotspots, and citizen engagement applications for a smart city.

13. List and briefly describe the layers of the ICT architecture for smart cities defined by ITU-T.

    - **Sensing layer:** Includes a collection of IoT-enabled devices with a variety of capabilities, including sensors, actuators, cameras, and RFID readers. Thus, this layer in fact encompasses more than sensing. The other main element of this layer is the capillary network, which is a local network that uses short-range radio-access technologies to provide local connectivity to things and devices.
    - **Network layer:** Consists of various networks provided by telecommunication operators, as well as other metro networks provided by city stakeholders and/or an enterprise private communication network.
    - **Data and support layer:** Provides support programs and databases for various city-level applications and services.
    - **Application layer:** Includes all applications that provide smart city services.

Chapter 6: Ultra-Reliable and Low-Latency Communications

1. What are the URLLC performance requirements listed in M.2410?

    - **User plane latency:** The minimum requirement (i.e., the maximum allowable value) is 1 ms, assuming unloaded conditions (i.e., a single user) for small IP packets (e.g., 0 byte payload + IP header), for both downlink and uplink.
    - **Control plane latency:** The minimum requirement is 20 ms.

- **Mobility interruption time:** The minimum requirement is 0 ms.
- **Reliability:** The minimum requirement is $1–10^{-5}$ success probability of transmitting a Layer 2 protocol data unit (PDU) of 32 bytes within 1 ms for the urban macro-URLLC test environment, assuming small application data (e.g., 20 bytes application data + protocol overhead).

2. What is the difference between user plane latency and control plane latency?

- **User plane latency:** This is the contribution by the radio network to the time from when the source sends a packet to when the destination receives it (in ms). It is defined as the one-way time it takes to successfully deliver an application layer packet/message from the radio protocol Layer 2/3 service data unit (SDU)[1] ingress point to the radio protocol Layer 2/3 SDU egress point of the radio interface in either uplink or downlink in the network for a given service in unloaded conditions, assuming that the mobile station is in the active state.
- **Control plane latency:** This refers to the transition time from a most battery-efficient state (e.g., Idle state) to the start of continuous data transfer (e.g., Active state).

3. Explain the four mobility classes.

- **Stationary:** 0 km/h
- **Pedestrian:** 0–10 km/h
- **Vehicular:** 10–120 km/h
- **High-speed vehicular:** 120–500 km/h

4. What are some examples of urgent healthcare use cases?

- **Remote patient monitoring:** This use case involves remote patient monitoring via communication with devices that measure certain health indicators, such as pulse, blood glucose, blood pressure, and temperature. On an individual basis, the data rate and latency requirements are modest. However, for this use case to become pervasive, 5G is needed to support the massive increase in the number of connections per square meter while still maintaining the requisite QoS.
- **Remote healthcare:** This use case provides for individualized consultation, treatment, and patient monitoring built on a video linkup capability. The video conferencing can be augmented with remote transfer of health-related data in real time. Treatment could also be offered using smart pharmaceutical devices that correctly administer approved dosages of a drug on a schedule specified by the physician or practitioner.

---

1. In a packet, the SDU is data that the protocol transfers between peer protocol entities on behalf of the users of that layer's services. For lower layers, the layer's users are peer protocol entities at a higher layer; for the application layer, the users are application entities outside the scope of the protocol layer model.

- **Remote surgery:** A demanding use case is remote surgery via control of robotic devices. This application area may be appropriate in ambulances, disaster sites, and remote areas. Important requirements are precise control and very low latency, very high reliability, and tight security.

5. Explain the difference between assisted driving, autonomous driving, and tele-operated driving.

   - **Assisted driving:** 5G enables the delivery of advanced driver assistance features that reduce fatal accidents and traffic congestion. These features include real-time maps for navigation, speed warnings, road hazards, vulnerabilities, heads-up display systems, and sensor data sharing. These features enable a vehicle to dynamically change course on the road under certain scenarios and conditions. Vehicle-to-network (V2N) communication is necessary for this use case. Information from the vehicle enables the remote application to perform short-range modeling and recognition of surrounding objects and vehicles as well as mid- to long-range modeling of the surroundings using information on the latest digital maps, traffic signs, traffic signal locations, road construction, and traffic congestion.

   - **Autonomous driving:** Fully autonomous driving involves the capability of a vehicle to sense its environment and navigate without human input under all scenarios and conditions. A 5G network with URLLC capability enables a number of necessary features, including using complex algorithms to distinguish between different cars on the road and to identify appropriate navigation paths, given obstacles and the rules of the road, and the exchange of information in real time between thousands of cars connected in the same area.

   - **Tele-operated driving:** This use case refers to the use of remote driver assistance in areas where automatic driving is not possible. Tele-operated driving can provide enhanced safety for disabled people, elderly populations, and those in complex traffic situations. Typical application scenarios include disaster areas and unexpected and difficult terrains for manual driving (e.g., in mining and construction). Tele-operated driving requires the wireless network to support V2N communication of video, sound feed information, and diagnostics from the vehicle, along with environmental information, to the remote driver. The network must support transmission of the control commands from the remote driver to the vehicle to maneuver the vehicle in real time.

6. What are mission-critical services?

   With mission-critical services, use cases involve communications requirements that are critical and need a higher priority compared to other communications in the networks, as well as some means of enforcing this priority.

7. What are cyber-physical systems?

   A cyber-physical system is a networked, interacting system of digital, analog, physical, and human components. Embedded computers and human operators monitor and control physical processes using feedback loops of sensors and actuators.

8. Summarize the characteristics of the four generations of the Industrial Revolution.

   - **First generation:** Mechanization through water and steam power
   - **Second generation:** Mass production and assembly lines using electricity
   - **Third generation:** Adoption of computers and automation
   - **Fourth generation:** Smart and autonomous systems fueled by data and machine learning

9. Explain the difference between smart factory, digital factory, and virtual factory.

   - **Smart factory:** Characterized by extensive use of IoT sensors in manufacturing machines to collect data on their operational status and performance. This extensive sensor network enables automated or human-supervised monitoring for signs that particular parts may fail, enabling preventive maintenance to avoid unplanned downtime on devices. Manufacturers can also analyze trends in the data to try to spot steps in their processes where production slows down or is inefficient in terms of use of materials. IoT networks of actuators can enable more flexible tailored manufacturing processes.
   - **Digital factory:** Refers to human-team agile exploitation/analysis of vast amounts of digital information, knowledge management, informed planning, complex simulation, and collaborative product-service engineering support. This includes product life cycle management, modeling, design, and optimization.
   - **Virtual factory:** Refers to the use of computers to model, simulate, and optimize the critical operations and entities in a factory plant. The main technologies used in the virtual factory include computer-aided design (CAD), 3D modeling and simulation software, product life cycle management (PLM), virtual reality, high-speed networking, and rapid prototyping. These technologies enable an organization to analyze the manufacturability of a part or product as well as evaluate and validate production processes and machinery and train managers, operators, and technicians on production systems.

10. What is the difference between IT and OT?

    - **Information technology (IT):** The common term for the entire spectrum of technologies for information processing, including software, hardware, communications technologies, and related services. In general, IT does not include embedded technologies that do not generate data for enterprise use.

- **Operational technology (OT):** Hardware and software that detects or causes a change through the direct monitoring and/or control of physical devices, processes, and events in the enterprise.

11. What is the difference between factory automation and process automation?

    - **Factory automation:** Factory automation refers to the automated control, monitoring, and optimization of processes and workflows within a factory. This includes aspects like closed-loop control applications (e.g., based on programmable logic or motion controllers), robotics, and aspects of computer-integrated manufacturing. Communication services for factory automation need to fulfill stringent requirements, especially in terms of latency, communication service availability, and determinism. Operation is limited to a relatively small service area, and typically no interaction is required with the public network (e.g., for service continuity and roaming).

    - **Process automation:** Process automation refers to the control of production and handling of substances such as chemicals, food, and beverages. Process automation improves the efficiency of production processes, energy consumption, and safety of the facilities. Sensors measuring process values, such as pressures or temperatures, work in a closed loop by means of centralized and decentralized controllers with actuators, such as valves, pumps, and heaters. Process automation also includes monitoring of attributes such as the filling levels of tanks, quality of material, or environmental data, as well as safety warnings or plant shutdowns. A process automation facility may range from a few hundred square meters to many square kilometers or may be distributed over a certain geographic region. Depending on the size, a production plant may have up to tens of thousands of measurement points and actuators. Autarkic (i.e., independent and self-sufficient) device power supply for years is needed in order to stay flexible and to keep the total costs of ownership low.

12. Explain the difference between a UAV, an autonomous drone, and a UAS.

    - **Unmanned aerial vehicle (UAV):** An aircraft operated without the possibility of direct human intervention from within or on the aircraft. Equivalent terms are remotely piloted vehicle (RPV), drone, unmanned aircraft (UA), and uncommanded aerial vehicle (UCAV).

    - **Autonomous drone:** A type of UAV in which communications management software, instead of a human, coordinates missions and pilots the aircraft.

    - **Unmanned aircraft system (UAS):** Refers to the entire system required for UAV operations, consisting of the UAV (including antenna, sensors, software, and power supply), the ground or airborne control system, and communication links and networks.

Chapter 7: Software-Defined Networking

1. List the key requirements for a modern networking approach.

   - **Adaptability:** Networks must adjust and respond dynamically, based on application needs, business policy, and network conditions.
   - **Automation:** Policy changes must be automatically propagated so that manual work and errors can be reduced.
   - **Maintainability:** Introduction of new features and capabilities (e.g., software upgrades, patches), must be seamless, with minimal disruption of operations.
   - **Model management:** Network management software must allow management of the network at a model level rather than implementing conceptual changes by reconfiguring individual network elements.
   - **Mobility:** Control functionality must accommodate mobility, including mobile user devices and virtual servers.
   - **Integrated security:** Network applications must integrate seamless security as a core service instead of as an add-on solution.
   - **On-demand scaling:** Implementations must have the ability to scale up or scale down the network and its services to support on-demand requests.

2. How does SDN split the functions of a switch?

   The SDN approach splits the switching function between a data plane and a control plane that are on separate devices. The application plane consists of apps that deal with an abstracted network.

3. Describe the elements of an SDN architecture.

   The data plane consists of physical switches and virtual switches. Both types of switches are responsible for forwarding packets. The control plane consists of SDN controllers implemented on a server or virtual server.

4. Explain the function of the northbound, southbound, eastbound, and westbound interfaces.

   The southbound interface (SBI) provides the logical connection between the SDN controller and the data plane switches. The northbound (NBI) interface enables applications to access control plane functions and services without needing to know the details of the underlying network switches. Eastbound interfaces are used to import and export information among distributed controllers. The westbound interface enables communication between an SDN controller and a non-SDN network.

5. What are the two main data plane functions?

   - **Control support function:** Interacts with the SDN control layer to support programmability via resource-control interfaces. The switch communicates with the controller, and the controller manages the switch via the OpenFlow switch protocol.

   - **Data forwarding function:** Accepts incoming data flows from other network devices and end systems and forwards them along the data forwarding paths that have been computed and established according to the rules defined by the SDN applications.

6. Explain the difference between an OpenFlow switch, an OpenFlow port, and an OpenFlow channel.

   - **OpenFlow switch:** A set of OpenFlow resources that can be managed as a single entity, including a data path and a control channel. OpenFlow switches connect logically to each other via their OpenFlow ports.

   - **OpenFlow port:** Where packets enter and exit the OpenFlow pipeline. A packet can be forwarded from one OpenFlow switch to another OpenFlow switch only via an output OpenFlow port on the first switch and an ingress OpenFlow port on the second switch.

   - **OpenFlow channel:** Interface between an OpenFlow switch and an OpenFlow controller, used by the controller to manage the switch.

7. Explain the difference between an OpenFlow physical port, an OpenFlow logical port, and an OpenFlow reserved port.

   - **Physical port:** Corresponds to a hardware interface of the switch. For example, on an Ethernet switch, physical ports map one-to-one to the Ethernet interfaces.

   - **Logical port:** Does not correspond directly to a hardware interface of the switch. Logical ports are higher-level abstractions that may be defined in the switch using non-OpenFlow methods (e.g., link aggregation groups, tunnels, loopback interfaces). Logical ports may include packet encapsulation and may map to various physical ports. The processing done by the logical port is implementation dependent and must be transparent to OpenFlow processing. Logical ports must interact with OpenFlow processing in the same manner as OpenFlow physical ports.

   - **Reserved port:** Defined by the OpenFlow specification and specifies generic forwarding actions, such as sending to and receiving from the controller, flooding, or forwarding using non-OpenFlow methods, such as "normal" switch processing.

8. Explain the difference between an OpenFlow flow table, an OpenFlow group table, and an OpenFlow meter table.

   A flow table matches incoming packets to a particular flow and specifies what functions are to be performed on the packets. There may be multiple flow tables that operate in a pipeline fashion. A flow table may direct a flow to a group table, which may trigger a variety of actions that affect one or more flows. A meter table can trigger a variety of performance-related actions on a flow.

9. Define flow.

A flow is a sequence of packets traversing a network that share a set of header field values.

10. Define action and action set.

Actions describe packet forwarding, packet modification, and group table processing operations. An action set is a list of actions associated with a packet that are accumulated while the packet is processed by each table and that are executed when the packet exits the processing pipeline.

11. Describe OpenFlow ingress and egress processing.

- **Ingress processing:** Ingress processing always happens, and it begins with table 0 and uses the identity of the input port. Table 0 may be the only table, in which case the ingress processing is simplified to the processing performed on that single table, and there is no egress processing.
- **Egress processing:** Egress processing is the processing that happens after the determination of the output port. It happens in the context of the output port. This stage is optional. If it occurs, it may involve one or more tables. The separation of the two stages is indicated by the numeric identifier of the first egress table. All tables with a number lower than the first egress table must be used as ingress tables, and no table with a number higher than or equal to the first egress table can be used as an ingress table.

12. List and briefly define the essential functions of an SDN controller.

- **Shortest path forwarding:** Uses routing information collected from switches to establish preferred routes.
- **Notification manager:** Receives, processes, and forwards to an application events such as alarm notifications, security alarms, and state changes.
- **Security mechanisms:** Provide isolation and security enforcement between applications and services.
- **Topology manager:** Builds and maintains switch interconnection topology information.
- **Statistics manager:** Collects data on traffic through the switches.
- **Device manager:** Configures switch parameters and attributes and manages flow tables.

13. Characterize the operation of an intent NBI.

    There are essentially two approaches for developing an NBI:

    - **Prescriptive:** The application specifies or constrains the selection and allocation, virtualization and abstraction, or assembly and concatenation of resources needed to satisfy the request.
    - **Nonprescriptive:** This is also referred to as intent NBI. The application describes its requirements in application-oriented language, and the controller becomes an intelligent black box that integrates core network services to construct network applications to serve users' requests.

14. List key reasons for using SDN domains.

    - **Scalability:** The number of devices an SDN controller can feasibly manage is limited. Thus, a reasonably large network may need to deploy multiple SDN controllers.
    - **Privacy:** A carrier may choose to implement different privacy policies in different SDN domains. For example, an SDN domain may be dedicated to a set of customers that implement their own highly customized privacy policies, requiring that some networking information in this domain (e.g., network topology) should not be disclosed to an external entity.
    - **Incremental deployment:** A carrier's network may consist of portions of legacy and nonlegacy infrastructure. Dividing the network into multiple individually manageable SDN domains allows for flexible incremental deployment.

15. Describe the application plane architecture.

    The application plane contains applications and services that define, monitor, and control network resources and behavior. These applications interact with the SDN control plane via application control interfaces in order for the SDN control layer to automatically customize the behavior and the properties of network resources. The programming of an SDN application makes use of the abstracted view of network resources provided by the SDN control layer by means of information and data models exposed via the application control interface.

16. What is the purpose of a network services abstraction layer?

    It provides service abstractions that can be used by applications and services.

17. List and briefly describe six network application areas.

    - **Traffic engineering:** Traffic engineering involves dynamically analyzing, regulating, and predicting the behavior of data flowing in networks with the aim of performance optimization to meet service-level agreements (SLAs).

- **Measurement and monitoring:** The area of measurement and monitoring applications can roughly be divided into two categories: applications that provide new functionality for other networking services and applications that add value to OpenFlow-based SDNs.
- **Security and dependability:** Applications in this area have one of two goals: address security concerns related to the use of SDN or use the functionality of SDN to improve network security.
- **Data center networking:** Data center networking involves local networking of servers to support cloud computing, big data, and large enterprise networks.
- **Mobility and wireless:** Network providers must deal with problems related to managing the available spectrum, implementing handover mechanisms, performing efficient load balancing, responding to QoS and QoE requirements, and maintaining security.
- **Information-centric networking:** ICN is aimed at providing native network primitives for efficient information retrieval by directly naming and operating on information objects.

Chapter 8: Network Functions Virtualization

1. Briefly describe type 1 and type 2 virtualization.

   A type 1 hypervisor is loaded as a thin software layer directly into a physical server, much as an operating system is loaded. Once it is installed and configured, usually within a matter of minutes, the server is then capable of supporting VMs as guests. A type 2 hypervisor functions as an application loaded on top of a system OS.

2. Briefly describe container virtualization.

   In this approach, software known as a virtualization container runs on top of the host OS kernel and provides an isolated execution environment for applications. Unlike hypervisor-based VMs, containers do not aim to emulate physical servers. Instead, all containerized applications on a host share a common OS kernel. This eliminates the resources needed to run a separate OS for each application and can greatly reduce overhead.

3. List the key reasons organizations use virtualization.

   - **Legacy hardware:** Applications built for legacy hardware can still be run by virtualizing (emulating) the legacy hardware, making it possible to retire the old hardware.
   - **Rapid deployment:** Whereas it may take weeks or longer to deploy new servers in an infrastructure, it may be possible to deploy a new VM in a matter of minutes. A VM consists of files, and by duplicating those files in a virtual environment, you get a perfect copy of the server.

- **Versatility:** Hardware usage can be optimized by maximizing the number of kinds of applications that a single computer can handle.
- **Consolidation:** A large-capacity or high-speed resource, such as a server, can be used more efficiently by sharing the resources among multiple applications simultaneously.
- **Aggregating:** Virtualization makes it easy to combine multiple resources into one virtual resource, as in the case of storage virtualization.
- **Dynamics:** With the use of virtual machines, hardware resources can be easily allocated in a dynamic fashion. This enhances load balancing, fault tolerance, and the ability to satisfy QoS requirements.
- **Ease of management:** Virtual machines facilitate deployment and testing of software at a faster rate than is possible with bare-metal servers.
- **Increased availability:** Virtual machine hosts are clustered together to form pools of compute resources. Multiple VMs are hosted on each of these servers, and in the event of a physical server failure, the VMs on the failed host can be quickly and automatically restarted on another host in the cluster. Compared with providing this type of availability for a physical server, virtual environments can provide higher availability at significantly less cost and with less complexity.

4. List three key NFV principles involved in creating practical network services.

   - **Service chaining:** VNFs are modular, and each VNF provides limited functionality on its own. For a given traffic flow within a given application, the service provider steers the flow through multiple VNFs to achieve the desired network functionality. This is referred to as service chaining.
   - **Management and orchestration:** This involves deploying and managing the life cycle of VNF instances. Examples of functions are VNF instance creation, VNF service chaining, monitoring, relocation, shutdown, and billing. MANO (management and orchestration) also manages the NFV infrastructure elements.
   - **Distributed architecture:** A VNF may be made up of one or more VNF components (VNFCs), each of which implements a subset of the VNF's functionality. Each VNFC may be deployed in one or multiple instances. These instances may be deployed on separate, distributed hosts in order to provide scalability and redundancy.

5. What are the NFV domains of operation?

   - **Virtualized network functions (VNFs):** A collection of VNFs, implemented in software, run over the NFVI.
   - **NFV infrastructure (NFVI):** The NFVI performs a virtualization function on the three main categories of devices in the network service environment: computer devices, storage devices, and network devices.

- **NFV management and orchestration:** This encompasses the orchestration and life cycle management of physical and/or software resources that support the infrastructure virtualization and the life cycle management of VNFs. NFV management and orchestration focuses on all virtualization-specific management tasks necessary in the NFV framework.

6. What types of relationships are supported between VNFs?

   - **VNF forwarding graph (VNF-FG):** Covers the case where network connectivity between VNFs is specified, such as a chain of VNFs on the path to a web server tier (e.g., firewall, network address translator, load balancer).
   - **VNF Set:** Covers the case where the connectivity between VNFs is not specified, such as a web server pool.

7. What is the difference between CapEx and OpEx?

   A CapEx is a business expense incurred to create future benefits. A CapEx is incurred when a business spends money either to buy fixed assets or to add to the value of an existing asset with a useful life that extends beyond the tax year. An OpEx is a business expense incurred in the course of ordinary business, such as maintenance and operation of equipment.

8. List and briefly describe the major parts of the NFV reference architecture.

   - **NFV infrastructure (NFVI):** The NFVI comprises the hardware and software resources that create the environment in which VNFs are deployed. The NFVI virtualizes physical compute, storage, and networking and places them into resource pools.
   - **VNF/EMS:** A collection of VNFs is implemented in software to run on virtual compute, storage, and networking resources, together with a collection of element management systems (EMSs) that manage the VNFs.
   - **NFV management and orchestration (NFV-MANO):** This framework for the management and orchestration of all resources in the NFV environment includes compute, networking, storage, and virtual machine (VM) resources.
   - **OSS/BSS:** Operational and business support systems are implemented by the VNF service provider.

9. What is meant by the term *reference point* in the context of NFV?

   Reference points constitute interfaces between functional blocks.

10. List and briefly describe the main reference points in the NFV architecture.

    - **Vi-Ha:** This reference point marks interfaces to the physical hardware. A well-defined interface specification enables operators to share physical resources for different purposes, reassign resources for different purposes, evolve software and hardware independently, and obtain software and hardware components from different vendors.

- **Vn-Nf:** These interfaces are APIs used by VNFs to execute on the virtual infrastructure. Application developers, whether migrating existing network functions or developing new VNFs, require a consistent interface that provides functionality and the ability to specify performance, reliability, and scalability requirements.
- **Nf-Vi:** These reference points mark interfaces between the NFVI and the virtualized infrastructure manager (VIM). The Nf-Vi reference points can facilitate specification of the capabilities that the NFVI provides for the VIM. The VIM must be able to manage all the NFVI virtual resources, including allocation, monitoring of system utilization, and fault management.
- **Or-Vnfm:** This reference point is used for sending configuration information to the VNF manager and collecting state information of the VNFs necessary for network service life cycle management.
- **Vi-Vnfm:** This reference point is used for resource allocation requests by the VNF manager and the exchange of resource configuration and state information.
- **Or-Vi:** This reference point is used for resource allocation requests by the NFV orchestrator and the exchange of resource configuration and state information.
- **Os-Ma:** This reference point is used for interaction between the orchestrator and the OSS/BSS systems.
- **Ve-Vnfm:** This reference point is used for requests for VNF life cycle management and exchange of configuration and state information.
- **Se-Ma:** This reference point is the interface between the orchestrator and a data set that provides information regarding the VNF deployment template, VNF forwarding graph, service-related information, and NFV infrastructure information models.

11. What is the NFV infrastructure?

    The heart of the NFV architecture is a collection of resources and functions known as the NFV infrastructure (NFVI).

12. List and briefly describe the three NFVI domains.

    - **Compute domain:** Provides COTS high-volume servers and storage.
    - **Hypervisor domain:** Mediates the resources of the compute domain to the virtual machines of the software appliances, providing abstraction of the hardware.
    - **Infrastructure network domain:** Comprises all the generic high-volume switches interconnected into a network that can be configured to supply infrastructure network services.

13. What is the difference between an NFVI container and a virtualization container?

An important concept in NFVI is the container interface. Unfortunately, the ETSI documents use the term *container* in a unique sense. The NFV infrastructure document states that *container interface* should not be confused with *container* as used in the context of container virtualization as an alternative to full virtual machines. Further, the infrastructure document states that some VNFs may be designed for hypervisor virtualization, and other VNFs may be designed for container virtualization.

14. What are the principal elements in a typical computer domain?

- **CPU/memory:** A multicore processor, with main memory, that executes the code of the VNFC.
- **Internal storage:** Nonvolatile storage housed in the same physical structure as the processor, such as flash memory.
- **Accelerator:** Accelerator functions for security, networking, and packet processing.
- **External storage with storage controller:** Access to secondary memory devices.
- **Network interface card (NIC):** An adapter circuit board installed in a computer to provide a physical connection to a network. It provides the physical interconnection with the infrastructure network domain, which is labeled Ha/CSr-Ha/Nr and corresponds to interface 14 in Figure 8.8.
- **Control and admin agent:** An agent that connects to the virtualized infrastructure manager (VIM).
- **Eswitch:** Server embedded switch. The eswitch function, described below, is implemented in the compute domain. However, functionally it forms an integral part of the infrastructure network domain.
- **Compute/storage execution environment:** The execution environment presented to the hypervisor software by the server or storage device ([VI-Ha]/CSr, interface 12 in Figure 8.8).

15. Explain the concept of eswitch.

In a virtualized environment such as NFV, all VNF network traffic would go through a virtual switch in the hypervisor domain, which invokes a layer of software between virtualized VNF software and host networking hardware. This can create a significant performance penalty. The purpose of the eswitch is to bypass the virtualization software and provide the VNF with a direct memory access (DMA) path to the NIC. The eswitch approach accelerates packet processing without any processor overhead.

16. What is an NFVI node?

    The documents define an NFVI-Node as a collection of physical devices deployed and managed as a single entity, providing the NFVI functions required to support the execution environment for VNFs.

17. List and briefly describe the types of compute domain NFVI nodes.

    - **Compute node:** A functional entity that is capable of executing a generic computational instruction set (with each instruction being fully atomic and deterministic) in such a way that the execution cycle time is on the order of tens of nanoseconds, regardless of what specific state is required for cycle execution. In practical terms, this defines a compute node in terms of memory access time. A distributed system cannot meet this requirement as the time taken to access state stored in remote memory cannot meet this requirement.

    - **Gateway node:** A single identifiable, addressable, and manageable element within an NFVI-Node that implements gateway functions. Gateway functions provide the interconnection between NFVI-PoPs and the transport networks. They also connect virtual networks to existing network components. A gateway may process packets going between different networks, such as removing headers and adding headers. A gateway may operate at the transport level, dealing with IP and data link packets, or at the application level.

    - **Storage node:** A single identifiable, addressable, and manageable element within an NFVI-Node that provides storage resources using compute, storage, and networking functions. Storage may be physically implemented in a variety of ways. It could, for example, be implemented as a component within a compute node. An alternative approach is to implement storage nodes independent of the compute nodes as physical nodes within the NFVI-Node. An example of such a storage node might be a physical device accessible via a remote storage technology, such as Network File System (NFS) or Fibre Channel.

    - **Network node:** A single identifiable, addressable, and manageable element within an NFVI-Node that provides networking (switching/routing) resources using compute, storage, and network forwarding functions.

18. What are the principal elements of the hypervisor domain?

    - **Compute/storage resource sharing/management:** This element manages these resources and provides virtualized resource access for VMs.

    - **Network resource sharing/management:** This element manages these resources and provides virtualized resource access for VMs.

    - **Virtual machine management and API:** This element provides the execution environment of a single VNFC instance ([Vn-Nf]/VM, interface 7 in Figure 8.8).

- **Control and admin agent:** This element connects to the virtualized infrastructure manager (VIM).
- **vswitch:** The vswitch function, described below, is implemented in the hypervisor domain. However, functionally it forms an integral part of the infrastructure network domain.

19. Explain the concept of vswitches.

    A vswitch is an Ethernet switch implemented by the hypervisor that interconnects virtual network interface cards (NICs) of VMs with each other and with the NIC of the compute node. If two VNFs are on the same physical server, they are connected through the same vswitch. If two VNFs are on different servers, the connection passes through the first vswitch to the NIC and then to an external switch. This switch forwards the connection to the NIC of the desired server. Finally, this NIC forwards it to its internal vswitch and then to the destination VNF.

20. Explain the difference between L2 and L3 virtual networks.

    Protocol-based virtual networks can be classified based on whether they are defined at protocol Layer 2 (L2), which is typically the LAN medium access control (MAC) layer, or Layer 3 (L3), which is typically the Internet Protocol (IP) layer.

21. List and briefly describe the VNF interfaces.

    - **SWA-1:** The interface enables communication between a VNF and other VNFs, PNFs, and endpoints. Note that an interface is to the VNF as a whole and not to individual VNFCs. SWA-1 interfaces are logical interfaces that primarily make use of the network connectivity services available at the SWA-5 interface.
    - **SWA-2:** This interface enables communications between VNFCs within a VNF. This interface is vendor specific and thus not a subject for standardization. This interface may also make use of the network connectivity services available at the SWA-5 interface. However, if two VNFCs within a VNF are deployed on the same host, other technologies may be used to minimize latency and enhance throughput.
    - **SWA-3:** This is the interface to the VNF manager within the NFV management and orchestration module. The VNF manager is responsible for life cycle management (creation, scaling, termination, etc.). The interface typically is implemented as a network connection using IP.
    - **SWA-4:** This is the interface for runtime management of the VNF by the element manager.
    - **SWA-5:** This interface describes the execution environment for a deployable instance of a VNF. Each VNFC maps to a virtualized container interface to a VM.

22. Define scale up, scale down, scale in, and scale out.

- **Scale up:** Expand capability by adding resources to a single physical machine or virtual machine.
- **Scale down:** Reduce capability by removing resources from a single physical machine or virtual machine.
- **Scale out:** Expand capability by adding additional physical or virtual machines.
- **Scale in:** Reduce capability by removing physical or virtual machines.

Chapter 9: Core Network Functionality, QoS, and Network Slicing

1. List and define the IMT-2020 network operational requirements.

- **Network flexibility and programmability:** The network should support a wide range of devices, users, and applications, with evolving requirements for each. Significant concepts in this regard are network functions virtualization (discussed in Chapter 8, "Network Functions Virtualization"), separation of user and control planes, and network slicing. The latter two concepts are discussed later in this chapter.
- **Fixed mobile convergence:** The focus of this requirement is to enable subscriber access through multi-access networks in seamless, integrated fashion.
- **Enhanced mobility management:** The network should support a wide variety of mobility options.
- **Network capability exposure:** The IMT-2020 network should provide suitable ways (e.g., via application program interfaces [APIs]) to expose network capabilities and relevant information (e.g., information for connectivity, QoS, and mobility) to third parties. This enables third parties to dynamically customize the network capabilities for diverse use cases within the limits set by the IMT-2020 network operator.
- **Identification and authentication:** There should be a unified approach to user and device identification and authentication mechanisms.
- **Security and personal data protection:** The IMT-2020 network must provide effective mechanisms to preserve security and personal data protection for different types of devices, users, and services, including rapid adaptation to dynamic network changes.
- **Efficient signaling:** There are two aspects to this requirement. The signaling mechanisms should be designed to mitigate risks of control and data traffic bottlenecks. Also, the network should provide lightweight signaling protocols and mechanisms to accommodate limited-resource devices.

- **Quality of service control:** The network should support different QoS levels for different services and applications.
- **Network management:** The network should provide a unified network management framework to support interworking of different providers and management of legacy networks.
- **Charging:** The IMT-2020 network needs to support different charging policies and requirements of network operators and service providers, including third parties that may be involved in a given IMT-2020 network deployment. The charging models to be supported include, but are not limited to, charging based on volume, time, session and application.
- **Interworking with non-IMT-2020 networks:** IMT-2020 networks should support user-transparent interworking with legacy networks.
- **IMT-2020 network deployment and migration:** The network design should accommodate incremental deployment with migration capabilities for services and related users.

2. List and define the basic network requirements for 5G specified by 3GPP.

   - **Network slicing:** Network slicing enables operators to customize their network for different applications and customers. A network slice can provide the functionality of a complete network, including radio access network (RAN) and core network functions.
   - **Efficiency:** This category covers four capabilities:
     - **Resource efficiency:** 5G networks need to be optimized for supporting diverse UE and services.
     - **Efficient user plane:** Cloud-based applications can involve substantial computation that occurs far from the end user device, with substantial or time-sensitive data transfers. Such cases require low end-to-end latencies and high data rates.
     - **Efficient content delivery:** Video-based services, such as live streaming and virtual reality, can place a considerable burden on the cellular network. To support such services, 5G networks emphasize caching content as much as possible near the end user, such as by using multi-access edge computing. In addition, 5G must support applications that involve a relatively small amount of data but have stringent latency requirements. Efficient delivery of such small packets requires the use of signaling protocols that do not require lengthy procedures and do not involve large amounts of control data.
     - **Energy efficiency:** 5G design must put minimal control signaling burden on constrained devices.
   - **Diverse mobility management:** 5G must support different mobility management methods that minimize signaling overhead. It optimizes access for user equipment with different mobility management needs.

- **Priority, QoS, and policy control:** 5G must support the provision of priority, QoS, and policy control features.
- **Connectivity models:** 5G needs to support the following connection models:
  - **Direct 3GPP connection:** An example is a sensor that communicates with an application server or with another device through a 5G network.
  - **Indirect 3GPP connection:** An example is a smart wearable that communicates through a smartphone to the 5G network.
  - **Direct device connection:** An example is a biometric device that communicates directly with other biometric devices or with a smartphone associated with the same patient.
- **Network capability/exposure and context awareness:** In order to allow third parties to access information regarding capabilities provided by the 5G network (e.g., information for connectivity, QoS, and mobility) and to dynamically customize the network capabilities for diverse use cases, the 5G network should provide suitable ways (e.g., via APIs) to expose network capabilities and relevant information to third parties. Applications may also provide the network with context awareness information.
- **Flexible broadcast/multicast service:** A broadcast/multicast service should allow flexible and dynamic allocation of resources between unicast and multicast services within a network, and it should also allow the deployment of standalone broadcast networks. It should be possible to stream multicast/broadcast content efficiently over wide geographic areas as well as target the distribution of content to very specific geographic areas spanning only a limited number of base stations.

3. What are the differences among priority, QoS, and policy control?

   - **Policy control:** The process by which network resources are controlled to implement a given policy for a given user.
   - **Priority:** A value assigned to specific packets transmitted to/from a user that determines the relative importance of transmitting those packets during the upcoming opportunity to use the medium.
   - **Quality of service:** The measurable end-to-end performance properties of a network service, which can be guaranteed in advance by a service-level agreement between a user and a service provider, to satisfy specific customer application requirements. These properties may include throughput (bandwidth), transit delay (latency), error rates, security, packet loss, packet jitter, and so on.

4. What is 5G tunneling?

   It is IP tunneling within a 5G network.

5. Define the two types of tunnels and explain the purpose of each of them.

   The CN tunnel is unidirectional, providing an uplink for the UE with a path from the gNB to the UPF. The AN tunnel provides a downlink path from the UPF to the gNB.

6. What are the differences among PDU session, QoS flow, and service data flow?

   - **PDU session:** This is a logical connection that carries all the communication between UE and a data network (DN).
   - **QoS flow:** This is the lowest level of granularity within the 5G system for defining policy and charging rules. A PDU session may contain multiple QoS flows.
   - **Service data flow (SDF):** An SDF provides an end-to-end packet flow between UE and a specific application at the DN. One or more SDFs can be transported in the same QoS flow if they share the same policy and charging rules.

7. What are the main purposes of QoS capabilities?

   - Enable networks to offer different levels of QoS to customers on the basis of customer requirements.
   - Allocate network resources efficiently, maximizing effective capacity.

8. What are the four stages in the QoS life cycle?

   QoS planning, QoS provisioning, QoS monitoring, QoS optimization

9. Define the common functions used to provide QoS in the data plane.

   - **Traffic classification:** Refers to the assignment of packets to a traffic class by the ingress router at the ingress edge of the network.
   - **Packet marking:** Encompasses two distinct functions. First, packets may be marked by ingress edge nodes of a network to indicate some form of QoS that the packet should receive. Second, packet marking can be used to mark packets as nonconformant, either by the ingress node or intermediate nodes, so that they can be dropped later, if congestion is experienced.
   - **Traffic shaping:** Controls the rate and volume of traffic entering and transiting the network on a per-flow basis.
   - **Congestion avoidance:** Deals with means for keeping the load of the network under its capacity so that it can operate at an acceptable performance level.
   - **Traffic policing:** Determines whether the traffic being presented is, on a hop-by-hop basis, compliant with prenegotiated policies or contracts.
   - **Queuing and scheduling algorithms, also referred to as queuing discipline algorithms:** Determine which packet to send next and are used primarily to manage the allocation of transmission capacity among flows.

- **Queue management algorithms:** Manage the length of packet queues by dropping packets when necessary or appropriate.
- **Admission control:** Determines what user traffic may enter the network.
- **QoS routing:** Determines a network path that is likely to accommodate the requested QoS of a flow.
- **Resource reservation:** Reserves network resources on demand for delivering desired network performance to a requesting flow.

10. Define the common functions used to provide QoS in the management plane.
    - A **service-level agreement (SLA):** Typically represents an agreement between a customer and a provider of a service that specifies the level of availability, serviceability, performance, operation, or other attributes of the service.
    - **Traffic metering and recording:** Concerns monitoring the dynamic properties of a traffic stream using performance metrics such as data rate and packet loss rate.
    - **Traffic restoration:** Refers to the network response to failures. This encompasses a number of protocol layers and techniques.
    - **Policy:** Is a category that refers to a set of rules for administering, managing, and controlling access to network resources.
11. What are the differences among QoS classification, marking, and differentiation?
    - **Traffic classification:** Grouping traffic into classes based on user-defined QoS values.
    - **User plane marking:** Marking packets to indicate to which QoS classification they belong.
    - **QoS differentiation:** The use of a different QoS set of values for different categories of traffic.
12. Define the QoS parameters specified in TS 23.501.
    - **5QI (5G QoS identifier):** This is an integer value used as a reference to a set of values assigned to QoS characteristics. Thus, a standardized combination of QoS characteristics can be preconfigured so that the AN and CN are informed of the QoS characteristics for a flow by means of the 5QI.
    - **ARP (allocation and retention priority):** The ARP consists of three attributes:
        - **ARP priority level:** Defines the relative importance of a QoS flow. The range of the ARP priority level is 1 to 15, with 1 as the highest level of priority. In cases of congestion, when all QoS requirements cannot be fulfilled for one or more QoS flows, the priority level determines for which QoS flows the QoS requirements are prioritized. In cases where there is no congestion, the priority level determines the resource distribution between QoS flows.

  - **ARP pre-emption capability:** Defines whether a QoS flow may get resources that were already assigned to another QoS flow with a lower ARP priority level. It is set as either enabled or disabled.
  - **ARP pre-emption vulnerability:** Defines whether a QoS flow may lose the resources assigned to it in order to admit a QoS flow with a higher ARP priority level. It is set as either enabled or disabled.

- **RQA (reflective QoS attribute):** Reflective QoS means that the UE uses the same QoS parameters on the uplink as obtained from the downlink QoS flow. RQA, when included, indicates that some (not necessarily all) traffic carried on this QoS flow is subject to reflective QoS.

- **Flow bit rates:** There are two categories of flow bit rates. A guaranteed bit rate (GBR) guarantees at least a minimum bit rate capacity for the flow. A non-GBR QoS flow does not guarantee the bit rate.

- **Notification control:** This parameter indicates whether notifications are requested from the NG-RAN when the GFBR can no longer (or can again) be guaranteed for a QoS flow during the lifetime of the QoS flow.

- **Aggregate bit rates:** Two parameters related to bit rates are associated with each UE:

- **Per session aggregate maximum bit rate (Session-AMBR):** For each PDU session of a UE, this parameter limits the aggregate bit rate that can be expected to be provided across all non-GBR QoS flows for a specific PDU session.

  - **Per UE aggregate maximum bit rate (UE-AMBR):** This parameter limits the aggregate bit rate that can be expected to be provided across all non-GBR QoS flows of a UE.

  - **Default values:** For each PDU session setup, these default values apply to one or more non-GBR flows.

- **Maximum packet loss rate:** This parameter indicates the maximum rate for lost packets of the QoS flow that can be tolerated in the uplink and downlink directions.

- **Wireline access network-specific 5G QoS parameters:** There are additional parameters applicable only to wireline access networks.

13. Define the QoS characteristics specified in TS 23.501.

    - **Resource type:** There are three resource types: GBR, delay-critical GBR, and non-GBR.

    - **Priority level:** This characteristic indicates a priority in scheduling resources among QoS flows.

    - **Packet delay budget (PDB):** This characteristic defines an upper bound for the time that a packet may be delayed between the UE and the UPF.

- **Packet error rate (PER):** This characteristic defines an upper bound for the rate of PDUs (e.g., IP packets) that have been processed by the sender of a link layer protocol but that are not successfully delivered by the corresponding receiver to the upper layer.
- **Averaging window:** This characteristic represents the duration over which the GFBR and MFBR are calculated (e.g., in the RAN, UPF, and UE).
- **Maximum data burst volume (MDBV):** This characteristic denotes the largest amount of data that the 5G-AN is required to serve within a period of 5G-AN PDB (i.e., the 5G-AN part of the PDB).

14. What is the difference between QoS parameters and characteristics?

Although there is some overlap, very broadly it can be said that the QoS parameters are used at configuration time to determine the network resources needed for creating a network slice for supporting this set of QoS parameter values. The QoS characteristics are more relevant to dynamic decisions made during the operation of the QoS flow, such as using the priority level as a tie-breaker when two flows compete for a resource.

Another perspective is based on the distinction between an application and a specific instance of an application. 3GPP has determined that the variables defined as characteristics are typically the same for a wide variety of instances of an application, and so it is efficient and useful to the user to provide standardized sets of values. Depending on the context, individual instances of an application may require different sets of values for some of the parameters, especially the flow bit rates (GFBR and MFBR) and the aggregate bit rates (session-AMBR and UE-AMBR).

15. List and define the service requirements for network slicing specified by 3GPP.

- Support is needed to provide connectivity to home and roaming users in the same network slice.
- In a shared 5G network configuration, operators can apply all the requirements to their allocated network resources.
- IMS needs to be supported as part of a network slice.
- IMS needs to be supported independent of network slices.

16. List and define the operational requirements for network slicing specified by 3GPP.

- The operator should be able to create, modify, and delete a network slice.
- The operator should be able to define and update the set of services and capabilities supported in a network slice.
- The operator should be able to configure the information that associates UE to a network slice.

- The operator should be able to configure the information that associates a service to a network slice.
- The operator should be able to assign UE to a network slice, to move UE from one network slice to another, and to remove UE from a network slice based on subscription, UE capabilities, the access technology being used by the UE, the operator's policies, and services provided by the network slice.
- A mechanism is needed for the VPLMN (visited public land mobile network), as authorized by the HPLMN (home public land mobile network), to assign UE to a network slice with the needed services or to a default network slice.
- A UE should be able to be simultaneously assigned to and access services from more than one network slice of one operator.
- Traffic and services in one network slice should have no impact on traffic and services in other network slices in the same network.
- Creation, modification, and deletion of a network slice should have no or minimal impact on traffic and services in other network slices in the same network.
- A network slice should be able to scale (i.e., adapt its capacity).
- The network operator should be able to define a minimum available capacity for a network slice. Scaling of other network slices on the same network should have no impact on the availability of the minimum capacity for that network slice.
- The network operator should be able to define a maximum capacity for a network slice.
- The network operator should be able to define a priority order between different network slices in the event that multiple network slices compete for resources on the same network.
- The operator should be able to differentiate policy control, functionality, and performance provided in different network slices.

17. What NFs support multiple slices for UE?

    - **Access and mobility management function (AMF):** Network slice instance selection is usually triggered as part of the registration procedure by the first AMF that receives the registration request from the UE. When UE accesses the network, the AMF provides functionalities to register and de-register the UE with the network, and it establishes the user context in the network. In the registration procedure, AMF performs (but is not limited to) network slice instance selection, UE authentication, authorization of network access and network services, and network access policy control. In addition, when a session establishment request message is received from UE, the AMF performs discovery and selection of the SMF that is the most appropriate to manage the session.

- **Network slice selection function (NSSF):** The AMF retrieves the slices that are allowed by the user subscription and interacts with the NSSF to select the appropriate network slice instance (e.g., based on allowed S-NSSAIs, 5G network ID, and other parameters). The NSSF responds with a message that includes the list of appropriate network slice instances for the UE. As a result, the registration process may switch to another AMF if needed.
- **Network repository function (NRF):** During the AMF-NSSF interaction, the NSSF may return the identity of one or more NRFs to be used to select NFs and services within the selected network slice instance(s).

18. What NFs support a single slice instance?

    - **Session management function (SMF):** The UE sends a message to the AMF, requesting that a PDU session be associated to one S-NSSAI and one data network (DN). The AMF selects the appropriate SMF, which manages the PDU session. The SMF sets up the PDU session for the UE and controls the user plane operation. The SMF selects the UPF and invokes enforcement of QoS and charging policies.
    - **User plane function (UPF):** Once a PDU session is established, QoS flows for this PDU session over this network slice pass through the UPF.
    - **Policy control function (PCF):** The SMF gets policy information related to session establishment from the PCF.
    - **Network repository function (NRF):** The SMF uses the NRF to discover the required NFs for the individual network slice.

Chapter 10: Multi-Access Edge Computing

1. Explain the difference between cloud computing, edge computing, cloud-edge computing, and MEC.

    - **Cloud computing:** A loosely defined term for any system providing access via the Internet (or other networks) to processing power, storage, network, software, or other computing services, often via a web browser. Often, these services are rented from an external company that hosts and manages them.
    - **Edge computing:** A strategy to deploy processing capability at the network edge, where end terminals are connected, and to perform the bulk of processing of data that is derived from and fed to the end terminals.
    - **Cloud-edge computing:** A form of edge computing that offers application developers and service providers cloud computing capabilities, as well as an IT service environment, at the edge of a network. The aim is to deliver compute, storage, and bandwidth much closer to data inputs and/or end users.

- **Multi-access edge computing (MEC):** Cloud-edge computing that provides an IT service environment and cloud computing capabilities at the edge of an access network that contains one or more types of access technology and in close proximity to its users. It is characterized by either ultra-low latency or high data rate capacity or both. For wireless access (e.g., in a radio access network), MEC provides real-time access to radio network information that can be leveraged by applications.
- **Mobile edge computing:** A term that was formerly used, with the acronym MEC, to denote what is now referred to as multi-access edge computing but limited to wireless access to a cellular network.

2. How does MEC support each of the three 5G application areas?

   - **Enhanced mobile broadband (eMBB):** eMBB applications require high data rates—both peak data rates and overall capacity rates. Moving much of the UE communications to a near edge relieves the core network of a significant burden. In particular, QoS flows supported by network slices generally follow a much shorter path, making satisfaction of QoS more achievable.
   - **Massive machine type communications (MMTC):** 5G must support massive Internet of Things (IoT) deployments with high connection density. An edge computing paradigm enables the collection of huge amounts of data at local edge processors. The edge processors can do some processing and consolidation of the data before transmitting the results across the 5G network to a central repository.
   - **Ultra-reliable and low-latency communications (URLLC):** URLLC applications by definition demand very low latencies. Low latency levels can be achieved only if the interaction between the UE and the URLLC application is local. Transmission across the breadth of a core network would prohibit providing the quality of service (QoS) required of such applications.

3. What is a local breakout?

   Local breakout is a concept in which the data plane traffic is routed locally to cloud services (compute and storage) without having to cross the breadth of the core network.

4. List and briefly explain the general design and development principles defined by ETSI.

   - **NFV alignment:** A MEC edge host system dynamically supports a number of applications. Thus, the resources required vary over time, and a virtualized environment provides the needed flexibility. The ETSI architecture employs NFV. If an edge system is part of the core network, this fits naturally with the NFV implementation of the core network. If the edge system is considered to be outside the core, the use of NFV enables hosting and management of virtual network functions (VNFs) using the same management infrastructure.

- **Mobility support:** A mobile user may require that support of edge-based applications move from one MEC system to another dynamically. Thus, the application needs to be implemented on a virtual machine such that the application can be moved from one virtual environment to another.
- **Deployment independence:** MEC capability should support deployment in variety of ways, including:
  - Deployment at the radio node.
  - Deployment at an aggregation point. An **aggregation point** is a location in a physical network deployment that is intermediate between the core network and a number of homogeneous or heterogeneous network termination points (base station, cable modems, LAN access points, etc.) and that can act as a location for a MEC host.
  - Deployment at the edge of the core network.
- **Simple and controllable APIs:** APIs should be easy to access and provide an effective means for controlling underlying resources to enable rapid development of applications.
- **Smart application location:** A MEC application needs to run at the appropriate physical location at any point in time, taking into account compute, storage, network resource, and latency requirements.
- **Application mobility to/from an external system:** The MEC architecture should support the movement of applications between a MEC host and an external cloud environment.
- **Representation of features:** The MEC architectural framework needs to support mechanisms to identify whether a specific feature is supported.

5. List and give a brief summary of each of the main elements of the MEC system reference architecture.
   - **Virtualization infrastructure manager (VIM):** The VIM corresponds to the VIM in the NFV architecture. It controls and manages the interaction of an app with the virtualized compute, storage, and network resources under its authority. It is responsible for allocating, maintaining, and releasing virtual resources of the virtualization infrastructure. The VIM also maintains software images for fast app instantiation.
   - **MEC host:** The mobile edge host is a logical construct that facilitates mobile edge applications (apps), offering a virtualization infrastructure that provides computation, storage, and network resources, as well as a set of fundamental functionalities (mobile edge services) required to execute apps, known as the mobile edge platform.
   - **MEC platform manager:** The MEC platform manager corresponds to the VNF manager in the NFV architecture. The MEC platform manager oversees the management of MEC applications and services.

- **MEC orchestrator:** The multi-access edge orchestrator corresponds to the Orchestrator in the NFV architecture.

6. What are the three main components of a MEC host?

   - **Virtualization infrastructure:** Corresponds to the NFV infrastructure (NFVI). The virtualization infrastructure includes a data plane that executes the traffic rules received by the MEC platform and routes the traffic among applications, services, the Domain Name System (DNS) server/proxy, the 5G network, other access networks, local networks, and external networks.
   - **MEC applications:** Virtual network functions that run on top of the virtual machines. These include all user and network applications that run on the edge host.
   - **MEC platform:** The collection of essential functionality required to run MEC applications on a particular virtualization infrastructure and enable them to discover, advertise, and consume edge services.

7. What functions are performed by the MEC platform manager?

   The MEC platform manager corresponds to the VNF manager in the NFV architecture. The MEC platform manager oversees the management of MEC applications and services.

8. What elements of a MEC architecture can be deployed as VNFs?

   - The MEC platform
   - All MEC applications
   - The data plane component of the virtualization infrastructure
   - Two components of the MEC platform manager: MEC platform element manager and MEC application rules and requirements management

9. How can MEC support network slicing?

   Each network slice is implemented as a separate MEC platform and MEC platform manager supporting one or more apps plus a data plane VNF that defines the user traffic rules and priorities. In the figure, the lighter and darker shaded boxes indicate dedicated MEC instances for the two slices. The MEC components in black boxes are shared across the two network slices. The white boxes are MEC components not directly involved in supporting different slices, although these components are slice aware.

10. List and briefly define the three categories of use cases specified by ETSI.

    - **Consumer-oriented services:** Services that directly benefit the end user (i.e., the user using the UE).

- **Operator and third-party services:** Services that take advantage of computing and storage facilities close to the edge of the operator's network. These services are usually not directly benefiting the end user, but can be operated in conjunction with third-party service companies.
- **Network performance and quality of experience (QoE) improvements:** Services aimed at improving performance of the network, either via application-specific or generic improvements. The user experience is generally improved, but these are not new services provided to the end user.

11. Describe the role for MEC in the factory of the future.

    Figure 10.10 illustrates a distributed MEC approach to satisfying the requirements for factory automation. In this example, some sensors are connected by fixed connections to a wireless LAN (WLAN) access point. A collection of sensors and actuators form an IoT that is connected to a field-level LAN by an IoT gateway. A MEC host at the field level provides processing and storage support for the sensor and actuator deployment. At a higher level, a MEC host could be deployed within the enterprise to support enterprise-wide applications and apps that support the operational domain (refer to Figure 10.9). This MEC host could also provide access to enterprise-wide databases and applications that are used to control actuators and consolidate and interpret sensor data.

12. Give some examples of the ways in which an application may target a group of subscribers.

    - Allowing an anonymous group of flat-rate billing subscribers access to content locally from the MEC host
    - Sending targeted advertising for a certain group of users within the mobile network
    - Providing content to a specific group of users that might be, for example, in the same club, association, or public service group
    - Providing enterprise services to company employees

13. Describe the role of video analytics in 5G use cases.

    - **Surveillance and public safety:** Processing live video streams almost instantaneously at the edge can lead to better surveillance and help in enforcing law and order. Two examples of this use case are face detection and incident identification and triggering, which allow law enforcement officers to take immediate actions involving an incident.
    - **Autonomous driving:** Real-time video of the scene as seen by a self-driving car needs be analyzed in a very short time to determine the actions to be taken by the car. A self-driving vehicle could already contain resources to process the scene instantaneously. Edge video analytics can help with processing (or preprocessing) further scenes or postprocessing video scenes for continual training and feedback.

- **Smart cities and IoT:** Video analytics at the edge is an important element in enabling smart cities. For example, traffic video analysis can be used to route traffic in the most efficient way. Fire or smoke detection in an area can be identified instantaneously to ensure that no traffic continues toward the danger zone; feedback can be sent to both the city infrastructure and connected cars in an area.
- **Enhanced infotainment services:** Video analytics at the edge can be used to enhance the real-life experience of event audiences such as those at sporting events, concerts, and other shows. Videos from different camera angles at an event can be analyzed and applied with AR/VR functions and presented to a live audience through large screens, smartphones, and VR devices.

14. How can MEC be used to support video analytics?

    Figure 10.12 illustrates a scheme in which video is first stored on a local MEC host in a video content cache. The video content is then processed by a video compression algorithm followed by a video analytics application. Typically the video compression scheme is that standardized by MPEG-4. This provides an ideal data representation for supporting indexing and retrieval schemes. It also simplifies the task of video structure parsing and keyframe extraction because many of the necessary content features (e.g., object motion) are readily available.

Chapter 11: Wireless Transmission

1. Define channel capacity.

   The maximum rate at which data can be transmitted over a given communication path, or channel, under given conditions is referred to as the channel capacity.

2. What is the difference between the theoretical channel capacity values derived from the Nyquist and Shannon formulas?

   The Shannon formula takes into account the SNR.

3. What key factors affect channel capacity?

   Modulation type and coding

4. What is the difference between optical and radio LOS?

   Radio LOS takes into account refraction.

5. What factors related to attenuation does a transmission engineer need to consider?

   - A received signal must have sufficient strength so that the electronic circuitry in the receiver can detect and interpret the signal.

- The signal must maintain a level sufficiently higher than noise to be received without error.
- Attenuation is greater at higher frequencies, causing distortion because signals are typically composed of many frequency components.

6. What variables are factors in free space loss?

   P_t = signal power at the transmitting antenna

   P_r = signal power at the receiving antenna

   λ = carrier wavelength

   f = carrier frequency

   d = propagation distance between antennas

   c = speed of light (3×10^8 m/s)

7. Define the path loss exponent.

   Both theoretical and measurement-based models have shown that beyond a certain distance, the average received signal power decreases logarithmically with distance according to a $10n\log(d)$ relationship, where n is known as the path loss exponent.

8. Name and briefly define four types of noise.

   - **Thermal noise:** Thermal noise is due to thermal agitation of electrons. It is present in all electronic devices and transmission media and is a function of temperature. Thermal noise is uniformly distributed across the frequency spectrum and hence is often referred to as white noise.
   - **Intermodulation noise:** When signals at different frequencies share the same transmission medium, the result may be intermodulation noise. Intermodulation noise produces signals at a frequency that is the sum or difference of the two original frequencies or multiples of those frequencies.
   - **Crosstalk:** Crosstalk is an unwanted coupling between signal paths.
   - **Impulse noise:** Impulse noise is noncontinuous noise, consisting of irregular pulses or noise spikes of short duration and of relatively high amplitude. It is generated by a variety of causes, including external electromagnetic disturbances, such as lightning, and faults and flaws in the communications system.

9. What is refraction?

   Radio waves are refracted (or bent) when they propagate through the atmosphere. The refraction is caused by changes in the speed of the signal with altitude or by other spatial changes in atmospheric conditions.

10. What is multipath fading?

Multipath fading refers to fading that occurs in any environment where there is multipath propagation. In a fixed environment, multipath fading is affected by changes in atmospheric conditions such as rainfall. But in a mobile environment, where one of the two antennas is moving relative to the other, the relative locations of various obstacles change, creating complex transmission effects.

11. What is the difference between diffraction and scattering?

Diffraction occurs at the edge of an impenetrable body that is large compared to the wavelength of the radio wave. When a radio wave encounters such an edge, waves propagate in different directions, with the edge as the source. Thus, signals can be received even when there is no unobstructed LOS from the transmitter.

If the size of an obstacle is on the order of the wavelength of the signal or less, scattering occurs. An incoming signal is scattered into several weaker outgoing signals. At typical cellular microwave frequencies, there are numerous objects, such as lamp posts and traffic signs, that can cause scattering. Thus, scattering effects are difficult to predict.

12. What is the difference between fast and slow fading?

As a mobile unit moves down a street in an urban environment, rapid variations in signal strength occur over distances of about one-half a wavelength. At a frequency of 900 MHz, which is typical for mobile cellular applications, a wavelength is 0.33 m. The rapidly changing waveform in Figure 11.10 shows an example of the spatial variation of received signal amplitude at 900 MHz in an urban setting. Note that changes of amplitude can be as much as 20 or 30 dB over a short distance. This type of rapidly changing fading phenomenon, known as fast fading, affects not only mobile phones in automobiles but even mobile phone users walking.

As a mobile user covers distances well in excess of a wavelength, the urban environment changes: The user passes buildings of different heights, vacant lots, intersections, and so forth. Over these longer distances, there is a change in the average received power level about which the rapid fluctuations occur. This is indicated by the slowly changing waveform in Figure 11.10 and is referred to as slow fading.

13. What is the difference between flat and selective fading?

Fading effects can also be classified as flat or selective. Flat fading, or nonselective fading, is the type of fading in which all frequency components of the received signal fluctuate in the same proportions simultaneously. Selective fading affects unequally the different spectral components of a radio signal. The term *selective fading* is usually significant only relative to the bandwidth of the overall communications channel. If attenuation occurs over a portion of the bandwidth of the signal, the fading is considered to be selective; nonselective fading implies that the signal bandwidth of interest is narrower than, and completely covered by, the spectrum affected by the fading.

14. What is the frequency range for millimeter wave?

    Traditionally, the term *millimeter wave* has been equated with the EHF (extremely high frequency) range of 30 to 300 GHz, with wavelengths between 10 mm and 1 mm, as shown in Figure 11.12. In a 5G context, millimeter waves refer to frequencies between 24 and 86 GHz, which encompass the frequency bands contemplated for 5G use.

15. List and briefly define the main sources of loss for mmWave transmission.

    - **Free space loss:** A transmitted signal attenuates over distance because the signal is being spread over a larger and larger area.
    - **Atmospheric gaseous losses:** This is the absorption of signal energy by molecules of oxygen, water vapor, and other gaseous atmospheric conditions.
    - **Rain losses:** Raindrops are roughly the same size as the millimeter radio wavelengths and therefore cause scattering of the radio signal.
    - **Foliage losses:** In the mmWave bands, foliage can introduce significant amount of loss.
    - **Blocking:** Blocking, or blockage, is related to the amount of loss sustained by a wireless signal due to a solid object on the LOS path.

Chapter 12: Antennas

1. Define adaptive equalization.

   Adaptive equalization can be applied to transmissions that carry analog information (e.g., analog voice or video) or digital information (e.g., digital data, digitized voice or video) and is used to combat intersymbol interference. The process of equalization involves some method of gathering the dispersed symbol energy back together into its original time interval.

2. What is the difference between space, frequency, and time diversity?

   - **Space diversity:** Some diversity techniques involve the physical transmission path and are referred to as space diversity, or spatial diversity.
   - **Frequency diversity**: The signal is spread out over a larger frequency bandwidth or carried on multiple frequency carriers.
   - **Time diversity:** Data is spread out over time so that a noise burst affects fewer bits.

3. What two functions are performed by an antenna?

   Radiating electromagnetic energy and collecting electromagnetic energy

4. What is an isotropic antenna?

   A point in space that radiates power in all directions equally

5. What information is available from a radiation pattern?

   When the radiated power of a transmitting antenna is measured around the antenna, a shape called the radiation pattern emerges. The pattern is a graphical representation of the radiation properties of an antenna as a function of space coordinates.

6. What is the advantage of a parabolic reflective antenna?

   Such surfaces are used in automobile headlights, optical and radio telescopes, and microwave antennas because of the following property: If a source of electromagnetic energy (or sound) is placed at the focus of the paraboloid, and if the paraboloid is a reflecting surface, the wave will bounce back in lines parallel to the axis of the paraboloid.

7. What factors determine antenna gain?

   Effective area and carrier frequency

8. What is meant by the term multiple-input/multiple-output antenna?

   If a transmitter and receiver implement a system with multiple antennas, this is called a multiple-input/multiple-output (MIMO) system.

9. What distinguishes single-user MIMO from multiple-user MIMO?

   With single-user MIMO (SU-MIMO), multiple streams using multiple antennas are directed to a single device. With multi-user MIMO (MU-MIMO), multiple streams using multiple antennas are directed to many devices.

10. Explain the two types of MIMO transmission schemes.

    - **Spatial diversity:** The same data is coded and transmitted through multiple antennas, which effectively increases the power in the channel proportionally to the number of transmitting antennas. This improves the signal-to-noise ratio (SNR) for cell edge performance. Further, diverse multipath fading offers multiple "views" of the transmitted data at the receiver, thus increasing robustness. In a multipath scenario where each receiving antenna would experience a different interference environment, there is a high probability that if one antenna is suffering a high level of fading, another antenna has sufficient signal level.

    - **Spatial multiplexing:** A source data stream is divided among the transmitting antennas. The gain in channel capacity is proportional to the available number of antennas at the transmitter or receiver, whichever is less. Spatial multiplexing can be used when transmitting conditions are favorable and for relatively short distances compared to spatial diversity. The receiver must do considerable signal processing to sort out the incoming substreams, all of which are transmitting in the same frequency channel, and to recover the individual data streams.

11. What are two applications of multiple-user MIMO?

- **Uplink—multiple access channel (MAC):** With MIMO-MAC, multiple end users transmit simultaneously to a single base station.
- **Downlink—broadcast channel (BC):** With MIMO-BC, the base station transmits separate data streams to multiple independent users.

12. What is beamforming?

Beamforming is a technique by which an array of antennas can be steered to transmit radio signals in a specific direction.

13. Describe the advantages of beamforming.

- **Higher SNR:** The highly directional transmission enhances the link budget, improving the range for both open-space and indoor penetration.
- **Interference prevention and rejection:** Beamforming prevails over cochannel interference (CCI), explained subsequently, by taking advantage of the antennas' spatial properties.
- **Higher network efficiency:** By substantially minimizing CCI, beamforming allows much denser deployments than are possible with single-antenna systems.

14. What is beam management?

Beam management refers to techniques and processes used to achieve the transmission and reception of data over relatively narrow beams.

15. Describe the four elements of beam management for downlink transmission.

- **Beam sweeping:** The base station antenna (i.e., the 5G radio access network node gNB) transmits beams in a predetermined sequence for beam measurement at the UE side.
- **Beam measurement:** The UE measures the characteristics of received beamformed signals.
- **Beam determination:** The UE selects the optimal beam. In essence, the UE isolates the receive beam, which affords the best reception. The best results are obtained when the transmitting and receiving beam pair is optimal for the location of the UE at the time.
- **Beam reporting:** The UE reports back to the gNB the information based on beam measurement.

16. What is FD-MIMO?

The term full-dimension MIMO (FD-MIMO), or 3D-MIMO, refers to a MIMO antenna system that is capable of varying the direction of a beam in both horizontal (azimuth) and vertical (elevation) dimensions.

17. Describe the benefits of an active antenna system compared to a passive antenna system.

    - The site footprint is reduced.
    - The distribution of radio functions to the individual antennas within the radome results in built-in redundancy and improved thermal performance.
    - Distributed transceivers support advanced beamforming features and enable FD-MIMO.
    - Integrating the active transceiver array and passive antenna array into one radome reduces cable losses.

18. What are the alternative duplexing approaches for implementing massive MIMO?

    FDD and TDD

Chapter 13: Air Interface Physical Layer

1. How are binary values represented in amplitude-shift keying, and what is the limitation of this approach?

   In ASK, the two binary values are represented by two different amplitudes of the carrier frequency. Commonly, one of the amplitudes is zero; that is, one binary digit is represented by the presence, at constant amplitude, of the carrier, and the other is represented by the absence of the carrier. ASK is susceptible to sudden gain changes and is a rather inefficient modulation technique.

2. How are binary values represented in frequency-shift keying?

   The most common form of FSK is binary FSK (BFSK), in which the two binary values are represented by two different frequencies near the carrier frequency.

3. How are binary values represented in phase-shift keying?

   In PSK, the phase of the carrier signal is shifted to represent data.

4. What is the difference between QPSK and offset QPSK?

   The difference is that a delay of one bit time is introduced in the Q stream for OQPSK.

5. What is the difference between data rate and modulation rate?

   Data rate, or bit rate, is the rate at which bits are transmitted, expressed in bits per second. Modulation rate is defined as the number of signal units per second, expressed in units of baud, or symbols per second.

6. What is the trade-off between bandwidth efficiency and error performance for MPSK?

   There is a trade-off between bandwidth efficiency and error performance: An increase in bandwidth efficiency results in an increase in error probability.

7. What is QAM?

Quadrature amplitude modulation (QAM) is a popular modulation technique that is used in a number of wireless standards. This modulation technique is a combination of ASK and PSK. QAM can also be considered a logical extension of QPSK. QAM takes advantage of the fact that it is possible to send two different signals simultaneously on the same carrier frequency, by using two copies of the carrier frequency, one shifted by 90° with respect to the other. For QAM, each carrier is ASK modulated. The two independent signals are simultaneously transmitted over the same medium. At the receiver, the two signals are demodulated, and the results are combined to produce the original binary input.

8. What is $\pi/2$-BPSK?

$\pi/2$-BPSK is a variation on BPSK in which the phase shift is $\pm\pi/2$.

9. Briefly explain OFDM.

OFDM, also called multicarrier modulation, uses multiple carrier signals at different frequencies, sending some of the bits on each channel. This is similar to FDM. However, in the case of OFDM, all of the subcarriers are dedicated to a single data source.

10. Define orthogonality in the context of OFDM.

Two signals, $s_1(t)$ and $s_2(t)$, are orthogonal if they meet this requirement:

Average over one bit time of $s_1(t)s_2(t) = 0$

11. What roles do FFT and IFFT play in implementing OFDM?

FFT converts source data into a set of frequency values. IFFT converts frequency values back to time domain values.

12. What is the purpose of the cyclic prefix?

- Additional time, known as a guard interval, is added to the beginning of the OFDM symbol before the actual data begins. This allows all residual ISI to diminish before it impacts the received data.
- This beginning time period is packed with data that is an actual copy of the data from the end of the OFDM symbol that is being sent. The effect of this is to isolate the parallel subchannels and allow for simple frequency-domain digital signal processing techniques.

13. Why is PAPR an important consideration in the design of OFDM schemes?

Due to the presence of a large number of independently modulated subcarriers in an OFDM system, the peak value of the system can be significantly higher than the average.

14. Why is intercarrier interference an important consideration in the design of OFDM schemes?

    Because OFDM frequencies are spaced as closely as possible, the frequency synchronization requirements are significantly more stringent. If they are not met, intercarrier interference (ICI) results.

15. What is the difference between OFDM and OFDMA?

    Like OFDM, OFDMA employs multiple closely spaced subcarriers, but the subcarriers are divided into groups of subcarriers. Each group is named a subchannel. The subcarriers that form a subchannel need not be adjacent. In the downlink, a subchannel may be intended for different receivers. In the uplink, a transmitter may be assigned one or more subchannels.

16. What is the difference between OFDMA and SC-OFDM?

    SC-FDMA performs a DFT prior to the IFFT operation, which spreads the data symbols over all the subcarriers carrying information and produces a virtual single-carrier structure. This then is passed through the OFDM processing modules to split the signal into subcarriers. Now, however, every data symbol is carried by every subcarrier.

Chapter 14: Air Interface Channel Coding

1. What is an error burst?

   A group of bits in which two successive erroneous bits are always separated by less than a given number x of correct bits. The number x should be specified when describing an error burst. The last erroneous bit in the burst and the first erroneous bit in the following burst are accordingly separated by x correct bits or more.

2. What is the difference between a BEC and a BSC?

   A binary symmetric channel (BSC) has input and output alphabets of {0, 1}. A binary erasure channel (BEC) has an input alphabet X of {0, 1} and an output alphabet Y of {0, 1, E}, where E is the erasure symbol.

3. In an (n, k) block error-correcting code, what do n and k represent?

   A k-bit block of data is mapped to an n-bit codeword.

4. What are the five possible outcomes of an FEC decoder?

   - **No errors:** If there are no bit errors, the input to the FEC decoder is identical to the original codeword, and the decoder produces the original data block as output.
   - **Detectable, correctable errors:** For certain error patterns, the decoder is able to detect and correct those errors. Thus, even though the incoming data block differs from the transmitted codeword, the FEC decoder can map this block into the original data block.

- **Detectable, not correctable errors:** For certain error patterns, the decoder can detect but not correct the errors. In this case, the decoder simply reports an uncorrectable error.
- **Detectable, falsely correctable errors:** For certain, typically rare, error patterns, the decoder detects an error but does not correct it properly. It assumes that a certain block of data was sent when in reality a different one was sent that differs in at least one bit position. The receiver cannot distinguish this case from the case of detectable and correctable errors unless another layer of error correction is applied.
- **Undetectable errors:** For certain even more rare error patterns, the decoder does not detect that any errors have occurred and maps the incoming n-bit data block into a k-bit block that differs from the original k-bit block. The receiver cannot distinguish this case from the case of no errors unless another layer of error correction is applied.

5. What is a parity-check code?

   An (n, k) parity-check code encodes k data bits into an n-bit codeword such that the codeword contains the original data bits plus $(n - k)$ check bits.

6. How many simultaneous equations define an (n, k) parity-check code?

   $m = n - k$ simultaneous linear equations

7. How do the equations mentioned in the preceding question relate to the parity-check matrix?

   - The $m \times n$ matrix $\mathbf{H} = [h_{ij}]$ is called the parity-check matrix. Each of the m rows of $\mathbf{H}$ corresponds to one of the individual equations. Each of the n columns of $\mathbf{H}$ corresponds to one bit of the codeword.

8. What is the relationship between the redundancy and code rate of an (n, k) code?

   The ratio of redundant bits to data bits, $(n - k)/k$, is called the redundancy of the code, and the ratio of data bits to total bits, k/n, is called the code rate.

9. What is the purpose of the syndrome in a parity-check code?

   The syndrome is a column vector that indicates which of the individual parity-check equations do not equal 0.

10. What properties determine a regular LDPC code?

    - Each code bit is involved with w_c parity constraints, and each parity constraint involves w_r bits.
    - Each row of $\mathbf{H}$ contains w_r 1s, where w_r is constant for every row.
    - Each column of $\mathbf{H}$ contains w_c 1s, where w_c is constant for every column.
    - The number of 1s in common between any two columns is zero or one.

- Both w_r and w_c are small compared to the number of columns (i.e., the length of the codeword) and the number of rows (i.e., $w_c << n$ and $w_r << m$).

11. What elementary function is the foundation of polar codes?

    Exclusive-OR

12. What is a synthetic channel?

    It is a virtual channel used in the mathematical construction and analysis of polar codes.

13. Explain the concept of polarization.

    Polarization involves deriving two synthetic channels that polarize the information capacity of the channels.

14. How does soft decision decoding improve HARQ?

    The decoding process can provide not just an assessment of a bit being 0 or 1 but also levels of confidence in those results.

Chapter 15: 5G Radio Access Network

1. List and define four types of RAN nodes.

Node Name	User Equipment (UE)	Core Network
gNB	5G	5G
ng-eNB	4G	5G
en-gNB	5G	4G
eNB	4G	4G

2. What is dual connectivity?

   Dual connectivity enables provision of better performance for 5G devices on a 4G core network. The en-gNB provides a 5G radio interface for UEs to connect to the 4G core network. A 4G eNB acts as a primary, or controlling, node that is in control of the radio connection with the UE, and the en-gNB is used as a secondary, or controlled, node.

3. List and define three core functional areas that are the primary means of interaction with a RAN.

   - **Access and mobility management function (AMF):** The AMF provides UE authentication, authorization, and mobility management services. The non-access stratum (NAS) is the highest protocol layer of the control plane between UE and the mobility management entity (MME) in the core network. The main functions of the protocols that are part of the NAS are the support of mobility of the UE and the support of session management procedures to establish and maintain IP connectivity between the UE and a packet data network gateway (PDN GW). The AMF is used to maintain continuous communications with the UE as it moves. In contrast, the access stratum is responsible for carrying information just over the wireless portion of a connection. NAS security involves IP header compression, encryption, and integrity protection of data based on the NAS security keys derived during the registration and authentication procedure. Idle state mobility handling deals with cell selection and reselection while the UE is in idle mode, as well as reachability determination.

- **Session management function (SMF):** The UE IP address allocation process assigns an IP address to the UE at session establishment. This ensures the ability to route data packets within the 5G system and also supports data reception and forwarding to outside networks and provides interconnectivity to external packet data networks (PDNs). In cooperation with the UPF, the SMF establishes, maintains, and releases a protocol data unit (PDU) session for user data transfer, which is defined as an association between the UE and a data network that provides a PDU connectivity.
- **User plane function (UPF):** UE mobility handling deals with ensuring that there is no data loss when there is a connection transfer due to handover that involves changing anchor points. Once a session is established, the UPF has a responsibility for PDU handling. This includes the basic functions of packet routing, forwarding, and QoS handling.

4. What functions are performed by Service Data Adaptation Protocol?

   SDAP supports the flow-based QoS model of the 5G core network. With this QoS model, the core network can configure different QoS requirements for different QoS flows of a PDU session. The SDAP layer provides mapping of IP flows with different QoS requirements to radio bearers that are configured appropriately to deliver that required QoS. The mapping between QoS flows and radio bearers may be configured and reconfigured by RRC signaling, but it can also be changed more dynamically without the involvement of RRC signaling through a reflective mapping process.

5. What functions are performed by the radio resource control?

   RRC is responsible for control and configuration of the radio-related functions in the UE. For each connection to UE, RRC operates using a three-state model. The RRC inactive state provides battery efficiency similar to RRC idle but with a UE context remaining stored within the NG-RAN so that transitions to/from RRC connected are faster and incur less signaling overhead. RRC supports an on-demand system information mechanism that enables the UE to request when specific system information is required instead of allowing the NG-RAN to consume radio resources to provide frequent periodic system information broadcasts.

6. What is the non-access stratum?

   The NAS consists of protocols between UE and the core network that are not terminated in the RAN. Specifically, NAS protocols terminate in the UE and the AMF of the 5G core network and are used for core network–related functions such as registration, authentication, location updating, and session management.

7. What functions are performed by GTP-U?

   GTP-U supports multiplexing of traffic of different PDU sessions (via N3) and carries QoS marking. GTP-U tunnels are used to carry encapsulated user plane PDUs between a given pair of GTP-U tunnel endpoints.

8. What are the chief differences between UDP and SCTP?

   - **User Datagram Protocol (UDP):** UDP is a connectionless transport layer protocol that provides for the exchange of transport layer datagrams without acknowledgments or guaranteed delivery. UDP adds a port-addressing capability to IP so that a specific source and destination application or service is designated.

   - **Stream Control Transmission Protocol (SCTP):** This is a reliable transport layer protocol. Like Transmission Control Protocol (TCP), SCTP ensures reliable transport of PDUs with congestion control. In contrast to TCP, SCTP allows delivery of out-of-order packets to applications; this type of delivery is more suitable for delay-sensitive applications. The multihoming feature of SCTP enables transparent handover over several heterogeneous overlapping wireless networks. One path with specified destination and source addresses plays the role of primary path. The remaining paths are secondary paths. SCTP can monitor, at runtime, delay and jitter on all active paths, and it makes the paths available to the application. Heartbeat messages are sent over secondary paths to collect required measurements. The collected key performance indicators (KPIs) are mapped to quality of experience (QoE) values using a suitable QoE/QoS mapping model. The path quality is compared at regular intervals, and the client decides whether a network switch is necessary according to its customized and internal policy.

9. What is the RAN transport network?

   The NG RAN transport network is a collection of communication links that interconnect nodes of the RAN and communication links that connect RAN elements to the 5G core networks.

10. Explain the differences between fronthaul, midhaul, and backhaul.

    - **Backhaul:** A network path between base station systems and a core network. The distance covered by a backhaul network between the core network and a base station could range from 1 km up to hundreds of kilometers.

- **Midhaul:** A network path between CUs and DUs that are physically separated. The typical range is 20 to 40 km.
- **Fronthaul:** A network path between centralized radio controllers and remote radio units of a base station function. The distance is in the range of less than 20 km.

11. Describe the air interface protocol architecture.

    The user plane consists of the following layers between UE and gNB: PHY, MAC, RLC, PDCP, SDAP. The control plane layers between UE and gNB are PHY, MAC, RLC, PDCP, and RRC; the NAS layer is between the UE and the core network AMF.

12. Describe the RAN–core interface protocol architecture.

    The user plane consists of the physical, data link, IP, UDP, GTP-U, and PDU session layers. The control plane consists of the physical, data link, IP, SCTP, and NGAP layers.

13. Describe the Xn interface protocol architecture.

    The user plane consists of the physical, data link, IP, UDP, GPT-U, and user plane PDUs. The control plane consists of the physical, data link, IP, SCTP, and XnAP layers.

14. Describe four possible deployment scenarios for transport networks.

    - **Independent RRU, CU, and DU locations:** In this scenario, there are fronthaul, midhaul, and backhaul networks.
    - **RU and DU integration:** In this scenario, an RU and a DU are deployed close to each other—perhaps hundreds of meters apart, such as in the same building. In order to reduce cost, an RU is connected to a DU just through straight fiber, and no transport equipment is needed. In this case, there are midhaul and backhaul networks.
    - **Collocated CU and DU:** In this scenario, the CU and DU are located together; consequently, there is no midhaul.
    - **RRU, DU, and CU integration:** This network structure may be used for small cell and hotspot deployments. There is only backhaul in this case.

15. Explain the difference between metro access, metro aggregation, metro core, and backbone.

    - **Metro access:** The access portion of a cellular network provides the last kilometer or last several kilometers of connectivity between UE and the radio unit. This corresponds to the air interface. For configurations with a remote RU, the term *metro access* roughly corresponds to the fronthaul network that connects the RUs to base stations or at least to the DU portion of a base station.
    - **Metro aggregation:** The metro aggregation network aggregates the traffic of progressively larger sets of different uses and transmits this aggregated traffic over increasingly higher-capacity facilities.

- **Metro core:** The metro core acts as a regional network providing connectivity between the various access and aggregation networks within a given metropolitan area and connecting to a larger backbone network.
- **Backbone:** This term generally refers to high-speed long-haul transmission links and networks that connect metropolitan area networks to more distant resources. The backbone encompasses the switched core network as well as the backhaul network.

16. What is a virtualized RAN?

A virtualized RAN (VRAN), formerly referred to as a cloud RAN, decouples hardware and software, allowing RAN functions typically run on a proprietary technology stack to exist as software workloads using commodity or custom hardware.

17. Explain the basic concept of IAB.

The key concept is to use the same wireless access technology that provides an air interface to user equipment (UE) for creating a backhaul link between 5G RAN nodes.

18. Explain the concepts of IAB donor, IAB node, parent node, and child node.

- **IAB donor:** A RAN node that provides connection to the core network for IAB nodes. It supports the CU function of the CU/DU architecture.
- **IAB node:** A RAN node that supports wireless relaying of NR access traffic from UEs via NR Uu backhaul links. It supports the UE function and the DU function of the CU/DU architecture.
- **Parent node:** The node closer to the core network of two nodes that are adjacent in an IAB backhaul transmission path from UE to the core network. The parent schedules the backhaul downstream and upstream traffic to/from the child node.
- **Child node:** The node farther from the core network of two nodes that are adjacent in an IAB backhaul transmission path from UE to the core network.

Appendix B

Glossary

access network A network that connects directly to the end user or customer.

actuator A device that accepts an electrical signal and converts it into a physical, chemical, or biological action.

air interface A wireless interface between user equipment and a base station, also called a *radio interface*. The air interface specifies the method for transmitting information over the air between base stations and mobile units, including protocols, frequency, channel bandwidth, and modulation scheme.

amplitude The size or magnitude of a voltage or current waveform.

amplitude modulation A form of modulation in which the amplitude of a carrier wave is varied in accordance with some characteristic of the modulating signal.

antenna The part of a transmitting or receiving system that is designed to radiate or to receive electromagnetic waves.

application programming interface (API) A language and message format used by an application program to communicate with the operating system or some other control program, such as a database management system (DBMS) or communications protocol. APIs are implemented by writing function calls in a program to provide linkage to the required subroutine for execution. An open or standardized API can ensure the portability of the application code and the vendor independence of the called service.

attenuation The reduction of strength of a signal as a function of distance traveled.

backhaul A network path between base station systems and a core network.

bandwidth The difference, in Hertz, between the limiting (upper and lower) frequencies of a spectrum.

base station A network element in a radio access network that is responsible for radio transmission and reception in one or more cells to or from the user equipment. A base station can have an integrated antenna or can be connected to an antenna by feeder cables. The base station interfaces the user terminal (through the air interface) to a radio access network infrastructure.

baseband The spectral band occupied by an unmodulated signal. Baseband transmission is usually characterized by being much lower in frequency than the signal that results if the baseband signal is used to modulate a carrier frequency.

broadband In general, wide bandwidth equipment or systems that can carry signals occupying a large portion of the electromagnetic spectrum. Typically, a broadband communication system can simultaneously accommodate voice, data, video, and other services. In digital transmission systems, the term connotes high data rate.

carrier frequency A continuous frequency capable of being modulated or impressed with a second (information-carrying) signal.

cellular network A wireless communications network in which fixed antennas are arranged in a hexagonal pattern and mobile stations communicate through nearby fixed antennas.

channel A single path for transmitting electric signals. Note: The word *path* is to be interpreted in a broad sense to include separation by frequency division or time division. The term *channel* may signify either a one-way or a two-way path.

circuit switching A method of communicating in which a dedicated communications path is established between two devices through one or more intermediate switching nodes. Unlike with packet switching, digital data is sent as a continuous stream of bits. Bandwidth is guaranteed, and delay is essentially limited to propagation time. The telephone system uses circuit switching.

cloud computing Any system providing access via the Internet to processing power, storage, software, or other computing services, often via a web browser. Often, these services are rented from an external company that hosts and manages them.

cloud-edge computing A form of edge computing that offers application developers and service providers cloud computing capabilities, as well as an IT service environment at the edge of a network. The aim is to deliver compute, storage, and bandwidth much closer to data inputs and/or end users.

code-division multiple access (CDMA) A multiplexing technique used with spread spectrum.

consortium A group of independent organizations joined by common interests. In the area of standards development, a consortium typically consists of individual corporations and trade groups concerned with a specific area of technology.

constrained device In an IoT, a device with limited volatile and nonvolatile memory, limited processing power, and a low-data-rate transceiver.

container Hardware or software that provides an execution environment for software.

container virtualization A technique in which the underlying operating environment of an application is virtualized. This environment is commonly the operating system kernel, and the result of virtualization is an isolated container in which the application can run.

control channel A logical channel that carries system control information.

core network A central network that provides networking services to attached distribution and access networks. Also referred to as a *backbone network*.

core router A router that resides within the middle of a network rather than at its periphery. The routers that make up the backbone of the Internet are core routers.

decibel A measure of the relative strength of two signals. The number of decibels is 10 times the log of the ratio of the power of two signals or 20 times the log of the ratio of the voltage of two signals.

diffraction Deviation of part of a beam, determined by the wave nature of radiation, and occurring when the radiation passes the edge of an opaque obstacle.

digital data Data consisting of a sequence of discrete elements.

digital signal A discrete or discontinuous signal, such as voltage pulses.

digital transmission Transmission of digital data, using either an analog or digital signal, in which the digital data is recovered and repeated at intermediate points to reduce the effects of noise.

edge computing A distributed information technology architecture in which client data is processed at the periphery of the network, as close to the originating source as possible.

end-to-end (E2E) Refers to communications between two endpoint devices or user equipment, across any arrangement of intervening administrative domains.

end-to-end (E2E) latency The time it takes from when a data packet is sent from the transmitting end to when it is received at the receiving entity (e.g., Internet server or other device). The measurement reference is the interface between Layers 2 and 3. Also referred to as one-trip time (OTT).

enhanced mobile broadband (eMBB) A 5G usage scenario characterized by high data rates for mobile devices.

error rate The ratio of the number of data units in error to the total number of data units.

error-correcting code A system of adding redundant data, or parity data, to a block of data such that the block can be recovered by a receiver even when a number of errors (up to the capability of the code being used) are introduced, either during the process of transmission or on storage.

error-detecting code A code in which each expression conforms to specific rules of construction, so that if certain errors occur in an expression, the resulting expression will not conform to the rules of construction, and thus the presence of the errors is detected.

fading The time variation of received signal power caused by changes in the transmission medium or path(s).

fixed wireless access (FWA) A connection that provides primary broadband access through wireless wide area mobile network–enabled customer premises equipment (CPE). This includes various form factors of CPE, such as indoor (desktop and window) and outdoor (rooftop and wall mounted).

flat fading That type of fading in which all frequency components of the received signal fluctuate in the same proportions simultaneously.

fog computing A scenario in which a massive number of heterogeneous, decentralized devices communicate with each other and with the network to perform storage and processing tasks without the intervention of third parties.

forward channel In a cellular network, the communications link from the base station to the mobile unit.

forward error correction Procedures whereby a receiver, using only information contained in the incoming digital transmission, corrects bit errors in the data.

free space loss The amount of attenuation of transmission energy on an unobstructed path between isotropic antennas. Basically, dilution of energy as the radio signal propagates away from a source.

frequency-division multiple access (FDMA) An access method at the data link layer based on FDM principles, in which different frequency bands are allocated to different data streams. The data link layer in each station tells its physical layer to make a bandpass signal from the data passed to it. The signal must be created in the allocated band. There is no multiplexer at the physical layer. The signals created at each station are automatically bandpass filtered. The signals are mixed when they are sent to the common channel. FDMA supports demand assignment, in which the assignment of frequency bands to users changes over time.

frequency-division multiplexing (FDM) A physical layer technique in which multiple baseband signals are modulated on different frequency carrier waves and added together to create a composite signal. The effect of FDM is to divide transmission bandwidth into multiple subchannels, each of which is dedicated to a particular baseband signal.

frequency modulation Modulation in which the frequency of an alternating current is the characteristic varied.

fronthaul A network path between centralized radio controllers and remote radio units of a base station function.

fundamental frequency The lowest-frequency component in the Fourier representation of a periodic quantity.

gain (of an antenna) The ratio of the radiation intensity, in a given direction, to the radiation intensity that would be obtained if the power accepted by the antenna were radiated isotropically.

ground wave A radio wave propagated over the earth that is ordinarily affected by the presence of the ground and troposphere. Notes: (1) The ground wave includes all components of a radio wave over the earth except ionospheric and tropospheric waves. (2) The ground wave is refracted because of variations in the dielectric constant of the troposphere.

handover The action of switching a call in progress from one cell to another (intercell) or between radio channels in the same cell (intracell) without interrupting the call. Handover is used to allow established calls to continue when mobile stations move from one cell to another (or as a method to minimize cochannel interference). Also called *handoff*.

haptic A sense perceived by touching an object. It involves tactile senses, which refers to the touching of surfaces, and kinesthetic senses, or the sensing of movement in the body.

hardware virtualization The use of software to partition a computer's resources into separate and isolated entities called virtual machines. It enables multiple copies of the same or different operating system to execute on the computer and prevents applications from different virtual machines from interfering with each other.

industry vertical An organization that provides products and/or services targeted to a specific industry, trade, profession, or other group of customers with specialized needs. A vertical might, for example, provide a range of products or services useful in the banking industry or healthcare. In contrast, a horizontal provides products or services that address a specific need across multiple industries, such as accounting or billing products and services.

information and communications technology (ICT) All devices, networking components, applications, and systems that together allow people and organizations to interact in the digital world.

information technology (IT) The common term for the entire spectrum of technologies for information processing, including software, hardware, communications technologies, and related services. In general, IT does not include embedded technologies that do not generate data for enterprise use.

Internet of Things (IoT) The expanding connectivity, particularly via the Internet, of a wide range of sensors, actuators, and other embedded systems. In almost all cases, there is no human user, and interaction is fully automated.

ionosphere The part of the earth's outer atmosphere where ionization caused by incoming solar radiation affects the transmission of radio waves. It generally extends from a height of about 50 km to about 400 km above the earth's surface.

isotropic antenna An antenna that radiates in all directions (about a point) with a gain of unity (not a realizable antenna, but a useful concept in antenna theory).

key performance indicator (KPI) Quantifiable measurements that reflect the critical success factors of a use case.

macro cell An outdoor cell with large cell radius, typically several kilometers.

massive machine type communication (mMTC) A 5G usage scenario characterized by the ability to support huge numbers of devices, such as in a large IoT deployment.

micro cell An outdoor or indoor cell with a small cell radius, typically less or much less than 1 km.

microwave Electromagnetic waves in the frequency range 1 to 40 GHz.

midhaul A network path between a centralized unit and distributed units of a base station that are physically separated.

millimeter wave (mmWave) An imprecise term for a system that typically operates in a region between 10 GHz (wavelength = 30 mm) and 300 GHz (wavelength = 1 mm).

mobile telecommunications switching office (MTSO) An office used by a cellular service provider for originating and terminating functions for calls to or from end user customers of the cellular provider. Also known as *mobile switching center (MSC).*

modulation The process of varying the amplitude, phase, and/or frequency of a periodic waveform, called the carrier signal, to convey analog or digital data. The modulating function, consisting of the analog or digital data, is called the baseband signal.

multi-access edge computing (MEC) Cloud-edge computing that provides an IT service environment and cloud computing capabilities at the edge of an access network that contains one or more types of access technology and in close proximity to its users. It is characterized by ultra-low latency and high data rate capacity. For wireless access (radio access network), MEC provides real-time access to radio network information that can be leveraged by applications. Also called *mobile edge computing*, although the term multi-access edge computing is preferred because it emphasizes that UE may connect through a means other than a cellular radio access network.

multipath A propagation phenomenon that results in signals reaching the receiving antenna by two or more paths.

multiplexing In data transmission, a function that permits two or more data sources to share a common transmission medium such that each data source has its own channel.

network function A processing function in a network that has defined functional behavior and interfaces. A network function can be implemented either as a network element on dedicated hardware, as a software instance running on dedicated hardware, or as a virtualized function instantiated on an appropriate platform (e.g., on a cloud infrastructure).

network functions virtualization (NFV) The virtualization of compute, storage, and network functions by implementing these functions in software and running them on virtual machines.

network operating system (NOS) A server-based operating system oriented to computer networking. It may include directory services, network management, network monitoring, network policies, user group management, network security, and other network-related functions.

noise Unwanted signals that combine with and hence distort the signal intended for transmission and reception.

packet A group of bits that includes data plus control information. Generally refers to a network layer (OSI Layer 3) protocol data unit.

packet switching A method of transmitting messages through a communication network in which long messages are subdivided into short packets. The packets are then transmitted as in message switching.

propagation delay The delay between the time a signal enters a channel and the time it is received.

protocol A set of semantic and syntactic rules that describe how to transmit data, especially across a network. Low-level protocols define the electrical and physical standards to be observed, bit and byte ordering, and transmission and error detection and correction of the bit stream. High-level protocols deal with data formatting, including the syntax of messages, semantics of messages, character sets, and sequencing of messages.

protocol architecture The software structure that implements the communications function. Typically, the protocol architecture consists of a layered set of protocols, with one or more protocols at each layer.

protocol control information Information exchanged between entities of a given layer, via the service provided by the next lower layer, to coordinate their joint operation.

protocol data unit (PDU) Information that is delivered as a unit between peer entities of a network. A PDU typically contains control information and address information in a header. The PDU may also contain data.

QoS flow The lowest granularity of a traffic flow where QoS and charging can be applied.

quality of experience (QoE) A subjective measure of performance in a system. QoE relies on human opinion and differs from quality of service (QoS), which can be precisely measured.

quality of service (QoS) The measurable end-to-end performance properties of a network service, which can be guaranteed in advance by a service-level agreement between a user and a service provider in order to satisfy specific customer application requirements. Note: These properties may include throughput (bandwidth), transit delay (latency), error rates, priority, security, packet loss, packet jitter, and so on.

radio access network (RAN) A network that connects radio base stations to the core network. The RAN provides and maintains radio-specific functions, which may be unique to a given radio access technology, that allow users to access the core network. RAN components include base stations and antennas, mobile telecommunications switching offices, and other management and transmission elements.

radio bearer An information transmission path with defined capacity, delay, bit error rate, and other parameters.

reference point A conceptual point at the conjunction of two non-overlapping functional groups.

reflection A process that occurs when an electromagnetic signal encounters a surface that is large relative to the wavelength of the signal; the angle of incidence equals the angle of reflection.

refraction The bending of a beam in transmission through an interface between two dissimilar media or in a medium whose refractive index is a continuous function of position.

reverse channel In a cellular or cordless network, the communications link from a mobile unit to the base station.

round-trip time (RTT) The time from when a data packet is sent from the source device until an acknowledgment or response is received from the destination device. The measurement reference is the interface between Layers 2 and 3.

scattering The production of waves of changed direction, frequency, or polarization when radio waves encounter matter.

selective fading Fading that affects unequally the different spectral components of a radio signal.

sensor A device that converts a physical, biological, or chemical parameter into an electrical signal.

service data unit (SDU) In a packet, data that the protocol transfers between peer protocol entities on behalf of the users of that layer's services. For lower layers, the layer's users are peer protocol entities at a higher layer; for the application layer, the users are application entities outside the scope of the protocol layer model.

sky wave A radio wave propagated obliquely toward, and returned from, the ionosphere.

smart city A municipality that uses information and communication technology to increase operational efficiency, share information with the public, and improve both the quality of government services and citizen welfare.

software-defined networking (SDN) An approach to designing, building, and operating large-scale networks based on programming the forwarding decisions in routers and switches via software from a central server. SDN differs from traditional networking, which requires configuring each device separately and relies on protocols that cannot be altered.

softwarization An overall approach for designing, implementing, deploying, managing, and maintaining network equipment and/or network components through software programming.

spectrum Refers to an absolute range of frequencies. For example, the spectrum of infrared is 3×10^{11} to 4×10^{14} Hz.

standard A document that provides requirements, specifications, guidelines, or characteristics that can be used consistently to ensure that materials, products, processes, and services are fit for their purpose. Standards are established by consensus among those participating in a standards-making organization and are approved by a generally recognized body.

standards-developing organization (SDO) An official national, regional, or international standards body that develops standards and/or that coordinates the standards activities of a specific country, region, or the world. Some SDOs facilitate the development of standards through support of technical committee activities, and some are directly involved in standards development.

time-division duplexing (TDD) A link transmission technique in which data are transmitted in one direction at a time, with transmission alternating between the two directions.

time-division multiple access (TDMA) An access method at the data link layer based on time-division duplexing principles. TDMA provides different time slots to different transmitters in a cyclically repetitive frame structure. For example, node 1 may use time slot 1, node 2 time slot 2, and so on until the last transmitter, when it starts over. TDMA supports demand assignment, in which the assignment of time slots to users changes over time.

time-division multiplexing (TDM) A physical layer technique of transmitting and receiving independent signals over a common signal path by means of synchronized switches at each end of the transmission line so that each signal appears on the line only a fraction of time in an alternating pattern. Thus, multiple stations may share the same frequency channel but use only part of its capacity.

time domain A characterization of a function or signal in terms of value as a function of time.

traffic channel A logical channel that carries user information. Also called a *data channel.*

transceiver A device that is capable of both transmitting and receiving information.

transmission medium The physical path between transmitters and receivers in a communications system.

transmission time interval or transmit time interval (TTI) An interval that relates to encapsulation of data from higher layers into frames for transmission on the radio link layer. TTI refers to the duration of a transmission on the radio link. The TTI is related to the size of the data blocks passed from the higher network layers to the radio link layer.

troposphere The portion of earth's atmosphere in which the temperature generally decreases with altitude, clouds form, and convection is active. The troposphere occupies the space from the earth's surface up to a height ranging from about 6 km at the poles to about 18 km at the equator.

ultra-reliable and low-latency communications (URLLC) A form of machine-to-machine communications that enables delay-sensitive and mission-critical services that require very low end-to-end delay, such as tactile Internet, remote control of medical or industrial robots, driverless cars, and real-time traffic control.

usage scenario A general description of the way in which an International Mobile Telecommunications network is used. A usage scenario dictates various performance and technical requirements. A wide but nevertheless constrained variety of use cases are encompassed by a usage scenario.

use case A specific application or way of using an International Mobile Telecommunications network. A use case dictates more specific and refined performance and technical requirements than the corresponding usage scenario.

virtual machine One instance of an operating system along with one or more applications running in an isolated partition within the computer. It enables different operating systems to run in the same computer at the same time and also prevents applications from interfering with each other.

virtual machine monitor (VMM) A system program that provides a virtual machine environment. Also called a *hypervisor*.

virtual network An abstraction of physical network resources as seen by some upper software layer. Virtual network technology enables a network provider to support multiple virtual networks that are isolated from one another. Users of a single virtual network are not aware of the details of the underlying physical network or of the other virtual network traffic sharing the physical network resources.

virtual private network (VPN) The use of encryption and authentication in the lower protocol layers to provide a secure connection through an otherwise insecure network, typically the Internet. VPNs are generally cheaper than real private networks using private lines but rely on having the same encryption and authentication system at both ends. The encryption may be performed by firewall software or possibly by routers.

virtualization A variety of technologies for managing computer resources by providing an abstraction layer between the software and the physical hardware. These technologies effectively emulate or simulate a hardware platform, such as a server, storage device, or network resource, in software.

wavelength The distance between two points in a periodic wave that have the same phase.

wireless Refers to electromagnetic transmission through air, vacuum, or water by means of an antenna.

Appendix C

Acronyms

3GPP 3rd Generation Partnership Project

5G NGN 5G NextGen

5G NR 5G New Radio

AMPS Advanced Mobile Phone System

API application programming interface

AR augmented reality

AuC Authentication Center

BPSK binary phase-shift keying

BSC base station controller

BTS base transceiver station

CDMA code-division multiple access

CS circuit switched

E2E end-to-end

EIR equipment identity register

eMBB enhanced mobile broadband

EN-DC E-UTRA-NR Dual Connectivity

EPC Evolved Packet Core

ETSI European Telecommunications Standards Institute

E-UTRAN Evolved Universal Terrestrial Radio Access Network

FDD frequency-division duplexing

FDM frequency-division multiplexing

FDMA frequency-division multiple access

FM frequency modulation

FWA fixed wireless access

GGSN Gateway GPRS Support Node

GMSC Gateway MSC

GSA Global Mobile Suppliers Association

GSM Global System for Mobile Communications

GSMA GSM Association

HARQ hybrid automatic repeat request

HLR home location register

IAB integrated access and backhaul

ICT information and communication technology

IEC International Electrotechnical Commission

IMT International Mobile Telecommunications

ITU International Telecommunication Union

ITU-R ITU Radiocommunication Sector

ITU-T ITU Telecommunication Standardization Sector

KPI key performance indicator

LDPC low-density parity check

LOS line of sight

mMTC massive machine type communications

MEC multi-access edge computing or mobile edge computing

mmWave millimeter wave

MSC mobile switching center

MTSO mobile telecommunications switching office

NF network function

NFV network functions virtualization

NFVI NFV infrastructure

NGMN Next Generation Mobile Networks

NG-RAN next-generation radio access network

O&M operations and maintenance

OFDM orthogonal frequency-division multiplexing

OFDMA orthogonal frequency-division multiple access

PAPR peak-to-average-power ratio

PLMN Public Land Mobile Network

PS packet switched

PSTN public switched telephone network

QAM quadrature amplitude modulation

QCI QoS class identifier

QoE quality of experience

QoS quality of service

QPSK quadrature phase-shift keying

RAN radio access network

RF radio frequency

RIT radio interface technologies

RNC radio network controller

RRC Radio Resource Control

RTT round-trip time

SC-FDMA single-carrier FDMA

SDU service data unit

SGSN Serving GPRS Support Node

SRIT set of radio interface technologies

TDD time-division duplexing

TDM time-division multiplexing

TDMA time-division multiple access

TRxP transmission and reception point

UMTS Universal Mobile Telecommunications System

URLCC ultra-reliable and low-latency communications

UTRAN Universal Terrestrial Radio Access Network

UE user equipment

V2X vehicle-to-anything

VLR visitor location register

VNF virtual network function

VR virtual reality

Wi-Fi Wireless Fidelity

Index

SYMBOLS

A

B

C

D

E

F

G

H

I

K

L

M

N

O

P

Q

R

S

T

U

V

W